HANDBOOK OF NONDESTRUCTIVE EVALUATION

To Jason Gerard
With my best wishes
Charles J. Hellier
July 12, 2001

HANDBOOK OF NONDESTRUCTIVE EVALUATION

Charles J. Hellier

McGRAW-HILL
New York   Chicago   San Francisco   Lisbon   London   Madrid
Mexico City   Milan   New Delhi   San Juan   Seoul
Singapore   Sydney   Toronto

Library of Congress Cataloging-in-Publication Data

Hellier, Charles J.
   Handbook of nondestructive evaluation / Charles J. Hellier.
     p. cm.
   ISBN 0-07-028121-1
   1. Nondestructive testing—Handbooks, manuals, etc. I. Title.
TA417.2 .H45   2001
620.1'127—dc 21   00-067564

McGraw-Hill

A Division of The **McGraw·Hill** Companies

Copyright © 2001 by The McGraw-Hill Companies, Inc. All rights reserved. Printed in the United States of America. Except as permitted under the United States Copyright Act of 1976, no part of this publication may be reproduced or distributed in any form or by any means, or stored in a data base or retrieval system, without the prior written permission of the publisher.

1 2 3 4 5 7 8 9 0   DOC/DOC   0 7 6 5 4 3 2 1

ISBN 0-07-028121-1

The sponsoring editor for this book was Kenneth McCombs and the production supervisor was Pamela A. Pelton. It was set in Times Roman by Ampersand Graphics, Ltd.

Printed and bound by R. R. Donnelley and Sons, Co.

This book was printed on recycled, acid-free paper containing a minimum of 50% recycled de-inked fiber.

McGraw-Hill books are available at special quantity discounts to use as premiums and sales promotions, or for use in corporate training programs. For more information, please write to the Director of Special Sales, Professional Publishing, McGraw-Hill, Two Penn Plaza, New York, NY 10121-2298. Or contact your local bookstore.

Information contained in this work has been obtained by The McGraw-Hill Companies, Inc. ("McGraw-Hill") from sources believed to be reliable. However, neither McGraw-Hill nor its authors guarantee the accuracy or completeness of any information published herein, and neither McGraw-Hill nor its authors shall be responsible for any errors, omissions, or damages arising out of use of this information. This work is published with the understanding that McGraw-Hill and its authors are supplying information but are not attempting to render engineering or other professional services. If such services are required, the assistance of an appropriate professional should be sought.

This handbook is dedicated to the memory of two pioneers in the field of nondestuctive testing:

Robert C. McMaster, 1913–1986

Samuel A. Wenk, 1914–1990

Their work served as the foundation for the technology that has become so important to industry. Without the contributions of these "giants," the world would not be the same.

CONTENTS

Contributors xi

Preface xv

Chapter 1 Introduction to Nondestructive Testing — 1.1
    I. What is NDT? — 1.1
    II. Concerns Regarding NDT — 1.2
    III. History of Nondestructive Testing — 1.3
    IV. Nondestructive versus Destructive Tests — 1.17
    V. Conditions for Effective Nondestructive Testing — 1.21
    VI. Personnel Considerations — 1.22
    VII. Certification Summary — 1.26
    VIII. References — 1.27

Chapter 2 Discontinuities—Origins and Classification — 2.1
    I. Primary Production of Metals — 2.2
    II. Castings — 2.4
    III. Cracks — 2.10
    IV. Welding Discontinuities — 2.11
    V. Discontinuities from Plastic Deformation — 2.15
    VI. Corrosion-Induced Discontinuities — 2.15
    VII. Operationally Induced Discontinuities—Fatigue Cracking — 2.18
    VIII. Operationally Induced Discontinuities—Creep — 2.19
    IX. Operationally Induced Discontinuities—Brittle Fracture — 2.20
    X. Geometric Discontinuities — 2.21
    XI. Summary — 2.23
    XII. Glossary of Metallurgy and Discontinuity Terms — 2.23
    XIII. Discontinuity Guide — 2.27

Chapter 3 Visual Testing — 3.1
    I. History and Development — 3.1
    II. Theory and Principles — 3.3
    III. Equipment and Accessories — 3.9
    IV. Applications and Techniques — 3.22
    V. Evaluation of Test Results — 3.45
    VI. Advantages and Limitations — 3.49
    VII. Glossary of Key Terms — 3.53

Chapter 4 Penetrant Testing 4.1

    I. Introduction   4.1
    II. History and Development   4.1
    III. Theory and Principles   4.4
    IV. Penetrant Equipment and Materials   4.6
    V. Penetrant Procedures   4.14
    VI. Techniques and Variables   4.18
    VII. Evaluation and Disposition   4.27
    VIII. Penetrant Testing Applications   4.29
    IX. Quality Control Considerations   4.29
    X. Advantages and Limitations   4.32
    XI. Glossary of Penetrant Testing Terms   4.32

Chapter 5 Magnetic Particle Testing 5.1

    I. History and Development   5.1
    II. Theory and Principles   5.2
    III. Equipment and Accessories   5.24
    IV. Techniques   5.30
    V. Variables   5.39
    VI. Evaluation of Test Results and Reporting   5.44
    VII. Applications   5.48
    VIII. Advantages and Limitations   5.50
    IX. Glossary of Key Terms   5.51
    X. References   5.53

Chapter 6 Radiographic Testing 6.1

    I. History and Development   6.1
    II. Theory and Principles   6.10
    III. Radiographic Equipment and Accessories   6.21
    IV. Variables   6.25
    V. Techniques and Procedures   6.39
    VI. Radiographic Evaluation   6.50
    VII. Applications   6.54
    VIII. Advantages and Limitations of Radiography   6.58
    IX. Compendium of Radiographs   6.60
    X. Glossary   6.68
    XI. Bibliography   6.70

Chapter 7 Ultrasonic Testing 7.1

    I. History   7.1
    II. Theory and Principles   7.4
    III. Equipment for Ultrasonic Applications   7.29
    IV. Techniques   7.52
    V. Variables   7.103
    VI. Evaluation of Test Results   7.105
    VII. Applications   7.107
    VIII. Advantages and Limitations   7.110
    IX. Glossary of Terms   7.111
    X. References   7.115

Chapter 8 Eddy Current Testing — 8.1

    8.1 History and Development — 8.1
    8.2 Theory and Principles — 8.3
    8.3 Alternating Current Principles — 8.10
    8.4 Eddy Currents — 8.17
    8.5 Test Equipment — 8.24
    8.6 Eddy Current Applications and Signal Display — 8.40
    8.7 Advantages and Limitations — 8.64
    8.8 Other Electromagnetic Test Techniques — 8.65
    8.9 Glossary of Key Terms — 8.67
    8.10 Suggestions for Further Reading — 8.70

Chapter 9 Thermal Infrared Testing — 9.1

    1. History and Development — 9.1
    2. Theory and Principles — 9.4
    3. Equipment and Accessories — 9.11
    4. Techniques — 9.16
    5. Variables — 9.23
    6. Data Storage — 9.26
    7. Applications — 9.27
    8. Advantages and Limitations — 9.43
    9. Glossary — 9.44
    10. Bibliography and References — 9.47

Chapter 10 Acoustic Emission Testing — 10.1

    1. History and Development — 10.1
    2. Principles of Acoustic Emission Testing — 10.2
    3. Advantages and Limitations of Acoustic Emission Testing — 10.36
    4. Glossary of Acoustic Emission Terms — 10.37
    5. References — 10.39

Basic Metric–English Conversions — B.1

Index — I.1

CONTRIBUTORS

Charles J. Hellier (primary author and reviewer, author of chapters 1, 4, and 6) was founder and is currently President of HELLIER (a division of Rockwood Service Corporation), a multidisciplinary organization offering a wide range of technical services throughout North America. He has over 40 years of experience in nondestructive testing, quality assurance, and inspection. He completed his formal education at Penn State and Temple Universities. He is a Registered Professional Engineer, a Board Certified Forensic Examiner, and holds Level III Certifications in five nondestructive testing methods. He also holds a Level III Certificate in five methods issued by the American Society for Nondestructive Testing (ASNT).

Mr. Hellier is past National President of ASNT and has been active in that organization for over 40 years, serving on many committees, boards, and various councils. He has presented many lectures and papers worldwide, and is widely published. Currently, he is the National President of the Nondestructive Testing Management Association (NDTMA) and holds memberships in ASNT (Fellow), ASME, ASTM, AWS, ASM, ABFE (Fellow), and NDTMA.

Michael W. Allgaier (author, Chapter 2) is presently the Manager NDE Instruction with Electric Power Research Institute (EPRI), supporting the NDE Center in Charlotte, NC. He has over 30 years of experience in the support of Navy nuclear program and the commercial nuclear power industry. He has provided technical and programmatic support as a manager, supervisor, technical analyst, and instructor in nondestructive testing, quality assurance and training programs. Mr. Allgaier attended Fairliegh Dickinson University, where he received a Bachelor of Science Degree in Business Management. He completed his Master of Science Degree at New Jersey Institute of Technology. His thesis was on the accreditation of Technical–Professional Personnel (NDE). He served General Public Utilities Nuclear (GPUN) as a NDE Level III in visual, liquid penetrant, magnetic particle, ultrasonic, and radiographic testing. He is active in the American Society for Nondestructive Testing. Previous service included six years on the National Certification Board. He has written several articles published in *Materials Evaluation* on visual testing and personnel certification. Mike was also the technical editor of Volume 8: *Visual and Optical Testing* of the *NDT Handbook* published by ASNT.

John Drury (co-author, Chapter 7) became involved with NDT when serving as an Engineer Officer in the Royal Air Force. After leaving the Forces, he continued his career in NDT, at first in the armaments and aerospace field and later in the steel, utilities, and petrochemical industries. In 1978 his book *Ultrasonic Flaw Detection for Technicians* was published and became standard reading for many certification schemes. Since 1983 he has run his own company, Silverwing (UK) Limited, specializing in ultrasonics, tube inspection, and magnetic flux leakage.

Richard D. Finlayson (author, Chapter 10) received his MSc from the Ohio State University with a major in Nondestructive Evaluation, A.Eng. from Canadian Forces School of Aerospace Training and Engineering, and BSEE from the Royal Roads and

Royal Military Colleges, Victoria, B.C. and Kingston, Ontario. He has had a military career that spanned 30 years and involved service in two different militaries. He served as the NDE and Condition Monitoring Program Manager for the Air Force. He was employed as Director of Marketing and Sales with Physical Acoustics Corporation for all NAFTA countries. His responsibilities also included development of Applications involving acoustic emission, submission of research and development proposals, and exploration of new markets, and he is now Director of Research, Engineering Applications, Certification, and Training and New Business Development at Physical Acoustics Corporation for all NAFTA countries.

Richard A. Harrison (author, Chapter 5) was born and educated in England and spent the first 20 years of his working life at British Aerospace, Military Aircraft, the latter 15 years in NDT, working both in a "hands on" and supervisory/management role in the major five nondestructive testing methods (UT, RT, ET, PT, and MT) plus VT. In 1995, after progressing to Senior Section Leader NDT at BAe, he left to begin a new role as General Manager and Senior NDT Instructor of Hellier NDT training school in California. For five years he taught NDT courses at Levels I, II, and III in six methods in addition to preparing, administering, and grading NDT examinations in all methods and also performing Level III outside agency services for numerous customers. In February 2000, he formed his own company, T.E.S.T. NDT, Inc., based in Southern California. He holds an ASNT Level III certificate in UT, RT, ET, PT, MT, and VT and is certified PCN (ISO 9712) Level III in PT, MT. RT, UT, and ET. He is currently the Secretary for the Greater Los Angeles ASNT Section, and a member of AWS, The British Institute for NDT, and is an Incorporated Engineer with the European Industry Council.

Robert B. Pond, Jr. (author, Chapter 2) received the Teacher of the Year Award in the Part Time Programs at Johns Hopkins University for 1995, the Instructor of Merit Award in 1996, and the Distinguished Educator Award in 1999 from American Society for Materials International. His professional affiliations have included: President, M-Structures, Inc., Baltimore, Maryland; Adjunct Faculty, The Johns Hopkins University, Baltimore, Maryland; Adjunct Faculty, The American Society for Materials International, Materials Park, Ohio; Associate Research Scientist, The Center for Nondestructive Evaluation, Johns Hopkins University, Baltimore, Maryland; Adjunct Faculty, The Society for Manufacturing Engineers; Adjunct Faculty, Loyola College, Baltimore, Maryland; Vice-President, Utility Operations, Karta Technology, Inc.; Principal Metallurgist, Baltimore Gas and Electric Company, Baltimore, Maryland; Member, The Off-Site Safety Review Committee for Calvert Cliffs Nuclear Power Plant; Assistant Professor, Department of Mechanical Engineering, The United States Naval Academy, Annapolis, Maryland; Consultant, Ballistics Research Laboratory, Aberdeen Proving Ground, Aberdeen, Maryland; Consultant, Department of Defense, Republic of South Korea; President, Windsor Metalcrystals, Inc., New Windsor, Maryland; Principal Researcher, Marvalaud, Inc., Westminster, Maryland.

His Professional affiliations include: American Society for Materials International, Materials Engineering Institute Committee; American Welding Society; Center for Non-Destructive Evaluation, The Johns Hopkins University, Representative for Baltimore Gas and Electric and Associate Research Scientist; Electric Power Research Institute's NDE Center, Chair of the Steering Committee; Edison Electric Institute's Materials and Processing Committee, Vice Chair.

Dr. Pond's areas of specialization include: materials engineering services and materials characterizations using metallography, fractography, hardness and microhardness testing; scanning electron microscopy with energy dispersive spectroscopy; servo-hydraulic mechanical testing; Charpy impact testing, macrophotography; nondestructive evaluation of materials by dye penetrant, magnetic particle, ultrasonic and isotope, X-ray, and microfocus X-ray radiographic examinations; heat treatment furnaces; and resources and experience for simulation testing. He is expert in evaluation of material failures and uses of materials in new applications.

George R. Quinn (author, Chapter 8) has over thirty years of experience in NDT training, problem solving, and marketing. He received his Bachelor of Arts degree in English from Saint Michael's College in Vermont. He then completed five years of service in the United States Air Force as an aircraft maintenance officer, attaining the grade of Captain. After completing military service, he bean his work in ultrasonic testing at Branson Instruments as Director of Training and served as Manager of Marketing Services for Krautkramer Branson. After almost twelve years with the Branson organization, Mr. Quinn formed his own NDT marketing company, with clients including the American Society for Nondestructive Testing and Hocking Electronics. For ten years, Mr. Quinn was Vice-President of Marketing at Hellier NDT. While at Hellier, Mr. Quinn wrote training manuals and developed specialized courses. He is now senior instructor in eddy current and ultrasonic testing at the Hellier division of Rockwood Service Corp. Mr. Quinn has lectured throughout North America, as well as Europe and Asia. He holds an ASNT Level III Certificate in the electromagnetic and ultrasonic test methods.

Michael Shakinovsky (co-author, Chapter 7) received his nondestructive testing training in England. Although ultrasonics is his specialty, he is an ASNT Level III certificate holder in Ultrasonics, Radiography, Magnetic Particle, and Penetrant Testing. After attending engineering college, he specialized in nondestructive testing and has been doing so for over 30 years. Transducer design, research and development, practical applications, and training have comprised a great part of his career. He has worked in many countries, setting up automated systems, conducting examinations, and teaching. He serves on both the ASTM and on the ASME national committees in the discipline of ultrasonics, is on the advisory board of one of the Connecticut State Community Colleges, and is often a guest lecturer.

John R. Snell, Jr. (co-author, Chapter 9) is a leader in the thermographic profession who first used thermal imaging equipment while providing energy consulting services with the Department of Energy Weatherization Assistance program and the Residential Conservation Service (RCS) program. In 1984 Mr. Snell established Snell Infrared to better serve the needs of his clients. Snell Infrared has since expanded their training services to many new clients and developed extensive on-site offerings. In 1992 the company began certifying thermographers and currently acts as the certifying agent for several large companies. Mr. Snell also continues to be professionally active. He has been on the Thermosense Steering Committee since 1990, was Chair for Thermosense XVI, and has worked on the standards development committee of the American Society for Nondestructive Testing. He has presented a postconference seminar on thermography and was the organizer and Track Chair for the T/IRT sessions of the ASNT Fall 1995 conference. In 1994 Mr. Snell had the honor of becoming the first ASNT Level III certificate holder in thermography in the United States. He is also currently working with three ASTM committees, as well as EPRI and IEEE, on standard written procedures and has authored numerous articles and professional papers. He volunteers in the local school systems of the City of Montpelier, is Chair of the Tree Board, and is on the Board of Directors of the Vermont Historical Society. Mr. Snell is a graduate of Michigan State University.

Robert W. Spring (co-author, Chapter 9) has been actively involved in the thermographic profession. In association with Snell Infrared he has provided thermographic training and inspection services to a broad range of industrial clients. His research on program development resulted in his co-authoring four professional papers for Thermosense. Mr. Spring maintains an active professional involvement in Thermosense, ASHRAE, and ASNT, where he serves on the Standards Development Committee for Thermographers. In 1995 Mr. Spring became a partner in Snell Infrared. From 1980 to 1995 Mr. Spring was a principal in a professional engineering consulting firm specializing in providing a broad range of energy management services to industry, utilities, and commerce. These services include technical analyses, project management, program development, and educational services. During this time he conceived of, developed, and presented a nationally recognized educational program to reduce institutional energy use. Mr. Spring's previous professional experiences include three years with the U.S. Public Health Service as a district engineer providing environmental health services to Native Americans in Alaska and the Eastern United States. While with the USPHS, he developed and presented a successful cross-cultural preventive maintenance training program for the operators of water and wastewater facilities in remote Alaskan villages. Mr. Spring also spent five years with the Army Corps of Engineers, where his duties included managing a large construction group and developing and presenting a human relations course to over 800 people. A graduate engineer of Norwich University, Mr. Spring is a Registered Professional Engineer, an ASNT NDT Level III certificate holder in thermography, as well as a Certified Energy Manager with the Association of Energy Engineers.

PREFACE

One may wonder why the title of this Handbook contains the word "evaluation" instead of the generic term "testing" that is usually used in connection with "Nondestructive." The *American Heritage Dictionary* properly defines "nondestructive" as "Of, relating to, or being a process that does not result in damage to the material under investigation or testing." The most appropriate definitions of the word "test(ing)" from the same source, are "to determine the presence or properties of a substance" and, "to exhibit a given characteristic when subjected to a test." There are also several other definitions that do not really apply. "Evaluate," on the other hand, has a definition that seems to be more fitting for the intent of this handbook: "To examine and judge carefully; appraise." "Evaluation," as defined in ASTM E-1316, is: "A review following interpretation of the indications noted, to determine whether they meet specified acceptance criteria."

In reality, these terms have been used interchangeably with other expressions such as "inspection," "examination," and "investigation." In general, all of these terms refer to the same technology, one that is still widely unknown or misunderstood by the general public. And the use of these different terms may have, in fact, contributed to this misunderstanding. Assuming it is acceptable to take some liberties with these definitions, I would like to suggest that the an appropriate definition of NDE, NDT, or NDI would be: "A process that does not result in any damage or change to the material or part under examination and through which the presence of conditions or discontinuities can be detected or measured, then evaluated."

It is the intent of this Handbook to introduce the technology of nondestructive testing to those who are interested in a general overview of the most widely used methods. There are many excellent reference books on the various methods that can provide additional in-depth information, if desired.

The key ingredient in the NDT process is the practitioner. Many times, NDT personnel are subjected to unfavorable environments and hazardous working conditions. These same individuals are required to complete extensive training programs and fulfill lengthy experience requirements as a prerequisite to becoming certified. And it doesn't stop there. Many codes and specifications require periodic retraining and recertification. Most inspectors/examiners are under constant scrutiny by client auditors or third party overseers. At times, travel to remote locations is required, resulting in extended periods away from home and long workdays. There should always be that desire to "do it right." Think of the consequences if a serious discontinuity is missed and some type of failure results. Conscientious examiners are concerned and caring individuals. In NDT, there is no room for those who are "just doing their job." It takes a special kind of dedicated person, but the rewards are great! The thought of helping mankind by being involved in a technology that is devoted to making this world a safer place is motivation for many. NDT is an honorable profession for those who are honorable. When NDT practitioners lose their ethics, they have lost everything!

This Handbook has been created by a group of professionals who all believe this to be true. It is our desire that it will be a source of knowledge and reference for many who are interested in this unique and challenging technology. The quest for excellence should be never-ending. As Robert Browning once wrote, "Ah! But a man's reach should exceed his grasp, or what's a heaven for?"

ACKNOWLEDGMENTS

This Handbook is the result of the combined efforts of many. Each contributor spent untold hours in the preparation of his segment and had to persevere through the many phone calls and e-mails received from the primary author. But this book would not have been possible without the support, encouragement, and dedication of Michael and Sheryl Shakinovsky. They did more than help. They worked, motivated; but mostly, they cared. I shall always be in their debt.

In addition to Mike and Sheryl, and the contributing authors, the efforts of the following added so much: Alice Baldi (tables and word processing), Christina Hellier (word processing and much encouragement), Lynne Hopwood (graphic design and illustrations), and William Norton (text review).

Finally, this Handbook would have taken much longer if it wasn't for the understanding, patience and support of the Rockwood Service Corporation management, especially Peter Scannell and James Treat. It seems that the word "thanks" just isn't enough.

HANDBOOK OF NONDESTRUCTIVE EVALUATION

CHAPTER 1
INTRODUCTION TO NONDESTRUCTIVE TESTING

I. WHAT IS NONDESTRUCTIVE TESTING?

A general definition of nondestructive testing (NDT) is an examination, test, or evaluation performed on any type of test object without changing or altering that object in any way, in order to determine the absence or presence of conditions or discontinuities that may have an effect on the usefulness or serviceability of that object. Nondestructive tests may also be conducted to measure other test object characteristics, such as size; dimension; configuration; or structure, including alloy content, hardness, grain size, etc. The simplest of all definitions is basically an examination that is performed on an object of any type, size, shape or material to determine the presence or absence of discontinuities, or to evaluate other material characteristics. Nondestructive examination (NDE), nondestructive inspection (NDI), and nondestructive evaluation (NDE) are also expressions commonly used to describe this technology. Although this technology has been effectively in use for decades, it is still generally unknown by the average person, who takes it for granted that buildings will not collapse, planes will not crash, and products will not fail. Although NDT cannot guarantee that failures will not occur, it plays a significant role in minimizing the possibilities of failure. Other variables, such as inadequate design and improper application of the object, may contribute to failure even when NDT is appropriately applied.

NDT, as a technology, has seen significant growth and unique innovation over the past 25 years. It is, in fact, considered today to be one of the fastest growing technologies from the standpoint of uniqueness and innovation. Recent equipment improvements and modifications, as well as a more thorough understanding of materials and the use of various products and systems, have all contributed to a technology that is very significant and one that has found widespread use and acceptance throughout many industries. This technology touches our lives daily. It has probably done more to enhance safety than any other technology, including that of the medical profession. One can only imagine the significant number of accidents and unplanned outages that would occur if it were not for the effective use of nondestructive testing. It has become an integral part of virtually every process in industry, where product failure can result in accidents or bodily injury. It is depended upon to one extent or another in virtually every major industry that is in existence today.

Nondestructive testing, in fact, is a process that is performed on a daily basis by the average individual, who is not aware that it is taking place. For example, when a coin is deposited in the slot of a vending machine and the selection is made, whether it is candy or a soft drink, that coin is actually subjected to a series of nondestructive tests. It is checked for size, weight, shape, and metallurgical properties very quickly, and if it passes all of these tests satisfactorily, the product that is being purchased will make its way through the dispenser. It is common to use sonic energy to determine the location of a stud behind a wallboard. The sense of sight is employed regularly to evaluate characteristics such as color, shape, movement, and distance, as well as for identification purposes. These examples, in a very broad sense, meet the definition of nondestructive testing—an object is evaluated without changing it or altering it in any fashion.

The human body has been described as one of the most unique nondestructive testing instruments ever created. Heat can be sensed by placing a hand in close proximity to a hot object and, without touching it, determining that there is a relatively higher temperature present in that object. With the sense of smell, a determination can be made that there is an unpleasant substance present based simply on the odor that emanates from it. Without visibly observing an object, it is possible to determine roughness, configuration, size, and shape simply through the sense of touch. The sense of hearing allows the analysis of various sounds and noises and, based on this analysis, judgments and decisions relating to the source of those sounds can be made. For example, before crossing a street, one can hear a truck approaching. The obvious decision is not to step out in front of this large, moving object. But of all the human senses, the sense of sight provides us with the most versatile and unique nondestructive testing approach. When one considers the wide application of the sense of sight and the ultimate information that can be determined by mere visual observation, it becomes quite apparent that visual testing (VT) is a very widely used form of nondestructive testing.

In industry, nondestructive testing can do so much more. It can effectively be used for the:

1. Examination of raw materials prior to processing
2. Evaluation of materials during processing as a means of process control
3. Examination of finished products
4. Evaluation of products and structures once they have been put into service

Nondestructive testing, in fact, can be considered as an extension of the human senses, often through the use of sophisticated electronic instrumentation and other unique equipment. It is possible to increase the sensitivity and application of the human senses when used in conjunction with these instruments and equipment. On the other hand, the misuse or improper application of a nondestructive test can cause catastrophic results. If the test is not properly conducted or if the interpretation of the results is incorrect, disastrous results can occur. It is essential that the proper nondestructive test method and technique be employed by qualified personnel, in order to minimize these problems. Conditions for effective nondestructive testing will be covered and expanded upon later in this chapter.

To summarize, nondestructive testing is a valuable technology that can provide useful information regarding the condition of the object being examined once all the essential elements of the test are considered, approved procedures are followed, and the examinations are conducted by qualified personnel.

II. CONCERNS REGARDING NDT

There are certain misconceptions and misunderstandings that should be addressed regarding nondestructive testing. One widespread misconception is that the use of nondestructive testing will ensure, to a degree, that a part will not fail or malfunction. This is not necessarily true. Every nondestructive test method has limitations. A nondestructive test by itself is not a panacea. In most cases, a thorough examination will require a minimum of two methods: one for conditions that would exist internally in the part and another method that would be more sensitive to conditions that may exist at the surface of the part. It is essential that the limitations of each method be known prior to use. For example, certain discontinuities may be unfavorably oriented for detection by a specific nondestructive test method. Also, the threshold of detectability is a major variable that must be understood and addressed for each method. It is true that there are standards and codes that describe the type and size of discontinuities that are considered acceptable or rejectable, but if the examination method is not capable of disclosing these conditions, the codes and standards are basically meaningless. Another misconception involves the nature and characteristics of the part or object being examined. It is essential that as much information as possible be known and understood as a prerequisite to establishing test techniques. Important attributes such as the processes that the part has undergone and the intended use of the part, as well as applicable codes and standards, must be thoroughly understood as a prerequisite to performing a nondestructive test. The nature of the discontinuities that are anticipated for the particular test object should also be well known and understood.

At times, the erroneous assumption is made that if a part has been examined using an NDT method or technique, there is some magical transformation that guarantees that the part is sound. Codes and standards establish minimum requirements and are not a source of assurance that discontinuities will not be present. There are acceptable and rejectable discontinuities that are identified by these standards. There is no guarantee that all acceptable discontinuities will not cause some type of problem after the part is in service. Again, this illustrates the need for some type of monitoring or evaluation of the part or structure once it is operational.

Another widespread misunderstanding is related to the personnel performing these examinations. Since NDT is a "hands-on" technology, the qualifications of the examination personnel become a very significant factor. The most sophisticated equipment and the most thoroughly developed techniques and procedures can result in potentially unsatisfactory results when applied by an unqualified examiner. A major ingredient in the effectiveness of a nondestructive test is the personnel conducting it and their level of qualifications. This will be addressed in greater detail later in this chapter.

III. HISTORY OF NONDESTRUCTIVE TESTING

Where did NDT begin? There are those who would answer this question by referring to the account of the creation of the heavens and the earth in *Genesis:* "In the beginning, God created the heavens and the earth and He *saw* that it was good" (Figure 1-1). This is a theme that has been used from time to time when discussing the history of nondestructive testing. Seeing that the "heavens and the earth were good" has been identified as the first nondestructive test—a visual test!

It is impossible to identify a specific date that would indicate exactly when nondestructive testing, as we know it today, began. In ancient times, the audible ring of a Damascus sword blade would be an indication of how strong the metal would be in combat. This same "sonic" technique was used for decades by blacksmiths (Figure 1-2) as they listened to the ring of different metals that were being shaped. This approach was also used by early bell-makers. By listening to the ring of the bell, the soundness of the metal could be established in a very general way. Visual testing, while not "offi-

FIGURE 1-1   Earth from Space. (Courtesy of Library of Congress.)

cially" considered a part of early NDT technology, had been in use for many years for a wide range of applications. Heat sensing was used to monitor thermal changes in materials, and "sonic" tests were performed well before the term "nondestructive testing" was ever used.

Table 1-1 lists some of the key events in the chronology of NDT and the individuals who were mostly responsible for these developments. Certainly there were many other individuals who have made significant contributions to the growth of NDT, but it is impossible to name them all.

From the late 1950's to present, NDT has seen unprecedented development, innovation, and growth through new instrumentation and materials. The ability to interface much of the latest equipment with computers has had a dramatic impact on this technology. The ability to store vast amounts of data with almost instant archival capability has taken NDT to a level once only imagined, yet NDT technology is still in its infancy. This chronology will continue to grow as exciting new challenges present themselves through technology expansion and unique material developments. The quest to detect and identify smaller discontinuities will not end until catastrophic failures can no longer be related to the existence of material flaws.

The roots of nondestructive testing began to take form prior to the 1920s, but the majority of the methods that are known today didn't appear until late in the 1930s and into the early 1940s. Much of the latter developments came about as a result of the tremendous activity during the Second World War. In the 1920s, there was an awareness of some of the magnetic particle tests (MT) and, of course, the visual test (VT) methods, as well as X-radiography (RT), which at that time was primarily being used in the medical field. In the early days of railroading, the forerunner of the present day penetrant test (PT), a tech-

FIGURE 1-2 Early blacksmith. (Courtesy of C. Hellier.)

nique referred to as the "oil and whiting test," had been widely used. And there were also some basic electrical tests using some of the basic principles of eddy current testing (ET). The sonic or "ringing" method, as well as some archaic gamma radiographic techniques using radium as the source of radiation, were both used with limited success. From these roots, NDT technology has evolved to encompass the many sophisticated and unique methods that are in use today. (See Table 1-2 for a comprehensive overview of the major NDT methods.)

Prior to World War II, design engineers were content to rely on unusually high safety factors, which were usually built or engineered into many products, such as pressure vessels and other complex components, of that time. As a result of the war effort, the relationship of discontinuities and imperfections relative to the useful life and application of a product or system became a concern. In addition, there were a significant number of catastrophic failures and other accidents relating to product inadequacies that brought the concern for system and component quality to the forefront. Some of the improvements in fabrication and inspection practices can be attributed to boilers (Figure 1-3) and some of their early catastrophic failures.

One such failure occurred on a sunny and unseasonably warm day in Hartford, Connecticut, in March of 1854. People were just returning to their offices and shops after

TABLE 1-1 Chronology of Early Key Events in NDT

BC (approx.)	Visual testing becomes the first NDT method when God creates the heavens and earth and "sees" that it is good!
1800	First thermography observations by Sir William Herschel
1831	First observation of electromagnetic induction by Michael Farraday
1840	First infrared image produced by Herschel's son, John
1868	First reference to magnetic particle testing reported by S. H. Saxby, by observing how magnetized gun barrels affect a compass
1879	Early use of eddy currents to detect differences in conductivity, magnetic permeability, and temperature initiated by E. Hughes
1880–1920	"Oil and whiting" technique, forerunner of present-day penetrant test used for railroad axles and boilerplates
1895	X-rays discovered by Wilhelm Conrad Roentgen
1898	Radium discovered by Marie and Pierre Curie
1922	Industrial Radiography for metals developed by Dr. H. H. Lester
1927–28	Electric current innduction/magnetic field detection system developed by Dr. Elmer Sperry and H. C. Drake for the inspection of railroad track
1929	Magnetic particle tests/equipment pioneered by A. V. deForest and F. B. Doane
1929	First experiments using quartz transducers to create ultrasonic vibrations in materials were conducted by S. Y. Sokolov in Russia
1930	Practical uses for gamma radiography using radium were demonstrated by Dr. Robert F. Mehl
1935–1940	Penetrant techniques developed by Betz, Doane, and DeForest
1935–1940's	Eddy current instrument developments by H. C. Knerr, C. Farrow, Theo Zuschlag, and Dr. F. Foerster
1940–1944	Ultrasonic test method developed in United States by Dr. Floyd Firestone
1942	First ultrasonic flaw detector using pulse-echo introduced by D. O. Sproule (United Kingdom)
1946	First portable ultrasonic thickness measuring instrument, the Audigage, was introduced by Branson
1950	Acoustic emission introduced as an NDT method by J. Kaiser
Mid 1950's	First ultrasonic testing immersion B and C scan instruments developed by Donald C. Erdman

lunchtime. At about two o'clock in the afternoon, a man stepped into the engine room of the Fales and Gray Car Works and began a conversation with the operating engineer. Just about that time, the boiler exploded with tremendous force (Figure 1-4). The explosion destroyed the boiler room and an adjoining blacksmith shop, and it severely damaged the main building. As a result of this dramatic boiler explosion, nine people were killed immediately and 12 died later. In addition, more than 50 were seriously injured. This boiler was almost new—in service for less than one month. It was manufactured by a reputable, well-experienced boiler manufacturer. It should be emphasized again that at this time, boilers were being made with unusually high safety margins. In fact, many of the early

TABLE 1-2 Major NDT Methods—A Comprehensive Overview

Method	Principles	Application	Advantages	Limitations
Visual testing (VT)	Uses reflected or transmitted light from test object that is imaged with the human eye or other light-sensing device	Many applications in many industries ranging from raw material to finished products and in-service inspection	Can be inexpensive and simple with minimal training required. Broad scope of uses and benefits	Only surface conditions can be evaluated. Effective source of illumination required. Access necessary
Penetrant testing (PT)	A liquid containing visible or fluorescent dye is applied to surface and enters discontinuities by capillary action	Virtually any solid nonabsorbent material having uncoated surfaces that are not contaminated	Relatively easy and materials are inexpensive. Extremely sensitive, very versatile. Minimal training	Discontinuities open to the surface only. Surface condition must be relatively smooth and free of contaminants
Magnetic particle testing (MT)	Test part is magnetized and fine ferromagnetic particles applied to surface, aligning at discontinuity	All ferromagnetic materials, for surface and slightly subsurface discontinuities; large and small parts	Relatively easy to use. Equipment/material usually inexpensive. Highly sensitive and fast compared to PT	Only surface and a few subsurface discontinuities can be detected. Ferromagnetic materials only
Radiographic testing (RT)	Radiographic film is exposed when radiation passes through the test object. Discontinuities affect exposure	Most materials, shapes, and structures. Examples include welds, castings, composites, etc., as manufactured or in-service	Provides a permanent record and high sensitivity. Most widely used and accepted volumetric examination	Limited thickness based on material. density. Orientation of planar discontinuities is critical. Radiation hazard
Ultrasonic testing (UT)	High-frequency sound pulses from a transducer propagate through the test material, reflecting at interfaces	Most materials can be examined if sound transmission and surface finish are good and shape is not complex	Provides precise, high-sensitivity results quickly. Thickness information, depth, and type of flaw can be obtained from one side of the component	No permanent record (usually). Material attenuation, surface finish, and contour. Requires couplant
Eddy current testing (ET)	Localized electrical fields are induced into a conductive test specimen by electromagnetic induction	Virtually all conductive materials can be examined for flaws, metallurgical conditions, thinning, and conductivity	Quick, versatile, sensitive; can be noncontacting; easily adaptable to automation and in-situ examinations	Variables must be understood and controlled. Shallow-depth of penetration, lift-off effects and surface condition
Thermal infrared testing (TIR)	Temperature variations at the test surface are measured/detected using thermal sensors/detectors instruments/cameras	Most materials and components where temperature changes are related to part conditions/thermal conductivity	Extremely sensitive to slight temperature changes in small parts or large areas. Provides permanent record	Not effective for detection of flaws in thick parts. Surface only is evaluated. Evaluation requires high skill level
Acoustic emission testing (AE)	As discontinuities propagate, energy is released and travels as stress waves through material. These are detected by means of sensors	Welds, pressure vessels, rotating equipment, some composites and other structures subject to stress or loading	Large areas can be monitored to detect deteriorating conditions. Can possibly predict failure	Sensors must contact test surface. Multiple sensors required for flaw location. Signal interpretation required.

FIGURE 1-3 Old boiler. (Courtesy of C. Hellier.)

boilers were fabricated before the principles of thermodynamics were fully understood. This boiler failure in Hartford, Connecticut, was ultimately determined to have been caused by an excessive accumulation of steam. Based on a hearing that was held to determine the cause and to establish blame, the jury offered suggestions as to what could be done to prevent or minimize such accidents in the future. Their suggestions included the following:

FIGURE 1-4 Boiler explosion. (Courtesy of C. Hellier.)

- Initiation of regulations to prevent careless or inexperienced people from being in charge of boilers
- Safety inspections to be made on a regular basis by authorized municipal or state representatives
- Boilers should be placed outside the factory buildings
- Boilers should be prohibited from operating at higher temperatures than would be consistent with safety

This was a significant turning point in the importance and progress of inspection and NDT. Ten years later, in 1864, the State of Connecticut passed a Boiler Inspection Law. This law required an annual inspection of every boiler and would result in the issuance of a certificate if the boiler was satisfactory or, if it weren't, the boiler would be retired from service. Another benefit that resulted from this early boiler explosion was the founding of the Polytechnic Club in 1857. Basically, twelve men who had an interest in boilers met periodically and studied the problems relating to steam boilers.

During those early days of boilers, there were many other dramatic failures. One of the most memorable in history involved a steamship named "Sultana," a Mississippi side-wheeler with two tall stacks (Figure 1-5). On April 27th, 1865, she was steaming along above Memphis when three of her four boilers exploded. The actual cause for this catastrophic explosion was never determined. The Sultana usually carried about 375 passengers, but that day the boat was jammed from stem to stern with almost 2200 passengers, mostly union soldiers who had just been released from confederate prisons following Lee's surrender at Appomattox. Eyewitness accounts of this disaster reported that the side-wheeler had burned to the water line within 15 minutes and the death toll, although not precise, was estimated to be between 1200 and 1600. Depending upon the exact number, this could have been the worst disaster in marine history. In fact, there is a possibility

FIGURE 1-5 The *Sultana*. (Courtesy of Library of Congress.)

that there were more lives lost as a result of this explosion than there were when the Titanic sank in 1912 with a loss of 1517 lives.

In the spring of 1866, the Connecticut legislature approved an act of incorporation of the Hartford Steam Boiler Inspection and Insurance Company (Figure 1-6). This is significant because the premise upon which this company was founded was to provide insurance for boilers. Inspection was required as a prerequisite to issuing an insurance policy. On February 14th, 1867, Policy Number 1 was written on three horizontal tubular boilers for a face value of $5000. The premium was $60. In 1911 the first Boiler Code Committee was formed and the first code was developed during the years of 1911 to 1914 and first published in 1915. Visual testing (VT) was the initial method of nondestructive testing. Surely, the introduction and application of the Boiler Code has had a major impact on the growth and application of nondestructive testing over the years since its inception and first publication.

In 1920 Dr. H. H. Lester, the dean of industrial radiography, began his work at the Watertown Arsenal in Boston, Massachusetts. Figure 1-7 illustrates the lead-lined exposure room of his original X-ray laboratory as it looked in 1922. Dr. Lester was directed to develop X-ray techniques for the examination of castings, welds, and armor plate to basically improve the quality of materials used by the Army. Even though William Conrad Roëntgen had discovered X-rays some 27 years earlier, not much had been accomplished in applying X-rays for materials evaluation, due primarily to the low energy of the early X-ray units. Dr. Lester's laboratory had equipment that was archaic by today's standards, but his work and that early equipment served as the foundation for future development of the radiographic test method using X-ray sources.

FIGURE 1-6 Hartford steam boiler advertisement. (Courtesy of C. Hellier.)

FIGURE 1-7 Dr. Lesters' X-ray laboratory. (Courtesy of C. Hellier.)

The next key development in the history of nondestructive testing was also due to a catastrophe—a major train derailment. This resulted in the electric current innduction/magnetic field detection system that was developed by Dr. Elmer Sperry and H. C. Drake (Figure 1-8). The primary use of this method was to detect discontinuities in railroad track. From this development came the basic principles upon which the Sperry Rail Service was founded. This ultimately resulted in more advanced railroad track test cars. Tracks are still being inspected today using similar principles. This makes Sperry Rail Service the oldest continuously operated NDT service group in the United States. In 1929, the Magnaflux Corporation was formed to promote the use of magnetic principles for industrial NDT applications. Magnetic particle testing (MT) principles came from early electromagnetic conduction and induction experiments performed by Professor A. V. deForest and F. B. Doane. Some of the early equipment was very limited in its applica-

FIGURE 1-8 Early Sperry inspection railcar. (Courtesy of C. Hellier.)

FIGURE 1-9 Early magnetic particle unit. (Courtesy of C. Hellier.)

tion, but at that time it was a unique and novel technique able to detect surface discontinuities in ferromagnetic materials. Some of the early wet horizontal units manufactured by Magnaflux, like the one illustrated in Figure 1-10, are still in use today.

Dr. Robert F. Mehl was instrumental in developing the practical industrial uses of radium for gamma radiography in the 1930s. He was instrumental in expanding the use of radium for the detection of discontinuities in materials that were not possible to be examined with the low energy X-ray equipment in use at that time. In fact, as a result of his early work in gamma radiography, Dr. Mehl gave the first honor lecture (later named in his honor) in 1941. The title of his lecture was, "Developments in Gamma Ray Radiography—1928 to 1941." An early illustration of the use of gamma radiography using radium is illustrated in Figure 1-11, which shows a practice that was commonly known at that time as the "fishpole" technique.

Coincidentally, 1941 was also the year that the American Industrial Radium and X-ray Society, the forerunner of the American Society for Nondestructive Testing as it is known today, was founded. Early X-ray units were only capable of producing low energy X-rays, making exposure times on structures as depicted in Figure 1-12 extremely long.

There were other significant new developments in the 1940s in the area of industrial radiography. The first million volt X-ray machines were introduced by General Electric, which provided higher energies to permit the examination of thicker material cross-sections.

Even though early penetrant testing (PT) techniques, known as "oil and whiting," had been in use since before the turn of the 20th century, the method was not widely used until the addition of visible dyes and fluorescent materials resulting from the research of Robert and Joseph Switzer in the late 1930s and early 1940s. Figure 1-13 illustrates an early fluorescent penetrant system.

FIGURE 1-10 Early "Magnaflux" unit. (Courtesy of CJH Collection.)

FIGURE 1-11 Radium "fishpole" technique. (Courtesy of C. Hellier.)

FIGURE 1-12 Early industrial X-ray unit. (Courtesy of C. Hellier.)

INTRODUCTION TO NONDESTRUCTIVE TESTING   **1.15**

FIGURE 1-13   Early fluorescent penetrant unit. (Courtesy of C. Hellier.)

Even though the principles of eddy current testing (ET) had their roots in 1831, when Michael Faraday discovered the principles of electromagnetic induction, it wasn't until the 1940s that the full potential of this method was realized. The first recorded eddy current test was performed by E. E. Hughes in 1879; he was able to distinguish the difference between various metals by noting a change in excitation frequency, which basically resulted from the effects of test material resistivity and magnetic permeability. But it was not until the year 1926 that the first eddy current instrument, which was used to measure material thickness, was developed. Through World War II and the early 1940s, further developments resulted in better and more practical eddy current instruments. Figure 1-14 illustrates an early instrument developed by the Foerster Institute. Notice that a "standard" was placed in the "primary" coil and compared to the response from the part in the "secondary" coil. In the 1950s, Forster also developed advanced instruments with impedance plane signal displays, which made it possible to discriminate between a number of parameters.

Since the beginning of time, it has been known that materials can emit certain noises when they are stressed. For example, when a piece of wood is bent, creaking or "crying" sounds can be heard. In fact, as the noise intensity increases, it can, in many cases, serve as a warning that the object is ultimately going to fail or break apart. In the 1950s, the first extensive study of this phenomenon, which we now call acoustic emission testing (AE), was reported by Dr. Joseph Kaiser in Munich, Germany, in his Ph.D. thesis. Basically, his studies demonstrated that acoustic emission events were caused by small failures in a material that was being subjected to stress. Much of the original work was done in the audible frequency sound range, but today, for obvious reasons, most acoustic emission monitoring is conducted at very high or ultrasonic frequencies. AE has grown significantly and this method has become a valuable NDT method for determining condition, behavior, and the in-service characteristics of many materials and structures.

The use of high frequency sound for the detection of discontinuities in materials was

FIGURE 1-14 Early eddy current unit. (Courtesy of C. Hellier.)

also introduced in the 1940s. The efforts of Dr. Floyd Firestone led to the development of an instrument called the supersonic reflectoscope, which was introduced in the United States by Sperry Products in 1944 (Figure 1-15). In other countries, similar efforts were being made. Equipment and instrumentation was also being developed in England, Russia, and Germany.

Since humans can sense variations in temperature, it has always been possible to observe thermal energy changes. The physics of thermography was observed as early as the 1800s. Late in the 19th century, heat radiation was observed and explored, and ultimately, instrumentation was actually developed to measure changes in radiant energy. John Herschel created the first thermal picture in 1840, but the real development of thermal imaging did not occur in major areas of industry until the 1950s and early 1960s. This unique nondestructive test method has seen phenomenal growth and expansion as more and more applications have been developed.

As mentioned earlier, a key period in the history and development of nondestructive testing came during and after the Second World War. Prior to the war, nondestructive testing was typically considered to be part of the inspection activities of various companies where it was being employed. As a recognized technology, it could be said that NDT began with the formation of the American Industrial Radium and X-ray Society (now the American Society for Nondestructive Testing) back in 1941. The evolution of nondestructive testing can be directly related to increased concern for safety, the reduction of safety factors, the development of new materials, and the overall quest for greater product reliability. The changes that have occurred in aerospace, nuclear power and space exploration have all greatly contributed to the exciting and dynamic changes that have been experienced in this technology.

FIGURE 1-15 Early Sperry ultrasonic unit. (Courtesy of C. Hellier.)

IV. NONDESTRUCTIVE VERSUS DESTRUCTIVE TESTS

Destructive testing has been defined as a form of mechanical test (primarily destructive) of materials whereby certain specific characteristics of the material can be evaluated quantitatively. In some cases, the test specimens being tested are subjected to controlled conditions that simulate service. The information that is obtained through destructive testing is quite precise, but it only applies to the specimen being examined. Since the specimen is destroyed or mechanically changed, it is unlikely that it can be used for other purposes beyond the mechanical test. Such destructive tests can provide very useful information, especially relating to the material's design considerations and useful life. Destructive testing may be dynamic or static and can provide data relative to the following material attributes:

- Ultimate tensile strength
- Yield point
- Ductility
- Elongation characteristics
- Fatigue life
- Corrosion resistance
- Toughness
- Hardness
- Impact resistance

One of the more common destructive testing instruments used for mechanical testing, which is capable of measuring characteristics such yield point, elongation, and ultimate tensile strength, is illustrated in Figure 1-16.

Other than the fact that the specimen being examined typically cannot be used after destructive testing for any useful purpose, it must also be stressed that the data achieved through destructive testing are specific to the test specimen. Another destructive test commonly used to measure a materials resistance to impact is the Charpy test. In this test, a specimen that is usually notched is supported at one end and is broken as a pendulum is released and impacts in the region of the notch. The measure of the material's resistance to impact (or notch toughness) is determined by the subsequent rise of the pendulum (See Figure 1-17).

Hardness is also an important material characteristic. The hardness test (See Figure 1-18) measures the material's resistance to plastic deformation. There has always been a minor dispute as to whether this test was nondestructive or destructive, since there usually is an indentation made on the surface of the material. If the hardness test is made with-

FIGURE 1-16 Typical tensile testing machine. (Courtesy of J. Devis Collection.)

FIGURE 1-17 Charpy impact tester. (Courtesy of J. Devis Collection.)

out indentation (as is the case when using eddy currents or ultrasonics), it can be considered truly "nondestructive."

Although it is assumed in many cases that the test specimen is representative of the material from which it has been taken, it cannot be said with 100% reliability that the balance of the material will have exactly the same characteristics as that test specimen. Key benefits of destructive testing include:

- Reliable and accurate data from the test specimen
- Extremely useful data for design purposes
- Information can be used to establish standards and specifications
- Data achieved through destructive testing is usually quantitative
- Typically, various service conditions are capable of being measured
- Useful life can generally be predicted

FIGURE 1-18 Typical hardness tester. (Courtesy of J. Devis Collection.)

Limitations of destructive testing include:

- Data applies only to the specimen being examined
- Most destructive test specimens cannot be used once the test is complete
- Many destructive tests require large, expensive equipment in a laboratory environment

Benefits of nondestructive testing include:

- The part is not changed or altered and can be used after examination
- Every item or a large portion of the material can be examined with no adverse consequences
- Materials can be examined for conditions internal and at the surface
- Parts can be examined while in service
- Many NDT methods are portable and can be taken to the object to be examined
- Nondestructive testing is cost effective, overall

Limitations of nondestructive testing include:

- It is usually quite operator dependent
- Some methods do not provide permanent records of the examination
- NDT methods do not generally provide quantitative data
- Orientation of discontinuities must be considered
- Evaluation of some test results are subjective and subject to dispute
- While most methods are cost effective, some, such as radiography, can be expensive
- Defined procedures that have been qualified are essential

In conclusion, there are obvious benefits for requiring both nondestructive and destructive testing. Each is capable of providing extremely useful information, and when used jointly can be very valuable to the designer when considering useful life and application of the part.

V. CONDITIONS FOR EFFECTIVE NONDESTRUCTIVE TESTING

There are many variables associated with nondestructive testing that must be controlled and optimized. The following are major factors that must be considered in order for a nondestructive test to be effective.

1. *The product must be "testable."* There are limitations inherent with each of the nondestructive test methods and it is essential that these limitations be known so that the appropriate method is applied based on the variables associated with the test object. For example, it would be very difficult to provide a meaningful ultrasonic test on a small casting with very complex shapes and rough surfaces. In this case, it would be much more appropriate to consider radiography. In another case, the object may be extremely thick and high in density, making radiography impractical. Ultrasonic testing, on the other hand, may be very effective. In addition to the test object being "testable," it must also be accessible.

2. *Approved procedures must be followed.* It is essential that all nondestructive examinations be performed following procedures that have been developed in accordance with the requirements or specifications that apply. In addition, it is necessary to qualify or "prove" the procedure to assure that it will detect the applicable discontinuities or conditions and that the part can be examined in a manner that will satisfy the requirements. Once the procedure has been qualified, a certified NDT Level III individual or other quality assurance person who is suitably qualified to properly assess the adequacy of the procedure should approve it.

3. *Equipment is operating properly.* All equipment to be used must be in good operating condition and properly calibrated. In addition, control checks should be performed periodically to assure that the equipment and accessory items are functioning properly. Annual calibrations are usually required but a "functional" check is necessary as a prerequisite to actual test performance.

4. *Documentation is complete.* It is essential that proper test documentation be completed at the conclusion of the examination. This should address all of the key elements of the examination, including calibration data, equipment and part description, procedure used, identification of discontinuities if detected, etc. These are all key elements. In

addition, the test documentation should be legible. There have been cases where the examination was performed properly and yet the documentation was so difficult to interpret that it cast doubt on the results and led to concerns regarding the validity of the entire process.

5. *Personnel are qualified.* Since nondestructive testing is a "hands-on" technology and depends greatly on the capabilities of the individuals performing the examinations, personnel must not only be qualified, but also properly certified. Qualification involves both formalized planned training, testing, and defined experience.

VI. PERSONNEL CONSIDERATIONS

Introduction

The effectiveness of a nondestructive test is primarily dependent upon the qualifications of the individuals performing the examinations. Most nondestructive tests require thorough control of the many variables associated with these examinations. The subject of personnel qualification has been an issue of much discussion, debate, and controversy over several decades. There are many different positions regarding the assurance that an NDT practitioner is qualified. The most common approach is to utilize some form of certification. The controversies and the different positions taken regarding this very emotional subject have resulted in the development of a number of different certification programs.

The term "qualification" generally refers to the skills, characteristics, and abilities of the individual performing the examinations, which are achieved through a balanced blend of training and experience. "Certification" is defined as some form of documentation or testimony that attests to an individual's qualification. Therefore, the obvious process involved in the attainment of a level of certification necessitates that the individual satisfactorily completes certain levels of qualification (training combined with experience) as a prerequisite to certification. In fact, a simple way to relate to this system would be to consider the steps involved in becoming a licensed driver. A candidate for a driver's license must go through a series of practical exercises in learning how to maneuver and control a motor vehicle and, in time, is required to review and understand the various regulations dealing with driving that vehicle. Once the regulations are studied and understood, and upon completion of actual practice driving a vehicle, the individual is then ready to take the "certification examination." Most states and countries require the applicant to pass both written and vision examinations, as well as to demonstrate their ability to operate and maneuver the motor vehicle. Once those examinations are completed, the candidate is issued the "certification" in the form of a driver's license. The mere possession of a driver's license does not guarantee that there will not be mistakes. It is obvious that there are individuals who carry driver's licenses but are not necessarily qualified to safely drive the vehicles. This is quite apparent during "rush hour" traffic time. It is unfortunate that the same situation occurs in nondestructive testing. Since individuals by the thousands are certified by their employers, there are major variations within a given level of certification among NDT practitioners. Those countries that have adopted some form of centralized certification do not experience these variations to the same degree as those who still utilize employer certification approaches.

History

One of the earliest references to any form of qualification program for NDT personnel was found in the 1945 Spring issue of a journal entitled "Industrial Radiography," which was published by the American Industrial Radium and X-ray Society. The name of this organization was eventually changed to the Society for Nondestructive Testing (SNT) and, ultimately, the American Society for Nondestructive Testing (ASNT). The original journal, *Industrial Radiography,* is now referred to as *Materials Evaluation.* An article in that 1945 issue entitled "Qualifications of an Industrial Radiographer" proposed that the Society establish standards for the "registration" of radiographers by some type of examination, which would result in a certification program. By the late 1950s, the subject of qualification or registration was being discussed more frequently. A 1961 issue of the same journal, which was renamed "Nondestructive Testing," contained an article entitled "Certification of Industrial Radiographers in Canada." Then, in 1963, at the Society for Nondestructive Testing's national conference, a newly formed task group presented a report entitled "Recommended Examination Procedure" for personnel certification. Finally, in 1967, ASNT published the first edition of a "Recommended Practice" for the qualification and certification of nondestructive testing personnel in five methods (PT, MT, UT, RT, and ET). This first edition was referred to as the 1968 edition of SNT-TC-1A. This practice, which was a set of recommendations, was designed to provide guidelines to assist the employer in the development of a procedure that the document referred to as a "Written Practice." This Written Practice became the key procedure for the qualification and certification of the employer's NDT personnel.

Today, SNT-TC-1A continues to be used widely in the United States as well as in many other countries and, in fact, it is probably the most widely used program for NDT personnel certification in the world. Over the years, it has been revised, starting in 1975, then again in 1980, 1984, 1988, 1992, 1996, and 2000. With this pattern of revisions, it is anticipated that there should be a new revision every four years. Coincidental with the different editions of SNT-TC-1A, the Canadian program was first made available in 1961. A brief description of some of the commonly used certification programs follows.

An Overview of SNT-TC-1A

This program—a "Recommended Practice"—provides for personnel certifications to three different levels. Individuals who are just beginning their NDT careers are usually referred to as "trainees." A trainee is one who is in the process of becoming qualified to be certified as an NDT practitioner. After completion of recommended formalized training and experience, the trainee is considered to be qualified. Upon satisfactory passing of the recommended examinations, the individual can then be certified as a Level I. The Level I is an individual who is qualified to perform specific calibrations, tests, and evaluations for acceptance or rejection in accordance with written instructions or procedures, and to record the results of those examinations. A certified Level I individual should receive instruction and technical supervision from an individual who is certified to a higher level. After completion of additional training, and experience, the Level I can take additional examinations and then be considered qualified to become certified as an NDT Level II.

By definition, a certified NDT Level II individual is qualified to set up and calibrate equipment and to interpret, as well as evaluate, the test results with respect to applicable codes, standards, and specifications. Generally, the NDT Level II will follow procedures that have been prepared and approved by the highest level in this system, a Level III. In addition, the Level II may be responsible for providing on-the-job training and guidance to the trainees and NDT Level I personnel. It is expected that the Level II also be thoroughly familiar with the various aspects of the method for which qualified. A very important qualification requirement for the Level IIs is that they should be able to organize and report the results of the NDT being performed.

The highest level of certification described in SNT-TC-1A is the NDT Level III. By definition, the Level III is a highly qualified individual, capable of establishing techniques, developing procedures and interpreting codes, standards, and specifications. The Level III should also be capable of designating the particular test method for a given application, as well as specifying the correct technique and procedure to be used. In general, the Level IIIs are responsible technically for the NDT operations for which they are qualified and assigned, and should also have the capability of interpreting and evaluating examination results in terms of existing codes, standards, and specifications. Further, the Level IIIs should have sufficient practical background in the materials, fabrication, and product technology applicable to their job function in order to establish techniques and to assist in establishing acceptance criteria where none are otherwise available. It is also acknowledged that the Level III should have a general familiarity with the other major NDT methods and be capable of training and examining those individuals who would be candidates for Level I and Level II certification.

To summarize, SNT-TC-1A is a set of guidelines designed to assist employers in the development of their own certification and qualification procedures. Since this program is a set of recommendations, it is intended to provide employers with wide latitude so their procedures can be tailored to the requirements of the products being manufactured and to meet the needs of their clients. The application of this program has resulted in some distinct benefits and significant shortcomings.

SNT-TC-1A Benefits

1. It provides for flexibility as interpreted and applied by each individual employer.
2. It requires employers to analyze their position on certification and document it through the preparation of their written practice.
3. It implies customer responsibility through their acceptance and evaluation of the written practice, which, in turn can assure the adequacy necessary to comply with the terms and conditions of the contract or purchase order.
4. Through the written practice or procedure, the employer has an implied responsibility to train and to assure that their personnel are experienced and competent.
5. It gives the employer excellent guidelines with respect to examinations that can be administered as part of the certification process.
6. It provides a common foundation that, even with its flexibility, provides an audit path.

SNT-TC-1A—Its Limitations

1. The lack of consistency that results from employers taking advantage of the latitude contained in this recommended practice.
2. The fact that employers certify their personnel assumes a high level of responsibility and competence on the part of the employer. If employers do not understand the recommendations, or do not apply the intent of the recommendations, the effectiveness of their program can be questioned.
3. There are several paths that an individual can take to achieve Level III certification through the employer, which again results in a lack of uniformity with an employer-based certification scheme. An individual can be designated by the employer as a Level III, examined by the employer to become a Level III, or can be examined through the use of an independent outside agency and, based on the results, become certified as a Level III by the employer.

Summary

In summary, the limitations and weaknesses of SNT-TC-1A, being an employer-based program which depends greatly upon the integrity of the employer, leaves a lot to be desired. The solution to this problem lies with either a standard or code that will establish minimum requirements for certification, or a fully centralized approach to the certification of NDT personnel.

ASNT Level III Certificate Program

In 1976, the American Society for Nondestructive Testing initiated a program to issue Level III certificates to individuals through a "grandfathering" process that provided for issuance of the certificates without an examination. A select committee of highly recognized NDT professionals was appointed to review the applications submitted by those who met certain minimum requirements that had been previously established by an ad hoc Level III committee. During a window of 6 months, September 1, 1976 to February 28, 1977, over 1300 applications were received and reviewed by the select committee. Upon completion of the review, a total of 713 individuals were granted a Level III certificate by grandfathering. Beginning in 1977, ASNT offered Level III examinations in PT, MT, RT, UT, and ET. Eventually, VT, AET, and TIR were added. Examinations now include Neutron Radiography (NRT), Leak Testing (LT), and Vibration Analysis (VA) Testing. (These latter three methods are not included in this Handbook.)

The examinations administered by ASNT are written and cover the fundamentals and principles for each method. There are also questions relating to the application and establishment of techniques and procedures. The subjects of interpreting codes and specifications as they apply to each method are also included. In addition to the method examination, a basic examination is also administered that covers topics such as materials, fabrication, product technology, certification programs, and general knowledge of the basic NDT methods. The basic examination is required to be taken and passed only once, and is a prerequisite to obtaining a certificate for any method. Once the Level III certificate is issued, the employer may have to administer additional examinations if required by their written practice. Once the conditions of the written practice are met, the employer can certify the individual. If the written practice states that the only condition for certification as Level III is the passing of the ASNT examinations, the employer may then certify with no further examinations other than a vision examination, which is a requirement for all levels. In this context, certification remains the responsibility of the employer.

Certification Program Overview

Some of the other commonly used certification programs are briefly described here. There are certain elements of each program that are unique, but a review of these programs points to a need for the uniformity and universal endorsement of a central program that will apply consistently to all NDT personnel.

ASNT/ANSI-CP189

This program is a standard for the qualification and certification of nondestructive testing personnel. Initially issued in 1991 and revised in 1995, it provides minimum requirements for personnel whose job functions require appropriate knowledge of the various principles and application of the nondestructive testing method for which they will be certified. There are similarities between SNT-TC-1A and CP189. The major difference is that

CP189 is a standard (not a recommended practice). Since it is still an employer-based program, it requires the employer to develop a procedure for the certification and qualification of their personnel. The possession of an ASNT Level III Certificate is a prerequisite to Level III Certification by the employer.

MILSTD-410
This military standard was published as MILSTD-410D, which was created in 1974 and was eventually superceded by MILSTD-410E in 1991. The program is primarily used by military agencies and government agencies, as well as by a number of prime aerospace companies and their contractors. It contains the minimum requirements for the qualification and certification for personnel in nondestructive testing and includes provisions for training, experience, and examination. This program has been replaced with an Aerospace Industry Association document referred to as NAS410. While MILSTD-410E continues to be employed in certain industries, it has been replaced by NAS410. The format of both MILSTD-410E and NAS410 are similar in many respects to SNT-TC-1A and CP189.

NAVSEA 250-1500
This program, which was initiated in the late 1950s, was specifically developed for use in the naval nuclear program. It is a form of central certification requiring candidates to go to an examination site to take examinations that consist of both written and practical parts. They are administered by an independent agency. It is generally believed that this program was developed as a result of the late Admiral Rickover's disdain for recommended practices in the areas of quality assurance and NDT.

ISO 9712
This international standard was first published in 1992 and revised in 1999/2000, and is believed by many to be an excellent format for a truly international central certification program. It establishes a system for the qualification and certification of personnel to perform "industrial" nondestructive testing. Certifications under this program are accomplished by a central independent body, which administers procedures for the certification of NDT personnel. There are many countries that have adopted ISO 9712 as a basis for their country's certification program, including Canada, Japan, most of the European countries, Kenya, and others.

The ASNT Central Certification Program (ACCP)
The ACCP was introduced in 1996 and provides for the certification of NDT personnel through the administration of both written and hands-on practical examinations. It is designed to provide a form of centralized certification to those industries and employers who believe there are many benefits to this approach as compared to an employer-based system. It has not been warmly endorsed or widely used at this time, and future implementation does not look promising.

Other Major Certification Programs
Certification Scheme for Welding and Inspection Personnel (CSWIP). This has been in use since the early 1970s and applies specifically to personnel involved with the examination of welds. Since its inception, over 20,000 individuals have been certified worldwide under this program. The administration of this program is the responsibility of The Welding Institute (TWI) in England.

Personnel Certification in Nondestructive Testing (PCN). This is a worldwide centralized certification scheme that was developed from guidelines established in the 1970's.

Since 1985, it has superceded certification programs operated by a number of other organizations and is now one of the most widely recognized schemes in the world. It addresses industry sectors such as aerospace, casting, welding, wrought metals, and railroads. Over 15,000 certificates have been issued globally. It is ISO 9712 compliant.

The Canadian Certification System (CSGB). This began in 1960 as a centralized certification program and has remained a basically third-party program administered by the Ministry of the Federal Government of Canada (Natural Resources Canada). This program has implemented the provisions of the Canadian National NDT Standards. Standard CAN/CGSB-48.9712-95 complied with ISO 9712, 1992 edition. In December 2000, it was replaced with CAN/CGSB-48.9712, 2000 edition, which complies with ISO 9712, 1999 edition, and EN 473, 2000 edition. Broad-sector certification is offered in five methods (RT, UT, ET, MT, and PT) to three levels. Three levels of certification in Radiography are also available for the aerospace industry.

Some countries, including France (COFREND), Japan (JSNDI), and others, have centralized certification programs that either comply with, or are patterned after, ISO 9712.

VII. CERTIFICATION SUMMARY

As evidenced by the number of diverse certification schemes that are in existence worldwide, some would say that NDTcertification is in a state of chaos. It has become more evident that a globally recognized, centralized certification program must be adopted and agreed upon by all countries utilizing NDT. SNT-TC-1A certification will most likely remain in use in certain industries, but the benefits of central certification far outweigh the benefits and possible cost-effectiveness of employer-based certification programs. The bottom line is that NDT practitioners must be qualified and there must be an independent, unbiased system for evaluating their qualifications and establishing their credentials. The technology of NDT and its predicted growth depends on this.

VIII. REFERENCES

1. Posakony, G. J. *"A Look to the Past—A Look to the Future."* Lester Honor Lecture, Presented at the National Fall Conference of the American Society for Nondestructive Testing, September 27–30, 1976 in Houston, Texas in honor of Dr. Horace H. Lester, NDT Pioneer. Published in the December 1976 issue of *Materials Evaluation.*
2. R. C. McMaster, *Nondestructive Testing Handbook.* Ronald Press, New York, 1959.

CHAPTER 2
DISCONTINUITIES — ORIGINS AND CLASSIFICATION

Structural materials are composed of atoms and molecules that ideally have material continuity extending down into the microscopic scale. Uniformity of material and material properties is desired for most engineering applications. Design engineers assume some level of structural continuity, homogeneity, and definition of material properties. However, absolute homogeneity and continuity never exist in any engineering component.

Spatially sharp departures from material homogeneity and continuity inside a component at any level of magnification are called discontinuities. Engineering materials always possess some discontinuities, although they may be very small and they may or may not be acceptable. Examples of these discontinuities include voids, inclusions, laps, folds, cracks, chemical segregation, and local changes in microstructure.

Sharp transitions in surface homogeneity, continuity, and contour are also considered to be "discontinuities" on component surfaces. Geometric surface discontinuities include sharp angles, notches, gouges, scratches, galling, fretting, pitting, and welding undercut.

Discontinuities in engineering structures are unacceptable when they degrade the performance or durability of the structure below the expectations of design and when they challenge the operability, reliability, and life of a component. The primary goal of nondestructive examination for discontinuities in engineering materials is to determine whether or not the continuity and homogeneity of a component are adequate for its use.

Identified discontinuities are evaluated as either rejectable or nonrejectable conditions in a part. An evaluation usually is made in reference to a design basis and may include a code or rule-based criteria for acceptance and rejection. The evaluation of a discontinuity generally requires an adequate measurement of its size and location and identification of its character. Discontinuities are evaluated completely by determining their location, number, shape, size, orientation, and type.

The origin and types of discontinuities depend primarily on the manufacturing processes and the service histories of engineering components. In some cases, the operational environment may induce the growth and development of preexisting discontinuities. Discontinuities in structures may originate at any manufacturing step and may be introduced during the component use, maintenance, and repair.

An understanding of the origin of discontinuities is useful in determining the type and features of discontinuities that may be expected in a component. Awareness of the characteristics, locations, and orientations of discontinuities is most helpful and sometimes critical in their detection and evaluation.

Discontinuities may be categorized by the stage in processing at which they are introduced. An "inherent discontinuity" is one that is generated in the original production of an alloy stock material. Discontinuities occurring in the first forming stages from a primary alloy are called "primary processing discontinuities," and any discontinuities that occur in subsequent forming and finishing steps are called "secondary processing discontinuities." The discontinuities that are created during the use of a component are called "service discontinuities." Discontinuities additionally may be categorized in terms of the forming process that caused them. For instance, discontinuities generated during welding are called "welding discontinuities." Any discontinuities caused by casting may be called "casting discontinuities," and discontinuities generated in forging obviously would be "forging discontinuities."

Most engineering structures and components are composed of alloys.[1] Alloys are made from metals that are refined from ores.[2] The metals are cast into primary ingots that are then remelted and mixed with other elements to make alloys.

Alloy ingots are reformed into shapes by melting and casting, mechanical forming, welding, consolidation of alloy powders, and machining. Combinations of forming processes may also be used to shape components. Discontinuities and inhomogeneities may be introduced in the initial winning of metals from ores, in the primary production of alloys, and during any of these subsequent forming steps.

Discontinuities induced during manufacture may propagate with the same or different character during operations and they may also be generated during repair and during use of a component. Discontinuities typically may arise from application and sequences of loads, reactions with chemical and electrochemical environments, or from other environmental conditions. Energetic inputs that create discontinuities in structures include applied and induced loads, ranging from thermal stresses to damage from neutron bombardment in nuclear power reactors. Service discontinuities also can be generated from the microstructural changes that occur over time within some alloys.

I. PRIMARY PRODUCTION OF METALS

Metals originate from ores that are generally rich in rock and relatively poor in metals. The extraction of metals from ores requires processes that often carry over some of the mineral impurities from the rocks and the chemical additions used in the refinement process. These impurities exist in large part as low-density nonmetallic slags that tend to float to the surface of the molten metal and are purposefully segregated and separated from the molten metal.

Small amounts of the slags are often retained within the metals during primary production and become incorporated in alloys. Slag impurities trapped in alloys are usually specified and controlled below allowable limits. However, occasionally the impurity content is outside of acceptable limits, and even if within proper amounts on the average, impurities may not be distributed homogeneously within a primary ingot. Slag concentrations are usually found in the pipe regions of ingots. This localization of slag may be carried over into secondary forming operations to generate stringers and laminations.

Slag inclusions within a structure are discontinuities, usually called nonmetallic inclusions. Even though there is usually material continuity throughout structures that contain inclusions, the physical properties of the inclusions are deficient compared to the alloy

---

1. "Alloy" is used in the classical sense to mean any mixture of elements that possesses metallic properties.
2. An "ore" is a mineral aggregate from which elements may be extracted at a profit. The extraction of metals is called "winning."

matrix. Inclusions are usually weak and brittle in ambient conditions and they may be soft and weak at high temperatures. Inclusions that are planar in shape may act as metallurgical notches within a material.

Inclusions in steel often are composed of the silicates and sulfides that come from the iron ore, the dissolution of refractory systems, and purposeful additions that are made to refine steel from its ores. Many of the nonmetallic inclusions are plastic at high temperatures and capable of being deformed during hot working of steels. When steel is shaped by being deformed at relatively high temperatures, these included particles will elongate in the direction of flow of the steel. Elongated inclusions in a steel structure are called stringers (Figure 2-1). Stringers in adequately high number will result in directionality, a property called "geometric anisotropy," in the steel. Such steel will have a fibrous structure like wood and will be stronger along the elongated fibrous structure and weaker across the elongated structure. Rupture and cracking along the weak planes of this structure during subsequent forming or use generates a discontinuity called "lamellar tearing." Lamellar tearing has the appearance of a split wood grain structure.

Manganese is added to steel during primary production, in part to react with excess sulfur that might be in the steel. Manganese has a greater chemical affinity for sulfur than iron. However, if manganese is not present in adequate quantity to balance out the sulfur, brittle iron sulfide will form in the steel at the boundaries of the first grains that form. Manganese sulfide is generally in the benign form of rounded particles. These spherically shaped discontinuities are accepted in controlled amounts because of their random distribution through the structure and the rounded nature of the phase boundaries.

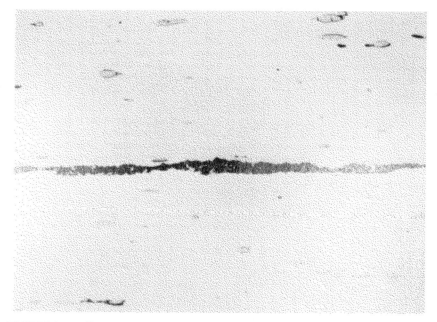

FIGURE 2-1 Silicate stringer in cold drawn bar of steel (235X). [From *The Making, Shaping and Treating of Steel*, 10th Edition, US Steel (1985).]

The inhomogeneities of inclusions common in steels and in other alloys are generally acceptable if the size and total inclusion content is relatively small and the distribution is adequately random.

II. CASTING

A metal or an alloy is transformed from a liquid to a crystalline solid by the extraction of thermal energy from the melt during casting in a mold. Casting is usually done in the primary production of metals and in the initial production of alloys, and is the starting point for many discontinuities in engineering components.

Casting of alloys entails the change of a metallic liquid solution into a crystalline solid alloy. Virtually all of the engineering alloys are crystalline solids.[3] All crystalline solids form inhomogeneously[4] from very small crystals on the relatively cold walls of a casting mold. The crystallization of a solid begins at small and discrete locations called nucleating sites, and alloy crystals grow from these sites into the alloy melt. These separated and individual crystals grow, consuming the melt, until they connect with each other and form a solid. Each of the crystals is called a grain and the place of the meeting of two grains is called a grain boundary. Adjacent grains of one solid phase differ from each other as crystals only in the orientations of their pattern of atoms. Their size, shape, and spatial and geometric disposition further characterize the grains in a structure.[5]

The growth of grains in castings is sometimes directional and the most rapid growth tends to proceed in the direction of heat dissipation. An extreme condition of this directionality in castings is called ingotism (Figure 2-2). Directionality of grain structure may create difficulties in ultrasonic examinations, as is experienced in cast stainless steel piping.

When metals and most alloys are solidified, the crystalline structures are denser than the liquids from which they came. Consequently, the casting volume is less than the melt volume after the liquid changes to a crystalline solid. This phenomenon is called solidification shrinkage. A natural consequence of solidification shrinkage occurs when liquid is entrapped within the crystalline shell of a solidifying casting. Unfilled spaces in the cast solid are created by the shrinkage of the included melt. These voids are called shrinkage voids or shrinkage porosity. A central depression called "pipe" is formed when the shrinkage is concentrated in the top of an ingot (Figure 2-3). Pipe often contains concentrations of slag and must be cropped from the ingot prior to secondary fabrication.

Solidification shrinkage may induce cracking. This cracking may occur at relatively high temperatures when the cast material is weaker. The cracking due to contraction of the cast solid also may occur at weak locations within the microstructure of an alloy, typically at grain and phase boundaries.

Forms of casting discontinuities include pipe, voids, porosity, microporosity, interdendritic porosity, slag, and cracking.

---

3. There is a class of engineering material called metallic glass that currently is used in special applications in relatively small amounts. These are alloys that have some characteristic metallic properties and an amorphous atomic structure.

4. Inhomogeneity of casting nucleation is the local and independent formation of small crystals starting at the liquidus temperature for most alloys and at the solidification temperature for metals.

5. The spatial disposition of grains includes on occasion the nonrandom distribution of orientations of the grains. This material property of a structure is called "texture" and it is a measure of the anisotropy or directionality of material properties in an alloy.

FIGURE 2-2 Ingotism in cross section of stainless steel billet. [From *Liquid Metals and Solidification*, ASMI (1958).]

FIGURE 2-3 A cross section through a billet showing pipe from solidification shrinkage. (Courtesy of C. Hellier.)

Shrinkage voids are simply unfilled macroscopic regions of a casting that have been created by the entrapment of relatively large amounts of melt within the casting. Voids may be found at the last regions of a casting to solidify, and occasionally they appear on the surfaces of castings. Shrinkage voids are potentially deleterious, weakening structures by reducing the continuity of load bearing material. These voids may cause or induce a breach in a boundary, thus allowing subsequent leakage through the wall of a component.

Shrinkage porosity is a distribution of a number of small voids. These small voids are usually localized in clusters and exist in the regions of final solidification. Shrinkage porosity extends from the macroscopic to microscopic scale. Porosity also reduces the load bearing section of a component. If porosity is adequately limited in amount and uniformly distributed, it may have a negligible effect on the strength of a component. Localized porosity in too great a quantity can degrade the mechanical properties of a component. Porosity also presents a potential problem when it provides a path for the breach of a component fluid boundary. Interconnected pores may readily provide leakage paths. Surface porosity may also reduce the fatigue resistance of a component.

Microporosity may be found in the specific form of interdendritic porosity. This occurs when the last fluid to solidify is contained between the arms of crystalline growth forms called dendrites. The dendrite arms are always spaced in a regular array and the arm spacing is dependent on the solidification rate. The dendritic forms are usually transient and the final form of the structure is a grain or multiphase structure. However, the interdendritic porosity will remain as a residue of the casting process. This form of discontinuity is rarely adequate in size, number, or distribution throughout a structure to challenge component integrity.

Gas evolution from melts may also cause porosity in cast structures. Gases are more soluble in liquid alloys than in their crystalline solids. During the crystallization process, the gas is released from the growing crystal into the remaining liquid. If the gas is released in the interior of the casting, it may be entrapped and exist in the form of gas voids or gas porosity (Figure 2-4). Gas porosity has some characteristics different from shrinkage porosity. Unlike shrinkage porosity, gas develops in the solidifying melt with pressure. The gases will try to move within the melt to lower-pressure regions. This occasionally leads to elongated void structures called wormholes. These elongated pores form as the gas tries to move away from the solid interfaces in the casting. In some cases, this elongated structure will provide a ready leakage path through the boundary of the casting. Gas porosity, like shrinkage porosity, will decrease the load bearing capacity of a component.

Dissolved gas emanating near the surface of a component may cause splitting and deforming of material near the surface. The appearance and name of this form of surface and near-surface discontinuity is "blistering."

There are occasions when a casting mold may contain regions that are not completely filled. These discontinuities in structure are called casting cavities. This condition is caused by blockage of the melt from the mold cavity by early solidification within a mold passage.

Shrinkage and the stresses arising from shrinkage after casting may be adequate to rupture a casting, as shown in Figure 2-5a. This is called solidification shrinkage cracking and it may occur through the crystal grains of an alloy or around the boundaries of grains and solid phases. Shrinkage cracks may later propagate in an alloy and cause failure during subsequent heating and mechanically forming. This damage is a form of hot cracking and hot tearing; see Figure 2-5b. Hot cracking usually is considered to be internal cracking and hot tearing involves cracks open to the surface of a casting. These solidification and hot cracking discontinuities are deleterious to service and are unacceptable.

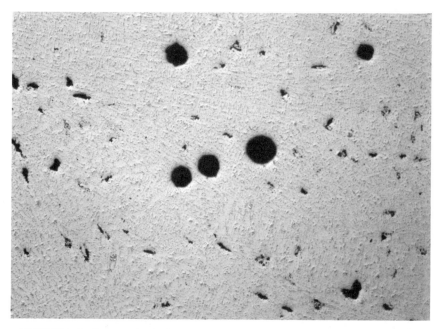

FIGURE 2-4 Gas porosity in alumiunum alloy die casting. (Courtesy of R. B. Pond, Jr.)

Inclusions of foreign objects in casting typically occur when pieces of refractory are broken off into the melt. The refractory may come from the primary production process or from refractories used in casting molds and pouring crucibles.

There is another form of discontinuity characteristic of casting called a cold shut. This is a discontinuity on the surface of a casting caused by a stream of liquid metal solidifying on and not fusing with a previously solidified part of the component. A cold shut may also refer separately to a plugging of a channel in a mold by early solidification, which then prevents the entire mold cavity from filling, resulting in casting cavities.

A scab is a surface discontinuity that has a rough and porous texture, usually with a cavity underneath caused by refractory inclusion near the surface. Scabs are more commonly found in thin sections of castings.

A mold parting line may give rise to a geometric discontinuity on the casting called a casting seam. A mold parting line seam is one indication of a casting process.[6]

An inherent cause of microscopic inhomogeneity in materials comes from phase separation in alloy solids. A phase is a region of matter that is homogeneous and physically distinct from its surroundings, independent of the size, form, or disposition of the phase. Most engineering alloys are composed of structures that are mixtures of numerous microscopic phases that are inhomogeneous in chemistry and/or crystal structure. This is a consequence of the natural separation of phases that occurs over discrete regions of temperature and chemistry, as shown in equilibrium phase diagrams.

---

6. There are casting processes, such as investment casting, which leave no mold parting lines. Parting lines may also be dressed out of a component after casting.

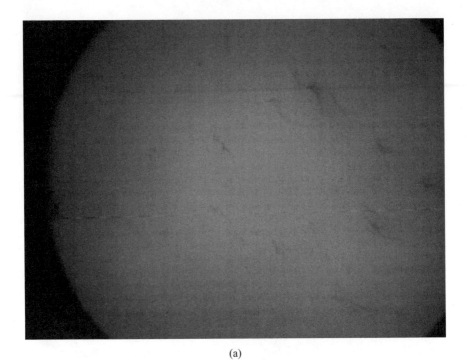
(a)

(b)

FIGURE 2-5 (a) Radiograph of shrinkage in a casting. (b) Radiograph of hot tear in a casting. (Courtesy of C. Hellier.)

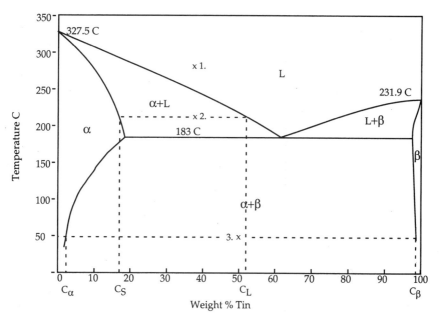

FIGURE 2-6 The equilibrium binary phase diagram for lead-tin. (Courtesy of R. B. Pond, Jr.)

In metallurgical systems, an equilibrium phase diagram is an empirical map of the phases in alloys that exist as a function of temperature and composition. Although most engineering alloys generally are composed of many different phases and characteristic geometric combinations of phases called constituents, the principle of chemical phase separation may be seen by example. A common type of binary phase diagram called a binary eutectic is found in the lead–tin system (Figure 2-6).

There are three points illustrated in the lead–tin diagram in Figure 2-6. Point one is in the single-phase liquid region, and the structure is a liquid solution with a uniform chemistry of lead and tin. At point two, the phase region is composed of two phases, a liquid and a solid. In this two-phase liquid and solid region, the liquid and the solid are not composed of the initial alloy chemistry and the points on the bounding lines at the temperature selected define these two chemistries. In the example, the solid is a lead-rich crystal structure with tin atoms replacing some of the lead atoms. This alloy solid is called a solid solution. In the mixture of the crystalline solid and the liquid alloy at point two, the liquid is richer in tin than is the solid. Neither of the phases is of the original alloy composition, but in combination they must yield the original composition.[7]

Point three of Figure 2-6 is in the two-phase solid region named alpha and beta (α + β). Here is found a mixture of two different crystalline solids. The alloy is composed of a lead-rich crystalline solid solution, with tin atoms replacing some of the lead atoms, and a tin-rich crystal, with lead atoms replacing some of the tin atoms. A micrograph of this structure is seen in Figure 2-7. Like most engineering alloys, this solid is inhomogeneous on the microscopic level.

[7]The average of the liquid and the solid composition *is* the alloy composition.

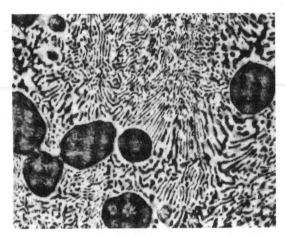

FIGURE 2-7 Microstructure of 50% Sn-Pb (400X). [From *ASM Handbook Volume 9, Metallography and Microstructures*, ASMI (1985).]

Microscopic mixtures of different solid phases are the rule for engineering alloys. The physical properties of the different phases are different. The separation of the different solids on a microscopic scale does not necessarily represent a structural discontinuity of concern. However, the sizes, shapes, and distributions of alloy solid phases characterize the microstructure, which is critical in determining the physical properties of the alloy.

Macroscopic segregation in the casting operation can give rise to inhomogeneities that are deleterious in an alloy. Macroscopic chemical segregation is a condition that often occurs in very large alloy castings, such as ingots. The segregated glassy material formed during casting is called slag. Relatively large local aggregates of slag challenge the mechanical strength of an alloy.

There are also changes in alloy microstructure that occur at relatively high temperatures where a nonequilibrium structure changes to an equilibrium structure. Graphitization of carbon and molybdenum steels is a good example of this phenomenon. The carbides in this alloy steel will decompose to iron and graphite at high temperatures and the location of the decomposition will favor highly stressed regions. Therefore, highly stressed regions of these steels will change their microstructure over time and generate weak particles of graphite. These graphitic regions may eventually initiate cracks.

III. CRACKS

A planar breach in continuity in a material is called a crack. Usually, a crack is envisioned as a physical breach of a material that had previously been continuous. Forces from forming operations or usage of components are generally the cause of cracks.

Cracking may occur during casting operations. Solidification cracking may occur when a cast component with reentrant angles solidifies and then cools and shrinks while it is restrained by the mold. Metallic materials are weaker at high temperatures and the forces from differential contraction may be enough to cause cracking.

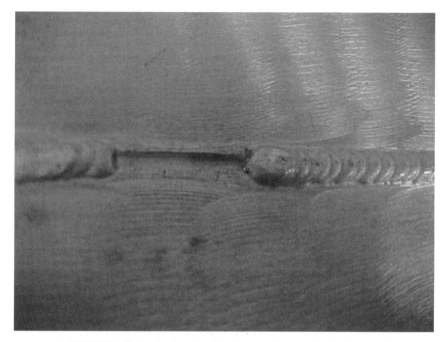

FIGURE 2-8 Lack of penetration in weld. (Courtesy of C. Hellier.)

Additionally, the evolution of gas in the interior of a casting at high temperatures may cause pressure that is adequate to crack the interior of the casting.

There are also crack-like discontinuities that are not formed by forces. These are disruptions in continuity of a component that are caused by the overlaying of two surfaces that are not joined, except along a boundary. An example of a crack-like discontinuity would be lack of fusion in a welding operation. The surface of the weldment is not fused to the base material substrate, and this discontinuity is geometrically similar to a crack. Lack of fusion, like a crack, can initiate fatigue cracking during component use.

Lack of penetration in a welding operation, generally at the root of a weld, may also provide a discontinuity in a structure that has crack-like geometry. For instance, incomplete penetration in an autogenous weld used to join pipe will yield two mating base material surfaces that are a crack-like discontinuity in the joint (Figure 2-8).

IV. WELDING DISCONTINUITIES

There are many welding processes, and each may give rise to discontinuities that are common to casting and occasionally unique to the welding process.

Inclusion of nonmetallic material is common to welding processes that use protective glasses called slags. Welding over slag covered surfaces may trap slag inside the weldment. This type of weld discontinuity occurs in shielded metal arc (SMAW), submerged arc (SAW), and flux core arc welding (FCAW) processes.

Slag inclusions in welding have the same effect as the slag inclusions in cast components. Slag weakens a structure by limiting its load-bearing capability. If rounded in shape, the major effect of slag is a reduction in load bearing material. Generally, there are allowable amounts of slag above which the slag condition is rejectable.

Welding discontinuities include solidification cracks that are caused by the casting operation. There is a stress condition that arises in welding operations due to the differential expansion and contraction of the base material. This is caused by the unrestrained expansion of the base material into the melt during welding. After solidification of the weldment, restraint against the contraction of the base material by the solid bridge across the joint induces stresses as the weld cools.

The differential expansion and subsequent restraint on cooling may give rise to forces that exceed the ultimate strength of either the base material, weldment, or heat-affected zone. This may cause deformation, cracking, fracture, and at the least, residual stresses. These effects are more pronounced in thick wall material and may be minimized by adequate preheating of the base material. The preheating reduces the differential in expansion and contraction during welding. Postweld heat treatment (PWHT) is sometimes additionally required to relieve the residual stresses in a weld and to make beneficial changes in microstructure.

Other cracks in welds are classified according to their location or the conditions of occurrence. For instance, cracking that is created in the base material adjacent to the fusion zone of a weld is called underbead cracking. Cracking occurring in the process of postweld heat treatment is called reheat cracking.

Discontinuities called lack of fusion are found in local unfused regions between beads in a weld or between the base material and the weld. Incomplete melting of a substrate in a welding process causes lack of fusion. These discontinuities are potential initiation sites for fatigue cracks and they are usually disallowed by code.

A microstructural change that occurs in welding is called the heat-affected zone (HAZ). The HAZ is the unmelted region of base material adjacent to the weldment that has a microstructure that is altered by the high temperature of the welding operation (Figure 2-9). This region has effectively been through a locally applied heat treatment during the welding process.

In the case of ferritic steels, the microstructural change is largely due to the allotropic transformation[8] of the steel. This transformation occurs as a function of temperature–time profiles in the HAZ and may be thought of as a local heat treatment of the steel. If the transformation occurs at too rapid a rate because of rapid cooling from the heat sink of the base material, then the heat-affected zone may form a metastable structure called martensite.

Martensite is hard and brittle and prone to crack under quenching stresses and operational loads. Martensite formation tends to occur in medium- and high-carbon steels and in high-strength ferritic alloy steels. Mitigation may require preheating the base material to reduce the rate of cooling so as to reduce or eliminate the martensite transformation. Proper postweld heat treatment will decompose the martensite into a benign microstructure of ferrite and carbides.

Other microscopic and material property changes may occur in the HAZ of different alloys. For instance, an alloy that is strengthened by work hardening will soften in the HAZ from a weld. Heat affected zone properties may result in some damage mechanisms being available. For instance, the HAZ of AISI 304 stainless steel has a microstructure that may be susceptible to stress corrosion cracking.

---

8. Allotropic transformation is the change in crystal structure, in this case, in a specific temperature range.

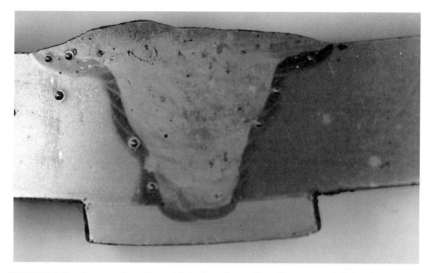

FIGURE 2-9 Cross section of a weld showing heat affected zone as a dark band adjacent to the weld. (Courtesy of R. B. Pond, Jr.)

Similar to casting operations, welding may introduce gas into the molten weldment that subsequently is released, resulting in gas wormholes and porosity.

There is also the possibility of damage due to incorporation of hydrogen atoms into a steel weldment. This typically occurs when water and organic molecules are split apart and disassociated in the high-energy welding process. The melt will incorporate hydrogen that after solidification will exist in nonequilibrium concentrations in the solidified weld.

The hydrogen atoms migrate in the crystal structure until by chance two come close to each other and combine, forming molecular hydrogen. This molecule will not be mobile and will act as a sink for other hydrogen atoms. After a time, there will be multiple small hydrogen bubbles formed in the interior of the crystalline solid. The hydrogen will typically precipitate at inclusion boundaries and grain boundaries.

If the alloy is relatively low strength and ductile, the material around the gas bubbles will plastically deform to accommodate the damage. If the alloy is very strong, it will resist the formation of the bubbles and the first accommodation to stress may be cleavage and fracture of the crystalline structure. This damage is called hydrogen cracking or delayed cracking and it occurs in the interior of the welded component, usually in the HAZ. This mode of damage requires time to occur and is the reason for mandatory requirements for a waiting period prior to nondestructive examination for some high-strength steels.

Because of the sensitivity of hydrogen damage in high-strength steels, sources of hydrogen must be controlled. Requirements are found for dryness and cleanliness of the base material and for use of low-hydrogen consumables for the filler materials and fluxes.

There are geometric discontinuities associated with welding. One of these is the formation of a ditching at the weld toe. The weld toe is the edge of the fusion zone on the surface of a base material. This discontinuity is called undercut and it has the effect of creating a "notch" effect on the surface (Figure 2-10).

Excessive crown reinforcement of a weldment usually results in a discontinuous geo-

FIGURE 2-10 Undercut at the edges of weld beads in a multiple pass weld. (Courtesy of C. Hellier.)

metric transition with the base material. The geometric discontinuity at the toe of the weld in this case will be a region of stress concentration. When this condition exceeds a design criterion it is called excessive convexity. The same type condition may occur when the weld crown is concave. The concave discontinuity is called "excess concavity" when it exceeds a design or code criterion.

A joint that is incompletely filled represents a geometric departure from design. This condition is called lack of penetration. This may provide a geometric discontinuity in addition to inadequate load bearing material in the weld joint. The condition of inadequate filling of a joint also exists in brazing, where it is called lack of fill. Brazing is a technique that joins components by capillary action of and metallurgical bonding with a lower melting point filler material that is drawn into a narrow space between the parent materials. The flow of the brazing alloy is dependent on the gap, surface cleanliness, and temperature of the base materials. Inadequate coverage of the brazing alloy in the joint weakens the joint.

There also are brazing conditions of inadequate bonding called unbond. Inadequate cleanliness and contamination of the base material surfaces, use of the incorrect brazing materials, and inadequate brazing temperature may cause unbond.

V. DISCONTINUITIES RESULTING FROM PLASTIC DEFORMATION

Forging is a process that forms an alloy by shaping it in a die under compressive loading. The flow of the metal is responsive to distributed loads and the ability of the metal to flow is dependent on its temperature. The higher the temperature, the easier the flow and the greater the ductility.

It is possible that the stress resulting from a forming operation may exceed the strength of the material, causing the material to break apart. In a forging operation, this load-induced cracking is called a burst. A burst may be entirely internal or the cracking may extend to the surface of the component (Figure 2-11).

The volume of a body is approximately constant when a component is mechanically formed. During the flowing of the alloy, material occasionally will lap over itself due to surface flow instability. The folded material will not fuse if the temperature is low and if the surface is contaminated with dirt. This type of discontinuity is called a fold or a lap. A long straight lap may be called a seam. Laps and folds have crack-like characteristics and they are considered to be rejectable discontinuities.

There are geometric discontinuities caused by pressing scale and debris left in forging dies into the surface of a component. These indentations are called scale pits.

A crack that transects a section is called a split. Splits may occur in forging and in other forming techniques. Rolling, swaging, spin forming, and extrusion may give rise to cracks, laps, burst, splits, scale pits, etc.

VI. CORROSION-INDUCED DISCONTINUITIES

Electrochemical corrosion of alloys may be thought of in terms of a battery. This form of corrosion requires an anode, a cathode, an electrolyte, and an electrical connection between the cathode and anode. The degradation generally occurs on the anode of the cathode–anode couple. Many operational environments provide the opportunity for electrochemical corrosion and resulting discontinuities.

A common surface discontinuity caused by corrosion is pitting (Figure 2-12). Pitting is a localized attack on a surface on which specific regions are anodic to the remaining surface and the attack, once started, persists at those areas. The pitting usually has associated debris of corrosion products that provides an environment that assists in the continued local attack. Pitting behavior is a product of the alloy and the chemistry of the electrolyte. There are many combinations of alloy and electrolyte that will not result in pitting.

FIGURE 2-11 Multiple forging bursts in a steel blank. (Courtesy of C. Hellier.)

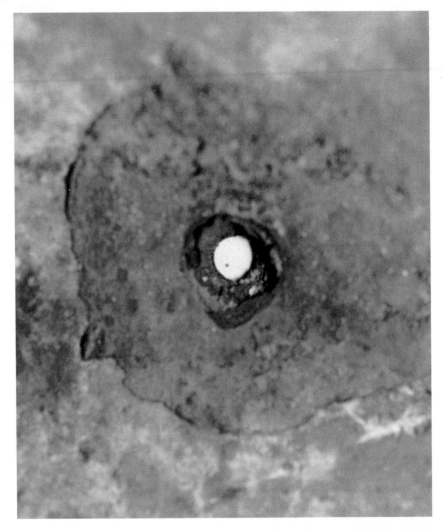

FIGURE 2-12 Corrosion pit that penetrated a pipe wall (5×). (Courtesy of R. B. Pond, Jr.)

The forms of pits are variable and range over a wide spectrum of geometry. Pits may be conical in form; they may mushroom beneath the surface or take tortuous paths (a form called wormhole corrosion).

Pitting damage is potentially deleterious. In addition to loss and penetration of the material, the rough surface of the pitting is a ready initiation site for fatigue cracking in components that are subjected to cyclic tensile loads. Additionally, the fatigue resistance of an alloy may be lowered in the electrolyte. Generally, an electrolyte that causes pitting will also be aggressive with respect to lowering the fatigue resistance of the component, and fatigue cracks may emanate from the pitted regions.

There are crack discontinuities that may be induced by corrosion. A cracking mechanism known as stress corrosion cracking (SCC) requires a metal or alloy within a range of microstructural condition to be under tensile stress in one of a limited number of specific electrolytes for it to exhibit this behavior. Any alloy or metal may be susceptible to stress corrosion cracking under certain conditions.

Stress corrosion cracks that run around the grain and phase boundaries of an alloy are known as intergranular stress corrosion cracking (IGSCC). They may also run across the grains (trans or intragranular cracking). The cracking may be mixed mode, consisting of inter- and transgranular cracking. In most cases, stress corrosion cracks have many branches (Figure 2-13). These cracks will begin on the surface and may be deleterious, causing penetration of the wall of the component and serving as a site of fatigue propagation in cyclic tensile loading conditions. This cracking condition is untoward and unacceptable.

A corrosion attack at the grain boundaries over broad areas is called intergranular attack (IGA). The character of IGA often includes a checkerboard pattern of cracking on a surface. Although IGA does not usually penetrate the wall of a component, it does provide a ready initiation site for fatigue.

There is a mode of cracking discontinuity that is induced at the cathode of a corrosion cell. This is called hydrogen embrittlement. The electrochemical reaction at the cathode

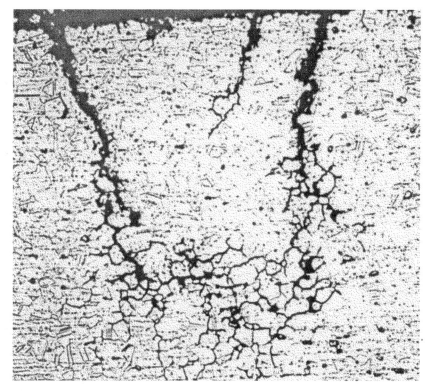

FIGURE 2-13 Stress corrosion cracking in nickel alloy (200×). (Courtesy of R. B. Pond, Jr.)

includes the reduction of hydrogen ions from the electrolyte. The atom of hydrogen that is formed on the surface of alloys is small enough to dissolve into the alloy. Over time, the hydrogen will migrate throughout the component. In the same way as embrittlement of high-strength material welds, the hydrogen will form molecular hydrogen at inclusions and grain and phase boundaries. The pressures associated with the included hydrogen gas are extremely high.

Ductile materials will plastically deform in the region of a bubble to accommodate its growth, but high-strength materials may resist plastic deformation and crack before they flow. This internal cracking may occur over substantial regions of a component and cause a weakening of the component as well as producing initiation sites for fatigue crack propagation. The original hydrogen-induced cracking is typically very bright and faceted. These small cracks are called flakes. There are occasions when the cracking will be on the boundaries of grains. Both forms of cracking result in a loss of strength of a component and sensitivity to catastrophic failure.

There are situations where corrosive attack will cause strips of material to disengage from a component surface. Elongation and segregation of the alloy microstructure cause the selective subsurface attack. This discontinuity is called delamination.

VII. OPERATIONALLY INDUCED DISCONTINUITIES—FATIGUE CRACKING

Alloys that are subjected to cyclic tensile loading may exhibit surface cracks after a critical number of load cycles. This damage may occur even though the component has maximum tensile stress far below the yield stress for the material.

Cracking will often initiate at a discontinuity in the form of a discontinuous geometric change or inhomogeneity in the alloy. Once begun, a fatigue crack usually will propagate irreversibly, a small amount with each cycle of tensile load. Eventually, the fatigue crack will grow large enough to cause a catastrophic failure of the component.

Fatigue cracks have unique characteristics. A fatigue crack will propagate on the plane(s) of maximum tensile stress. Often, the distributed tensile load is perpendicular to the planar surface of the material. In these situations, the fatigue cracking will be relatively flat. Usually, each cycle of cracking induces a variation in stress state local to the tip of the crack. Large excursions in load cause the crack to run slightly out of plane during the stress cycle. The consequence of this is a sequence of concentric ridges that are called beach marks (Figure 2-14). The beach marks radiate from the initiation site of the cracking and extend to the point where the crack runs catastrophically due to ductile failure or brittle failure.

Another characteristic of fatigue is macroscopic plastic deformation with no change of shape associated with the crack propagation. For this reason, fatigue cracking preserves the shape of the component and fatigue cracks are often hard to see, requiring augmented inspection techniques for detection.

Conditions of thermal cycling may cause stresses due to differential contraction and expansion of the cooled and heated surface compared to the bulk of a component. There is the possibility in this condition of the creation of thermally induced fatigue cracking. These cracks usually occur over an area and they are often in a checkerboard pattern. Thermal cracks are initiators of crack propagation due to other operational stresses.

In grinding operations on steels it is possible to raise the temperature of a surface layer high enough to cause an allotropic transformation to generate martensite. The associated thermal stresses can give rise to a cracking pattern that is called crazing. This is often

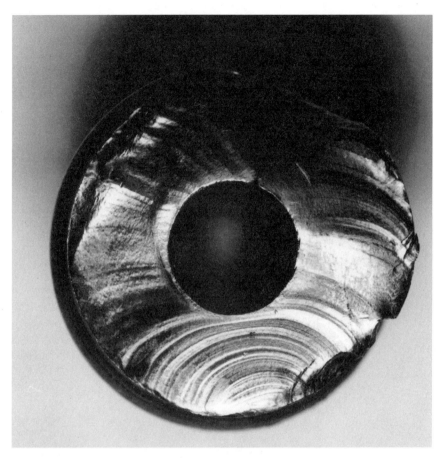

FIGURE 2-14 Beach marks in a drill rod fractured by fatigue. (Courtesy of R. B. Pond, Jr.)

associated with an oxide coloration due to the high-temperature condition. Craze cracks are also initiation sites for fatigue.

VIII. OPERATIONALLY INDUCED DISCONTINUITIES—CREEP

High-temperature operation of alloys for long periods of time may give rise to an operationally induced cracking called creep cracking. This condition may occur even if the stresses are relatively low. The cracks are usually preceded by discontinuities in the form of creep voids, which are small and distributed. The creep voids grow and then cracking links these voids. The catastrophic failure due to high-temperature creep is usually a relatively low ductility failure characterized by thick lip rupture (Figure 2-15).

FIGURE 2-15 Thick lip fracture from creep in a power plant superheater tube. (Courtesy of R. B. Pond, Jr.)

IX. OPERATIONALLY INDUCED DISCONTINUITIES—BRITTLE FRACTURE

Brittle fracture is usually catastrophic in nature and nearly always emanates from a discontinuity. Some materials such as glasses are inherently brittle and some alloys may be made brittle due to their process history and environment.

The formation of martensite in steel is an example of embrittlement caused by microstructural changes in an alloy. This is usually caused by phase transformation at high

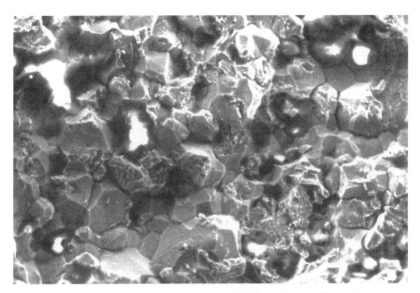

FIGURE 2-16 "Rock candy" fracture surface of a grade 8 bolt cracked by hydrogen embrittlement. (Courtesy of R. B. Pond, Jr.)

temperature followed by rapid cooling and phase transformation at relatively low temperature. There are conditions of precipitation of phases at grain boundaries of alloys that cause the grain boundaries to be brittle, and brittle fracture in those cases will tend to be confined to the envelope of the grains. Fracture about the grain boundaries has a characteristic appearance called a "rock-candy" fracture surface (Figure 2-16).

In other cases, the brittle cracking occurs through the grains on crystallographic planes of weakness. This gives a specular, faceted appearance to the brittle cracked surface (Figure 2-17).

Ferritic steels will change toughness and ductility as a function of temperature, becoming "glass-like" below a critical transition temperature. The critical temperature is a function of steel alloy chemistry and microstructure and it ranges from far below the freezing point of water to hundreds of degrees centigrade.

In all cases, brittle fracture will begin at a discontinuity in a structure. For a given discontinuity in a material there is a given static stress that must be exceeded before a crack may propagate. This is the basis for categorizing crack-like discontinuities as critical in size.

X. GEOMETRIC DISCONTINUITIES

Geometric discontinuities are often involved as sites of cracking and failure. These discontinuities may be created by deficiencies in the design, manufacture, operation, and repair of a component. Sharp transitions in surface slope will be locations of stress concentration, and under cyclic loading conditions these locations will be the first to crack due to fatigue.

FIGURE 2-17 Brittle fracture surface of fastener. Bright flecks are planar cleavage facets of ferrite in the steel. (Courtesy of R. B. Pond, Jr.)

An open and sharp relief in the surface of a component is called a notch. The dimensions and the slope transition of the sides of the notch are critical characteristics of the extent of stress concentration that will exist under operational loads.

Gouges are usually considered to be relatively shallow grooves or cuts on a surface. Gouges may be caused by plastic deformation or by cutting action. A cut in a surface is a discontinuity similar to a gouge, except that it is usually caused by the removal of material from the surface by interaction with other components, similar to machining. Water and high-velocity gas flow (erosion), abrasive particle wear, mechanical wear, local plastic deformation, and local chemical action may cause gouges.

Galling is a surface condition caused by the metallurgical bonding of surfaces under pressure and the local tearing out of surface material during motion between the surfaces.

There are other machining-related discontinuities that have specific and unique characteristics. These include excessive undercut, sharp radius, scoring, scratching, and burring.

Fretting is a wear condition that occurs during operations when two surfaces repeatedly rub in a reciprocating motion. The debris from the surface interactions and oxidation products are included between the surfaces and act as a grinding medium. The result of this motion is a roughened surface that invariably will have reduced fatigue resistance.

Surface finish is not usually considered to be a discontinuity. However, surface finish strongly influences the fatigue resistance of a material, and for given environmental conditions, smoother surfaces provide a greater resistance to fatigue cracking.

XI. SUMMARY

Challenges to engineering-material integrity largely involve the discontinuities in components. Disruptions in continuity may be either internal or on the surface. Discontinuities may be macroscopic or microscopic and they may limit the strength, ductility, toughness, and endurance of a component.

A primary responsibility of the examiner is to detect and characterize discontinuities in a component. The location and type of discontinuity is dependent on the fabrication and operation history of a component. An examiner is given an advantage when he or she understands the relationship of product form and history to the consequent discontinuities in a component.

Understanding the origin of discontinuities in engineering components should result in efficiencies in nondestructive examinations and enhancements in quality of examination results.

XII. GLOSSARY OF METALLURGY AND DISCONTINUITY TERMS

Anisotropy The material condition in which material vector properties change with direction.

Annealing Any treatment of metals and alloys at relatively elevated temperatures for the purpose of altering the properties of the material. The types of changes include softening, reducing residual stress, and recrystallizing.

Artifact An indication originating from a source other than a discontinuity that resembles an indication from a discontinuity.

Billet A solid, semifinished, round or square product that has been formed by hot working (e.g., forging, rolling, and extrusion).

Blister A surface or near-surface discontinuity in metal and cast alloy components caused by gas evolution during casting. Small blisters are called "pinheads" or "pepper blisters."

Blowhole A discontinuity in the form of a hole in a casting or a weld that is caused by gas entrapment during solidification.

Brazing   Joining of metals and alloys by bonding; the alloys have liquidus temperatures above 800° F (below the liquidus temperatures of the materials being joined).
Brittle Cracking   The propagation of a crack that requires relatively little energy and results in little or no plastic deformation.
Brittleness   The material quality that leads to fracture after little or no plastic deformation.
Burr   Ragged edge on a part usually caused by a machining or grinding process.
Burst   Fissure or rupture caused by improper rolling or forging.
Capillary Action   The movement of liquid within narrow spaces that is caused by surface tension between the liquid and the substrates. The mechanism that is used to fill or penetrate a joint in soldering and brazing.
Cast Structure   The microscopic and macroscopic distribution of grains and chemistry that is characteristic of a casting.
Casting Shrinkage   The reduction in component volume during casting. The reductions are caused by 1) liquid shrinkage, which is the reduction of volume of the liquid as it cools to the liquidus temperature; (2) solidification shrinkage, which is the total reduction of volume of the alloy through solidification; (3) the shrinkage of the casting as it cools to room temperature.
Casting Discontinuties   Discontinuities that are generated in casting operations.
Charpy Test   An impact fracture test that characterizes the energy required to break a standard specimen. The test machine is a pendulum hammer.
Chatter   The vibration of a tool in a grinding or cutting operation. Also, the wavy surface on a machined component caused by chattering of the working tool.
Checks   Multiple small cracks on the surface of components caused during manufacturing.
Cleavage   The splitting of a crystal on a crystallographic plane of relatively high density.
Constituent   A characteristic geometric arrangement of microscopic phases.
Cold Shut   A discontinuity on the surface of a casting caused by the impingement of melt (without fusion) within a part of a casting.
Cold Working   Plastically deforming a material at relatively low temperatures, resulting in creation of dislocation defects in the crystal structure.
Columnar Crystal Structure   Elongated grains that are perpendicular to a casting surface.
Corrosion   The electrochemical degradation of metallic materials.
Corrosion Fatigue   An acceleration of fatigue damage caused by a corrosive environment.
Crack   A breach in material continuity in the form of a narrow planar separation.
Crater   A local depression in the surface of a component caused by excessive chip contact in machining or arc disturbance of a weldment.
Creep Cracks   Cracking that is caused by linking of creep voids at the end of tertiary creep.
Creep Voids   Small voids that form in the third stage of creep.
Crevice Corrosion   The loss of surface material in a crevice subjected to an electrolyte.
Decarburization   The loss of carbon from the surface of a ferrous alloy because of high temperature oxidation of carbon.
Defect   A component discontinuity that has shape, size, orientation, or location such that it is detrimental to the useful service of the part.
Dendrite   A tree-like crystal structure that forms in some casting and vapor deposition crystallization.
Ductility   The ability of a material to deform plastically without fracturing.

Exfoliation   A corrosion degradation mode that causes layers parallel to the surface of an alloy to be separated and elevated due to the formation of corrosion product.

Fatigue   Progressive cracking leading to fracture that is caused by cyclic tensile loading in the range of elastic stress, eventually initiating small cracks that sequentially and irreversibly enlarge under the action of fluctuating stress.

Flakes   Short internal fissures in ferrous materials caused by stresses produced by evolution of hydrogen after hot working. Fractured surfaces containing flakes with bright and shiny surfaces.

Folds   Discontinuities composed of overlapping surface material.

Forging Discontinuties   Discontinuities that are created in forging operations.

Fracture   A break, rupture, or crack large enough to cause a full or partial separation of a component.

Fretting   Low-amplitude reciprocal motion between two component surfaces under pressure causing surface roughness.

Gas Holes   Holes created by gas escaping from molten metal during solidificaton.

Gas Porosity   Minute voids distributed in a casting that are caused by the release of gas during the solidification process.

Geometric Discontinuity   A sharp change in surface configuration that may be specified or an unplanned consequence of manufacture.

Gouge   A groove cut in a surface caused by mechanical, thermal, or other energy sources.

Grain   An individual crystal in a polycrystalline material.

Grain Boundary   The narrow zone of material between crystals of differing orientation.

Grinding Cracks   Shallow cracks formed in a surface of relatively hard materials because of excessive grinding heat.

Gross Porosity   Pores, gas holes, or globular voids in weldments or castings that are larger and in greater number than is acceptable in good practice.

Heat-Affected Zone   The portion of base material that was not melted during brazing, cutting, or welding whose microstructure and physical properties have been altered by the heat of the joining operation.

Hot Cracks   Cracks that are caused by differential contraction between a casting and its mold. These may be branched and scattered through the interior and on the surface of the casting.

Hot Tear   A relatively large fracture formed in the interior or on the surface of a cast component due to restricted contraction.

Hydrogen Embrittlement   Low ductility caused by cracking of the interior of a component due to the evolution and precipitation of hydrogen gas.

Inclusion   Usually a solid foreign material that is encapsulated in an alloy. Inclusions comprised of compounds such as oxides, sulphides, or silicates are referred to as nonmetallic inclusions.

Incomplete Fusion   Failure of weldment to fuse with the base material or the underlying or adjacent weld bead in a weld.

Incomplete Penetration   Fusion into a joint that is not as full as dictated by design.

Inherent Discontinuity   A discontinuity that is generated in the primary production of a material.

Lack of Fusion   Failure of a weldment to fuse with the base material in a weld.

Lamination   Separation or structural weakness, usually in plate that is aligned parallel to the surface of a component. This may be caused by the elongation during plastic forming of segregates that are the result of pipe, blisters, seams, and inclusions.

Lamellar tearing   The typical rupture of a material that is weakened by elongated slag inclusion. The crack surfaces have the appearance of wood fracture.

Lap A surface discontinuity appearing as a seam caused by the folding over of hot alloy fins, ears, or corners during forging, rolling, or other plastic forming without fusing the folds to the underlying material.
Macroshrinkage Casting discontinuity that is detectable at magnifications less than ten times and that is the result of voids caused by solidification shrinkage of contained material.
Metallurgical Notch A material discontinuity, usually involving hardness, that has the geometric characteristics of a crack.
Microfissure A microscopic crack.
Microsegregates Segregates that are microscopic in size.
Microshrinkage Cracking Microscopic cracks that are caused by solidification shrinkage.
Microshrinkage Porosity Microscopic pores that are caused by solidification shrinkage.
Nonmetallic Inclusion A slag or glass-like inclusion in an alloy.
Notch A sharp reentrant on a surface of a component that causes a local concentration of stress.
Phase Diagram A descriptive map that shows the existance of equilibrium phases, usually as a function of alloy composition and temperature.
Phase Diagram—Binary A phase diagram between two components.
Phase Diagram—Ternary A phase diagram between three components; the representation uses a three dimensional representation of data. The interior of an equilateral triangle base providing unique location representing all the compositions possible between the three components.
Pitting Forming of small cavities in a surface by corrosion, arcing, wear, or other mechanical means.
Primary Processing Discontinuities Discontinuities that are generated in the first forming steps of an alloy.
Scratches Small grooves in a surface created by the cutting or deformation from a particle or foreign proturberance moving on that surface.
Secondary Processing Discontinuities Discontinuities that are generated in the secondary and finishing forming steps.
Segregation Nonuniform distribution of alloying elements, impurities, or phases in alloys.
Service Discontinuities Discontinuities that are generated by service conditions.
Slag An aggregate of nonmetallic glasses that is found in primary metals production and in welding processes. Slag may be used as a covering to protect the underlying alloy from oxidation.
Split A rupture in a component, usually open to the surface.
Stringers Nonmetallic inclusions that are elongated by a hot forming process.
Scale Pits Surface discontinutities in the form of local indentations caused by the capture and rolling of scale into the surface of an alloy.
Voids A cavity inside a casting or weldment that is usually caused by solidification shrinkage.
Welding Discontinuities Any breaches in material continuity that are generated in welding processes.
Weld Undercut A groove at the toe or root of a weld caused by melting of the base material.

XIII. DISCONTINUITY GUIDE

The following table provides examples of discontinuities. These examples are designed as a reference and to provide the reader with a representation as to how some of the conditions described herein and in other chapters can appear. All discontinuities are different. When there is doubt regarding the classification and type of discontinuities, this guide may help. Following the table is a collection of figures that illustrate some of the discontinuities (Figures 2-18 to 2-29).

Discontinuity Guide

Nondestructive Test Methods	Visual	Fluorescent Penetrant	Visible Penetrant	Wet D.C. Magnetic Particle	Dry D.C. Magnetic Particle	Dry A.C. Magnetic Particle	Eddy Current	Thermal Infrared	Radiography	Straight Beam Ultrasonics	Angle Beam Ultrasonics
Types of Discontinuities		Surface and Near Surface Methods								Subsurface	
Process Category											
Inherent											
Laminations	P	U	U	U	U	U	U	U	U	A(1)[c]	U
Pipe[d]	U	U	U	U	U	U	U	U	A(2)	A(1)	A(3)
Seams	P	A(4)	A(5)	A(1)	A(2)	A(3)	P	P	U	U	P[b]
Stringers	U	U	U	A(1)	A(2)	A(3)	P	U	P	P	P[a]
Mechanical Forming Processes											
Flare or Split	U	U	U	U	U	U	U	U	U	A(1)	A(2)
Forging Bursts	P	A(2)	A(3)	A(1)	A(4)	A(5)	P	P	P	P	P
Forging Laps	P	A(2)	A(3)	A(1)	P	P	P	U	U	P	P
Rolling Lap (Seam)	P	A(2)	A(3)	A(1)	P	P	P	P	U	P	P
Casting Process											
Casting Cold Shuts	P	P	P	P	P	P	P	U	A(1)	P	P
Casting Shrinkage Cracks	U	U	U	P	P	U	U	U	A(1)	A(2)	P
Gas Porosity	U	U	U	U	P	U	P	U	A(1)	A(2)	P
Hot Tears	U	U	U	P	P	U	P	P	A(1)	A(2)	P
Scabs	P	A(3)	A(4)	A(1)	A(2)	A(5)	P	U	U	U	P
Shrinkage Porosity	P[e]	P[e]	P[e]	U	U	U	U	U	A(1)	A(2)	P
Inclusions	U	P[e]	P[e]	P[e]	P[e]	P[e]	U	U	A(1)	A(2)	P
Secondary Process											
Grinding Checks	P	A(2)	A(3)	A(1)	P	P	P	U	U	U	P
Machining Tears	P	A(2)	A(3)	A(1)	U	U	U	U	U	U	U
Plating Cracks	P	A(2)[e]	A(3)[e]	A(1)	P	P[e]	P	U	U	U	U
Quench and Heat Cracks	P	A(2)	A(3)	A(1)	A(4)	A(5)	P	P	P	U	P
Welding/Joining											
Crater Cracks	P	A(2)	A(3)	A(1)	P	P	P	U	U	P	P
Dense Inclusion	U	U	U	U	P	U	U	U	A(1)	P	U

Excessive Concavity	A(1)	U	U	U	U	U	U	U
Excessive Convexity	A(1)	U	U	U	U	U	U	U
Incomplete Penetration	P	U	U	U	U	U	A(1)	A(2)
Lack of Fusion	U	P	P	P	U	U	A(2)	A(1)
Porosity	P	P	P	P	U	U	A(1)	U
Slag Inclusion	U	U	U	A(2)	U	U	A(1)	P
Subsurface Cracks	U	P[e]	P	U	U	U	A(1)	A(1)
Surface Cracks	P[e]	P[e]	U	P	U	U	U	A(1)
Underbead Cracks	U	U	U	U	U	U	A(1)	A(1)
Undercut	A(1)	U	U	U	U	U	U	P
Brazing								
Lack of Fill	U	U	U	U	U	U	A(1)	U
Unbond[f]	U	U	U	U	U	U	A(1)	U
Service								
Brittle Overload Crack	P	A(5)	A(4)	A(2)	A(1)	P	P	P
Corrosion Pitting	P	A(1)	A(2)	U	U	U	P	U
Fatigue Cracking	P	A(5)	A(4)	A(2)	A(1)	P	P	P
Fretting	A(1)	U	U	U	U	P	U	U
Graphitization	U	U	U	U	U	U	P	A(1)
Metallurgical Notch	U	U	U	U	U	U	P	P
Overload Cracking	P	A(5)	A(4)	A(2)	A(1)	P	P	P
Segregation	U	U	U	U	U	U	A(1)	P
Stress Corrosion Cracking	P	A(1)	A(2)	P	U	A(3)	A(2)	A(3)
Wear	A(1)	U	U	U	U	U	U	U

Key:
U = Unsatisfactory; P = Possible; A(1) = First order of preference; A(2) = second order of preference; A(3) = third order of preference; A(4) = fourth order of preference

[a] Inspection of billet possible.
[b] Possible using surface waves.
[c] Assuming it is completely internal.
[d] May be seen on ends by penetrant or magnetic particle.
[e] If open to surface.
[f] Thermal tests possible.

Note: Acoustic emission testing (AET) has not bee included in this guide because this method applies only to those discontinuities that propagate under applied loads.

FIGURE 2-18 Lamination in plate edge.

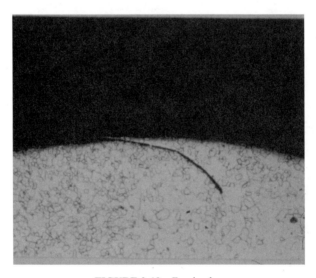

FIGURE 2-19 Forging lap.

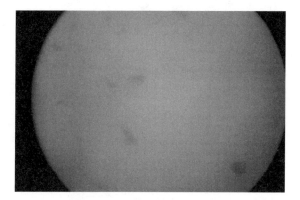

FIGURE 2-20 Sand and slag inclusions in a casting.

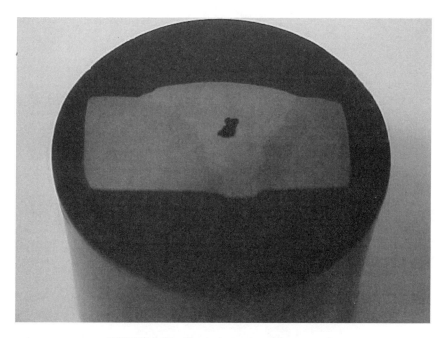

FIGURE 2-21 Slag inclusion in weld cross section.

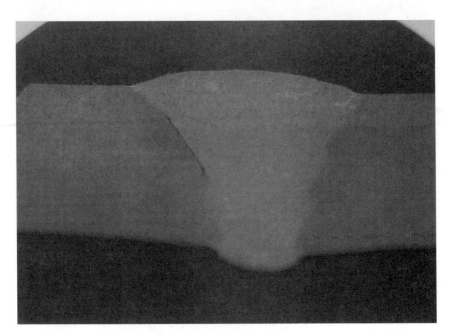

FIGURE 2-22 Lack of fusion.

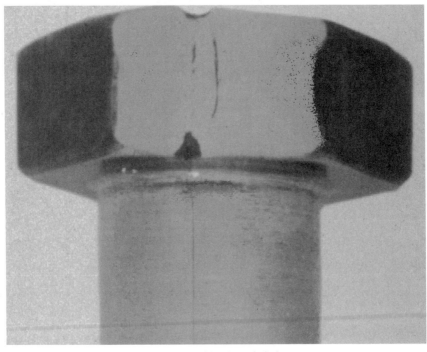

FIGURE 2-23 Seam in bolt.

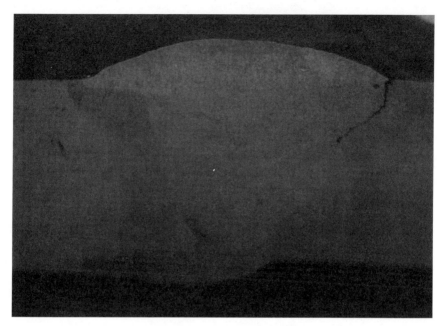

FIGURE 2-24 Toe crack in weld.

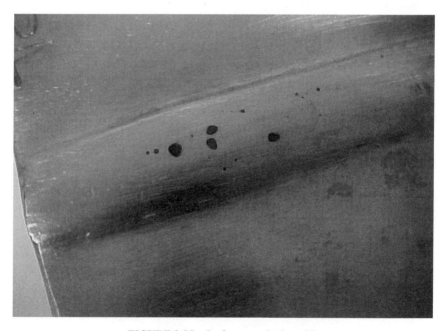

FIGURE 2-25 Surface porosity in weld.

FIGURE 2-26 Crack between two welds.

FIGURE 2-27 Burst in bar.

FIGURE 2-28 Crack in bolt.

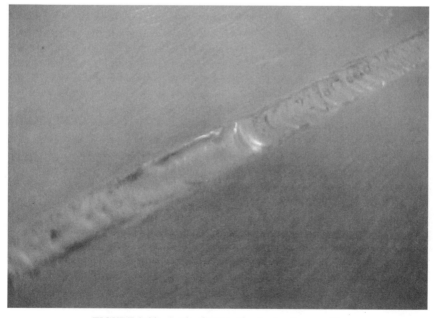

FIGURE 2-29 Lack of penetration in aluminum weld.

CHAPTER 3
VISUAL TESTING

I. HISTORY AND DEVELOPMENT

Visual examination or testing (VT) is a method of nondestructive testing that has been neglected for much of its industrial application life. VT was the first nondestructive test (NDT) method used in the nondestructive testing industry, but was last method to be formally acknowledged. Development of the visual method as an independent entity was fostered by the Electric Power Research Institute (EPRI) Nondestructive Examination (NDE) Center in the early 1980s. This was the result of the development of a training program for visual examination technology that included 120 hours of formal training. The need was prompted by the American Society of Mechanical Engineers, specifically Section XI—Rules for Inservice Inspection of Nuclear Power Plant Components. The program was designed to qualify personnel as visual examiners. Examination personnel scrutinizing the general condition of components were to comply with the requirements of the American Society for Nondestructive Testing, Recommended Practice No. SNT-TC-1A. Visual examination of components for general mechanical and structural conditions was to satisfy the requirements of ANSI N45.2.6. This standard was codified via the U.S. Federal Regulations, "Title 10, Code of Federal Regulations, Part 50," requiring nuclear power plants to meet certain requirements for licensing. ASME sectored the visual examination into four categories based on the scope of inspection. The categories are classed as VT-1, VT-2, VT-3, and VT-4. VT-1 addresses the condition of a component, VT-2 the location of evidence of leakage, VT-3 the general mechanical and structural conditions of components and their supports, and VT-4 (which has been eliminated) focussed on the conditions relating to the operability of components or devices.

Performance requirements for NDT are referenced in ASME Boiler and Pressure Vessel Code, Section V—Nondestructive Examination. Direct and remote visual testing is described in Article 9 of Section V—Visual Examination. Direct visual testing is defined as using "visual aids such as mirrors, telescopes, cameras, or other suitable instruments." Direct visual examination is conducted when access allows the eye to be within 25 inches (610 mm) of the surface to be examined, and at an angle not less than 30° to the surface to be examined. This is illustrated in Figure 3-1.

Remote visual testing is divided into three categories: borescopes, fiberscopes, and video technology. These have been developed chronologically. "Borescopes," also referred to as "endoscopes," were originally used to inspect the bores of rifles or cannons utilizing a hollow tube and mirror. The second generation of the endoscopes included a relay lens system in a rigid tube. This upgraded the image. Due to its rigid structure, endoscopes are limited to straight-line access, as is depicted in Figure 3-2. Later innovations corrected this limitation by providing flexibility to the endoscopes. By 1955, the introduction of glass fiber bundles and fiber optic image transmission enabled the development of the fiberscope. Medical researchers experimented with different techniques in fiber optic image transmission during this period.

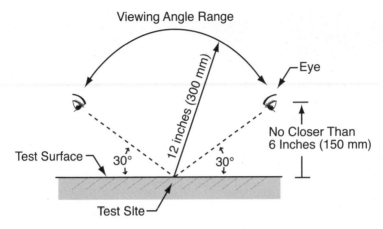

FIGURE 3-1 Minimum angle for typical visual testing.

FIGURE 3-2 Rigid borescope. (Courtesy of Olympus Industrial, with permission.)

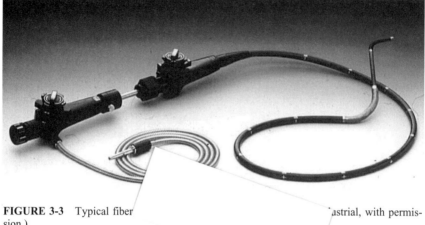

FIGURE 3-3 Typical fiber ... ...ustrial, with permission.)

Imaging with fiber optic bundles decreased the clarity o... ...age transmission compared with the rigid lens systems of borescopes; however, this was a small price to pay for the opportunities it presented. The flexibility of the bundle opened up previously inaccessible areas to remote visual inspection, providing a more versatile tool to be used in industrial situations. This often eliminated the need to dismantle equipment for inspection. A typical fiberoptic borescope is illustrated in Figure 3-3.

The evolution of the endoscope continued as the problems of eye fatigue associated with the use of endoscopes and fiberscopes prompted the development of various "add-on" cameras or closed circuit TV cameras that allowed for the display of images on a monitor. The first of such innovations was the tube-type camera. Many add-on camera systems are still presently in use, but due to their bulky exterior, smaller, solid-state imaging sensors, some of which are known as charge-coupled devices (CCDs), are replacing them. Figure 3-4 provides a basic demonstration of how a charge coupled device (CCD) works.

This new generation of CCDs stimulated a new wave of videoendoscope technology. Small in diameter with high-resolution images, this new technology increased the range of industrial endoscopy applications. The physical size of the CCD as well as its ability to allow for electronic image processing and its other advantages broaden the application possibilities. One technological aspect worthy of mention is the CCD's ability to record images. Whether the camera is orthicon or vidicon tube technology or CCD technology, the present systems can record the images on videotape. With the advent of digital storage technology, the recording of images on other permanent media enhances the system's versatility.

II. THEORY AND PRINCIPLES

Object Factors

In order to understand the physics of vision, it is necessary to first consider the characteristics of the eye. The eye can be compared to a radiation detector. Different wavelengths of light travel through the lens and reach the retina, which is located at the back of the

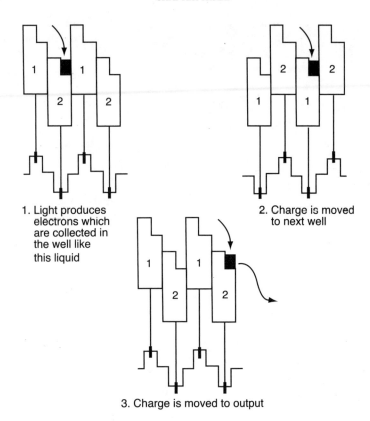

FIGURE 3-4 How charge coupling works.

eye. The rods and the cones of the retina in the human eye can sense wavelengths from about 400 nm up to approximately 760 nm. The eye performs the function of a spectrum analyzer that measures the wavelengths and intensity, as well as determining the origin of the light (from the sun or an artificial source). The light strikes the object to be viewed and is reflected towards the eye, through the lens and onto the retina as an image. The brain analyzes this image. The retina is similar to an array of tiny photosensitive cells. Each of these elements (cells) is connected to the brain through individual optic nerves. The optic nerves linking the eye to the brain can be compared to a bundle of electric cables. The major parts of the eye are shown in Figure 3-5.

The iris opens and closes, thus varying the amount of light reaching the retina. The light then passes through the lens, which by changing shape, focuses the light and produces the image on the retina at the rear of the eye. Here a layer of rods and cones are found. The neurological connection from the rods and the cones pass through the rear of the eye via the optic nerve, which transmits the neurological signals to the brain. The brain processes the signals as perceptions of colors and details that vary in light intensity and color. It is necessary for a certain minimum level of light to be present before the eye can produce an image. This level is known as the "intensity threshold." Contrast is some-

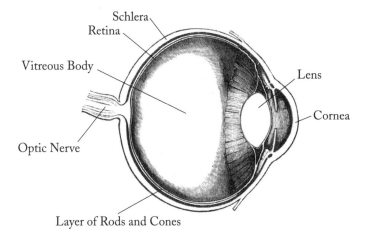

FIGURE 3-5 Key components of a human eye.

thing that shows differences between images placed side by side. Lighting requirements are frequently expressed in terms of ratios, due to the eye's ability to perceive a percentage of change rather than an absolute change in brightness. The retina will only retain an image for a certain amount of time. This varies according to the size of the object and speed at which it is moving. The main limitations of vision are intensity threshold, contrast, visual angle, and time threshold.

Visual acuity is the ability to distinguish very small details. For example, as the distance from the eye to the object increases, two lines that are close together appear as one heavy, dark line. The normal eye can distinguish a sharp image when the object being viewed subtends an arc of one-twelfth of a degree (five minutes), irrespective of distance from the eye to the object. Practically speaking, a person with "normal" vision would have to be within eight feet of a 20-inch TV monitor to resolve the smallest detail displayed.

White light contains all colors. Newton proved that color is not a characteristic of an object but, rather, various wavelengths of light that are perceived as different colors by the eye. Color can be described as having three measurable properties: brightness, hue, and saturation. The color of an object, ranging from light to dark, emitting more or less light, is known as brightness. Different wavelengths give us different perspectives of colors; this is known as hue. How green something is as opposed to white, is how saturated it is with green.

In the United States NDT environment, visual acuity examinations are a requirement for certification. The visual inspector's natural visual acuity must be examined. The Jaeger (J) test is used in the United States for near-distance visual acuity. It consists of an English language text printed on an off-white card. The parameters for near-distance visual acuity are described in personnel certification and qualifications programs. Visual acuity requirements will vary depending upon the needs of specific industries.

Light

For decades, light wavelengths have been measured in angstrom units (10^{-10} meters) and currently are also measured using the nanometer (nm), which is 10^{-9} meters. Brightness of light is an important factor in test environments. The apparent brightness of a test surface is dependent on the intensity of the light and the reflectivity of the surface reflecting the light to the eye. Excessive brightness will interfere with the ability to see and will "white out" the object. Inadequate light can cause excessive shadows and result in insufficient light being reflected from the surface, preventing observation of the surface attributes. Some codes require a minimum intensity of 15 foot candles (fc) for general visual testing and a minimal 50 fc for critical and fine detail viewing. The Illumination Engineering Society requires 100–300 fc for critical work. The inverse square law governs the intensity of light noted or measured. It states that illuminance (E) at a point on a surface varies directly with the luminous intensity of the source (I) and inversely as the square of the distance (d) between the surface and the source (see Equation 1 and Figure 3-6).

$$E = \frac{I}{d^2} \qquad (3\text{-}1)$$

where:
E = luminance
I = intensity of the source
d = distance between the source and the surface

This equation is accurate within 0.5% when d is at least five times the maximum dimension of the source, as viewed from the point on the surface.

Cleanliness
The amount of light that reaches the eye from an object is dependent on the cleanliness of the reflecting surface. In visual testing, the amount of light may be affected by distance, reflectance, brightness, contrast or the cleanliness, texture, size, and shape of the test object. Cleanliness is a basic requirement for a successful visual test. Opaque dirt can mask or hide attributes, and excessively bright surfaces cause glare and prevent observation of the visual attributes.

Brightness
Excessive brightness within the field of view can cause an unpleasant sensation called glare. Glare interferes with the ability to see clearly and make critical observations and judgments.

Surface Condition
Scale, rust, contaminants, and processes such as milling, grinding, and etching may affect the ability to examine a surface. This will be further discussed in the section relating to tools, equipment, and accessories.

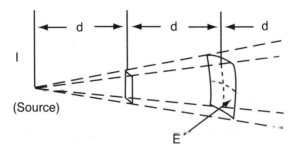

FIGURE 3-6 Inverse square law.

VISUAL TESTING   **3.7**

Shape
The shape of an object can influence the amount of light reflected to the eye, due to various angles that can determine the amount of light that will be reflected back to the eye.

Size
The size of an object will determine the type of scan pattern that may be used to view 100% of the object or it may determine that some magnification is necessary to get a closer view of details otherwise unobservable.

Temperature
Excessive temperature may cause distortion in viewing due to the heat wave effect. Most are familiar with the heat waves coming off a desert, resulting in a mirage; this is known as "heat wave distortion." In a nuclear reactor, underwater components are frequently distorted due to heat waves rising from the core that can interfere with the view from an underwater camera as it scans a particular component during visual examination.

Texture and Reflectance
One of the greatest variables in viewing an object is the amount of light that is reflected from it, and the angle at which the light strikes the eye. Excessive rust or roughness can cause diffusion of the light and limit the light returning to the eye. This can easily be corrected by increasing the amount of light or improving the surface condition of the object under examination.

Human Factors

Environmental
The amount of light required for a visual test depends on several factors, such as speed or accuracy, reflection from background, and other inspection variables. These inspection variables include physiological processes, psychological states, and the inspector's experience, health, and fatigue. All of these factors contribute to the accuracy of a visual inspection. One of the key factors in viewing a lighted object is the difference (contrast) between the light on the object and the background. A contrast ratio of 3:1 between the test object and the background is desirable. If the background is dark, a ratio of 1:3 is recommended between the test object and lighter surroundings, 3 being the most intense light in both cases. Psychological factors can also affect a visual inspector's performance. Surrounding colors and patterns can have an effect on the inspector's attitude. Dark walls can absorb up to 50% of the light. A high contrast on the pattern being inspected can cause eye fatigue. It is recommended that blue colors be utilized and brilliant colors avoided.

Physiological
The act of seeing something is not a passive activity. The observer must be active in keeping track of what is going on. Constant eye shifting back and forth from one location to another or scanning a large area at a rapid speed causes the muscles in the eye to fatigue. If the eye isn't focusing quickly when changing directions, the image can be lost altogether. Any fatigue on the part of the observer can result in reduced efficiency and accuracy in interpreting the visual data.

Psychological
Individuals can be in various psychological states. They can be suffering from tensions, emotions, and other influences. These may influence the appraisal and ability to visualize

an object and may also influence performance of a visual task. The intention of a viewer may affect perception. A great deal of information is potentially available immediately after viewing, but if one does not expect to find certain attributes, one may well overlook the physical evidence that has been viewed and not perceive them. One of the ways to overcome this is to know ahead of time what to expect, the attributes of what is to be seen, and what the greatest possibility is of these attributes existing.

Perception
The ability of the eye to sense a variety of views is not constant. If one is well rested in starting out an inspection activity, perception can be greater than when one is fatigued. Grossly changing light levels may cause painful glare. After a long period of relative darkness, normal light may seem painful. (Normal light is that which is comfortable to the eyes.) Sudden exposure to full sunlight can cause discomfort. It may take up to 30 minutes for the iris to adjust and regain normal vision. As the iris becomes tired and the muscles that adjust the lens become fatigued due to age, overuse, drugs, disease, or emotions, vision can be greatly affected. Vision examinations are usually administered annually to provide assurance that the inspector meets the requirements for performing VT. Daily influences external to the inspector, such as emotions, drugs, excessive light, inadequate sleep, etc., can cause temporary loss of visual acuity.

Another influence on perception is appearance. The two lines represented in Figure 3-7 appear to have different lengths due the perceptions created by the "Vee" extensions on each end. In fact, the two lines are identical in length; the brain perceives the lengths differently.

Visual Angle and Distance
Visualize two pencils representing two parallel lines. If these are perpendicular to the incident angle of view, two pencils can be clearly seen. As the two pencils rotate and become relatively parallel to the incident view of the eye, the two pencils appear to become one, since one is now behind the other. The quantitative ability of a person to resolve objects is determined in a practical manner from the distance of the object to the eye and the angle of separation of two points that can be resolved by the eye. This is known as resolving power. For the average eye, the minimal resolvable angular separation of two points on an object is about one minute of arc (1/60th of one degree). This means that at about 12 inches (300 mm) from the test surface, the best resolution to be expected is 0.0035 inches (0.09 mm); at 24 inches (600 mm), the best anticipated resolution is about 0.007 inches (0.18 mm). The best result from a visual test is obtained

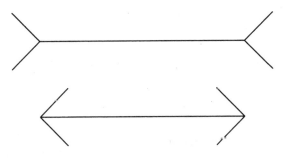

FIGURE 3-7 Which line is longer?

when the object is brought close to the eye and a large visual angle is obtained. There is a limitation to distance; the eye cannot sharply focus if it is nearer to 10 inches (250 mm) to an object. Therefore, direct visual tests are best performed at a distance of 10–24 inches (250–600 mm). Also of importance is the viewing angle between the line of site that the eye makes with the test site and the plane of the test surface. As described above, two pencils can become one as the viewing angle of the eye changes. For practical consideration, this angle should not be less than 30° of view off the plane of the surface under inspection, as shown in Figure 3-1.

III. EQUIPMENT AND ACCESSORIES

Direct Visual

Eyes
The most important instrument in visual testing is the human eye (refer to Figure 3-5). As the camera replicates the action of the human eye, it can serve as an model for discussing the eye's components. Light emanates from a source and strikes object. Light reflects off the object and travels through a medium, usually air or other transparent or translucent materials such as water, glass, crystals, etc. The light incident on the human eye enters the lens, after which it is focused on the retina. There the cones and rods transform electromagnetic radiation energy into neurological signals that are transmitted to the brain through individual optic nerves. Analogous to focusing the lens of a camera by changing the depth of field and angle, the lens in the eye is elastic. It changes shape as the eye muscle contracts or relaxes. The iris opens (like the aperture opening in a camera) to allow different quantities of light into the eye. The light first passes through the lens and is focused on the retina. The image focus depends on the shape of the lens as it is changed by the variations in tension of the eye muscle. The light rays then pass through the fluid in the eye and strike the rear of the eyeball at the retina (properly focused). This image-recording medium (the retina) is equivalent in function to the film in a camera. The image is temporarily retained (up to 1/6 of a second) before being passed to the brain. The neurological connection from the rods and cones in the retina goes through the rear of the eye via the optic nerve. The optic nerve transmits the neurological signals to the brain, which processes them as perceptions of colors and details that vary in light, intensity, and hue. Through this neurological connection, the brain translates the signals into three-dimensional perception.

The unique "equipment" known as the eye can suffer damage and deterioration over time. Or it can be inherently defective. The lens can become inflexible with age and not be able to change shape so as to focus. The lens itself can become defective due to cataracts, stigmas, scars, or scratches that distort or block light transmission. The recording medium (the retina) can become detached due to sharp blows, genetic predispostition, or disease. The ability of the eye to perceive color may have been limited or abnormal since birth (color blindness). The optical nerve system can become damaged like a wire that is shorted or broken. Chemical influences such as the effects of alcohol or hallucinogens, brain cell deterioration, or neurological impairment can affect the processing system in the brain.

The process of seeing is even more analogous to a charged coupled device, in which light striking small chips, causes electrons to accumulate, resulting in the transmission of signals to a computer where they are reassembled onto chip arrays that yield an image for the human eye to perceive. This is discussed in more detail later in this section.

To summarize, the eyes are the most important and essential components in the entire VT system. It would be impossible to perform examinations without them. They also play a key role in all other methods of NDT, including evaluation prior to, during, and most importantly, after completion of the performance of the examination.

Direct Visual Aids

The eye may need assistance when visualizing the detail of the vast variety of surfaces that are normally accessible to the direct view. Enhancement of the view can be achieved through magnification. A change of angle can be achieved through reflective mirrors. Both of these aids can enhance views, and their removal restores direct view again. Hence, their use and the option to return to unaided viewing results in "direct viewing with an aid."

Magnifier
The magnifier can increase the image size of the viewed object. The power of the magnification is expressed as follows:

$$\text{Magnifying power} = 10 \div \text{focal length (in inches)}$$

The following is an example of determining the magnification power of a lens: A piece of paper is held in one hand and a magnifying lens in the other. The image of a light source (e.g., a candle) is visible on the paper. After focusing the image on the paper by moving the lens toward and away from the paper, measure the distance from the center of the lens to the paper. This distance is the focal length of the lens. Divide ten by the focal length distance in inches. The resultant quotient is the magnifying power of the lens. For example, assume the focal distance was 2 inches; the power of the lens is ten divided by 2 inches, which equals a magnification power of five ($5\times$).

Another unit of measurement for the magnifier is the diopter. This is a measurement of the refractive power of lenses equal to the reciprocal of the focal length in meters. A "five diopter" lens has the magnification power of five times. The principal limitation to the amount of magnification is the depth of field. As magnification increases, the depth of field decreases. This is why a microfocused view of a small object such as a bee on a flower yields an image with the bee in focus and an out of focus background. It may be so out of focus that it is sometimes blurred out and not distinguishable.

Two common visual magnification devices are hand-held lenses and pocket magnifiers or microscopes. Hand-held lenses with a frame and handle may contain one lens or multiple fold-out lenses. They are generally plastic (acrylic) or glass. Normal sizes range from one half inch to six inches. The other common magnifying device is the pocket microscope. Small diameter tubes (usually 6 inches in length) are fitted with half-inch diameter lenses. Light is available through cut outs in the tube or through translucent tube ends. Due to the higher magnification ranges of $25\times$ to $60\times$, the depth of field and field of view are extremely limited. The larger the diameter of the lens, the lower the magnifying power.

Light Sources (Direct)
When employing magnification devices, additional light sources are usually required. Several lighting devices are available that permit light to be concentrated on a small spot. The most common of these is the hand-held flashlight. This light source is usually held at some angle to, and within inches of, the specimen to be examined. The actual light intensity (foot candles or lumens) at the surface to be examined is dependent on distance, light

angle, light bulb wattage, and battery strength. Another common source of auxiliary light is the "drop light," which is usually a 100-watt light bulb encased in a protective cage with a reflective shield connected to a length of electric cable. Again, the actual light intensity is dependent upon angle, distance, and wattage. A light meter can be placed on the specimen surface and the actual light intensity measurement can be made. As discussed, when the distance from the light source to a surface is doubled, the light intensity is decreased to one fourth of the original intensity. This is seen in the following equation, which is expressed differently than Equation 3-1:

$$\frac{I_1}{I_2} = \frac{D_2^2}{D_1^2} \qquad (3\text{-}2)$$

where:
 I_1 = intensity at first location
 I_2 = intensity at second location
 D_1 = distance to first location
 D_2 = distance to second location

Measuring Devices

There are a multitude of measuring devices available for different applications (See Figure 3-8). For the sake of brevity, only a few will be discussed here. The direct visual inspection method is frequently augmented by the use of several common tools used to measure dimensions, discontinuities, or range of inspection. Among these are linear measuring devices, outside diameter micrometers, ID/OD calipers, depth indicators, optical comparators, gauges, templates, and miscellaneous measuring devices.

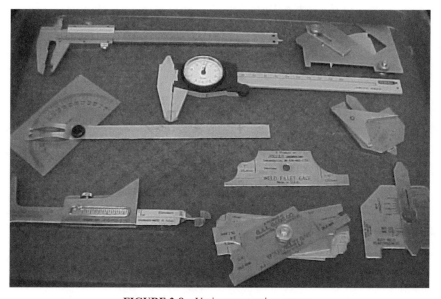

FIGURE 3-8 Various measuring gauges.

The most common linear measuring device is the straightedge scale. Typical scale lengths are 6 inches (15 cm) and 12 inches (30 cm) long. Additionally, 25-foot to 50-foot tape measures are frequently utilized. The 6-inch pocket scale is frequently misused by placing the thumb on one edge of the specimen and abutting the end of the scale against the soft and pliable thumb flesh. This is inherently inaccurate when a movable surface is the beginning point for a measurement. Preferably, the first whole unit of measurement (metric or imperial) should be aligned with one edge of the specimen and the other edge read for total distance. *Caution:* One must then remember to subtract the first unit of measurement from the total measurement noted.

Micrometers are used to measure outside or inside diameters. They are very accurate measuring devices commonly utilized to measure to the nearest one thousandth of an inch. Micrometers are available that can be used to measure to an accuracy of one ten thousandth of an inch. The outside diameter (OD) micrometer is made up of several parts, including the anvil, stirrup, spindle, sleeve, thimble, internal screw, ratchet, and knurled knob. It is worth noting that the essential component is the internal screw. It has been turned to have 40 threads per inch. One divided by 40 results in 0.025 thousandth of an inch spindle movement per 360° turn of the thimble. The micrometer may be outfitted with attachments for different applications. For example, a 0.2" ball may be attached to the spindle to accommodate the inside radius of a pipe when measuring pipe wall thickness. Additionally, a pointed attachment may be utilized when measuring pits or very localized variations in wall thickness.

The vernier caliper is a variation of the basic caliper. A caliper is usually used to measure the outside diameter of round objects. A common application is the measurement of remaining stock on a part in a machine lathe. If transfer calipers are used, the calipers are placed with the two contact points touching opposite sides of a round object. The calipers are removed and compared to the distance between the contact points as measured on a linear scale. Vernier calipers have inside diameter (ID) and OD capabilities, depending on which set of points or jaws are utilized. The handle that forms part of the stationary jaw has one or two inscribed scales. There is also a vernier slide assembly that is integral with the movable jaw. The correct OD or ID set of calipers or jaws makes contact with the specimen. The user reads the vernier scale zero mark upon the stationary scale for the whole unit measurement. The vernier scale marks are matched with the stationary scale marks and the best match-up of lines are utilized to read the linear distance to the nearest one thousandth of an inch or millimeter as the case may be. These two basic forms of calipers have since been replaced with direct read-out dial or digital calipers. Some digital gauges can be interfaced with a laptop or hand-held computer so that a large number of readings may be stored.

Depth indicators (dial indicators) are frequently used to measure surface discontinuity depths. Examples of discontinuities are pits, corrosion, or wastage. Verification of dimensions can also be conducted. In either case, zero depth must first be verified. The dial indicator or the digital read-out indicator is placed on a flat surface and zeroed. It is then moved over the depression. The spindle movement indicates the depth of the depression. When using the older dial indicator, the sweep hand rotations must be counted if the movement exceeds one 360° sweep. If a digital indicator is used, one must note the minimum and maximum point to which the indicator or its extensions are to range, i.e., 0"to 1"or 4"to 5", etc. It is a common mistake not to count the number of rotations or not to note the beginning and end range of the indicator, resulting in only the last increment of measurement being accurate.

Optical comparators are frequently used in machine shops where close tolerance measurements are desired. A comparator produces a two-dimensional enlarged image of the object on a large smoked (frosted) glass screen. The reflected light or background lighting is utilized to cast a magnified image onto the glass screen. This image is compared with a template of the object to check for dimensional accuracy.

The welding, fabrication, and construction industries use templates extensively to measure fillets, offset, mismatch, weld reinforcement, undercut, and other dimensional attributes. A common application is a template with minimum and/or maximum dimensions made from sheetmetal stock. The actual weld or base metal is compared to the template to determine "go" or "no-go" status. Accurate and actual measurements would still require linear measuring devices.

Miscellaneous measuring devices come in many sizes and configurations. Snap gauges, feeler gauges, radius gauges, temperature gauges, pitch gauges and diameter gauges are just a few. Some applications could have a gauge custom made for a specific need or requirement.

Remote Visual (Indirect)

Whenever the eye cannot obtain a direct, unobstructed view of the specimen test surface without use of another tool, instrument, or device, a "remote visual" examination is performed. Recall that a direct visual examination is an examination that can usually be made when access is sufficient to place the eye within 24 inches (610 mm) of the surface to be examined and at an angle not less than 30° to the surface to be examined. Figure 3-1 illustrates this definition. Most codes permit the use of mirrors and magnifying lens to assist with direct visual examinations. A remote visual examination can be defined as an examination that uses visual aids such as mirrors, telescopes, borescopes, fiber optics, cameras, or other suitable instruments.

Borescopes

One of the oldest applications of remote visual examination is the inspection of gun barrels. The failure of a gun barrel, be it the earliest mortar siege gun of medieval times or the precision machined barrel of a modern rifle, is catastrophic to the gunner at the very least. Two obstacles had to be overcome in the inspection of gun barrels: access to the area to be inspected, and a provision of a light source to provide adequate illumination in order to see the conditions of interest on the inside surface. Small rigid borescopes containing a series of lenses provided the answer (see Figure 3-9).

Small lamps were used at the far end to provide sufficient light. Since the original applications utilized a lens train to access the bore of a rifle, the term "borescope" was the original and lasting term for the device. From this technology evolved the glass fiber bundle (referred to as a fiberoptic borescope). This device transmits both the light to illuminate the inspection surface as well as the light reflected off the object back to the viewing end (see Figure 3-10).

Both the rigid and the fiberoptic borescopes are capable of providing access to small openings. The rigid borescope utilizes a lens train; this is called the *lens optic* technique. The fiberoptic borescope utilizes a fiber bundle with a lens at each end; this is called the *fiber optic* technique.

The lens optic technique brings the object image to the eye using an objective lens, sometimes a prism, a relay lens, and an eyepiece lens. The eyepiece lens allows each inspector to adjust the focus as needed with the use of a "focus ring."

The advent of miniature lamps the size of a grain of wheat was the reason that the light sources used in these devices were called "wheat lamps." These wheat lamps provided limited light and burned out quickly. A more practical means of transferring light to the examination surface is with the use of a light source transmitted through a fiber bundle.

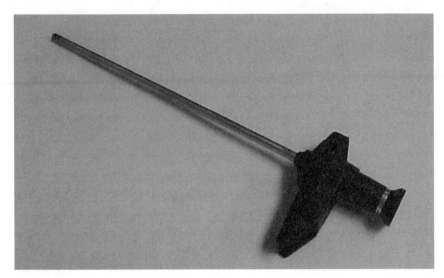

FIGURE 3-9 Small rigid borescope. (Courtesy of Olympus Industrial, with permission.)

FIGURE 3-10 Fiberoptic borescope. (Courtesy of Olympus Industrial, with permission.)

The reflective image from the object is returned to the eye via a second "image" bundle. Both bundles are made up of individual fibers that are "clad" with a glass material with a different refractive value sufficient to reflect the light from the outside diameter of each fiber back through the core and down the length of the glass fiber.

Both methods of transmitting the light and image down and back in a bore result in variations in the direction that the view can be delivered. Looking straight ahead is known as the *direct* view. An angle off the straight-ahead but still forward-looking is known as the *fore-oblique* view. A sideways look is known as the *side* view. Anything more than 90° past the straight-ahead view is known as the *retrospective* view. Different manufacturers will designate the forward, straight-ahead view as 0° or alternatively as 90°. In any case, the included angle of the resultant view is the "field of view."

A recent development is the combination of the lens-optic and fiberoptic techniques. The object image is transmitted to the eye along a rigid tube containing the lens-optic train. But the light is transmitted to the object through fiberglass surrounding the tube, bringing light from the external source of light to the object.

Yet another version of the borescope is the miniborescope. A single solid fiber diffuses ions in a parabolic arc from the center of the rod to the periphery of the rod, with a graded index of refraction. This still captures the light within the rod and passes it down the length of the rod. But the light actually bends down the length of the rod, forming an image at specific intervals. Since the lens aperture is so small, the lens has an infinite depth of field, eliminating the need for a focusing mechanism.

An interesting physical fact can be observed when using borescopes. A wide field of view reduces the magnification but results in greater depth of field. This can be observed when looking down a great distance during a tube inspection. The image less than a quarter inch away will be magnified. But the image from a quarter inch to one foot away, and theoretically to infinity, may be in focus but greatly reduced in image size. Conversely, a narrow angle field of view produces higher magnification but results in a shallow depth of field. The image at one point may be in focus, but at a short distance in front and behind that point, the image is out of focus.

Fiber Cameras

Like the fiberoptic borescope, the fiber camera uses light from an external source brought to the test site via the fiberoptic bundle (light guide). This fiber bundle that delivers the external light to the object site is limited in its length. An articulated (four-directional tip) bundle can be manufactured up to 40 feet long. Fiberglass bundles up to 45 feet can be made without articulating tips. Unlike the fiberoptic borescope, the light waves from the object pass through the objective lens and strike a charged coupled device (CCD). The light waves are then converted to electronic signals and are transmitted to the processor in the central control unit (CCU). The signals are processed before being viewed on the monitor. Another name for this equipment is "video borescope" (see Figure 3-11).

Charged Coupled Device (CCD)

The workings of a charged coupled device are illustrated in Figure 3-4. Electromagnetic radiation in the visible light wave spectrum and some infrared radiation reflect off the object to be examined. The light waves pass through the lens, which can be focused through the mechanical means of moving the lens distance relative to the solid-state silicon chip or CCD. The CCD comprises thousands of light-sensitive elements arrayed in a geometric pattern of rows and columns (a matrix). Each single microchip contains a single point or picture element known as a pixel. Each chip is struck by light waves. The intensity or quantity of light striking the chip determines the amount of electrons that are generated in

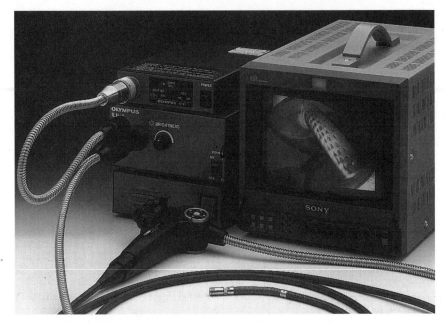

FIGURE 3-11 Video borescope. (Courtesy of Olympus Industrial, with permission.)

the silicon sandwich (chip). This accumulation of electrons in proportion to the amount of light that strikes that pixel is periodically passed on as packets. These packets of electrons are transferred from one pixel site to its adjacent site, just as water is moved from one bucket to the next in a bucket brigade.

The force required to move these electrons is produced in the pixels by means of a voltage that alternately turns on and off. As this "clock voltage" alternates, the electrons move from one site to the other. After the last pixel, the electrons are converted to a voltage proportionate to their number. This voltage is amplified, filtered, and then applied to the input of a video monitor. The differential in electrical charge is what is processed down the wire at the other end, where the processor is located. By assignment, the pixel is coded as one of the primary colors (red, green, or blue) and is reassembled as electronic signals that are sent to a color monitor. The monitor reads the electronic signals and through electron beams scanning the interior of the vacuum tube (cathode ray tube) causes visible light of the appropriate color to be displayed on the screen.

The application of CCD images to flexible endoscopes during the 1970's permitted development of a solid-state scope that relied on electronics rather than fiber optics for image transmission. The CCD image processor produces a brighter, higher-resolution image than does the conventional fiberscope. This advanced microelectronic technology enables the CCD to function as a miniature TV camera able to record and display images with great clarity on a video monitor.

Miniature Cameras
Another variation of the fiberoptic camera is the miniature camera developed for the examination of boiler tubes (see Figure 3-12). The limiting factor in fiberoptic cameras is

FIGURE 3-12 Miniature cameras. (Courtesy of Visual Inspection Technologies, with permission.)

the distance the fiber bundle can be extended. There is also a practical limit as to how long the fiber bundle can be manufactured. Also, the longer the fiber bundle, the greater the possibility of breaks and disconnections. The solution to these problems is to place the light source at the object end via electrical wire. Since boiler tubes have inside diameters of several inches, the logistics of assembling a ring of light can be realized and cable lengths of up to 100 feet in length can be attained. The same elements of a lens and CCD transmitting the signal back to the CCU (processor) and on to the monitor remain, but now multiple wires traverse the length of the cable—one with power for the light source and another to transfer signals from the CCD to the processor. No fiber bundles are necessary.

A recent variation of the miniature camera is to replace the CRT monitor with a liquid crystal display (LCD) monitor. This is less expensive and results in less resolution. If the application is less critical, e.g., checking for clogged drains or major leaks in boiler tubes rather than looking for cracks or minor erosion, this loss of resolution may be acceptable.

Miniature Camera with Pan, Tilt, and Zoom Lens

The previous examples of fiberoptic miniature cameras with external light sources or miniature cameras with a light source at the object end still have some commonality. Both use a simple lens and adjustment apparatus to focus the object onto the CCD a very short distance away. Adding a zoom lens to focus on objects very far away requires great versatility in adjusting the lens assembly. Focusing at greater distances creates more problems to be overcome. First, light, which decreases to one-fourth in intensity for each doubling of the distance, must be greatly increased. Second, motor controls for the remote movement of the more complicated zoom lens assembly must be dealt with. Third, motor controls for panning and tilting must be designed to cover a large inspection area with the possibility of limited camera movement (see Figure 3-13).

The lights now have to target two possible zones of illumination due to the longer viewing distances. Spot lamps or flood lamps could be used. The typical light source is two 35 Watt bulbs with reflectors. The spot lamp will produce 8500 candle power with a 10° beam spread. The flood lamp will provide 15,000 candle power with a 38° beam spread. These can become very hot and dangerous in hazardous environments.

The motor controls for the zoom lens are bulkier because the lens assembly is now more complicated. The lens components in these assemblies are moved to change the relative physical positions, thereby varying the focal length (fl) and angle of view through specified ranges of magnification. Zooming is a lens adjustment that permits seeing detailed close-up images of a subject or a broad overall view. The angular field of view (FOV) can be made narrower or wider, depending on the setting. This ability to zoom in or out gives the impression of camera movement, even though there is no movment at all. This is done with variable focal length and multiple lens assemblies that go from wide angle to telephoto while the image remains in focus. To achieve this effect, three groups of lens elements are needed:

FIGURE 3-13 Miniature camera with pan, tilt, and zoom lens. (Courtesy of Visual Inspection Technologies, with permission.)

1. Front-focusing objective group
2. Movable zoom group
3. Rear stationary relay group.

If the zoom lens is well designed, a scene in focus at the wide-angle (short fl) setting remains in focus at the narrow angle (telephoto) setting and everywhere in between.

Pan and tilt mechanisms allow the camera to be pointed in almost any direction. The pan and tilt controller has a two-axis drive capability for panning (side to side) and tilting (up and down). The addition of the pan and tilt gears and drives make for a heavy head that needs additional strength in the support fixture. To assure spark-free and low or noncombustible situations, the components must all be housed in an inert gas environment. Additionally, the internal pressure of the fixture housing must exceed the external pressure to assure that explosive gases do not enter the housing (where sparks are possible). The housing must be airtight to prevent sparks from exiting the housing into combustible environments.

The zoom lens with pan and tilt capabilities offers a great advantage over the fixed focal length lens in that many different fields of views can be obtained with one lens. An additional point of interest regarding the zoom lens is that the typical magnification limit via optical adjustment is 12:1. Commercially available units can attain magnification up to 24:1. This doubling of magnification is attributable to computer enhancement. It can be noted during zooming that the magnification is smooth up to the 12:1 point, and then a slight "jolt" is noted as the computer enhancement takes over. At the maximum magnification, slight pixel squares may be barely visible to the eye when viewed on the monitor.

The final variation is the assortment of fixtures that provide the camera with a platform from which to operate. The "snaking" means of locomotion is satisfactory for short distances with minimal friction. Inserting the "distal" end or objective end into a pipe, tube, internal component, etc., is acceptable if the rigidity of the cable is adequate for traversing the distance in question. If not, the camera or flexible borescope can be affixed to a fiberglass rod to add rigidity. Additional mobility can be gained by mounting cameras onto robotic self-propelled platforms or crawlers. A power wheeled system or tractor with sufficient weight can overcome great amounts of friction from the cable, but can only enter an opening larger than the tractor. The larger the tractor, the longer the distance that can be traveled. However, access to smaller openings is limited. Smaller tractors can access smaller openings but cannot travel as far. This is discussed further in the applications section.

Video Cassette Recorders and Video Tape Recorders

Regardless of the camera, monitor, or fixturing mechanisms, the image is usually recorded on videotape for review, evaluation, and record retention. The medium of the future may well be the digital laser disc or digital tape medium. At this time, the most common methods are the videocassette recorder (VCR) or video tape recorder (VTR).

This is one area of technology where the consumer market has driven the commercial market. In a race to produce an acceptable format, Video Home System (VHS) and BETA formats were promoted. It is widely accepted that the most common format for consumer use is the VHS format. The commercial world has accepted the Super VHS (SVHS) format for technical applications.

The VTR consists of the scanner, transport system, servo controls, frequency response, and signal processing components. The word "heads" is often heard when we purchase our own home system. The more heads, the better the system. Either the heads are laying down magnetic tracks when the recording is being made or they are picking up the tracks of magnetic signals when the tape is being played for viewing. The video signal is a series of magnetic tracks laid down diagonally across the tape, which is coated with a ferromagnetic material that holds a residual magnetic field. This field can be picked up later when the tape is played. The transport system scans the tape across the face of the head. There is no actual contact, but rather a thin film of air upon which the tape glides. This prevents or reduces the wear on the head and tape. The tape is guided through its path automatically when recorded or played back. Servo controls assure that precise angles for recording and playing back of the videotape are maintained. The positions of the revolving heads must be precisely locked in phase with the incoming video signal, so that alternate A and B heads contact the tape at the right moment. The heads are adjusted by the servo controls.

In analog recording techniques, a Mylar base tape is coated with a magnetic oxide layer. The magnetic head (which is an electromagnet) records magnetic values onto the tape to record vertical and horizontal signals that produce the video image. The more lines of information available, the better the resolution of the image. VHS has a maximum resolution of 240 horizontal lines. SVHS has a maximum resolution of 400 horizontal lines. Since the SVHS format will produce fairly high quality results at a reasonable cost it, is frequently used today. The horizontal resolution provided in the half-inch formats for SVHS are clearly below the standard camera and monitor resolution requirements [525 for commercial broadcast and 900–1000 for closed circuit television systems (CCTV)]. That is why if the maximum observable resolution is desired, the image should be evaluated in "real time" on the monitor. Because of the reduction from 900 lines horizontal resolution in CCTV to 400 lines horizontal resolution for SVHS, there is a 50% loss of resolution.

In the not-too-distant future, video signals will be real-time recorded in digital format on ultra-high-capacity random-access media. At this time, digital video requires large amounts of memory. One frame of video requires about 1 megabyte of storage. Technology is advancing, with memory now available in gigabytes, signal compression, and increased speed of processing (1 GHz in the consumer PC market).

Monitors

The best camera will only produce the number of lines of resolution that the monitor is capable of displaying. The National Television System Committee (NTSC) defines resolution as the extent to which details can be distinguished in an image. Video resolution is most commonly measured using "lines" as the measuring criterion. Horizontal resolution is a measure of the number of evenly spaced vertical lines across the entire picture that the video system can resolve. The NTSC calls for 525 lines per frame. Two alternating scans of 262.5 lines, interlaced, equals one image of 525 lines per frame. The standard of the Electronic Industries Association (EIA) measures the apparent resolution on a monitor with the use of the Standardized EIA Resolution Test Pattern (see Figure 3-14). This consists of nonparallel lines that begin converging towards the center of the test pattern. The converging lines are repeated in each corner of the pattern, since the resolution is not the same across the entire face of the monitor. When the pattern is displayed on the monitor, the human eye discerns where the converging lines merge and are not separated as four distinct lines. At this location along the four lines is a scale of resolution values. The number nearest the converging four lines is noted. This is the resolution value. Remember that the human eye, with its limited resolution, is part of the visual equipment system. Different patterns are more pleasing to some than others. Humans do not have linear sensitivity to all patterns.

Monitors are available with picture diagonals ranging from an inch to 25 inches and even larger. However, at least a 5 inch monitor is needed to resolve 512 TV lines, and a

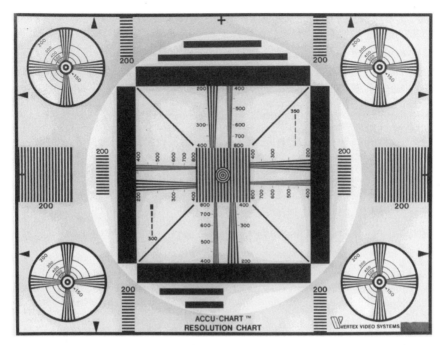

FIGURE 3-14 Standard EIA test pattern. (Courtesy of Visual Inspection Technologies, with permission.)

13-inch monitor to resolve 1100 TV lines horizontally per picture height. The size and shape of the electron beam focused on the phosphor screen is also a factor in determining resolution. An individual cannot "see" any detail smaller than the electron beam. Electron optics and the quality of their voltage and current supplies all govern the shape of the landing electron beam.

Digital Cameras
Digital cameras use the charged coupled device (CCD) as the means of capturing images. These images are stored temporarily on floppy discs or memory chips, or are archived permanently. The bitmap or other file format can then be downloaded directly to a personal computer, attached to e-mail, stored on web pages, or otherwise transferred for viewing. Some cameras utilize removable media cards for short-term or permanent archiving.

Typical resolution may vary. CCD pixel arrays with 640 × 480 to 1280 × 960, or 1280 × 1024 up to 1536 × 1024 are available. Some cameras feature LCD screens built into the camera for on-the-spot viewing of the image, thus allowing retakes as required. Reusable storage cards can provide image capturing capability limited only by the number of storage cards available.

Once an image is downloaded to a personal computer, various image programs allow image manipulations. Enlargements, segmenting, framing, changing brightness or color, as well as other options allow a wide range of changes to be made. Image printouts on color printers are producing near-photographic quality prints.

IV. APPLICATIONS AND TECHNIQUES

General Applications

Visual examinations and other nondestructive test methods cover the spectrum of examining materials from raw product form to the end of their useful lives. Initially, when raw material is produced, a visual examination is conducted to locate inherent discontinuities. As the material is further transformed through the manufacturing process, a product results. At this stage, the visual examination method is used to find discontinuities that are produced during the primary processing steps. When the product is further developed into its final shape and appearance, the secondary processes that give the product its final form can also introduce new discontinuities. Finally, the product is placed into service and is subject to stresses, corrosion, and erosion while performing its intended function. The process concludes when the material has reached the end of its useful life and is removed from the source. At every stage, the visual examination method is applied using various techniques to ascertain the physical condition of the material that became the component, system, or structure serving the needs for which it was intended.

After material is produced, visual examination is used to assure that a product will meet the specification requirements prior to processing into a product form for use in its intended service.

The technology associated with visual testing (VT) and remote visual testing (RVT) includes a spectrum of applications, including various products and industries such as:

- Tanks and vessels
- Buildings
- Fossil-fuel power plants
- Nuclear power plants
- Turbines and generators
- Refinery plants
- Aerospace

Tanks and vessels usually contain fluids, gases, or steam. Fluids may be as corrosive as acid or as passive as water, either of which can cause corrosion. Tank contents are not always stored at high pressure. Conversely, vessels usually contain substances under substantial pressure. This pressure, coupled with the corrosive effects of fluids and thermal or mechanical stresses, may result in cracking, distortion, or stress corrosion of the vessel material.

Buildings also serve as a source for a myriad of RVT applications. These applications include location of clogged piping; examination of heating and cooling (HVAC) heat exchangers; and looking for cracking, pitting, blockages, and mechanical damage to the components. Structural damage that may be present in the support systems, beams, flooring, or shells, such as cracking, corrosion, erosion, or warpage can also be detected.

Fossil-fuel power plants have piping, tubing, tanks, vessels, and structures that are exposed to corrosive and erosive environments as well as to other stresses. These components may require RVT.

Turbines and generators, existing at both fossil-fuel and nuclear power plants, are vulnerable to damage due to high temperatures, pressures, wear, vibration, and impingement of steam, water, or particles. Accessing the small openings and crevices to reach damaged turbine blades becomes a very tedious job and a serious challenge, but the effort of performing remote inspections through limited access ports reduces the need and cost of downtime and disassembly of major components.

VT and RVT technologies and techniques are used in nuclear power plants as well. Water used for shielding and cooling is exposed to both ionizing radiation and radioactive surface contamination. The use of water as a coolant and radiation shield in a nuclear environment places additional requirements on RVT evaluation. The equipment must not only be waterproof, but also tolerant of radioactive environments.

Due to process requirements in refineries, the containment of pressure and temperature is a necessity of paramount importance, as is the containment of hazardous materials. These same materials can be a source of corrosion to piping, tanks, vessels, and structures, all of which are in constant need of monitoring.

Standards, Codes, and Specifications

The above applications all require VT and RVT to detect surface anomalies. A common source for material specifications is the American Society for Testing and Material (ASTM) standards. ASTM was founded in 1898 and is a scientific and technical organization formed for "the development of standards on characteristics and performance of materials, products, systems and services; and the promotion of related knowledge." At this time, the annual ASTM Standards fill 75 volumes and are divided into 16 sections. An ASTM standard represents a common viewpoint of producers, users, consumers, and general interest groups intended to aid industry, government agencies, and the general public. Two metal material sections are Section 1—Iron and Steel Products and Section 2—Nonferrous Metal Products. These standards provide guidance on the material conditions that must exist in order to be considered satisfactory for use.

Additionally, when material is fabricated and formed into a product, other standards, specifications, and codes delineate visual testing requirements. The American Society of Mechanical Engineers (ASME) publishes the ASME Boiler and Pressure Vessel Code, Sections I through XI. The Material section is Section II. The Design sections are I, III, IV, VIII, and X. Section V contains the methodology for nondestructive examination, and Section IX addresses welding and brazing. Section VI deals with heating boilers, Section VIII, pressure vessels, and Section XI, in-service inspection of nuclear power plant components. Scattered throughout these sections are visual examination requirements.

The American National Standards Institute (ANSI) is a consensus-approval-based organization. ANSI's B31.1—Power Piping and B31.7—Nuclear Power Piping also provide visual examination requirements for materials, fabrication, and erection.

The American Petroleum Institute (API) has developed approximately 500 equipment and operating standards relating to the petroleum industry that are used worldwide. An example is the API standard for Welding of Pipelines and Related Facilities (API 1104).

The aerospace and military standards are being replaced with the more commonly accepted industry codes and standards.

Visual Detection of Discontinuities

Chapter 2 addresses and illustrates many of the following discontinuities in greater detail.

Inherent Discontinuities

The VT technique most often used to detect inherent discontinuities is direct visual. The human eye assisted by measuring devices, auxiliary light sources, visual aids (e.g., magnifiers and mirrors), and the recording media of photographs and sketches are most commonly used for this application.

Applications. Raw material may be checked for inherent discontinuities and compared with the requirements of material specifications for acceptance. Frequently, the first examinations performed on new materials and products are dimensional checks. Attributes such as thickness, diameter, out of round, roughness or smoothness, etc., are examples of the first round of visual checks made on new materials. Checking for discontinuities that interrupt the normal physical structure of the part is also conducted at the material fabrication stage. The inspector might look for blemishes or imperfections that are generally superficial and could be located by observing a stain, discoloration, or other variations on the surface. Discontinuities on the surface could be indicators of other anomalies in the material. Discontinuities or lack of some element that is essential to the usefulness of the part should be of importance to the inspector. Any of these attributes would be evaluated against the specified code, acceptance criteria, standard, or customer requirements. The following is a typical sequence of observation steps an inspector or examiner might follow for a general condition visual examination.

1. The observation of an anomaly—something different, not normal, or irregular.
2. The visual indication is evaluated and determined to be relevant or not.
3. The relevant indication is compared to the appropriate standard, code, specification, or customer requirements.
4. The relevant indication is judged acceptable, not recordable, recordable, or unacceptable and rejected.

Discontinuities may start out as something small. They may result from corrosion, a slight scratch, or other discontinuities inherent in the material such as porosity, cracks, or inclusions. These seemingly harmless anomalies may develop into cracks due to stress concentration under varying loads and propagate with time. Eventually, there may no longer be sufficient solid material left to carry the load.

A defect is a discontinuity of such size, shape, type, number, or location that results in a condition that will cause the material to fail in service or not be used for its intended purpose.

Ingots. Common discontinuities detected visually in the ingot are cracks, scabs, pipe, and voids.

Cracks may occur longitudinally or in the transverse plane in the ingot. Although cracks may be observable in the ingot stage, it is more likely that they will appear in subsequent processing operations.

Scabs are surface conditions on ingots caused by the hot molten metal splashing on the mold surface, solidifying and oxidizing such that it will not re-fuse to the ingot. Again this condition may be more detrimental to the product at later stages of processing.

Pipe is a cavity formed by shrinkage during the last stage of solidification of the molten metal. It is an elongated void with a "pipe like" shape and occurs in the center of the ingot. This discontinuity can result in a lamination after the ingot is rolled into plate or sheet. If ingot is processed into bars, blooms, or billets, the pipe will become elongated as a result of the processing and generally remain in the center.

Voids are produced during the solidification of the ingot, caused by insufficient degassing. The gas, which becomes entrapped, often forms spherical cavities.

Nonmetallic inclusions in ingots are chiefly due to deoxidizing materials added to the molten steel in the furnace, ladle, or ingot mold. Oxides and sulfides constitute the bulk of nonmetallic inclusions.

All of these dicontinuities can occur during the pouring of molten metal into the ingot mold. The resulting discontinuities are the same as those found in the processing stage when pouring metal into a mold and creating a casting.

Techniques. The types of anomalies found in ingots are typically gross in nature. These can generally be found with direct visual examination techniques. A magnifying lens, auxiliary light, and possibly a mirror should suffice for this application.

Primary Processing Discontinuities

The visual technique most often used to detect primary processing discontinuities is the direct visual technique. Measuring devices, auxiliary light sources, visual aids (e.g., magnifiers and mirrors), along with the recording media of photographs and sketches, are the most common techniques used for this application.

Applications

Materials are formed into a final or near-final shape as a result of the various primary processes. The forming processes of wrought products include forging, rolling, drawing, extruding, and piercing. Discontinuities caused by these methods of forming the product shape are known as primary processing discontinuities. When steel ingots are worked down into usable sizes and shapes such as billets and forging blanks, some of the inherent discontinuities described earlier may appear. These primary-processing discontinuities may ultimately become defects. Working the metal down into useful forms such as bars, rods, wire, and forged and cast shapes will either change the shape of existing discontinuities inherent to the product or possibly introduce new ones.

Forging. Typical forging discontinuities include forging bursts, laps, and cracks. Bursts are internal or external forging discontinuities. Forging bursts appear as scaly, ragged cavities inside the forging. (Note: Internal bursts are more likely to be detected with the use of volumetric inspection methods such as radiography or ultrasonics. For additional information refer to Chapter 2.)

Forging laps are folded flaps of metal that have been forced into the surface of the part during the forging process. The lap can vary from shallow to very deep on the forging surface. A lap can appear to be tight and irregular, straight, crack-like and linear, or even wide and U-shaped. Forging laps are difficult to detect without magnification.

Cracks in a forging are usually referred to as "forging bursts." Many forging bursts open to the surface are detectable visually; however, the smaller ones may require other methods such as PT or MT.

Technique. The direct visual technique is best suited for this application. This can be best accomplished by examination using a 5× to 10× magnifier with an auxiliary light source held at an angle so as to cast a shadow on any irregularities.

Rolling. Rolling of product forms can create a variety of discontinuities. The more common ones are seams, stringers, cracks, and laminations. These can appear in product forms such as bars, rods, channels, beams, angle iron, and strip metal.

Seams and stringers generally appear on the outside surface of the product parallel to the direction of rolling and are very difficult to detect visually. Seams contain nothing other than entrapped oxides. A seam is usually produced when two unbonded surfaces are pressed together during the rolling process. Stringers are inclusions that get "strung out" during rolling and may be observable if exposed to the surface. Complex structures may be formed from plate or rolled shapes, making some surface indications difficult to detect. Additionally, internal discontinuities such as laminations are best detected using UT.

Techniques. The direct visual technique is best suited to detect these conditions in this application. This can be best accomplished by examination using a 5× to 10× magnifier with an auxiliary light source held at an angle so as to cast a shadow on any irregularities. Indirect visual examination with a borescope or fiberscope may be appropriate for viewing internal areas of complex product forms. The limiting factors will be access and distance.

Drawing, Extruding, and Piercing. Drawn, extruded, or pierced products may exhibit similar types of surface discontinuities. Any irregularities on the surface that are gross in nature could result in problems during the forming process. Common observable anomalies may be scabs, slugs, or scoring.

Scabs are loosely adhering material that can be removed from the final product. These can result from dirty dies that press in the loose foreign material during the drawing process or from oxide layers that are not bonded to the material.

Slugs are usually larger pieces of material that are firmly impressed into the product form and may not come off easily.

Scoring appears as deep scratching of the surface of the product by foreign material that may be adhering to the inside of the die or piercing mandrel during processing.

Technique. The direct visual technique is best suited to detect these conditions. This can be best accomplished by examination using a 5× to 10× hand-held magnifying lens with an auxiliary light source held at an angle so as to cast a shadow on any irregularities.

Castings. Casting is a primary process. Castings may exhibit various types of discontinuities that are inherent in the metal forming process, including nonmetallic inclusions (slag), hot tears, gas, unfused chills and chaplets, cold shuts, and misrun. Some of these discontinuities may also occur internally. The casting process can produce some very complex shapes. Access for visual examination can be a real challenge.

Inclusions exposed to the surface may be akin to the "tip of the iceberg" in that the visible area may be merely a small percentage of the inclusion contained within the casting. The visual examiner can only detect and measure the exposed area of the discontinuity.

Hot tears are more likely to be visible on the surface, since they occur at locations where changes in thickness occur. Hot tears are the result of different coefficients of expansion and contraction at a change of thickness. The abrupt change from thick to thin material sections provides a stress riser. The different rates of cooling (contraction) could cause a rupture of the metal under stress at these junctions.

Gas is a spherical void condition. It may occur both internally and at the surface of the casting. Surface gas typically has a hemispherical shape, since if exposed to the surface an incomplete sphere is trapped in the casting. There may be one or many of these anomalies distributed randomly or in clusters at any location.

Unfused chaplets and chills will be visible only if they protrude through the surface. Remember that the chaplet is usually metal with a low melting point that is used to support the internal mold cores until the molten metal enters the mould and supposedly melts the chaplet. The examiner may note the "footprint" of the metal form that is exposed to the surface. The chill can be likened to a nail with a head. The "head" may be all that the examiner observes, with the remainder of the "nail" encased in the casting.

Cold shuts could result from balls of molten splattered metal that adhere to the mold wall prior to the remainder of the molten metal rising over it and entrapping it on the inside of the mold and thus the outside of the casting. It can appear as a circle around a ball of trapped metal. Cold shuts also result when two regions of the casting do not metallurgically fuse; one surface begins to solidify before the remainder of the molten metal forms around or over the already solidified metal. This condition is generally not detectable by standard VT techniques.

Misrun is a surface condition in which the casting mold does not completely fill with molten metal, thereby leaving a noticeable depression, which is readily detectable with VT.

Techniques. Either the direct visual or remote visual techniques are necessary for examining castings for exposed inclusions—hot tears, surface voids, unfused chaplets or chills, cold shuts, and misrun. Accessible surfaces can best be examined by using a 5× to 10× hand-held magnifying lens with an auxiliary light source held at an angle so as to cast a shadow on any irregularities. The surfaces that are internal to the casting may require examination using a rigid borescope or flexible articulating end fiberoptic borescope, the latter being more versatile. Either one would bring the remote surface under close examination and provide light to the examination site. Extra care is required to assure full coverage and proper orientation when performing internal exams. Knowing where the discontinuity is located and accurately measuring its dimensions are requirements for proper evaluation of the indication. If a visual record is desired, a fiberoptic video camera may be appropriate. The distance to the area of interest as well as the size of the access port are important factors to be considered when choosing the proper instrument. Additionally, it is difficult to obtain accurate measurements at the remote site. Some instruments have been outfitted with detachable scales that are visible in the field of view. Other instruments have remote measuring devices that rely on shadow casting and geometric calculation techniques. These devices require calibration on surfaces similar to the orientation plane of the examination surface in order to be accurate.

Metal Joining Processes

Metal joining processes include a number of welding and allied processes. Each major process must be considered for unique as well as common discontinuities. Metalworking industries generally use soldering, brazing, and a broad variety of welding processes. A generic definition of the basic welding process is "a materials joining process that produces coalescence of materials by heating them to suitable temperatures, with or without the use of filler metal."

Soldering utilizes filler material that becomes liquid at temperatures usually not exceeding 450° C (±840° F). The solder is heated and distributed between two surfaces of a joint by capillary action.

Brazing produces coalescence of materials by heating them to a suitable temperature and using a filler metal that becomes liquid above 450° C (±840° F). The filler metal is distributed between the closely matching surfaces of the joint by capillary action. In either case, when completed, the only visible signs of completed wetting, and hence bonding, is the micro fillet at the intersection of the two materials. The normal configuration for joining is a lap joint of two materials overlapping each other. When heated and wetted with filler material, the resultant fillet should be uniform the entire length of the joint.

Arc welding produces coalescence of metals by heating them with an arc, with or without the application of pressure, and with or without the use of filler metal. Shielded metal arc welding (SMAW; Figure 3-15), submerged arc welding (SAW; Figure 3-16), gas tungsten arc welding (GTAW; Figure 3-17), and gas metal arc welding (GMAW; Figure 3-18) all have two things in common. Each utilizes an electric arc as the heat source for melting the base and filler metal, and each provides a means of shielding the arc to block out harmful elements found in the air (oxygen).

SMAW and SAW both use the combustion of flux as the means of consuming oxygen at the weld site, thus preventing the oxidation (burning) of the metals being joined. The common by-product of burning flux is slag.

GTAW and GMAW both utilize inert gas cupped around the weld site to eliminate the oxygen portion of the fire triangle (oxygen, heat, and fuel) and thus prevent the metals from burning at high temperatures. The electrode bearing the heat for GTAW is the tung-

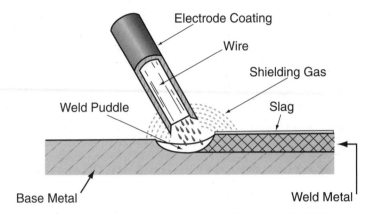

FIGURE 3-15 Shielded metal arc welding process (SMAW).

sten electrode. In this process, the electrode is not consumed. However, deposits of tungsten may flake off and remain in the weld. This is visible only if they remain on the surface of the weld during welding. GMAW uses a consumable electrode of filler metal. Therefore, tungsten cannot be deposited in the weld.

Welding processes are akin to a microcasting process. Therefore, many discontinuities found in the primary casting process can also be found in the welding processes. Porosity, nonmetallic inclusions, and cracks are common to both casting and welding processes.

Porosity may occur internally or at the surface of the weld. As with the casting process, gas porosity may have a hemispherical shape if it is open to the surface. There may be from one to a number of these, randomly distributed or in clusters, at any location. Variations in the shape of the porosity may be called by descriptive names such as worm-

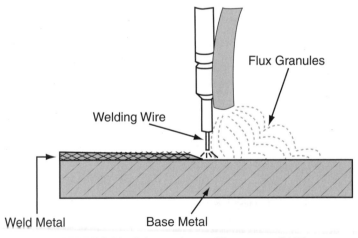

FIGURE 3-16 Submerged arc welding process (SAW).

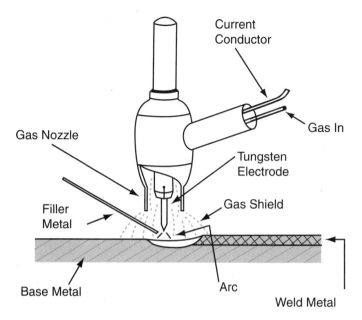

FIGURE 3-17 Gas tungsten arc welding process (GTAW).

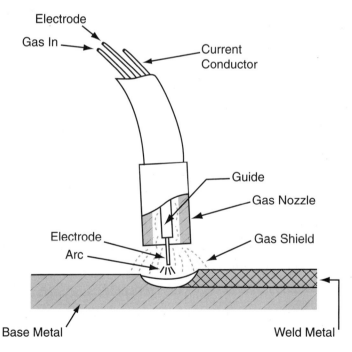

FIGURE 3-18 Gas metal arc welding process (GMAW).

3.29

hole porosity, cluster, and random or linearly aligned porosity. They all originate from trapped gases as the liquid metal solidifies upon cooling, thus forming spherical voids in the metal.

Nonmetallic inclusions commonly referred to as slag (burned flux in the case of welding processes) may be trapped internally or at the surface of the weld face. This condition may be sporadic, isolated, or clustered in groups. Slag is something included in the metal that is not metallic.

Note: Only the processes that utilize flux can have significant slag residue from the burning of the flux. Any base metal may have nonmetallic inclusions or impurities that may rise to the surface of a weld site during the melting of the metal. However, this quantity of "slag" is slight and not to be confused with the major source of nonmetallic inclusions, which is burned flux.

Cracking, with all its possible shapes, forms, and degrees of severity will occur whenever metal relieves itself of stresses by rupturing. The various orientations, locations, and appearances cause the cracks to have many names that uniquely identify them. Crater cracks are found in the depressed cup of the terminal puddle made at the last arc location of the weld bead. Transverse cracks are oriented perpendicular to the longitudinal axis of the weld. Longitudinal cracks are parallel to the long axis of the weld. Microcracks could occur anywhere but are very small by definition. Heat affected zone (HAZ) cracks occur at the toe of the weld and may exist in the weld and/or the base metal, but are limited to the zone adjacent to the toe of the weld. All cracks are characterized as being crack-like in appearance; that is to say, irregular, jagged, and linear in overall disposition.

Each weld process may have weld discontinuities unique to that process. Since only GTAW has a tungsten nonconsumable electrode, it is the only process that can have tungsten inclusions. The other weld processes use consumable material as filler.

Techniques. The direct visual technique is best suited for detecting most external weld discontinuities. This can be best accomplished by examination using a 5× to 10× magnifier with an auxiliary light source held at an angle so as to cast a shadow on any irregularities. Additionally, linear measuring devices are commonly employed to measure each discontinuity. Dimensional and profile checks are the most common direct visual examinations for welds. These visual examinations would be conducted at fit-up, in process, and after welding.

During prewelding and fit-up, many dimensions should be checked. Root opening, which is the separation between the members to be joined, and root face lands should also be checked. Groove face surfaces on each member of a butt joint must be checked. The single bevel angle on one member of the two members and the groove angle between the two members to be joined have tolerances and limits that require verification. The weld joint may have penetration minimums that necessitate confirmation. The members to be joined typically have minimum thickness measurement requirements. The inside diameter of butt pipe joints has diameter tolerances, land dimensions, and counterbores that must also be verified.

During welding, controls are frequently evaluated by the visual inspector; e.g., temperature control with the use of a pyrometer, temperature crayons, or other temperature indicators. Distortion or movement tolerances are checked during the welding process. Welding process controls such as amperage, temperature, weld wire, material certifications, and welder qualification etc. may also be performed by the visual examiner, the quality control inspector, quality verifier or, in this era of Total Quality Management (TQM) and other quality control philosophies, by the welders themselves. Other weld attributes that can be evaluated visually in the process include bead and weave pattern. Excessive weave patterns may induce uneven heat sinks, creating a skin effect that could lead to brittle layers of weld metal that may crack in service. Interpass cleaning is necessary to prevent entrapment of nonmetallic inclusions. Porosity in the intermediate weld beads may propagate throughout the weld cross section with subsequent bead deposits. Another imperfection that may occur during the welding process prior to completion is lack of fusion. Since lack of fusion is defined as nonfusion between weld beads or between the weld and the parent metal, it is not visible to the eye in the final weld. During deposit of each weld layer, sharp crevices that may lead to lack of fusion on the next weld deposit can be detected and either removed or ground into a radius, or more heat can be applied to melt to the bottom of the crevice. Detection of lack of sidewall fusion usually requires a volumetric test (i.e., UT or RT).

When dealing with completed welds, many dimensions and weld attributes are inspected for compliance with the weld specification. Final cleaning, weld size, and profile dimensions are also checked. Weld face, length, width, and thickness must meet dimensional requirements found on the drawing. The weld should also be examined to detect possible surface discontinuities. Convexity and concavity affect the weld thickness. The legs and actual throat dimensions of fillet welds must be measured. Welds that are accessible from both sides should be checked for adequate penetration. Partial penetration welds cannot be adequately detected by visual means unless cross-sectioned, which defeats the nondestructive aspect of visual examination. Stringer weld beads can be counted, lengths measured, distances between welds measured, and occurrence of surface discontinuities observed.

Final weld examinations include visual inspection for the following dimensional and profile attributes:

Misalignment—excessive offset observed on the inside or outside diameters of a pipe or the misalignment of the near or opposite sides of butt welds.

Underfill—insufficient weld filler material to fill the weld groove face. The weld surface is below the base metal surface (see Figure 3-19).

Undercut—base metal melting without sufficient metal deposit to bring the weld metal level with the base metal. Emphasis is on the melting of base metal and insufficient filler metal being added (see Figure 3-20)

Overlap—weld metal extends over the base metal at the toe of the weld without fusion to the base metal (see Figure 3-21).

Excessive reinforcement—weld reinforcement that exceeds the specification limits (see Figure 3-22).

Concavity—a depression on the inside diameter (root concavity) of a pipe, or near the far side of plate that results in less than specified weld thickness and is less that the base material thickness of the thinnest member (see Figure 3-23).

Insufficient throat in a fillet weld—a depression on the face of a fillet weld that results in the actual throat being less than that specified.

Insufficient leg in a fillet weld—less leg dimension in the fillet weld than that specified.

Incomplete penetration—this is the absence of penetration of the root pass. It is only observable by VT when the opposite side from the weld is accessible. It is, in fact, the original edge of the plate or pipe to be joined (at least one side) still visible and not totally fused or melted (see Figure 3-24).

Burn-through—excessive heat resulting in the loss of some metal, leaving behind a cavity that can vary in size and can extend from one side of the weld to the other.

Weld craters—depressions at the end of the weld, representing the last puddle in the weld bead (see Figure 3-25).

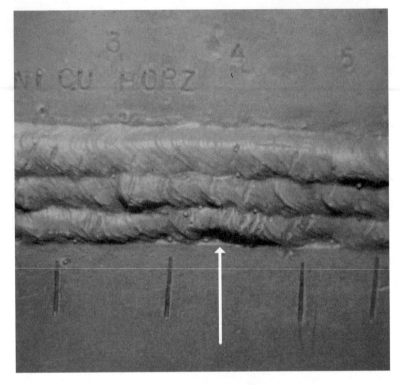

FIGURE 3-19 Underfill.

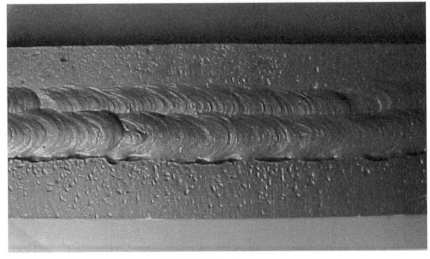

FIGURE 3-20 Undercut.

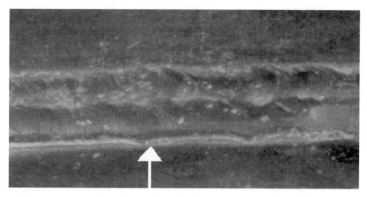

FIGURE 3-21   Overlap (or rollover).

Crater cracks—as mentioned above, cracks that occur in the weld crater. These cracks are due to thinner portions of the weld in the crater cooling first and becoming rigid. The main weld cools later and shrinks away from the crater portion of the weld. This results in residual stresses in the weld relieving themselves via rupture of metal or cracking. The crater crack often has a star-like appearance and is not easily detected with conventional VT techniques.

FIGURE 3-22   Excessive reinforcement.

FIGURE 3-23 Concavity.

FIGURE 3-24 Incomplete penetration.

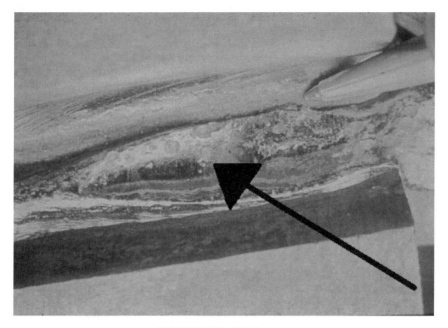

FIGURE 3-25 Weld crater.

Arc Strikes—arc welding that strikes the base metal, usually in an effort to initiate an arc to start welding another bead. These arc strikes melt localized portions of the base metal to shallow depths. The differentials in material hardness frequently result in stress risers, yielding microcracks at the time of welding or thereafter in service.

Spatter—globules of material expelled from the weld puddle or from the consumable electrode during welding. The localized melting of the skin of the base material contacted by the molten balls of excess metal can also cause local stress risers. Again, these differentials in material hardness frequently result in stress risers, yielding microcracks at the time of welding or later in service.

Techniques. Visual examination of welds and other metallurgical joining processes can best be performed with the direct visual technique. This application requires a variety of mechanical aids, such as templates, gauges, and measuring tools, to maximize efficiency and assure consistency of the results. The first overview should be accomplished by examination using a 5× to 10× magnifier with an auxiliary light source held at an angle in order to cast a shadow on any irregularities. Additionally, linear measuring devices need to be employed to measure each dimensional or profile attribute and the size and number of discontinuities. The root area of a pipe weld is occasionally examined for complete fusion or melting in the case of critical welds. This application requires a flexible fiberoptic borescope that is positioned in the area of the root weld. This visual examination can assure that the penetration or melting of the root is adequate, in lieu of performing radiography. On some occasions, the root insert can be observed during real-time welding to observe welder progress and adequacy. This is hazardous to the remote visual instrument, as the heat can destroy the fibers or lens if they are too close to the weld root while welding occurs.

Measuring Devices for Welding Process Visual Examination
Measurements and examinations of welds are conducted before, during, and after welding with the use of a large array of mechanical aids, including, but not limited to, the items listed in Table 3-1.

Service-Induced Discontinuities—Applications And Techniques

Service-induced discontinuities are the results of material deterioration during use. Wear, erosion, corrosion, and loss of integrity through fatigue all may lead to the failure of a component while performing its intended function. Wastage or general material loss can occur when the size of a component has been reduced such that the cross-sectional area is no longer able to support the designed load. Mechanical fatigue due to cyclic loading and thermal fatigue due to temperature fluctuations frequently lead to excessive stresses, cracking, and loss of integrity. Stress combined with a corrosive environment frequently results in stress corrosion cracking.

Wear
Wear can be defined as the undesired removal of material caused by contacting surfaces through mechanical action. In many respects, wear is similar to corrosion. The difference is that wear is mechanical in origin and corrosion is chemical in origin. The different types of wear are abrasive, erosive, grinding, gouging, adhesive, and fretting.

Abrasive wear has the appearance of a multitude of small scratches or grit marks that generally result in a local uniform reduction of material thickness. It occurs when one surface rolls or slides under pressure against another surface.

Erosive wear occurs when particles in a fluid or other carrier slide and roll at a relatively high velocity against a surface. Each moving particle contacting the surface removes a minute portion of the surface. The cumulative effect over a long period of time is

TABLE 3-1 Mechanical Aids Listed by Functional Category

Linear Measuring	General Gauges	Weld Gauges	Optical Aids	Image Recorders	Miscellaneous	Light Source
Steel rule and scale	Depth gauge	Cambridge weld gauge	Reticule Magnifier	Polaroid camera	Combination square	Flashlight
Tape	Thread gauge	Fillet weld gauge	Microscope (pocket)	35mm film camera	Temperature gauge	Extension light
Vernier caliper	Pitch gauge	Template gauge	Telescope	Video camera	Level and plumb line	High-intensity spotlight
Dial caliper	Thickness gauge	Hi-lo gauge	Optical comparator	Digital camera	Surface comparator	
Digital caliper	Diameter gauge	Weld height gauge				
Dial indicators	Snap gauge					
Micrometer	Radius gauge					

significant and can produce a high degree of material loss. A "scalloped" appearance with wave patterns over a general area can indicate erosive wear.

Grinding wear occurs primarily when particles under high pressure cut very small grooves at relatively low speed across a metal surface. High-pressure, low-speed operation is characteristic of metal turning against a dull cutting edge. A series of parallel scratches, either curved or straight, resulting in reduced material thickness, would indicate grinding wear.

Gouging wear is caused by high pressure that lifts fragments from a metal surface. When hard abrasive products are crushed, battered, or pounded under high pressure, rapid deterioration of the contact surfaces can be expected. Grooves with a buildup of material at the deeper end of the trough are good indications of gouging.

Adhesive wear can be characterized as microwelding. The microwelding or adhesive property causes a stationary component to lose particles of material that adhere to a moving component. Other terms for adhesive wear are scuffing, galling, and seizing. Adhesive wear frequently starts out on a small scale but rapidly escalates as the two sides alternately weld and tear metal from each other's surfaces.

Fretting wear is similar to adhesive wear in that microwelding occurs on mating surfaces. The difference is that adhesive wear is related to moving surfaces and fretting wear is related to stationary surfaces that unintentionally move due to applied stresses. Minute motions do occur, but the cyclic motion is extremely small in amplitude. It is, however, enough to cause microwelding and enough to cause fretting on both surfaces. These "welded" areas do not remain joined, and with subsequent movement of the surfaces, they break off. The particle debris resulting from this process causes wear. Other terms for fretting wear is fretting corrosion, false brinnelling, friction oxidation, chafing fatigue, and wear oxidation.

Visual techniques for examining materials for wear include magnification, auxiliary light, and surface comparators. The surface patterns, depth of wear, knowledge of the part's function, and hardness all play a role in effective evaluation. The biggest obstacle to overcome is to gain access to the surfaces where the wear is occurring. By definition, two surfaces in close contact could cause wear. Therefore, disassembly is required for direct visual examination.

Corrosion and Erosion

Corrosion can be defined as the deterioration of a metal resulting from electrochemical reactions with its environment. There are many types of corrosion (at least two may be acting simultaneously).

Galvanic corrosion is caused by physical differences between contacting metals or a metal and its environment. The corroding material is known as the "anode" the noncorroding complimentary material is known as the "cathode". When corrosion takes place, three conditions are necessary for the electrochemical reaction to occur. 1) Two different materials are needed. 2) They must meet in electrical contact with an electrolyte or electrically conductive liquid or paste. 3) When the anode corrodes, such as in the operation of a battery, hydrogen is released and energy in the form of an electric current is present. The continued electrochemical reaction causes more galvanic corrosion and additional material loss. To break the cycle, one of these three elements must be removed.

Uniform corrosion is the most common form. In the case of iron or steel, it is better known as rusting. Uniform corrosion is the result of microscopic galvanic cells in the surface of the metal. Because of local chemical differences, impurities, and alloying intermetallics within the metal, there are microscopic anodes and cathodes ready to cause corrosion if an electrolyte such as moisture is introduced. The corrosion is uniform only on the macroscopic level.

Crevice corrosion is a kind of galvanic corrosion that is caused by differential oxygen concentration cell corrosion, or what is also known as "poultice corrosion." Normally, metal may be expected to corrode when exposed to high levels of oxygen. However, a crevice between two surfaces or metal under a poultice of moist debris is more likely to corrode than the exposed metal. This occurs because there is little oxygen within the crevice or under the poultice. The metal there is anodic and it corrodes. Areas exposed to higher oxygen content are cathodic and do not corrode. Concealed metal at the edge of a joint or under debris tends to pit and eventually becomes perforated through the metal thickness. The gas bubbles under a coated (painted) surface of steel can be attributed to crevice corrosion.

An **erosion corrosion** form of material loss is particularly common in pumps or impellers. This form is known as **cavitational erosion.** A liquid stream can destroy the protective surface film on a metal. This is done through a change in pressure on the trailing edge of pump or impeller blades with an associated creation of gas bubbles. The subsequent collapse or implosion of the bubble can cause a great deal of accelerated metal loss over time. The collapsing cavities implode on the metal surface with compressive stresses estimated at thousands of pounds per square inch. With constant repetition of the pounding, the fatigue mechanism progresses until pits form in the material surface at these locations. Corrosion may enter the picture if surface films are formed on virgin metal in the pit when the system is at rest. The surface film is rapidly destroyed by the high compressive forces when the motion resumes and the cycle starts again.

Fatigue Cracking

Loss of integrity through fatigue cracking can be caused by many mechanisms. Fatigue can cause the failure of a material or component under repeated, fluctuating stresses having a maximum value less than the tensile strength of the material. In general, fatigue strength is a function of the ultimate tensile strength. There are three stages of fatigue: 1) initial damage, 2) crack propagation in the component, and 3) failure resulting from sectional reduction. The cracking mechanisms of mechanical fatigue and thermal fatigue are as follows.

Mechanical fatigue depends on the nature of the stress, how the stress is applied, the rapidity of the stress fluctuation, the stress magnitude, the fabrication procedures utilized, and the environment in which the item is used. The higher the tensile strength, the more susceptible the part is to mechanical failure. Surface finish is important when considering mechanical fatigue. The more polished the surface, the more fatigue-resistant the item is. When fatigue cracks occur, they characteristically are relatively straight and nonbranching. The cracks are transgranular (across the grains) when viewed metallographically.

Thermal fatigue is differentiated from mechanical fatigue by the nature of the load causing the fatigue. In thermal fatigue, there are temperature differentials within a component, which induce significant strains and stresses. Differential expansion may be caused by parts with different coefficients of expansion, i.e., parts within one component having different temperatures or connected to components having different temperatures. The cyclic rate of thermal fatigue is typically low. Thus, it is often but not exclusively a high-temperature phenomenon. High thermal stresses can lead to failure in a few cycles. Thermal fatigue is often initiated where there is a change in section thickness. This is due to different amounts of heat that is absorbed and the resultant temperature gradient. Examples of components that may suffer thermal fatigue are superheater tubes, pump casings, turbine blades, and piping. Failure is often caused by a single crack propagating through the cross section of the part. Occasionally, the failed piece may display a crack that is partially through the sectional area. This will exhibit a color resulting from oxidation. This occurs prior to the break, which exhibits a newer cracked surface.

Techniques. In general, fatigue cracks are open to the surface. Examining components such as pumps, valves, piping, vessels, and heat exchangers for wear, erosion corrosion, and loss of integrity through fatigue can best be performed with the direct visual technique. Access may frequently be limited and remote visual techniques may be required. The first overview should be attempted by direct visual examination using a 5× to 10× hand-held magnifying lens with an auxiliary light source held at an angle so as to cast a shadow on any irregularities. Additionally, linear measuring devices may be needed to measure each observable discontinuity. Internal examinations of valves, pumps, or components may require flexible fiberoptic borescopes, zoom cameras, crawler devices with miniature cameras, and hand-held or guided fixture devices.

Corrosion, Erosion, and Fatigue Applications

The locations of corrosion and erosion are usually inside pressure containers. Common examples are pumps, valves, piping, vessels, and heat exchangers.

Pumps function to move liquid from one place to another. This is achieved by imparting energy to the medium, generally by means of converting electrical to mechanical energy. In the process, resistance is encountered in the form of friction. Friction occurs from both the medium moving in the pressure boundary and from the moving parts of the pump. Gravity is also a force that has to be partially overcome.

In many cases, the medium being moved may have corrosive properties. Erosive wear is common in all cases. All of these factors lead to deterioration of the internal pressure boundary and the internal parts of a pump. Access to the areas to be examined usually requires some disassembly, if not the total breakdown of the pump for inspection. Frequently, the primary access to pumps "in situ" is either through the suction or discharge ends of the pump. Occasionally, disassembly provides access through the pump shaft opening.

Techniques. A common access technique is with the use of a crawler, which can be a tractor motor device with mounted camera and trailing electrical wire. This approach would be acceptable if the suction or discharge pipe were nearly horizontal. If the access pipe is vertical or very steep and the gravity force exceeds the friction of the tractor wheels, a different mechanism is needed. It may be a centering device with expanding and contracting spider arms. Sequential expansion and contraction of alternating sets of centering arms may be mechanically or hydraulically operated to allow the device to climb up or down a pipe (see Figure 3-26).

Once access to the impeller blade or pump vanes is achieved, a zoom camera or a camera with an articulating head can be manipulated to view the area of interest. The leading and trailing edges of the blades and vanes are very important areas to inspect. Additionally, the pump casings should be examined for signs of wash-out (erosion) or pitting. Pits or ridges can reduce efficiency and accelerate casing wear. Impellers should be checked for cracking. Diffuser vanes should be examined for wear or erosion. Shaft sleeves should be checked for scoring. Bearings should be checked for wear or damage. Shafts should be checked for corrosion, wear, and signs of washout.

Valves function to stop or start flow, regulate flow, and control the direction of flow. Short of catastrophic failure, the common problems with valves are erosion and corrosion (as with pumps). Different types of valves have different locations that should be focused on for examination.

Gate valves use linear movement to start or stop flow. The upstream seat of the wedge (gate) and downstream body seat are the most likely places for erosion and wear. The sealing surfaces behind removable seat rings are good locations to expect erosion. The

FIGURE 3-26 Video crawler system. (Courtesy of Visual Inspection Technologies, with permission.)

walls of the packing and stuffing box is a probable location of corrosion. The stem should have a smooth surface to allow sealing contact with the stuffing material. Scores, scratches, pitting, and erosion could all allow leakage.

Globe valves regulate flow and employ rotational movement. The sealing surfaces of the disc and body of the globe valve should be checked for cracks, scratches, galling, pits, indentations, or erosion that may prevent disc-to-body sealing. Examination of these areas would require either upstream or downstream pipe openings and an articulating fiberscope or video camera head, assuming that the valve and its openings are large enough to accommodate access. The valve stem should have a smooth surface to allow sealing contact with the stuffing material. Checking the stuffing box wall would require stem and stuffing removal.

Swing check valves function to control the direction of flow, and as the name implies, the disc "swings." As fluid pushes against the disc, it opens to allow flow. When backpressure is present, the disc reverses direction and seals against the body seal, preventing reverse flow. The areas to be inspected are the seating surfaces of the disc and seat ring. Additionally, the swinging gate rotates on a hinge. If access is possible, the hinge pin should be checked for wear or distortion, as should the seating surfaces. Swing check valves frequently have access port or bonnet caps that provide access from the top of the valve.

An articulating fiberscope is the most appropriate instrument for the examination of valves. Limited success could be achieved with a rigid borescope. Marginal success may be achieved using a flashlight and mirror. Only very large swing check valves could accommodate a miniature camera in conjunction with a video monitor or LCD.

Ball and plug valves function to start or stop flow. They have rotational movement. Openings from the up or downstream side could allow access for a rigid or flexible borescope to check the ball and seating surfaces for physical damage, wear, or corrosion that may cause leakage. As with other valves with stems, the stuffing box and stem should be checked for scratches, pits, wear, or erosion that might cause leakage.

An articulating fiberscope would be an effective device for the examination of ball and plug valves. Limited examination may be achieved with the use of a rigid borescope. Some success may be achieved with the use of a flashlight and mirror.

Butterfly valves function to start and stop flow in addition to regulating flow. The movement of the valve is rotational. The primary area for signs of physical damage, wear, or corrosion is the horizontal plane at the centerline. It is here that the plastic seal may be subjected to high flow rates and the most wear and erosion. As with the other valves with stems, the stuffing box and stem should be checked for scratches, pits, wear, or erosion that might cause leakage. Limited view may be achieved with the use of a flashlight and a mirror. An articulating fiberscope would be an appropriate instrument to use. A limited examination may be achieved with the use of a rigid borescope.

Diaphragm valves function to control or regulate flow. Their movement is linear. The sealing surface is a flexible diaphragm with a wear-resistant coating or layer sealing against the body of the valve. Most importantly, the diaphragm must be checked for evidence of aging and cracking. An articulating fiberscope would be the most appropriate instrument. Limited success may be achieved with a rigid borescope. A flashlight and a mirror may also be useful.

Most of the discussion regarding piping systems to this point has been related to the pumps and valves in the piping system. This is where the energy is imparted to the component or changed in the liquid medium. It is in these areas where energy is transformed or transferred that damage to the internals of the component or part within the system usually occurs. Evidence of the failure to contain a liquid can be observed externally by signs of leakage. Piping systems are designed to contain specified pressures during normal operation. Testing these systems usually requires achieving certain pressures at certain temperatures and for specified amounts of time. The more critical the function of the system, the higher the safety factors that are used. When holding times and pressures have been determined, a pressure test can be conducted. First, the inspector needs to be familiar with the complete scope of the impending test. Pumps, pressure gauges, lighting, drawings, temperature gauges, and layout familiarity are necessary. A checklist for the inspector would comprise a list of possible locations where evidence of leakage may be found. Exposed surfaces, surrounding areas and components, floor conditions, conditions of equipment or components, and drainage systems should be part of the checklist. Evidence of leakage could range from drips of water to a spraying torrent. Access to "out of the way" piping systems is needed to prevent any zone of interest from being overlooked.

Remote Visual Techniques

Leakage. Remote visual examination with the use of binoculars and a spotlight may allow access to areas not easily observed by direct visual examination. Evidence of leakage may be noted by observing stains. These stains may be red or rust discoloration due to oxidation of ferrous materials, white crystals from liquid containing boric acid, yellow stains from sulfur-containing liquid, and green stains from copper-containing liquid. Gravity will take water to the lowest point on vertical piping. Horizontal runs of pipe could drip at any low point. If the piping is covered with insulation, time must be allowed to pass to let the leakage penetrate through the insulation. A drop of water is very uniform in size. Since it is the smallest unit of evidence of leakage, being able to see a drop of water at the examination location would be the indication upon which to qualify the remote (using binoculars) visual method. The drop of water should be visible at the maximum distance to be inspected with a given light source. Light intensity is most often measured with a calibrated light meter.

In summary, leakage examination should be systematic and scheduled. The normal system pressure should be achieved or exceeded for the examination. A minimum amount of time should pass to allow evidence of leakage to be observed. Leak testing provides a good technique to discover a mechanical problem and has the advantage of testing the entire system at one time.

Component Supports
Component supports have different configurations. Hangers, snubbers, restraints, and other component supports are all designed to support piping systems. Detailed knowledge of each component is required prior to performing a thorough examination. A brief description of each type of component support will be followed by some general inspection techniques, as well as a discussion of the conditions that should be observed in order to ensure proper functioning of the supports.

Hangers usually consist of a rod attached to a component by a welded "ear" or clamp. They usually are simple in design, with few parts and connections.

Spring can supports can be of the variable spring-loaded type or the constant-weight type. They usually carry the weight from above and support the pipe in tension.

Pipe supports typically support the pipe from below. They can be in contact with the pipe in the form of a saddle, clamp, or tee. As with overhead supports, they can be variable spring load or constant load. They are installed vertically, carry the weight from below, and support the pipe in compression.

Restraints typically are rigid and anchor the piping system to a building structure. The purpose of a restraint is to prevent movement in one axis. Multiple restraints can prevent movement in several directions. They are typically installed horizontally to vertical runs of pipe but may be oriented in any direction. They are usually attached to the pipe by means of clamps. The other end of the restraint is anchored to the concrete or welded to I-beams. They prevent severed pipes from whipping and rupturing other pipes.

Snubbers are vibration dampers that are installed on components to minimize the effects of dynamic loading. They are attached to the piping by clamps and to a building or structure by anchor bolts or welded connections. The snubber provides structural stability through mechanical or hydraulic action. The snubber allows normal movement during operations but restrains, restricts, or prevents movement during dynamic loading events.

Techniques. The examination of hangers, snubbers, restraints, and component supports generally is done by direct visual examination. On occasion, indirect or remote visual examination is required due to access limitations. In general, the inspector is looking for signs of mechanism failure or future failure. This could include inadequate construction practices, physical damage in service or since installation, signs of overload, signs of corrosion, or signs of fatigue.

Generally, hangers, snubbers, restraints, and component supports are checked to determine their general mechanical and structural conditions. The inspector checks for loose parts, debris or abnormal corrosion products, wear, erosion, corrosion, and the loss of integrity at bolted or welded connections.

Specific conditions that should be observed include:

Inadequate construction practices that may be evidenced by discontinuities in the base material may be in the form of roughly cut or deformed threads in rods, bolts, or nuts; bolts, nuts, and rods not tightened or torqued properly; wrong-sized parts; missing washers, cotter pins, or locking nuts; and rough, excessive, or inadequate welds.

Physical damage may be evidenced by bent, twisted, or deformed clamps, rods, or beams; warped, discolored, or burned parts (due to welding); missing nuts, pins, load scales, or identification numbers; cracked, sheared, or broken pins, rods, bolts, or welds, chipped concrete; and elongated bolt holes, eyes, swivels, or turnbuckles.

Overload may be evidenced by cracked welds, broken pins, sheared rods, or stripped threads; deformed mounting or saddle plates; cracked concrete at embedded bolts and plates; twisted or bent parts; and elongated bolt or pin holes.

Corrosion may be evidenced by light or heavy layers of rust on exposed surfaces; shallow or deep pits; build-up of corrosion products or loss of material at openings and crevices; loss of rivets on load scales; frozen rollers or bearings; loss of threads on bolts, rods, and nuts; and areas of eroded material.

Fatigue can be evidenced by cracked or broken parts. These could appear as thin lines or gaps running through or adjacent to welds, gaps or openings in threaded roots, and gaps or openings in bars, rods, or pipes adjacent to an encircling clamp.

Pressure Vessels

Vessels or tanks are designed to contain liquids or gases while maintaining their integrity at specified design pressures. Visual examinations of vessels are performed to determine the condition of the critical inner surfaces, such as the high-stress points at the junction of the vertical and horizontal surfaces, at nozzle penetrations, at construction weld joint intersections, and transitions between thin and thick sections. Less apparent is the importance of the intersection of air and product, where crevice corrosion (discussed earlier) could be found. Direct visual inspection may be performed where access is possible. The typical large vessel contains product. Internal equipment or components may block access, as may high-pressure gases or liquids, or hazardous materials that offer biological or radioactive dangers to humans. Hence, it is seldom possible to examine the surfaces of pressure vessels by direct visual means within 24 inches and at an angle greater than 30° from the plane of inspection. More commonly, remote visual is performed with binoculars, periscopes, closed circuit television, fiberscopes, or zoom cameras mounted on fixtures that aid access.

Nuclear reactor vessels are inspected in accordance with the Internal Vessel Visual Inspection (IVVI) program. Simply stated, the internal components of the reactor vessel are inventoried. The type of inspection required for each component and surface area is identified as to its importance and the type of visual examination to be performed. For the more important structures, components and stress points receive close inspection with sensitive calibration requirements. For example, component integrity is checked for fine cracks. A 0.001 inch diameter wire is used to prove adequate visual system resolution. For a less stringent examination, the observance of a 1/32 inch black line on the 18% gray card or a sheet of metal can be used as verification of resolution. An example of a gray card is illustrated in Figure 3-27.

The intent of this inspection is to look for general integrity of the part or component. Location of distortion, missing parts, debris, general wear, major cracking, and discoloration are the main objects of this examination.

Techniques. **Nuclear reactor vessels** are inspected using a vidicon CCTV camera that can tolerate high levels of radiation without burning out. The CCD camera has become more radiation-tolerant for use in high radiation environments. At this time, the vidicon camera is still attached to long cables that are dangled from a bridge into 40 feet of water. The water provides radiation protection for examination personnel. With an extra control line attached, the skilled operator can access very restricted areas. Like the strings used by a puppeteer, the extra line articulates the small-diameter camera into very tight spaces, allowing awkward but necessary angles of view to be achieved. Alternatively, zoom cameras are attached to long aluminum poles that are placed in central areas to provide a stable base from which the camera can zoom in and out with minimal movement and maximum flexibility.

Storage tanks for petrochemical products are frequently inspected to assess their con-

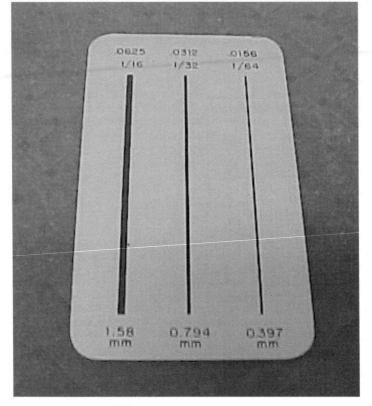

FIGURE 3-27 Example of gray card.

dition for continued service. Tanks that are several stories tall and hundreds of feet in diameter are inspected through access ports at the top. Smaller tanks on railroad cars and highway tractor-trailers are also inspected for the same reasons. In these cases, a camera on a pole may be attached to a gimble apparatus from the tank access port and manipulated to view as much area as possible. Alternatively, if the tank is free of product, the camera may be mounted on a tripod atop the tank and lowered into the tank. If access to the bottom of the tank is possible, it may be placed on the floor. From there it can be panned, tilted, and zoomed to achieve maximum coverage for inspecting the physical condition of the tank or vessel.

Another remote technique is to send in a crawler robot with the pan, tilt, and zoom camera attached. Tank or vessel configuration is very important for determining the specific technique most suitable for remote visual examination.

Heat exchangers, boilers, and steam generators provide more unique inspection challenges. The open space usual in a tank or empty vessel is now replaced with tubes, baffles, tube supports, deflector plates, and other structural supports. Lens optic borescopes, fiberoptic borescopes, and crawlers with pan, tilt, and zoom cameras may enable access to the heat exchanger internals. Boilers are often the victims of overheat failures. Short-term, severe overheating often leads to distortion and failures. Less extreme overheating leads to accelerated scaling or corrosion until failure results. Toward the end of their useful life, tubes may show signs of creep cracks by observable changes in dimensions. Distorted boiler tubes and enlarged diameters may be early indicators of immanent creep failure.

V. EVALUATION AND REPORTING OF TEST RESULTS

Evaluating and reporting of visual test results is dependent on the applicable codes, standards, specifications, contract requirements, and customer needs. Some general criteria apply to all examinations. The first requirement to be established is to determine the scope of the inspection. The second requirement is to decide "how" to inspect an object. The third requirement is to evaluate the results in accordance with a known standard. The fourth requirement is to accurately and legibly report the results.

Visual examination has long been an integral part of construction, fabrication, and manufacturing. As mentioned in Chapter 1, the *Book of Genesis* in the Bible mentions the first nondestructive test, and in particular, the first visual examination. This belief can be documented in the quote, ". . . and God *saw* that is was good." Apparently the acceptance criterion was in the eye of the beholder, in this case, the Creator. Industrial applications can be categorized into various groups, such as energy, petrochemical, aerospace, transportation, and construction.

The American Welding Society's *Structural Welding Code* for steel requires that welds meet certain workmanship standards. Welded structures such as buildings, tubular structures, and bridges must be welded in accordance with specific requirements regarding the material, welding processes, and joint designs. Once the application is established, welding parameters are specified in the workmanship requirements. The weld quality requirements are described in the workmanship section. The section on Inspection spells out the qualifications of Certified Weld Inspectors (CWIs). The tasks of the CWI include assuring that the materials, weld procedures, welder qualification and records of inspection all meet code requirements. The inspector keeps a record of qualifications of all welders, welding operators, and tackers; all procedure qualifications or other tests that are made; and other information that may be required. Generally, the CWI performs weld inspections in accordance with a procedure that contains the AWS Code requirements.

The American Society for Mechanical Engineers (ASME), Boiler and Pressure Vessel Code, Section V, Nondestructive Examination, Article 9, Visual Examination lists the following elements to be included in a visual examination procedure:

1. How visual examination is to be performed
2. Type of surface conditions available
3. Method or tools for surface preparation, if any
4. Whether direct or remote viewing is used
5. Special illumination, instruments, or equipment to be used, if any
6. Sequence of performing examination, as applicable
7. Data to be tabulated, if any
8. Report forms to be completed

The procedure should also contain or reference a report of the approach used to demonstrate that the examination procedure was adequate (procedure qualification).

In general, a fine line, 1/32 in. (0.8 mm) or less in width, or some other artificial flaw located on the surface to be examined or a similar surface is considered satisfactory evidence that the procedure is adequate. To prove the procedure, the line or artificial flaw should be located in the most unfavorable location on the area being examined. It is common to list acceptance criteria for various product forms in the procedure. The product forms should be specified in order to reduce the number of different visual procedures needed. In any case, substituting equipment with that of a different manufacturer or changing the details of the test parameters may require procedure requalification.

Reports

Depending on specific code requirements, a written report may be necessary. The report may contain the following for general applications:

1. Date of test
2. Examination procedure used
3. Illuminators and illumination requirements
4. Instruments, equipment, and tools
5. Test results
6. Inspector's signature and certification level

When specific product forms are expected to be examined repeatedly, it may be appropriate to design a report with attributes specific to the examination.

A report of the results of a weld examination may include the following items:

1. Surface finish
2. Discontinuities
3. Undercut
4. Overlap
5. Concavity
6. Convexity
7. Weld dimensions
8. Alignment
9. Evidence of mishandling
10. Arc strikes
11. Other appropriate observations

A report of the results of bolting examinations may include the following items:

1. Discontinuities
2. Necking down
3. Erosion
4. Crevice corrosion

A report of the results of valve examinations may include the following items:

1. Surface markings
2. Stem conditions
3. Seating surface cracks, scratches, galling, etc.
4. Guide surface clearances, wear, galling, corrosion, misalignment
5. Internal body surface erosion, wear, steam cutting
6. Stem-to-wedge connection for wear and corrosion
7. Diaphragms (if applicable) checked for aging, damage, cracking
8. Ball joints (if applicable) should be checked at bearing surfaces and position stops for wear
9. Hinge pins (if applicable) should be checked for wear, clearances, and misalignment
10. Stuffing boxes (if applicable) should be checked for steam cutting, corrosion, and mishandling.

A report of the results of pump exams may include the following items:

1. Erosion
2. Corrosion
3. Pitting
4. Leaching
5. Broken or lose parts
6. Debris
7. Shaft alignment

A report of the results of hanger, support, snubber, and restraint examinations may include the following items:

1. Galled or scraped areas of pipe surface (wear)
2. Loose nuts, pins, bolts, and locking pins
3. Oversized and undersized gaps and clearances
4. Distortion of the restraint
5. No freedom of movement or locked-up conditions
6. Extensive or deep corrosion
7. Cracked welds or broken parts
8. Stripped threads
9. Elongated bolt holes
10. Necked-down bolts or rods
11. Sheared rods
12. Debris
13. Loose parts
14. Erosion

   Note: Snubbers have special performance requirements unique for each application. These functional requirements exceed the normal visual examiner's scope of inspection.

A report of the results of vessel exams may include the following items:

1. Loose parts
2. Debris
3. Abnormal corrosion products
4. Wear
5. Erosion
6. Corrosion
7. Loss of integrity (broken or cracked)

Reports of visual examination results may be supplemented by special recording media or other permanent electronic record forms. Whenever questionable conditions or indications are encountered, some form of a permanent record should be considered. Standard or digital cameras with sufficient resolution to discern and record the attribute or condition of interest are excellent means of accomplishing this. Digital cameras usually include special computer programs that allow the image to be imported directly into the electronic report. Videotape recordings are useful for large areas that are being examined. This approach is especially helpful in providing evidence of coverage and the examination technique applied. Video camera recordings may also permit a picture frame to be "captured" and recorded directly on the electronic media report. Historically, sketches are simple and can be very helpful in recording examinations and providing location data regarding various discontinuities. Sketches have the disadvantage of incorporating the inspector's perception and subsequent misinformation that would not be in a photograph. Sketches made manually also may not be legible. Photographs and electronic images should include, where possible, a scale in order to permit the comparison of size and distance. Any disorientation of angle between the scale and the line of sight (sometimes referred to as "parallax"), could distort the true dimensions of an attribute, discontinuity, or condition. The use of a scale will usually provide a better perspective and result in better accuracy.

Acceptance Criteria

Acceptance criteria for welds will be as designated by the applicable code or specification and will usually include the following discontinuities:

1. Cracks
2. Incomplete penetration
3. Crater pits and cracks
4. Arc strikes
5. Undercut (dimension will be specified)
6. Surface porosity (usually defined by a maximum single size or some formula of aggregate amount in a total length of weld)
7. Slag (surface)
8. Spatter
9. Burn-through or melt-through
10. Overlap and rollover

11. Lack of fill
12. Excessive reinforcement

Specific acceptance criteria can be found in various codes, specifications, and standards. In many cases, they will be specified as part of the contract requirements.

Records of inspections or examinations should also include statements indicating part acceptance (if applicable), and include appropriate documentation to support findings. Inspection report forms should be complete, legible, concise, and signed and dated. If corrections are necessary, they should be made with a single line through the entry error, initialed, and dated. All reports, results, and/or drawings, sketches, photos, electronic files, digital video discs (DVD), or videotapes should be listed and retained in accordance with contractual requirements or codes and included with the data sheet.

VI. ADVANTAGES AND LIMITATIONS

Direct Visual Testing

As has been discussed, direct visual testing is a test in which the eye must be within 24 inches of the test surface and the angle of the line of site must be no less than 30° from the plane of the test surface. Direct visual examination has an advantage over other techniques in that it generally affords the clearest view of the inspection surface. The direct view will result in the processing of the full spectrum of light wavelengths that are available to the eye and the brain to form an image. All the color, shadows, textures, and visual attributes that are generally in existence are observed directly by the inspector. The limitation is that artificial enhancement of the image is not always possible and access restrictions may limit the view. Environmental extremes that are hazardous to inspection personnel may limit the viewing time or visualization of the image altogether.

Photographic Cameras
An advantage of photography is that it is an excellent record-keeping technique. It can be used very effectively in conjunction with written records or a voice recorder. The most commonly used devices may still be the 35 mm single lens reflex, the Polaroid™, and macro cameras. Digital cameras provide exceptional resolution and detail. The digital camera allows multiple images to be stored and downloaded to another camera, TV monitor, or computer storage media. Once filed, images can be retrieved and imported into a report. The limitation may be the quality of the image, i.e., the degree of resolution and clarity, depending upon the equipment used. There may also be limitations to the available storage space or memory. This may limit the quality or quantity of the stored images.

Film-Based Records
Resolution, contrast, and light sensitivity of film-based photographic records all depend on the grain size of the emulsion. In order to achieve greater detail, the use of film that has a finer grain size will be necessary. The decrease in grain size results in a film having a slower speed. This means that the film will require an increased exposure to light to record the image. Exposure is a function of light and time. A higher-speed film will have larger grains. This results in the reduction of the amount of light that is necessary to turn the film base into a latent image. This equates to a reduction in exposure. The latent image will appear as a negative image after it is chemically processed. The negative image is

transferred to light-sensitive paper as a positive image. The result is a positive paper-based photograph. The American and international standard for film speed is expressed in ASA numbers. The European or German standard is expressed in the DIN number. Color film comparisons are listed in the Table 3-2.

The three most common film speeds used for nonprofessional photographs are ASA 100, 200, and 400. The finest grain and slowest speed combination for snapshot photography with ASA 100 is considered as medium speed in the range of available film. Microphotography requires more light than is available from a consumer type flash attachment and a much finer-grain film than that which is used by the amateur photographer.

CCD Based Records
Section III discussed the concept of the charged coupled device (CCD). This is reviewed briefly in the following paragraph.

Photons (packets of energy that behave like particles), also known as light, are reflected from an object. They then pass through the lens of the camera and fall into the collection region of the picture element or pixel. Here electrons are freed from the semiconductor's crystal lattice. Thus, the more light that falls on the pixel, the more electrons that are liberated. Electrical fields trap and isolate the mobile electrons in one pixel from those of other pixels until the electrons can produce an indication of the amount of light that fell on the pixel. Electrons are passed from one CCD to another. This is analogous to a bucket brigade moving water. A tiny color filter over each pixel allows photons of only one color to pass through into the pixel's collection region. This pixel is assigned that color during the reconstruction of the image on the LCD monitor. The array of light-sensitive picture elements, or pixels, each measuring five to 25 microns across, make up the array matrix. The low-end digital camera for nonprofessional use typically has an array of 640 × 480 pixels. At the time of writing, arrays up to 1536 × 1024 are commercially available. Top of the line cameras are available for professional use. These comprise arrays that have millions of pixels.

Photographic Techniques
Whether the recording medium is emulsion-coated acetate-based film or CCD matrix arrays, the lens creates an image by focusing the light rays from an object on a plane behind the lens. An important concept to understand is depth of field. Depth of field can be defined as the overall range of focus apparent in a photograph. The principal plane of focus is the single plane through the subject that is actually in focus. The distance in front of and behind the principal plane of focus that is in focus is known as the depth of field. When working at higher magnifications, this effect becomes even more significant. For this application, the lens diaphragm is used to control the principal plane of focus or depth of field.

If the lens diaphragm opening is reduced, portions of the image that were blurred appear sharper because areas of the object both in front and behind of the principle plane become focused. If the opening is further reduced, this effect is greater. Increasing or reducing the diaphragm opening controls the depth of field. The offsetting factor when increasing the depth of field in this manner is that there is less light reaching the film. A

TABLE 3-2 Comparative Color Film Speeds

Slow	Medium	Fast
< 32 ASA (16 DIN)	64–125 ASA (19–22 DIN)	160 > (23 DIN)

balance between adequate light and optimum depth of field must be achieved. When using a typical 55 mm lens, the depth of field or area of focus extends farther behind the principal plane of focus than it does in front of it. Therefore, the best depth of field can be obtained by focusing at approximately one-third into the region or area of the object or area of interest. This way, the two-thirds region behind the optimum focus will also be in focus. As magnification is increased (by using 90 mm or 120 mm lenses), the reverse is true. This is why macrophotography, e.g., of a bee on a flower, may yield a sharply focused image of the flower or the bee while the background may appear out of focus.

Fiberoptic Cameras and Video Borescopes
Video Borescopes with charged coupled device (CCD) cameras are available with diameters as small as a quarter inch and pixel arrays with 410,000 pixels coupled with microlens technology. The monitor can provide 470 lines visible in the super-video home system (SVHS) format. The four-way articulating tip is available in up to 20 feet lengths. Without the articulating tip, lengths of 45 feet can be achieved. The camera control unit (CCU) can fit under the 7-inch diagonal color monitor and can control light intensity, gain, white balance, and shutter speed. The light source can be provided with 150 Watt halogen lamps or optionally with 300 Watt Xenon lamps.

This instrument is excellent for examining the inside surfaces of tubing found in heat exchangers, feedwater heaters, condensers, and steam generators. It would be especially useful for short distances in small-orifice access scenarios, such as with pumps, valves, engines, small-diameter pipes, and short-distance views of turbines. It would be a poor choice for large-space inspections, such as tanks, vessels, large-diameter pipes, or anywhere where detail at some distance was desired and an additional light supply could not be provided. The small diameter of the lens can yield near-infinite focus, but inadequate light would limit the view.

Miniature Cameras
Camera equipment specifically developed for boiler tube inspection application (See Figure 3-12) utilizes a half-inch color CCD with 420,000 pixels, a micro lens, a 460 horizontal video line (HVL) monitor for SVHS video, and a light source made from various light head options. These light attachments can produce from 9.6 Watts to 80 Watts of illumination. The elimination of the fiber bundle to transmit light to the site makes the electrical wire and frequency of the electric current the limiting factor for length rather than the limits of how long a fiber bundle can be manufactured. The half-inch diameter cable can be manufactured in 100 foot long lengths. Depending on light head diameters, tube diameters down to a half inch can be accessed. This type camera is best suited for boiler tubes but can do an adequate job on larger-diameter pipes or small vessels. The larger pipe diameters reduce the ability of the camera head to rest anywhere but on the bottom of the pipe, restricting the viewing options. Additionally, pushing a cable does not work well for any distance if the space is sufficient to allow snaking and curling of the cable. Therefore, centering devices and push rods are necessary to alleviate these problems. There remains the problem of articulation. The wide angle of view, as would be achieved with a "fisheye" lens, allows for a large downfield view and limited sidefield examination. This makes it suitable for many applications where examination for general conditions is the goal. Drain lines, oil lines, steam and process lines, headers, and vessels can all be internally inspected with the miniature camera with light head to lengths of 100 feet.

Miniature Camera with Pan, Tilt, and Zoom Lens
The pan, tilt, and zoom (PTZ) camera, as illustrated in Figure 3-13, is designed for multipurpose applications that require objects at major distances to be brought under closer ex-

amination. A one-third inch CCD chip is joined with a 12:1 or 24:1 lens train for zoom capabilities. The 12:1 lens can yield greater than 460 horizontal video lines of image. The 24:1 lens can yield more than 380 horizontal video lines of image. The 12:1 lens and the 24:1 lens both utilize variable lens diameters from 0.216 inches (5.4 mm) to 2.6 inches (65 mm).

A spot or flood lamp is needed for adequate light at the greater distances for which these cameras are employed. The standard cable length is 100 feet, but lengths up to 500 feet can be manufactured. A pan range of ± 175° is available. A tilt range of 253° is achievable. This type of camera and light arrangement can be mounted on a tripod, tractor crawler, push pole, or any other suitable platform to examine a wide variety of surface configurations.

This visual examination system is particularly useful in hazardous environments and confined spaces. Tank and vessel applications are probably the most common. Inspections of steam headers, sumps, manifolds, pipe system supports and large-diameter piping (including sewer and water utility lines) are also common, as are petrochemical, refinery, power generation, and engineering services industry applications. The transportation industry requires infrastructure inspection of structures such as bridges, shafts, and tunnels. The nuclear power and medical industries may need contaminated or hostile environment areas surveyed. Other industrial surveillance and law enforcement applications may include security and terrorist bomb investigations using remote means.

Other applications in the nuclear industry include the inspection of vast surface areas of reactor containment structures made of concrete or steel. A single location of a camera with a 24:1 zoom lens can inspect concrete walls and roofs for spalling, cracking, defoliation, and other concrete deterioration anomalies much more efficiently than placing an inspector hundreds of feet in the air and covering thousands of square feet of surface manually.

It is interesting to note the difference between 12:1 and 24:1 zoom capability. Moving the lens assembly optically performs the zoom effect up to 12:1 magnification. The zoom effect from 12:1 up to 24:1 is achieved by computer enhancement. Portions of the image are expanded and digitally magnified to create enlargements, providing greater detail for viewing and interpretation. When observing the zoom effect of this camera, the transition is smooth and seamless up to 12:1. Beyond 12:1 magnification, a slight "jiggle" in the image may be observed. From 12:1 to 24:1 the appearance of squares for each pixel are barely visible, but detectable. This is due to the digital magnification process.

Delivery Systems
Previous discussions of remote visual systems have made some reference to delivery systems. The following paragraphs list these by category. Some typical means of moving a camera from one point to another include fiberglass push rods, small-, medium-, and large-wheeled tractors, mobile platforms with tracks, and various crawlers with retrieval attachments. Applications can vary considerably with regard to the access size or opening, length of travel, environment, magnification, and lighting requirements. The basic elements of a remote visual delivery system are the camera, light, tractor, cable, mobility control unit, pan, tilt, and zoom controls, and monitor.

A small camera can be affixed to push rods. Limitations may be encountered when multiple changes in direction are required; i.e., the push rod soon binds up against multiple turns in the pipe, shaft, or opening.

A steerable-wheeled tractor can negotiate many turns with a camera attached for showing the way. The smaller the opening in which the tractor must operate, the greater the limitation to the payload that can be delivered. For example, a two-inch diameter pipe-crawling tractor may only weigh 3.5 pounds, thereby gaining access to a three-inch nominal pipe size. But it may only be able to drag along 100 meters of quarter-inch diameter cable. Multiple friction drag or contact points may reduce this to a much shorter distance. Additionally, inclined angles of the shaft, pipe, or structure would increase the load and decrease the access distance. Remote-controlled delivery systems may eliminate the need for cable but would require radio contact at all times.

A larger tractor that would fit into a 4 inch diameter pipe may be able to carry a 16 pound load. This may include 100 to 200 meters of cable. Larger tractors may carry a 24 pound load and 200 or more meters of cable. The advantage of larger tractors and accompanying larger wheels would be the ability to traverse adverse terrain and steeper inclines. The unmanned rovers that landed on Mars in the mid 1990s demonstrated that multijointed and complex designs can be designed to fit into small packages. They may appear frail but they were able to cross over rocks approaching a 2:1 ratio. Generally, the heavier the structure, the more friction that can be generated; thus, the heavier the load, the rougher the terrain that can be negotiated.

Any of these remote systems can incorporate a retrieval system. The "three pronged fork and tine" is one approach. Two pronged pincers resembling a crab claw is another. Each additional function requires more cables, lines, and articulating capabilities.

Monitors

Video monitors are the last link in the video system chain. They are the display devices. In simple terms, the monitor reverses the electronic coding of the video camera and returns the signals to visible displays. The standard monitor since the invention of TV has been the cathode ray tube (CRT). Solid-state and liquid crystal displays (LCD) have mostly replaced the CRT, allowing for thinner and smaller monitors. In particular, the miniature camera with lens and CCD designed for boiler tubes has been outfitted with LCDs. The detail and nature of the indications being sought in tubes, pipes, and small vessels do not necessarily require higher resolution. The LCD display is less expensive and more rugged in the field.

Conclusion

Generally, the main limitation to visual testing is access. The image of the object must be delivered to the eye. That image is always of the surface of an object. Visual testing is capable of examining the surface of an object unless the material is translucent. Remote visual testing advances are being driven today, as in recent years, by consumer demand and improvements in video technology. The challenge remains to understand fully "what" the inspector is examining and "how" the image is delivered to the eye. As designers make the image-gathering package smaller and smaller, the limitations of access will be further reduced. Applications in the field of medicine have been influencing the industrial field for years. Military applications including drones and robotic devices should continue to bring innovations to the technology of remote visual testing.

VII. GLOSSARY OF KEY TERMS

This section provides definitions of the terms associated with visual examination or testing and the nomenclature associated with visual equipment. Most terms relating to discontinuities are defined in Chapter 2 and within the text of this chapter.

Angle of Field—the greatest angle in between two rays coming from the object through the objective lens and into the optical system.

Angstrom—unit of length, equal to 0.1 nanometer.

Aspect Ratio—in television, the ratio of the frame width to the frame height.

Borescope—an industrial scope used to transmit images from inaccessible interiors for visual testing. Borescopes are so called because they were originally used in machined apertures and holes such as gun barrel bores. There are both flexible and rigid, fiberoptic, and geometric light (lens optic) borescopes.

Brightness—the attribute of visual perception in accordance with which an area appears to emit more or less light.

Burned-in Image—an image that persists in a fixed position in the output signal of a camera tube after the camera has been turned to a different scene.

Charged Coupled Device (CCD)—a solid state image sensor. CCDs are widely used inspection systems because of their accuracy, high speed scanning, and long service life. Semiconductors, when struck by photons, accumulate electrons that are passed on from one charged coupled device to another. Each pixel represents one color or another. Each pixel and its color is reconstructed on a monitor to form an image.

Code—a standard enacted or enforced as a law.

Color—sensation by which humans distinguish light of different intensities (brightness) and wavelengths (hue).

Cone—in biology, a retinal receptor that dominates the retinal response when the luminance level is high and provides the basis for the perception of color.

Contrast—the range of light and dark values in a picture or the ratio between the maximum and minimum brightness values.

Corrosion—loss or degradation of metal as a result of chemical reaction.

Crevice Corrosion—a kind of galvanic corrosion caused by differences in metal ion concentrations in neighboring portions of the corrodent.

Depth of Field—in photography, the range of distance over which an imaging system gives satisfactory definition when its lens is in the best focus for a specific distance. (Also see *Depth of Focus*.)

Depth of Focus—the region in front of and behind the focused distance within which objects still produce an image of acceptable sharpness or resolution when viewed by an observer having normal vision, or an observer whose vision is corrected to normal. In the case of a fixed-focus system, this parameter is often called the Depth of Field (DOF).

Direct Viewing—viewing of a test object in the viewer's immediate presence. The term is used in the fields of robotics and surveillance to distinguish conventional from remote viewing.

Distal—in a manipulative or interrogation system, of or pertaining to the end opposite from the eyepiece and farthest from the person using the system. Also objective or tip.

Dustproof—so constructed or protected that dust will not interfere with successful operation.

Dust-tight—so constructed that dust will not enter the enclosing case.

Endoscope—device for viewing the interior of objects. From the Greek words for inside view, the term endoscope is used mainly for medical instruments.

Erosion—loss of material or degradation of surface quality caused by friction or abrasion from moving fluids or particles.

Explosion-proof, Intrinsically Safe, Purged and Pressurized—constructed in a manner to prevent the surrounding atmosphere from being exploded by the operation of, or the results from, operating the item so classified.

Fiber Optics—an array of flexible glass or plastic fibers that has the capability of transmitting light (random array) or an image (coherent array) axially through the fiber bundles.

Flux (luminous)—the intensity of light per unit area of its source.

Foot-candle—a unit of illuminance when the foot is taken as the unit of length. It is the illuminance on a surface one square foot in area on which there is a uniformly distributed flux of one Lumen, or the illuminance at a surface, all points of which are at a distance of one foot from a uniform source of one Candle.

Fovea Centralis—a small depression near the center of the retina, constituting the area of most acute vision.

Frame (in Television)—the total area, occupied by the picture, that is scanned while the picture signal is not blanked.

Geometric Distortion—any aberration that causes a reproduced picture to be geometrically dissimilar to the perspective-plane projection of the original scene.

Horizontal (Hum) Bars—relatively broad horizontal bars, alternately black and white, that extend over the entire picture. They may be stationary or may move up or down. Sometimes referred to as a "Venetian-blind" effect, they are caused by the approximate 60-cycle interfering frequency or one of its harmonic frequencies.

Image—a reproduction of an object produced by light rays. An image-forming optical system gathers a beam of light diverging from an object point and transforms it into a beam that converges toward another point. If the beam converges to a point, a real image is produced, which can be projected upon a screen, a film plane, video camera tube, or semiconductor array.

Interlaced Scanning—a scanning process in which the distance from center to center of successively scanned lines is two or more times the nominal line width, and in which the adjacent lines belong to different fields.

Lumen—the unit of luminous flux. It is equal to the flux through a unit solid angle (steradian) from a uniform point source of one Candle, or to the flux on a unit surface, all points of which are at unit distance from a uniform point source of one Candle.

Object—the figure seen through or imaged by an optical system that may contain natural or artificial structures, or may be the real or virtual image of an object formed by another optical system.

Peripheral Vision—the seeing of objects displaced from the primary line of sight and outside the central visual field.

Pixel—a lighted point on the screen of a digital image. The image from a conventional computer monitor is an array of over 256,000 pixels, each of which has a numerical value. The higher the number for a pixel, the brighter it is. Formerly called picture element.

Raster—a predetermined pattern of scanning lines that provides substantially uniform coverage of an area.

Reference Standard—work piece or energy source prepared according to precise instructions by an approved agency for tests and calibrations requiring precise and consistent measurements. (In the case of photometry, the standard is a light source.)

Reflection—a general term for the process by which the incident flux leaves a surface or medium from the incident side, without change in frequency. Reflection is usually a combination of regular and diffuse reflection.

Remote Viewing—viewing of a test object not in the viewer's immediate presence. The word remote previously implied either closed-circuit television or fiberoptic systems remote enough so that, for example, the eyepiece and the objective lens could be in different rooms. High-resolution video and digital signals can now be transmitted around the world with little loss of image quality. Compare with *Direct Viewing*.

Resolution—an aspect of image quality pertaining to a system's ability to reproduce objects, often measured by resolving a pair of adjacent objects or parallel lines. See also *Resolving Power*.

Resolution (Horizontal)—the amount of resolvable detail in the horizontal direction in an image. It is usually expressed as the number of distinct vertical lines, alternately black and white, that can be seen in a distance equal to image height. This information usually is derived by observation of the vertical wedge of a test pattern. An image that is sharp and clear and shows small details has good or high, resolution. If the picture is soft and blurred and small details are indistinct, it has poor or low resolution. Horizontal resolution depends upon the high-frequency amplitude and phase response of the system, the transmission medium, and the image monitor, as well as the size of the scanning spots.

Resolution (Vertical)—the amount of resolvable detail in the vertical direction in a picture. It is usually expressed as the number of distinct horizontal lines, alternately black and white, that can be seen in a test pattern. Vertical resolution is fundamentally limited by the number of horizontal scanning lines per frame. Beyond this, vertical resolution depends on the size and shape of the scanning spots of the pickup equipment and picture monitor and does not depend upon the high-frequency response or bandwidth of the transmission medium or picture monitor.

Resolution Threshold—minimum distance between a pair of points or parallel lines when they can be distinguished as two, rather than one; expressed in minutes of arc.

Retina—the tissue at the rear of the eye that senses light.

Resolving Power—the ability of vision or other detection system to separate two points. Resolving power depends on the angle of vision and the distance of the sensor from the test surface. Resolving power is often measured using parallel lines. Compare with *Resolution*.

Retained Image (Image Burn)—a change produced in or on the target that remains for a large number of frames after the removal of a previously stationary light image and yields a spurious electrical signal corresponding to that light image.

Rod—retinal receptor that responds at low levels of luminance, even down below the threshold for cones. At these levels, there is no basis for perceiving differences in hue and saturation. No rods are found in the fovea centralis (see definition).

Simple Magnifier—a device having a single converging lens.

Specification—a set of instructions or standards invoked by any organization to govern the results or performance of a specific set of tasks or products.

Standard—document to control and govern practices in an industry or application, applied on a national or international basis and usually produced by consensus. See also *Reference Standard*.

Tip—in casual usage, the distal or objective end of a borescope.

Trace—line formed by electron beams scanning from left to right on a video screen to generate a picture.

Video—pertaining to the transmission and display of images in an electronic format that can be displayed on a cathode ray screen.

Videoscope—jargon for video borescope. See *Borescope, Video*.

Visibility—the quality or state of being perceivable by the eye. In many outdoor applications, visibility is defined in terms of the distance at which an object can be just perceived by the eye. In indoor applications, it usually is defined in terms of the contrast or size of a standard test object, observed under standardized viewing conditions.

Vision—perception by eyesight.

Vision Acuity—the ability to distinguish fine details visually. Quantitatively, it is the reciprocal of the minimum angular separation in minutes of two lines of width subtending (extending under or opposite to) one minute of arc when the lines are just resolvable as separate.

Visual Angle—the angle subtended by an object or detail at the point of observation. It usually is measured in minutes of arc.

Visual Field—the locus of objects or points in space that can be perceived when the head and eyes are kept fixed. The field may be monocular or binocular.

Visual Perception—the interpretation of impressions transmitted from the retina to the brain in terms of information about the physical world displayed before the eye. Visual perception involves any one or more of the following: recognition of the presence of something (object, aperture, or medium); identifying it; locating it in space; noting its relation to other things; identifying its movement, color, brightness, or form.

Visual Testing—method of nondestructive testing using electromagnetic radiation at visible frequencies.

Waterproof—so constructed or protected that water will not interfere with successful operation. This does not imply submersion to any depth or pressure.

Watertight—provided with an enclosing case that will exclude water applied in the form of a hose stream for a specified time as stated in the following note.

Note: a common form of specification for watertight is: "So constructed that there shall be no leakage of water into the enclosure when subjected to a stream from a hose with a one-inch nozzle and delivering at least 65 gallons per minute, with the water directed at the enclosure from a distance of not less than 10 feet for a period of 5 minutes, during which period the water may be directed in one or more directions as desired."

BIBLIOGRAPHY

IES Lighting Handbook: Reference Volume. New York: The Illuminating Engineering Society of North America (1984).

1999 Annual Book of ASTM Standards, Section 3, *Metals Test Methods and Analytical Procedures,* Volume 03.03, *Nondestructive Testing.* Philadelphia: The American Society for Testing and Materials (1999).

Anderson, Robert Clark. *Inspection of Metals,* Vol. 1 *Visual Examination.* American Society for Metals (1983).

Inoue, Shinya. *Video Microscopy.* New York: Plenium Press (1989).

Lorenz, Peter G., *The Science of Remote Visual Inspection, Technologies, Applications, Equipment.* New York: RVI Olympus Corporation.

Lorenz, Peter G. Expanding Technology Adds Value to Visual Testing and Remote Visual Inspection Procedures, *Materials Evaluation,* September (1997).

Nondestructive Testing Handbook, 2nd edition: Vol. 8, *Visual and Optical Testing.* American Society for Nondestructive Testing (1993).

Remote Visual Testing Manual. Visual Inspection Technologies, Inc. (1998).

Visual Examination Technology Containment Inspection, Level II IWEMIWL, 2nd edition. American Society for Metals (1997).

Welding Inspection Handbook, 3rd edition. American Welding Society, Florida (2000).

CHAPTER 4
PENETRANT TESTING

I. INTRODUCTION

Penetrant testing (PT) is one of the most widely used nondestructive testing methods for the detection of surface discontinuities in nonporous solid materials. It is almost certainly the most commonly used surface NDT method today because it can be applied to virtually any magnetic or nonmagnetic material. PT provides industry with a wide range of sensitivities and techniques that make it especially adaptable to a broad range of sizes and shapes. It is extremely useful for examinations that are conducted in remote field locations, since it is extremely portable. The method is also very appropriate in a production-type environment where many smaller parts can be processed in a relatively short period of time. This method has numerous advantages and limitations that can be found in Section X of this chapter.

II. HISTORY AND DEVELOPMENT

Although the exact date of the first "penetrant" test is unknown, it is generally believed that the earliest tests were performed in the late 19th century and were primarily limited to the examination of various railroad parts, such as axles and shafts. Even though it was a very rudimentary method, it was capable of revealing fairly large cracks in metallic parts by using what is referred to as the "oil and whiting" method. It is interesting to note that the oil and whiting method employed the same processing steps that are in use today with current penetrant testing procedures. In this early penetrant method, the part to be examined was cleaned and then submerged in dirty engine oil. The oil that was used in those days came from large locomotive engines and was very heavy. It was generally diluted with kerosene or alcohol so that it would be thin enough to penetrate surface discontinuities. It seemed that the dirty oil worked the best in the presence of a discontinuity, since it provided a dark, oily stain on the test surface. After saturation with oil, the part was allowed to drain. During the draining time, known today as the "dwell" time, the thinned oil would penetrate surface cracks. After the excess oil was removed from the surface with a solvent, the part was coated with "whiting," which consisted of a chalk-like powder suspended in alcohol. The oil that was entrapped in the void would then "bleed" out into the whiting and a dark, oily stain indicated the presence and location of a discontinuity.

Sometimes the whiting used was a whitewash material similar to that used to wash and paint fences, hence the expression "oil and whiting" method. It is also believed that talcum powder was used in those early days as a developer to help bring the oil back out to the surface so it could be observed.

As can be expected, there were many problems with this early technique. There was a general lack of consistency, since there were no established procedures or standards and the dwell and development times were pretty much left up to the judgment of the user. The materials varied in content and use and only the grossest type of discontinuities could be detected.

The oil and whiting method had a place in the early examinations of railroad parts, but its use began to diminish with the introduction of magnetic particle testing (MT) in the 1930s. Its decline was most notable in the years from 1925 into the mid-1930s. At this time, many parts that were considered to be critical were made of ferromagnetic materials and the advent of MT techniques provided a much more reliable and repeatable method for the detection of surface discontinuities. By the mid-1930s, the use of aluminum and other materials that could not be magnetized was increasing and it was quite apparent that there had to be another nondestructive test method for detecting discontinuities in these nonferromagnetic materials. Certainly, at that time, internal discontinuities could be detected using x-ray techniques, but many of the x-ray techniques in use were not capable of revealing smaller, tight discontinuities at the surface. It follows that there was an obvious need for a method that would be sensitive enough to detect these small surface discontinuities.

The early pioneers Carl Betz, F. B. Doane, and Taber deForest, who worked for the Magnaflux Corporation at the time, were experimenting with many different types of liquids and solvents that might fulfill this need for surface discontinuity detection. These early techniques that were tried used brittle lacquer, electrolysis, anodizing, etching, and various color-contrast penetrants. Some of these early approaches were discarded, with the exception of the anodizing process. During this period, the anodizing process was used for detecting cracks in critical aluminum parts, generally associated with aircraft. This resulted in the publication of a military specification, MIL-I-8474, on September 18th, 1946. The title of this specification was "Anodizing Process for Inspection of Aluminum Alloys and Parts."

Years before the publication of this military standard, Robert Switzer had been working with fluorescent materials, primarily for producing fluorescent advertisements for movie theaters. He and his younger brother, Joseph, were pioneers in the early use and development of various fluorescent materials. As early as 1937, Robert Switzer became aware of a problem that a local casting company was having with parts for the Ford Motor Company. There was a large batch of aluminum castings that were found to contain a number of discontinuities that were not observed until after the surfaces of the castings had been machined. Switzer recalled how the different fluorescent materials he was familiar with were capable of clinging to surfaces and fluorescing when observed using ultraviolet (black) light. He thought that this material would be appropriate for detecting the surface cracks that had been uncovered by machining. He was able to obtain a number of samples and began experimenting with the various fluorescent pigments that he was developing. He found that although these pigments were unique in clinging to a person's hands and other porous-type materials, they were not very effective for the detection of very fine surface cracks. Switzer continued to try different combinations and, eventually introduced them into various liquids that would be used to carry the fluorescent pigment into the discontinuities. His work was ultimately successful and he applied for a patent in August of 1938. It is interesting to note that he asked his brother, Joseph, to share the patent-filing fee, which amounted to $15.00, and Joe declined. Robert then proceded to file the patent on his own and, ultimately, found information about the Magnaflux Corporation, which at that time was still developing and pioneering the magnetic particle inspection method. He decided to investigate the company and contacted one of their sales representatives in New York City to discuss the possibility of expanding their surface discontinuity efforts to include a fluorescent penetrant.

Some of the early attempts to demonstrate the fluorescent penetrant method to Carl Betz and others from the Magnaflux Corporation were not successful. Lack of success in these early demonstrations was probably due to the fact that the castings being examined were supposed to have discontinuities but those discontinuities probably did not exist. One of the observers, A. V. deForest, who was a pioneer in MT, happened to have a specimen containing known discontinuities that he had used to demonstrate MT, and a fluorescent penetrant test was performed on this specimen. The indications were quite apparent. Naturally, all of the observers were very impressed that the known discontinuities appeared, In addition, other discontinuities that A. V. deForest was not aware of were also observed.

Subsequent to this demonstration, it was found that the castings that were first examined had been peened, thus closing the discontinuities that were at the surface. This demonstration proved unique. The year 1941 became memorable, not only for the patent that was awarded to Switzer that summer, but because it also marked the beginning of the Second World War. This new test would be widely used in supporting the different products that would be used by the military.

Early Penetrant Techniques

Some of the early penetrant processes were quite similar to those in use today. The part would be cleaned with a strong solvent and then, after drying, it was immersed in the penetrant for about 10 minutes. After this penetration, or dwell time, the penetrant was removed, usually with a strong solvent, and the part was wiped until dry and clean. The removal step was usually performed under a black light. According to early accounts, the parts would then be struck with a hammer, which would cause the entrapped penetrant materials to "bleed out" to the surface, at which time the part would be examined under a black light. These early techniques were still quite archaic and the test results were not very consistent. During the following years, in order to achieve the level of consistency that is essential for quality assurance, many different types of materials were tried and a variety of techniques were attempted. These led to a most unique development: the water-washable or water-removable (perhaps a more appropriate term) penetrant.

Water-rinsable penetrant materials and related equipment were first offered around June of 1942. There was much interest in them and a number of companies started to use the water-rinsable technique. As a matter of record, the first purchaser of the water-removable penetrant equipment was Aluminum Industries of Cincinnati, Ohio, which would use the equipment for the examination of aluminum castings. Many other applications and uses followed, including the testing of propeller blades, pistons, valves, and other critical aircraft parts. It should be noted that a major step forward in the application of PT occurred when it was included in the maintenance and overhaul programs for aircraft engines. As a result of this application, the PT method of nondestructive testing was on its way to gaining wide acceptance. A patent on the "water-washable" technique was applied for in June 1942 and was issued in July 1945. The early developer compounds were yellowish materials consisting mainly of talc. The wet developer technique began to be used in late 1944 and early 1945.

One of the problems associated with the water-removable technique was the potential for removal of entrapped penetrant from discontinuities as a result of a vigorous water rinse step; this led to concerns about "overwashing." This overwashing was the result of an emulsifier that had been mixed in with the penetrant. In fact, some early penetrants were marketed as promoting "super washability," which resulted in extremely clean surfaces but also produced a greater danger of the penetrant being removed from surface discontinuity openings. The solution to this overwashing concern was to remove the emulsifier from the penetrant and to apply it later. This finally occurred in 1952 when this process was referred to as the postemulsification, or PE, technique.

The first postemulsification penetrants were introduced in 1953. With this technique, the emulsifier was applied after the penetrant dwell time and carefully controlled, so that the penetrant in the discontinuities was not emulsified. It would remain in the discontinuity even after the emulsified surface penetrant was removed with a water rinse. Although this added an additional step and an additional liquid to the process, it did provide a higher level of control over the detection of small, shallow discontinuities that may not have detected with the use of a typical water-removable technique containing emulsifiers.

Visible Penetrants

The penetrant techniques described to this point were of the fluorescent type. The fluorescent penetrant technique required tanks, a water supply, electricity for the black lights, and a darkened area for the evaluation of indications. In order to permit penetrant tests to be performed in the field and to provide portability, a simpler, visible dye had to be developed. In the 1940s, a Northrop metallurgist named Rebecca Smith developed a visible dye penetrant approach. Rebecca Smith, who would be later known as Becky Starling, collaborated with Northrop chemists Lloyd Stockman and Elliot Brady, who also assisted in the development of a visible dye penetrant. This was considered necessary for examining critical jet engine parts outdoors, where creating a darkened area necessary for the use of fluorescent penetrants was inconvenient. The development of the visible dye penetrant technique would take several years; Stockman applied for a patent in March 1949. By this time, there were several choices of penetrants: fluorescent water removable, fluorescent emulsifiable, and a visible dye.

Other Developments

In the late 1950s and early 1960s, much work was done to quantify and analyze the various penetrants that were available. By the early 1960s, a variety of techniques were being used with a range of sensitivities that would satisfy the demanding requirements of many industries. Many of the variables associated with the penetrant tests were evaluated and the entire process was improved, so that consistency and sensitivity would be an intrinsic part of the process. One of the most widely used military standards, MIL-I-6866, was issued in 1950 and is still in use in a number of industries today, with very few changes from the original document. Currently, there has been a shift away from military standards and, eventually, all military standards will be replaced with standards developed by the American Society for Testing and Materials (ASTM). The remainder of this chapter will focus on the materials and techniques that are in use at the present time and will describe the many applications for which this unique NDT method can be effectively applied.

III. THEORY AND PRINCIPLES

The basic principle upon which penetrant testing is based is that of capillary "attraction" or "action." Capillary action is a surface tension phenomenon that permits liquids to be drawn into tight openings as a result of the energies that are present at the surfaces of the openings. In most high school physics classes, the principle of capillary action is demon strated by placing a glass straw into a beaker filled with colored water. The surface tension associated with the opening of the glass straw, or capillary, causes the liquid level to move to a higher level inside that capillary than the level of the liquid in the beaker. A simple demonstration of capillary action using two glass panels clamped together is illustrated in Figure 4-1a and b.

One can consider that discontinuities open to the surface behave in much the same fashion as shown by the glass panels in Figure 4-1b. The liquid used in this example is a

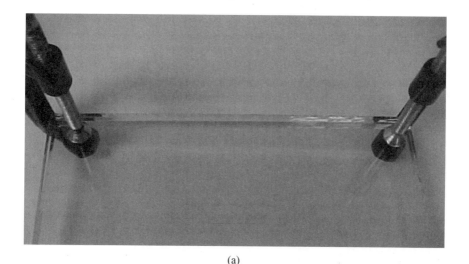

(a)

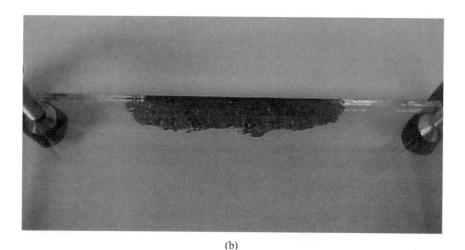

(b)

FIGURE 4-1 Demonstration of capillary action. (All illustrations in this chapter courtesy C. J. Hellier.) (a) Glass panels clamped together. (b) Visible color contrast penetrant applied to edge of panels.

typical visible contrast penetrant. The capillary action forces are very strong and, in fact, if a penetrant test were being performed on a specimen in an overhead position, the penetrant would be drawn into the opening, against the force of gravity. The capillary force is much stronger than gravity and the discontinuities will be detected even though they may be in an overhead specimen.

IV. PENETRANT EQUIPMENT AND MATERIALS

Penetrant Equipment

Penetrant systems range from simple portable kits to large, complex in-line test systems. The kits contain pressurized cans of the penetrant, cleaner/remover, solvent, and developer and, in some cases, brushes, swabs, and cloths. A larger fluorescent penetrant kit will include a black light. These kits are used when examinations are to be conducted in remote areas, in the field, or for a small area of a test surface. In contrast to these portable penetrant kits, there are a number of diverse stationary-type systems. These range from a manually operated penetrant line with a number of tanks, to very expensive automated lines, in which most steps in the process are performed automatically. The penetrant lines can be very simple, as illustrated in Figure 4-2.

In this particular system, there is a tank for the penetrant, a tank for the water rinse, a drying oven, and a developer station. The final station is the examination area, which includes a black light. This manually operated system is a typical small water-removable penetrant line. The steps in the testing process would be: cleaning of the parts, application of the penetrant, removal of the penetrant with a water spray, drying, application of the developer, and finally, inspection. This entire process is covered in much greater detail in Section V, Techniques.

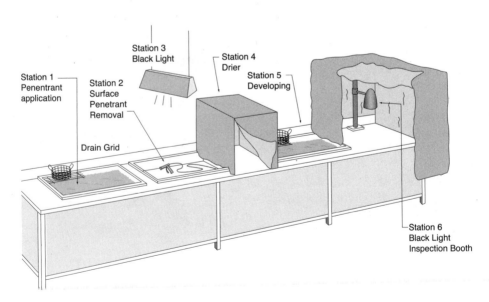

FIGURE 4-2 Typical fluorescent penetrant line arrangement.

If a postemulsifiable penetrant is to be used, the manually operated line will require an additional tank. This tank will contain an emulsifier that will render the surface penetrant removable after a specified emulsification time. Again, this technique will be covered in much greater detail later in this chapter. The automatic penetrant lines in use today vary from small, rather simple systems to very large complex lines that are computer controlled. Figure 4-3 illustrates a large automatic penetrant line.

Although the steps in an automated penetrant system have been somewhat mechanized, it is interesting to note that the examinations still must be conducted by inspectors who have been trained and are qualified in the process. The arrangement of these large automated penetrant lines vary with different layouts to permit the most flexibility from the standpoint of processing the parts. Normally, the systems will be arranged in a straight line; however a U shape or other configuration may be used to provide more effective use of floor space.

Other Equipment

The *black light* (see Figure 4-4) is an essential accessory for fluorescent penetrant inspection. Black lights used in penetrant testing typically produce wavelengths in the range of

FIGURE 4-3   Automated fluorescent penetrant line.

FIGURE 4-4 Black light.

315 to 400 nm (3150–4000 Angstrom units) and utilize mercury vapor bulbs of the sealed-reflector type. These lights are provided with a "Woods" filter, which eliminates the undesirable longer wavelengths. Black light intensity requirements will range from 800–1500 microwatts per square centimeter ($\mu W/cm^2$) at the test surface. Specific requirements will vary, depending upon the code or specification(s) being used. Recent developments in black light technology provide lights that can produce intensities up to 4800 $\mu W/cm^2$ at 15" (38.1 cm).

Light intensity meters are used to measure both white light intensities when visible PT is used and black light intensities for fluorescent penetrant techniques. This measurement is necessary to verify code compliance and to assure that there is no serious degradation of the lights. Some meters are designed to measure both white light and black light intensities.

Test panels, including comparator blocks, controlled cracked panels, tapered plated panels, and others for specific industries such as the TAM panel (typically used in aerospace), are employed to control the various attributes of the PT system. They also provide a means for monitoring the materials and the process. This is discussed in Section X, Quality Control Considerations.

In summary, the equipment used will be greatly influenced by the size, shape, and quantity of products that are to be examined. If there are large quantities involved on a continuing basis, the use of an automated system may be appropriate, whereas with small quantities of parts, the use of penetrant kits may be more suitable. The size and configuration of the part will also influence the type of penetrants that will be most appropriate.

Penetrant Materials

The various materials that will be used in the different penetrant processes must exhibit certain characteristics. Above all, these materials must be compatible with each other and collectively provide the highest sensitivity for the application. The term "penetrant family" is sometimes used to indicate a group of materials all from the same manufacturer. The intent is to provide a degree of assurance that the different materials will be compatible with each other. There are usually some provisions for using materials outside the "family" if the combination of the different materials can be proven compatible through qualification tests.

The materials used in the penetrant process are classified into four groups. The characteristics for each will be presented in detail. The first group of materials that are essential for a penetrant test are precleaners. The second group of materials, which has the greatest influence on sensitivity, are penetrants. The third group comprises the emulsifiers and solvent removers, and the fourth group the developers.

Precleaners

Precleaning is an essential first step in the penetrant process. The surface must be thoroughly cleaned to assure that all contaminants and other materials that may prohibit or restrict the entry of the penetrant into surface openings are removed. Thorough cleaning is essential if the examination results are to be reliable. Not only does the surface have to be thoroughly cleaned, but openings must be free from contaminants such as oil and water, oxides of any kind, paint or other foreign material which can greatly reduce the penetrant sensitivity.

Typical cleaners include the following.

Solvents are probably the most widely used liquids for precleaning parts in penetrant testing. There are a variety of solvents that can be effective in dissolving oil, films, grease, and other contaminants. These solvents should be free of any residues that would remain on the surface. Solvents cannot be used for the removal of spatter, rust, or similar materials on the surface. These must be removed by some type of a mechanical cleaning process (see Section V, Prerequisites).

Ultrasonic Cleaning. Of all the precleaner materials and processes, ultrasonic cleaning is probably the most effective. Not only will the contaminants be removed from the surface, but also if there are entrapped contaminants in discontinuities and other surface openings, the power that is generated in the ultrasonic cleaning process will usually be effective in breaking up and removing them. A typical small ultrasonic cleaner is illustrated in Figure 4-5.

Alkaline cleaning. Alkaline cleaners used for precleaning are nonflammable water solutions that, typically, contain specially selected detergents that are capable of removing various types of contamination.

Steam Cleaning. In some rare instances, steam may be used to remove contaminants from the surface. Although very effective in removing oil-based contaminants, this is not a widely used technique.

Water and detergent cleaning. There are various devices that utilize hot water and detergents to clean part surfaces. This technique depends largely upon the type of contamination that is present on the test surfaces. Usually, if parts are covered with oil or grease, the contaminants will not be satisfactorily removed from the surface with this cleaning technique.

Chemical cleaning. Chemical cleaning techniques usually involve etchants, acids, or alkaline baths. This precleaning approach is primarily confined to softer materials, such as aluminum and titanium, where prior mechanical surface treatments, such as machining

FIGURE 4-5 Ultrasonic cleaner.

or grinding, could possibly have smeared metal over discontinuity openings. Both acid or alkaline liquids are usually effective in the removal of rust and surface scale; however, a slight amount of the surface material is also removed, so this process must be very carefully controlled. Steps must be taken to assure the complete removal of these liquids from all surface openings.

Penetrants
The most important characteristic that affects the ability of a penetrant to penetrate an opening is that of "wetability." Wetability is a characteristic of a liquid and its response to a surface. If a drop of water is placed on a very smooth, flat surface, a droplet with a very pronounced contour will result, as shown in Figure 4-6(c). Although water is a liquid and is "wet," its wetting characteristics are not good enough to make it an effective penetrant.

The "contact angle," θ, is measured from a line drawn in an arc from its junction point with the surface to the opposite surface (see Figure 4-6). If that same droplet of liquid is emulsified, such as would be the case with the addition of a small amount of liquid soap, the droplet will tend to flatten out and the contact angle will be somewhat decreased (see Figure 4-6(b). In the case of a liquid penetrant, its wetting properties are so great that it will, essentially, lie almost flat on a smooth surface and the contact angle will be very low, as in Figure 4-6(a). Therefore, the penetrants with the lowest contact angles will have the best wetability and provide good penetrability for a given material. Two other important characteristics of the penetrant are the dye concentrate and the viscosity. Dye concentrate has a major and direct influence on the "seeability" or sensitivity of the pene-

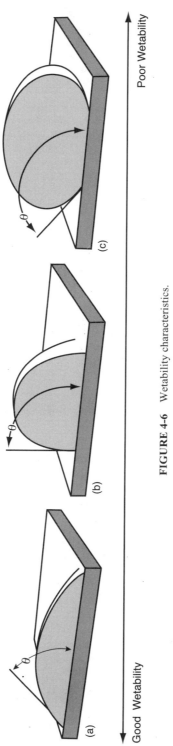

FIGURE 4-6 Wetability characteristics.

4.11

trant material, assuming that the wetting properties and the penetrability are of a very high level. The dye concentrate makes the penetrant more visable to the human eye. In some early penetrants, different colored dyes were tried, including blue, yellow, and green, but it seemed that the red dye resulted in the best response to visible observation. In fact, the term "contrast ratio" is generally used to express the seeability of a penetrant. If a white penetrant were used on a white developed background, the contrast ratio would be one-to-one. In other words, there would be no contrast between the penetrant and the background. In the case of a red penetrant, the contrast ratio is said to be six-to-one, making that contrast very noticeable on a white background surface. The dye concentrate is also important in a fluorescent penetrant. The contrast ratio of a fluorescent penetrant is said by many references to be forty-to-one, compared to the six-to-one of a visible red dye. In reality, the contrast ratio of the fluorescent penetrant being viewed under a black light in a virtually dark room would essentially be the same as would exist if there were a single candle in a perfectly pitch-black room. The contrast ratio in this case would be exceptional and, for this reason, the fluorescent penetrant produces a much higher degree of seeability or sensitivity as compared to the visible dye penetrants.

Viscosity is defined as the state or quality of being viscous. Liquids with higher viscosity values are thicker than those with lower ones. Although viscosity is an important characteristic of a penetrant, it is not as influential as the wetting characteristics and dye concentrate. If the penetrant has good wetting characteristics and exceptional dye concentration, it will still provide a meaningful examination, even if the viscosity is high. The difference in viscosity will influence the actual dwell time, or the amount of time that it will take for the liquid to effectively penetrate a given surface opening.

There are other characteristics that an effective penetrant should possess. In addition to being able to penetrate small surface openings, they must also:

- Be relatively easy to remove from the surface during the removal step
- Be able to remain in the discontinuities until they are withdrawn during the development step
- Be able to bleed from the discontinuities when the developer interacts with it and have the ability to spread out in the developer layer
- Have excellent color and the ability to be displayed as a contrasting indication in order to provide the sensitivity that is necessary
- Exhibit no chemical reaction between the penetrant materials and the test specimen
- Not evaporate or dry rapidly

In addition, they should be nonflammable, odorless, and nontoxic; possess stability under conditions of storage; and be cohesive, adhesive, and relatively low in cost.

In summary, the most important characteristics of a penetrant when performing a penetrant test are (1) capillary action, (2) wetting characteristics, (3) dye concentrate, and (4) viscosity.

Emulsifiers/Removers
The purpose of the emulsifiers used in penetrant testing is to emulsify or break down the excess surface penetrant material. In order for these emulsifiers to be effective, they should also possess certain characteristics, including:

- The reaction of the emulsifier with any entrapped penetrant in a discontinuity should be minimal in order to assure that maximum sensitivity is achieved.
- The emulsifier must be compatible with the penetrant.

- The emulsifier must readily mix with and emulsify this excess surface penetrant.
- The emulsifier mixed with the surface penetrant should be readily removable from the surface with a water spray.

Solvent Removers
Solvent removers are used with the solvent removable technique and must be capable of effectively removing the excess surface penetrant. There are a number of commercially available solvents that make excellent removers. These solvents should readily mix with the penetrant residues and be capable of removing the final remnants from the surface. They should also evaporate quickly and not leave any residue themselves. It is essential that the removers not be applied directly to the surface, since they are also good penetrants. Spraying or flushing the part with the solvent during the removal step is prohibited by many specifications. Even so, there are still users who insist on performing this unacceptable practice in order to "thoroughly" remove the surface penetrant. When using the visible color contrast penetrants, a slight trace of pink on the cloth or paper towel will indicate that the removal is adequate. For fluorescent penetrants, slight traces of the fluorescent penetrant as observed under the black light will also indicate the proper level of removal.

Developers
There are four basic types of developers:

1. Dry developer
2. Solvent-based developers, also referred to as "spirit" or nonaqueous
3. Wet developers suspended in water
4. Wet developers that are soluble in water

Developers have been described as "chalk" particles, primarily because of their white, chalk-like appearance.

In order for the developers to be effective in pulling or extracting the penetrant from entrapped discontinuities, thus presenting the penetrant bleed-out as an indication that can be evaluated, they should possess certain key characteristics. They should:

- Be able to uniformly cover the surface with a thin, smooth coating
- Have good absorption characteristics to promote the maximum blotting of the penetrant that is entrapped in discontinuities
- Be nonfluorescent if used with fluorescent penetrants
- Provide a good contrast background that will result in an acceptable contrast ratio
- Be easily applied to the test specimen
- Be inert with respect to the test materials
- Be nontoxic and compatible with the penetrant materials
- Be easy to remove from the test specimen after the examination is complete

There are other types of developers that are used on rare occasions. These are referred to as strippable, plastic film, or lacquer developers. They are typically nonaqueous suspensions containing a resin dissolved in the solvent carrier. This developer sets up after application and is then stripped off the surface, with the indications in place. It can then be stored and maintained as part of the inspection report.

V. PENETRANT PROCEDURES

Selecting the correct technique for penetrant testing is very important. Prior to performing the examination, a procedure should be developed and qualified. When preparing the procedure, the following should be considered:

- The requirements of the code, specification, or contract
- The type and size of discontinuity that is anticipated
- The surface condition of the test specimen
- The configuration of the part
- The quantity of parts to be examined
- Systems and equipment that are available

Prerequisites

Prior to any penetrant test, there are certain prerequisites that have to be addressed.

Temperature
Penetrant materials are influenced by temperature variations. Most codes and specifications require that the test part and the penetrant materials be within a specified temperature range, typically between 40 °F (4.4 °C) up to as high as 125 °F (51.6 °C). The part and the penetrant materials must fall within the specified temperature range. If the test part or the penetrant is extremely cold, the penetrant becomes very thick and viscous, which will affect the time it will take to penetrate the discontinuities. If the test surface or penetrants are high in temperature, some of the more volatile constituents may evaporate from the penetrant, leaving a thick residue that will not effectively penetrate the discontinuities.

Environmental Considerations
Since some of the solvent cleaners and removers used with penetrant testing can be somewhat flammable, it is essential that the penetrant test be performed in an area where there are no open flames or sparks that may tend to cause the penetrant materials to ignite. Typically, penetrant materials have relatively high flash points, but some of the cleaner/remover solvents could ignite when exposed to sparks or open flames. Also, some of the solvents may give off fumes. Therefore, penetrant testing should be performed in an area where there is adequate ventilation.

Lighting
There must be adequate lighting in the examination area, especially during the time when the evaluation is performed.

Surface Condition Considerations
Surfaces to be examined having coatings such as paint or plating, or extremely rough conditions, must be addressed. If the surface contains scale and rust, some type of strong mechanical cleaning process is required. Many codes and specifications do not permit the use of some mechanical cleaning techniques, such as shot-blasting, shot-peening, or sandblasting, since these processes tend to peen over the test surface, potentially closing a crack or other surface discontinuity. If wire brushing is used to remove scale or rust, it should be

done with extreme care for the same reason. If extreme pressure is applied to a grinding wheel or power wire brush, it is possible to cause a smearing of the metal on the surface.

The Penetrant Procedure

Precleaning
After addressing the prerequisites, it is necessary to remove all contaminants from the surface and, after the surface has been cleaned, all evidence of any residues that may remain. (See Figure 4-7) After precleaning, it is essential that the precleaners evaporate and that the test surface be totally dry prior to application of the penetrant. This will prevent contamination or dilution of the penetrant in the event that it interacts and becomes mixed with the precleaner.

Penetrant Application
The penetrant can be applied to the surface of the test part in virtually any effective manner, including brushing (Figure 4-8), dipping the part into the penetrant, immersion, spraying, or just pouring it on the surface. Figure 4-9 shows a water-removable fluorescent penetrant being applied with an electrostatic sprayer. The key is to assure that the area of interest is effectively wetted and that the penetrant liquid does not dry during the penetration or *dwell time,* which is the period of time from when the penetrant is applied to the surface until it is removed. The codes and specifications give detailed dwell times that must be followed. It is quite common to have a dwell time of 10 to 15 minutes for many applications.

FIGURE 4-7 Precleaning of a weld surface.

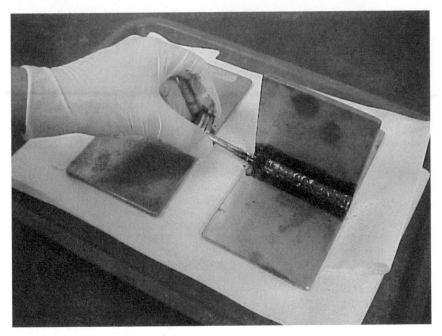

FIGURE 4-8 Application of a visible penetrant.

FIGURE 4-9 Water-removable fluorescent penetrant being applied with electrostatic spray.

Penetrant Removal

In this step the excess surface penetrant is removed from the test specimen surface; the method of removal depends on the type of penetrant that is being used. There are three techniques for excess surface penetrant removal: water, emulsifiers, and solvents. Figure 4-10 illustrates excess surface visible contrast penetrant being removed with a solvent-dampened cloth. Removal of fluorescent penetrants is usually accomplished under a black light. This provides a means of assuring complete removal of the excess surface penetrant while minimizing the possibility of overremoval.

Application of Developer

The type of developer to be used will be specified in the penetrant procedure. As mentioned above, the four types of developers are dry, nonaqueous, aqueous suspendable, and aqueous soluble. The entire test surface or area of interest must be properly developed, although there are rare applications where developers are not used. A nonaqueous developer is applied by spraying (See Figure 4-11). It must be applied in a thin, uniform coating. Thick layers of developer, whether nonaqueous, dry, or aqueous, can tend to mask a discontinuity bleed-out, especially if that discontinuity is small and tight.

Development Time

The developer must be given ample time to draw the entrapped penetrant from the discontinuity out to the test surface. Many codes and specifications will require a development time from 7 to 30 minutes and, in some cases, as long as 60 minutes. Development is de-

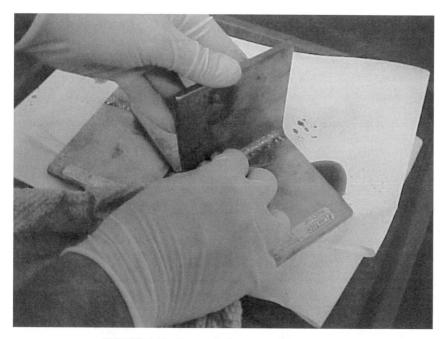

FIGURE 4-10 Removal of excess surface penetrant.

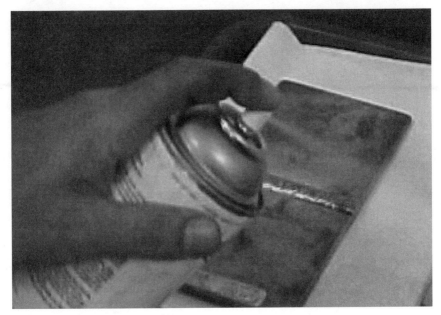

FIGURE 4-11 Application of a nonaqueous developer.

fined as the time it takes from the application of the developer until the actual evaluation commences. It is recommended that the surface be observed immediately after the application of the developer to assist in the characterizing and to determine the extent of the indication(s).

Interpretation
Upon completion of the development time, the indications from discontinuities or other sources that have formed must be interpreted. A visible contrast penetrant bleedout is illustrated in Figure 4-12 and fluorescent penetrant indications are shown in Figure 4-13. Bleedouts are interpreted based primarily on their size, shape, and intensity (see Section VII).

Postcleaning
After the part has been evaluated and the report completed, all traces of any remaining penetrant and developer must be thoroughly removed from the test surface prior to it being placed into service or returned for further processing.

VI. TECHNIQUES AND VARIABLES

It should be apparent by now that there are a number of PT techniques that can be used with the different materials described. A summary of these techniques is listed in Table 4-1. A detailed description of each technique follows. (*Note:* The technique and process designations in this section are for simplification and do not directly relate to code or specification classifications.)

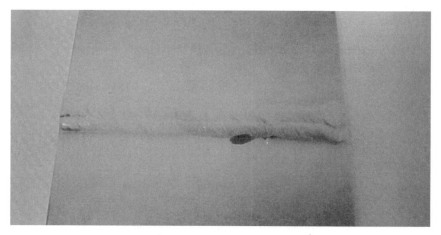

FIGURE 4-12 Indication of a bleedout.

Technique I, Process A (I-A)

Technique I Process A uses a fluorescent water-removable penetrant that can be used with either dry, aqueous, or nonaqueous developers (see Figure 4-14). This technique is generally used for the following applications:

1. When a large number of parts or large surface areas are to be examined
2. When discontinuities that are not broad or shallow are anticipated
3. When parts to be examined have complex configurations such as threads, keyways, or other geometric variation

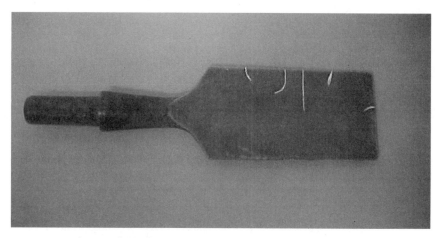

FIGURE 4-13 Fluorescent penetrant bleedouts.

TABLE 4-1 Penetrant Technique Classification Summary

Technique	Process	Materials
I (Fluorescent)	A (Figure 4-14)	Water-removable penetrant; dry, aqueous, or nonaqueous developer
	B (Figure 4-15)	Postemulsifiable penetrant; lipophilic emulsifier; dry, aqueous, or nonaqueous developer
	C (Figure 4-16)	Solvent-removable penetrant; solvent cleaner/remover; dry or nonaqueous developer
	D (Figure 4-17)	Same as I B except the emulsifier is hydrophilic
II (Visible, color contrast)	A (Figure 4-14)	Water-removable penetrant; aqueous or nonaqueous developer
	B (Figure 4-15)	Postemulsifiable penetrant, emulsifier, and aqueous or nonaqueous developer
	C (Figure 4-16)	Solvent-removable penetrant; solvent cleaner/remover, aqueous or nonaqueous developer

4. When the parts to be examined have surfaces that are rough, such as with sand castings or as-welded conditions

Advantages:

1. Higher sensitivity
2. Excess penetrant is easily removed with a coarse spray
3. Easily adaptable for large surfaces and large quantities of small parts
4. The cost is relatively low

Limitations:

1. A darkened area is required for evaluation
2. Under- or overremoval of penetrant material is possible
3. Water contamination can degrade the effectiveness of the penetrant
4. Not effective for broad or shallow discontinuities
5. Dryers are required (usually) when using developers
6. This technique is usually not portable

Technique I Process B (Lipophilic) and Process D (Hydrophilic)

Technique I, Processes B and D use a fluorescent postemulsifiable penetrant, a lipophilic (L) or hydrophilic (H) emulsifier, and dry, aqueous, or nonaqueous developers. (Figures 4-15 and 4-17 illustrate overviews of Processes B and D, respectively.) The materials used are very similar to those described for Technique I Process A, except that these penetrants are not water-removable without emulsification. A lipophilic or hydrophilic emul-

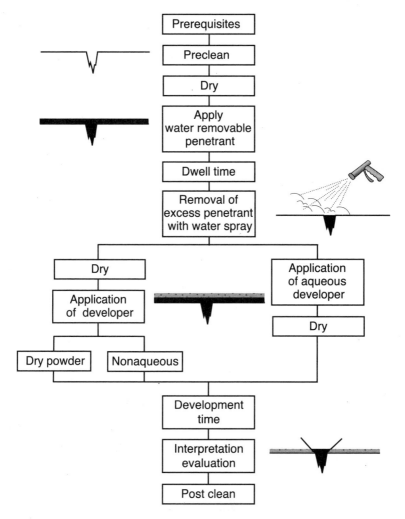

FIGURE 4-14 Water-removable technique (Process I-A or II-A).

sifier must be used after the dwell time has expired. This technique is generally used in the following situations:

1. When a large quantity of parts must be examined
2. When discontinuities that are broad and shallow are anticipated
3. For the detection of stress cracks or intergranular corrosion
4. For the detection of small discontinuities such as grinding cracks
5. Applications requiring higher-sensitivity techniques

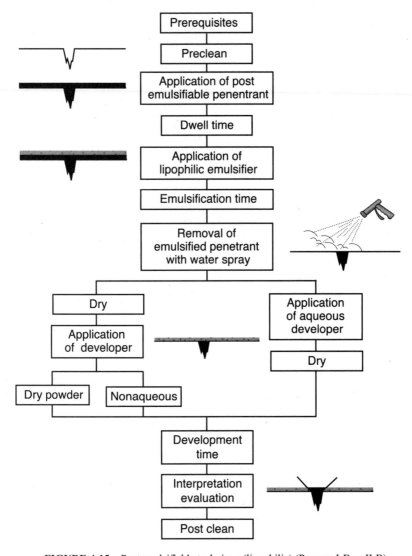

FIGURE 4-15 Postemulsifiable technique (lipophilic) (Process I-B or II-B).

Advantages:

1. High sensitivity for the detection of smaller discontinuities
2. For broad or shallow discontinuities (when they are expected)
3. Adaptable for high-quantity testing
4. Not easily affected by acids
5. Less susceptible to overremoval than Technique I-A

PENETRANT TESTING

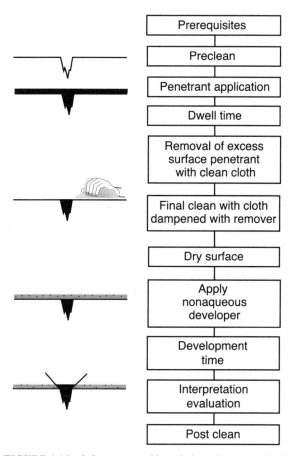

FIGURE 4-16 Solvent-removable technique (Process I-C or II-C).

Limitations:

1. This technique has an additional step, which requires an emulsifier. Therefore, more time and material is necessary.
2. It is not as effective for parts with complex shapes (e.g., threads) or rough surfaces, as is Technique I-A.
3. The emulsification time must be closely controlled.
4. As with Technique I-A, it requires drying prior to the application of dry or nonaqueous developers.
5. It is usually not portable.

Technique I, Process C (I-C)

Technique I Process C uses a fluorescent penetrant, which is solvent-removable, a solvent cleaner/remover, and a nonaqueous developer. The excess surface penetrant is first re-

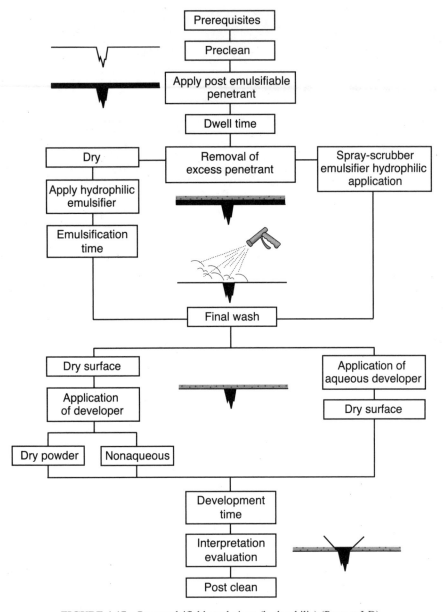

FIGURE 4-17 Postemulsifiable technique (hydrophilic) (Process I-D).

moved with a dry cloth, followed by cleaning with a cloth dampened with a solvent remover. (Figure 4-16 illustrates an overview of Technique I-C.) This process is generally used when removal with water is not desirable due to part size, weight, surface condition, water availability, or when a heat source is not readily available for drying.

Advantages:

1. Can be used for spot examinations on large parts
2. Effective when water removal is not feasible

Limitations:

1. The use of solvent for removal limits this technique to smaller areas
2. A black light and darkened area are required
3. The sensitivity can be reduced if excessive remover is applied
4. A "background" may occur with this technique, which could affect the contrast ratio, especially with rougher surfaces

Technique II Process A (II-A)

Technique II Process A uses a visible color-contrast, water-removable penetrant and an aqueous or nonaqueous developer. Dry developer is not usually used with Technique II penetrants. Some specifications, in fact, do not permit the use of dry developers with Technique II penetrants. Figure 4-14 illustrates Technique II-A.

The penetrant contains an emulsifier, making it water-removable. This technique is generally used for the following applications:

1. Examinations of a large quantity of parts or large surface areas
2. For discontinuities that are generally tight
3. For the examination of parts with threads, keyways, and other complex geometries
4. For parts with generally rough surfaces

Advantages:

1. No blacklight or darkened area is required for evaluation.
2. It is relatively quick and inexpensive.
3. The excess penetrant is easily removed with a coarse water spray.
4. It is effective for the examinations of a large quantity of parts.
5. It can be used for rough surfaces, keyways, threads, and other complex geometries.

Limitations:

1. Its sensitivity is inferior to Technique I-A.
2. Penetrant can be overremoved.
3. Water contamination can degrade the effectiveness of the penetrant.
4. It is not usually effective for the detection of broad or shallow discontinuities.

Technique II Process B (II-B)

Technique II Process B uses a visible color-contrast, postemulsifiable penetrant, an emulsifier, and an aqueous or nonaqueous developer. The materials used (except for the penetrant) are very similar to those described for Technique I Process B (as illustrated in Figure 4-15). An emulsifier (usually lipophilic) is applied to the surface penetrant after the dwell time to make it water-removable. Technique II Process B is generally used for the following applications:

1. When a large quantity of parts must be examined
2. Whenever lower sensitivity than that achieved with Technique I is acceptable
3. When broad and shallow discontinuities are anticipated

Advantages:

1. No black light or darkened area for evaluation is required.
2. Broad or shallow discontinuities may be detected.
3. Useful when there are large quantities of parts to be examined.
4. This technique is not as susceptible to overremoval, as are the Process A penetrants.

Limitations:

1. The additional step of an emulsifier requires more time and additional material.
2. It is not as effective for parts with a complex geometry (e.g., threads), as is Process A.
3. The emulsification time is very critical and must be closely controlled.
4. Drying is required if nonaqueous developers are used.

Technique II, Process C (II-C)

Technique II Process C uses a visible, color-contrast, solvent-removable penetrant, a solvent cleaner/remover, and an aqueous or nonaqueous developer. Figure 4-16 illustrates this technique.

The excess penetrant is not water-removable and must be removed with a solvent remover. This technique is widely used for field applications when water removal is not feasible, or when examinations are to be conducted in a remote location.

Advantages:

1. This technique is very portable and can be used virtually anywhere.
2. It can be used when water removal is not possible.
3. Black lights or darkened evaluation areas are not required. Evaluation is done in visible light.
4. It is very adaptable for a wide range of applications.

Limitations:

1. The use of solvent to remove excess surface penetrant limits the examinations to smaller areas and parts without a complex geometry.
2. Sensitivity is reduced when an excessive amount of remover is used during the removal step.
3. Excess penetrant removal is difficult on rough surfaces, such as sand casting and as-welded surfaces, and usually results in a "background."

4. This technique has a lower level of sensitivity compared to Technique I penetrants.
5. It is more "operator-dependent" due to the variables involved in the removal step.

Summary

Of all the techniques described in Section VI, the most widely used fluorescent penetrant technique is I-A (water-removable). Technique II-C is the most widely used visible color-contrast penetrant.

VII. EVALUATION AND DISPOSITION

After the penetrant process has been completed and the indications are observed and recorded, the final step will be to establish whether or not these conditions are acceptable or rejectable. The size of the indication can usually be related to the amount of penetrant entrapped in the discontinuity. The larger the discontinuity volume, the greater the amount of penetrant that will be entrapped and, therefore, the larger the bleed-out after development. The shape of the indication is important because it relates to the type or nature of the discontinuity; e.g., a crack or lack of fusion will show up as a linear bleed-out rather than a rounded one. A linear indication, by most codes and specifications, is defined as a bleed-out whose length is three times or greater than its width. The intensity of the bleed-out gives some evidence as to how tight the discontinuity is. A broad shallow-type discontinuity will tend to be diluted, to an extent, by the remover liquid, and not be as brilliant as a bleed-out from a very tight discontinuity. It is essential that corrective action be taken to remove or repair the discontinuity if it is deemed to be rejectable. In most cases, a crack or other serious discontinuity will be cause for the rejection or scrapping of the part. Repairs to discontinuities will often be accomplished by grinding. A recommended technique to assure complete removal of an indication after grinding is to merely reapply the developer. This usually verifies whether the discontinuity has been removed, since the bleedout will reappear if it has not (see Figure 4-18).

This process should be repeated, i.e., grinding, reapplication of developer, and then grinding again until no further bleedout occurs. At that point, the area that has been ground out must be reexamined, following the penetrant procedure from the beginning, to assure that in fact, the discontinuity has been totally removed. It is further recommended that the grinding be performed in the same direction as the longest dimension of the discontinuity. This is to minimize the possibility of smearing the material over the discontinuity. After the repair is completed, the repaired surface must be reexamined. During the evaluation process, it is necessary that a suitable light source be used. For visible penetrants, a light source of 100 foot-candles is common, although codes and specifications may require different light intensities. A black light is used for evaluation of fluorescent penetrants, as seen Figure 4-19. An intensity of between 800 to 1200 μW per square centimeter is typically required.

Penetrant indications must be recorded. For recording purposes, a number of satisfactory techniques can be used, including photographs, hand sketches, and the transparent tape lift-off technique. Photographic techniques employed for recording visible penetrant indications are quite standard. When photographing indications under black light conditions, specialized exposures and filters may be necessary when using photographic film. Digital cameras are usually quite exceptional for recording both visible and fluorescent

FIGURE 4-18 Initial grindout of an indication.

penetrant indications. In addition, the digital image can be observed immediately after exposure. Hand drawings, when used with test reports, should be prepared with as much accuracy and detail as possible. The transparent tape lift-off technique is usually very effective for visible dye penetrants, since the tape can be placed directly on the developed indication, lifted off, then placed on the test report. This will provide a means of displaying the actual bleed-out indication with its true size and intensity. This technique is not

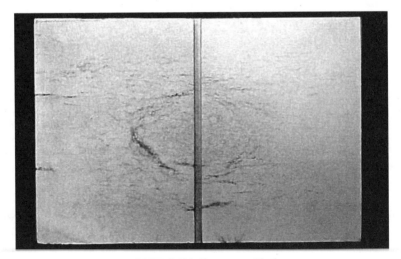

FIGURE 4-19 Comparator block.

usually effective with fluorescent penetrants, unless the test report is to be evaluated under a black light.

VIII. PENETRANT TESTING APPLICATIONS

Penetrant testing is extremely versatile and has many applications. It is used in virtually every major industry and for a wide variety of product forms. Industries that widely use penetrant testing techniques include:

- Power generation, both fossil- and nuclear-fueled
- Petrochemical
- Marine/Shipbuilding
- Metalworking, including foundries and forging shops
- Aerospace
- Virtually all of the various welding processes and metals-joining industries

Another unique application of penetrant testing is for the detection of through-wall leaks. With this application, penetrant is applied to one surface, for example, of a tank, and developer applied to the opposite surface. If there is an unimpeded through leak, the penetrant will follow that path and be exposed on the opposite developed surface as an indication. Some leak tests, such as those for complex components and piping systems that must be evaluated to determine the source and location of a leak path, are conducted using fluorescent tracers.

IX. QUALITY CONTROL CONSIDERATIONS

In order to control the many variables associated with penetrant testing, there are a number of quality control (QC) checks that should be made periodically. The applicable requirements will specify which ones must be made and how often. There are three major areas that include the various QC issues, and once these are identified as essential or required, they should be added to the procedure. The three major categories are:

1. Material checks
   - New
   - In-use
2. System checks
3. Equipment checks

Material Checks—New

All incoming or new penetrant materials should be verified to assure compliance with specifications. Some codes and specifications contain unique requirements regarding the presence of contaminants in the PT materials such as chlorine, sulfur, and halogens. Some codes and standards include analytical tests to determine the amount of these contaminants but most of the penetrant manufacturers will provide material certifications that specify the amount in that particular batch of materials. Each can or container of the penetrant will have the batch number clearly marked to make it easy to correlate with the certifications. These certifications should be maintained in the QA files for future reference and for ready access in the event of an audit.

Material Checks—In-Use

It is good practice, and in some cases a requirement, that penetrant materials be checked periodically to assure that they have not degraded and are still performing satisfactorily. These checks may include but are not limited to the following:

1. Contamination of the penetrant
2. Water content in the penetrant (for water-removable penetrants)
3. Water content in lipophilic emulsifier
4. The condition of the dry developer (fluorescent penetrant carry-over, etc.)
5. Wet developer contamination
6. Aqueous developer concentration (using a hydrometer)
7. Hydrophilic emulsifier concentration

The results of these checks should be entered in a logbook to permit periodic reviews of the condition and the trends of the materials as they are being used.

System checks

Probably the most effective overall checks of the performance of penetrant systems involve the periodic use of panels or samples containing known discontinuities (see Figure 4-20). As mentioned earlier, there are a number of cracked panels available for this purpose, including plated, single thickness; plated, tapered to provide for a range of crack depths; and industry-specific panels. Prior to use, these panels should be thoroughly cleaned, preferably in an ultrasonic cleaner to ensure that there are no residual penetrants remaining that could affect their response.

There are other systems checks that may be waived if the system performance check with the panels is satisfactory. They include tests for:

1. Penetrant brightness
2. Penetrant removability
3. Penetrant sensitivity
4. Emulsifier removability

Equipment Checks

There are routine checks that should be made on the equipment that is used in the PT process. They include, but are not limited to, the following:

1. Black lights—for intensity, filter condition, bulb, and reflector.
2. Light meters, black and white—for functionality and calibration status. Some codes require that these meters be periodically calibrated; this is usually done by the meter manufacturer.

FIGURE 4-20   Cracked plated panels.

TABLE 4-2   Recommended Tests and Frequency of Performance

Test	Frequency
System performance	Daily
Penetrant contamination	Daily
Developer contamination (aqueous: soluble and suspension)	Daily
Developer condition (dry)	Daily
Water spray pressure	Each Shift
Water spray temperature	Each Shift
Black light intensity	Daily
Black light reflectors, filters, and bulbs	Daily
Inspection area cleanliness	Daily
Emulsifier concentration (hydrophilic)	Weekly
Penetrant sensitivity	Weekly
Fluorescent brightness	Quarterly
Penetrant removability	Monthly
Emulsifier removability	Monthly
Emulsifier water content (lipophilic)	Monthly
Drying oven temperature	Quarterly
Light meter calibration	Semiannually

3. The temperature of the drying oven—to assure that it is within the specified range.
4. The water used for removal—temperature and pressure.
5. Pressure gauges, when compressed air is used for application of the penetrant, emulsifiers, and developers.

X. ADVANTAGES AND LIMITATIONS

The major advantages of penetrant testing include:

    Portability
    Cost (inexpensive)
    Sensitivity
    Versatile—virtually any solid nonporous material can be inspected
    Effective for production inspection
    Nondestructive

The limitations include:

    Only discontinuities open to the surface of the test specimen can be detected
    There are many processing variables that must be controlled
    Temperature variation effects
    Surface condition and configuration
    Surface preparation is necessary
    The process is usually messy

XI. GLOSSARY OF PENETRANT TESTING TERMS

Adhesion—The tendency of the penetrant to adhere to the surface
Angstrom unit (Å)—A unit of length that may be used to express the wavelength of electromagnetic radiation (i.e., light). One angstrom unit is equal to 0.1 nanometers (1 nm = 10^{-9}m).
Background—The surface of the test part against which the indication is viewed. It may be the natural surface of the test part or the developer coating on the surface.
Black light—Electromagnetic radiation in the near-ultraviolet range of wavelength (315–400 nm, 3150–4000 Å).
Black light filter—A filter that transmits near-ultraviolet radiation while absorbing other wavelengths.
Bleedout—The surfacing of penetrant entrapped in discontinuities to form indications.
Blotting—Developer soaking up penetrant from a discontinuity to accelerate bleedout.
Carrier—A liquid, either aqueous or nonaqueous, in which penetrant examination materials are dissolved or suspended.
Clean—Free of contaminants.
Cohesion—The intermolecular action by which the elements of a penetrant are held together.

Contaminant—Any foreign substance present on the test surface or in the examination materials that adversely affects the performance of penetrant materials.

Contrast—The difference in visibility (brightness or coloration) between an indication and the background.

Contrast ratio—The ratio of the amount of light reflected or emitted between a penetrant and a background (usually the test surface). The contrast ratio for visible color contrast penetrants is 6:1; it is 40:1 for fluorescent penetrants.

Developer—A material that is applied to the test surface to accelerate bleedout and to enhance the contrast of indications.

Developer, aqueous—A suspension of developer particles in water.

Developer, dry powder—A fine, free-flowing powder used as supplied.

Developer, liquid film—A suspension of developer particles in a vehicle that leaves a resin/polymer film on the test surface after drying.

Developer, nonaqueous—Developer particles suspended in a nonaqueous vehicle prior to application.

Developer, soluble—A developer completely soluble in its carrier (not a suspension of powder in a liquid) that dries to an absorptive coating.

Development time—The elapsed time between the application of the developer and the examination of the part.

Drain time—That portion of the dwell time during which the excess penetrant or emulsifier drains from the part.

Drying oven—An oven used for increasing the evaporation rate of rinse water or an aqueous developer vehicle from test parts.

Drying time—The time required for a cleaned, rinsed, or wet developed part to dry.

Dwell time—The total time that the penetrant is in contact with the test surface, including the time required for application and the drain time.

Emulsification time—The time that an emulsifier is permitted to remain on the part to combine with the surface penetrant prior to removal.

Emulsifier—A liquid that interacts with an oily substance to make it water-removable.

Emulsifier, hydrophilic—A water-based liquid used in penetrant examination that interacts with the penetrant oil, rendering it water-removable.

Emulsifier, lipophilic—An oil-based liquid used in penetrant testing that interacts with the penetrant oil, rendering it water-removable.

Etching—The removal of surface material by chemical or electrochemical methods.

Evaluation—A review to determine whether indications meet specified acceptance criteria.

False indication—A response not attributed to a discontinuity or test object condition (usually caused by faulty or improper NDT processing).

Family—A complete series of penetrant materials required for the performance of a penetrant examination; usually from the same manufacturer.

Foot-candle (fc)—The illumination on a surface 1 ft^2 in area, on which is uniformly distributed a flux of 1 lm (lumen). It equals 10.8 lm/m^2.

Immersion rinse—A means of removing surface penetrant, in which the test part is immersed in a tank of either water or remover.

Nonrelevant indication—An indication caused by a condition related to the test object shape or design that entraps penetrant but is not rejectable.

Overemulsification—Excessive emulsification time that may result in the removal of penetrant from discontinuities.

Overremoval—Too long and/or too vigorous use of the remover liquid that may result in the removal of penetrant from discontinuities.

Penetrant—A solution or suspension of dye (visible or fluorescent) that is used for the detection and evaluation of surface-breaking discontinuities.

Penetrant comparator block—An intentionally flawed specimen having separate but adjacent areas for the application of different penetrant materials so that a direct comparison of their relative effectiveness can be obtained.

Penetrant, postemulsifiable—A penetrant that requires the application of a separate emulsifier to render the excess surface penetrant water-washable.

Penetrant, solvent removable—A penetrant so formulated that most of the excess surface penetrant can be removed by wiping, with the remaining surface penetrant traces removable by further wiping with a cloth or similar material lightly moistened with a solvent remover.

Penetrant, visible—A penetrant that is characterized by an intense color, usually red.

Penetrant, water-removable—A penetrant with a built-in emulsifier.

Postemulsification—The application of a separate emulsifier after the penetrant dwell time.

Postcleaning—The removal of residual penetrant test materials from the test part after the penetrant test has been completed.

Precleaning—The removal of surface contaminants from the test part so that they will not interfere with the examination process.

Solvent remover—A volatile liquid used to remove excess penetrant from the surface being examined.

TAM Panel—(Tool Aerospace Manufacturing) A controlled test panel for checking system performance

Viscosity—The property of a fluid that presents a resistance to shearing flow (measured in centistokes).

Wetability—The ability of a liquid to spread over and adhere to solid surfaces.

CHAPTER 5
MAGNETIC PARTICLE TESTING

I. HISTORY AND DEVELOPMENT

Magnetism Discovered

The ancient Greeks, originally those near the city of Magnesia, and also the early Chinese knew about strange and rare stones (possibly chunks of iron ore struck by lightning) with the power to attract iron. A steel needle stroked with such a "lodestone" became "magnetic" and around the year 1000 AD, the Chinese found that such a needle, when freely suspended, pointed north–south.

Early Physicists Develop the "Basics" of Magnetism

In the late 1700s and the early 1800s, several exiting new discoveries were made in the field of Physics that paved the way for today's magnetic particle testing (MT) technology.

In the 1700s, Charles Coulomb, a French physicist, discovered that "the magnetic forces of attraction and repulsion are directly proportional to the strength of the poles and inversely proportional to the square of the distance from them," (the inverse square law). He also invented the magnetoscope and magnetometer, which are devices for measuring the Earth's magnetic field strength.

Until 1821, only one kind of magnetism was known; the one produced by iron magnets. Then, Danish scientist Hans Christian Oersted, while demonstrating to friends the flow of an electric current in a wire, noticed that the current caused a nearby compass needle to move. Physicist Andre-Marie Ampere, who concluded that the nature of magnetism was quite different from what everyone had believed, studied this new phenomenon in France. It was basically a *force between electric currents:* two parallel currents in the same direction *attract*, in the opposite direction they *repel*. This proved that magnetic fields exert an influence on current flow.

Phenomenon Discovered Leading to Inspection Principles

In the late 1800s it was observed that a compass needle deflected when it passed over a crack in a magnetized cannon barrel. In the 1920s it was discovered that iron filings on parts held in a magnetic machining chuck produced patterns showing the magnetic lines of flux. Closer investigation revealed patterns that corresponded to discontinuities within the parts.

This is probably the earliest documented application of the magnetic particle method for detecting discontinuities in engineering components. This event was recorded in the journal, "Engineering."

Development of Current MT Techniques

Development of the means to establish the magnetic flux in test objects ran in two distinctly separate paths on either side of the Atlantic Ocean. Most American developments used DC storage batteries to produce magnetism in components, whereas the European developments were based on alternating current. These initial preferences are still present in many of the standards and codes, although the DC storage batteries have been replaced with full-wave rectified current.

Advances were also made in the detection media. Instead of using iron filings, iron oxides and iron powders became much more popular, and colored coatings were added to increase the visibility of the indications formed. Later still, fluorescent coatings were applied to the particles, which greatly increased the sensitivity of the inspection process when viewed under "black light" or ultraviolet illumination.

The Present

Today's range of magnetic particle testing techniques is vast and encompasses portable, transportable, fixed, and semiautomatic inspection devices. These use permanent magnets, electromagnets using either AC or DC, or a combination of both. The detection media are available as dry powder or as a suspension in a liquid carrier. They are supplied in many colors so as to provide a contrast with the test surface background color and are also available as fluorescent particles for maximum sensitivity.

II. THEORY AND PRINCIPLES

Introduction

Magnetic particle testing (MT) is a nondestructive testing (NDT) method for detecting discontinuities that are primarily linear and located at or near the surface of ferromagnetic components and structures. MT is governed by the laws of magnetism and is therefore restricted to the inspection of materials that can support magnetic flux lines. Metals can be classified as ferromagnetic, paramagnetic, or diamagnetic.

Ferromagnetic metals are those that are strongly attracted to a magnet and can become easily magnetized. Examples include iron, nickel, and cobalt.

Paramagnetic metals such as austenitic stainless steel are very weakly attracted by magnetic forces of attraction and cannot be magnetized.

Diamagnetic metals are very slightly repelled by a magnet and cannot be magnetized. Examples include bismuth, gold, and antimony.

Only those metals classified as ferromagnetic can be effectively inspected by MT. In order to understand MT, one should have a basic understanding of magnetism and electromagnetism.

Principles of Magnetism

Polarity
Many of the basic principles of magnetism can be deduced by simple observation of the behavior of a magnetized rod and its interaction with ferromagnetic materials, including other magnetized rods. If the rod is suspended at its center, it will eventually align itself with the Earth's magnetic field so that one end points to geographic north and the other end to the south. If the north-pointing end is identified, it will be found that it is always this end that points north. By convention, this end of the rod is called the "north-seeking pole," usually abbreviated as "north pole," and the other end is called the "south pole."

Magnetic Forces
When the north pole of one magnetized rod is placed close to the south pole of another, it will be observed that they attract one another. *The closer they come together, the stronger the force of attraction.* Conversely, if two north poles or two south poles are placed close together, they will repel each other. This can be summarized as *"like poles repel, unlike poles attract."* One way of defining the phenomenon of magnetism could be: "a mechanical force of attraction or repulsion that one body has upon another," especially those that are ferromagnetic

Magnetic Field
The simple observations of attracting and repelling indicate that some force field surrounds the magnetized rod. Although invisible, this force field is clearly three-dimensional because the attraction or repulsion can be experienced all around the rod. A two-dimensional slice through this field can be made visible by placing a sheet of plain white paper over a bar magnet and sprinkling ferromagnetic particles onto it. The particles will collect around the lines of force in the magnetic field, producing an image such as the one shown in Figure 5-1.

This image is called a "magnetograph" and the lines of force are referred to as lines of "magnetic flux." Lines of magnetic flux will flow in unbroken paths that do not cross each other. Each line of force forms a closed loop that flows through and around the magnet. The word "flow" suggests some sort of direction of movement and by convention this direction is said to be the direction that would be taken by a "unit north pole" placed at the north pole of a bar magnet. A unit north pole is an imaginary concept of a particle with no corresponding south pole. Such a particle would be repelled by the magnet's north pole and attracted to the south pole. In other words, magnetic flow is from its *north pole to its south pole through the air* around the magnet and in order to complete the magnetic circuit, flow will be from the south pole to the north pole within the magnet.

Flux Density
The flowing force of magnetism is called "magnetic flux." The magnetograph image does not show the direction of flux flow, but it can be seen from the magnetograph that the *area of maximum flux concentration (flux density) is at the poles.* Flux density is defined as "the number of lines of force per unit area." The unit area referred to is a slice taken perpendicular to the lines of force. Flux density is measured in Gauss or Tesla, the Tesla being the current unit, and flux density is given the symbol "β" (beta).

Magnetizing Force
The total number of lines of force making up a magnetic field determines the strength of the force of attraction or repulsion that can be exerted by the magnet and is known as the "magnetizing force" and given the symbol "H."

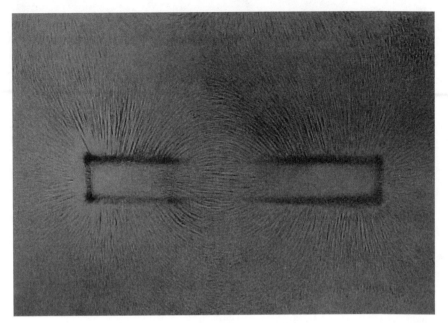

FIGURE 5-1 Magnetograph.

Magnetic Permeability

A German physicist, Wilhelm Weber, postulated a theory about a material's ability to generate or concentrate a magnetic flux field. This theory became known as the magnetic domain theory and relies on an assumption that a magnetic domain is the smallest independent particle within a material that still exhibits a north and south pole (i.e., it has polarity). In an unmagnetized material, these magnetic domains are arranged in a random (haphazard) direction such that their magnetic fields cancel each other out when considering the total magnetic field exhibited by the material.

When a magnetic force is applied to the material, the domains will tend to align themselves with the magnetizing force so that the domains' north poles point in one direction while the south poles point in the opposite direction. The material will now exhibit an overall polarity, which equates to the sum of all of the magnetic domains combined. A flux field will exist around and through the material, as depicted in the sketch in Figure 5-2.

The ease with which the domains align themselves is called "permeability," which is expressed with the symbol "μ" (mu). Materials in which the domains align easily under low magnetizing forces are said to have high permeability. To determine absolute values for permeability, the flux density produced is divided by the magnetizing force applied. Stated mathematically this becomes

$$\mu = \frac{\beta}{H}$$

For practical applications, however, it is far easier to use a comparative measure of permeability, which is determined simply by comparing a material's permeability to that of a

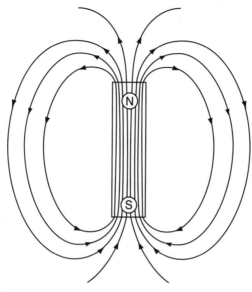

FIGURE 5-2 Magnetograph sketch showing polarity and flux direction.

vacuum. This produces a response that is referred to as "relative permeability" and is given the symbol, "μ_r."

Permeability in magnetic theory can be compared with "conductivity" in electrical theory. Relative permeability can then be compared with the IACS (International Annealed Copper Standard) conductivity scale, in which all conductive materials have their conductivity expressed as a percentage of that of copper. Ferromagnetic materials will have relative permeability values far in excess of one, which is the base value for a vacuum (sometimes referred to as "unity"). They can have values of several hundred times a vacuum's permeability.

Paramagnetic materials in which the domains resist alignment even under high magnetizing forces are said to have low permeability. Paramagnetic materials will have a relative permeability value of slightly greater than 1; for example, 1.003 for some stainless steels.

The third class of metals consists of the diamagnetic materials. The magnetic domains of diamagnetic materials will rotate to positions that are 90° to the direction of the applied magnetizing force, and will create a very slight repulsion to the magnetizing force.

Diamagnetic materials will have a relative permeability value slightly less than one; for example, 0.9996. Paramagnetic and diamagnetic materials are usually referred to as "nonmagnetic" because of their slight or absence of reaction to a magnet.

Magnetic Reluctance

"Reluctance" in a magnetic circuit is the equivalent of resistance in an electrical circuit. The magnetic flux will always follow the path of least magnetic reluctance, which is usually through a ferromagnetic material. The factors affecting reluctance are:

1. The length of the magnetic circuit (λ)
2. The cross-sectional area of the magnetic circuit (A)
3. The permeability of the magnetic circuit (μ)

The reluctance, R, of a given magnetic circuit can be described mathematically as

$$R = \frac{\lambda}{\mu A}$$

Another way of referring to the phenomenon of a material having low permeability is by stating that the material has high reluctance (the domains are reluctant to align themselves).

Magnetic Saturation

The lines of force in a magnetic field repel adjacent lines flowing in the same direction. As the flux density increases, the force of repulsion increases. For a given material there is a maximum value for flux density that can be sustained. Upon reaching this value, the material is said to be "saturated." As the flux density is increased towards saturation, the reluctance of the material increases and the permeability decreases towards that of a vacuum. At saturation, any further increase in magnetizing force finds that the path of least reluctance is now through the air surrounding the material and the excess flux extends out into the air.

Hysteresis

"Hysteresis" describes how the flux density (β) in a magnetic material varies as the magnetizing force (H) is varied. A graphical representation of how flux density increases with an increase in the magnetizing force is shown in Figure 5-3.

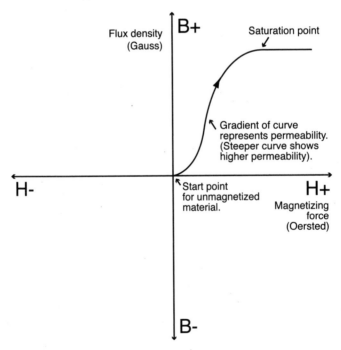

FIGURE 5-3 Hysteresis curve—flux density due to initial magnetization of ferromagnetic material.

From the graph it can be seen that a ferromagnetic material produces a steep initial "virgin curve" due to the relatively small amount of magnetizing force required to produce a high flux density. The flux is being concentrated by the material's permeability. The steepness of the initial "virgin curve" is therefore a measure of the material's permeability.

Also noticeable from the graph is the fact that the ferromagnetic material curve changes in gradient toward the top and will actually flatten out completely at one point. From this point onward, any increase in magnetizing force will not produce an increase in flux density. As mentioned above, this condition is called "saturation." The reason for the change in gradient is due to the reduced amount of magnetic domains available for alignment. The total number of domains present is fixed; therefore, as more of these become locked in alignment due to the magnetizing force, there will be fewer "fluid" domains to create the permeability effect. The relative permeability of a material becomes less as it becomes more magnetized.

The graph shown in Figure 5-3 displays a positive magnetizing force that produces an increasing positive flux density. If the effects of reducing the magnetizing force and plotting the resultant reduction in flux density is considered, even more information about the properties of ferromagnetic materials is gained, as shown in Figure 5-4.

As seen in Figure 5-3, increasing the magnetizing force will increase the flux density until the saturation point is reached. If the magnetizing force is decreased, it will produce a corresponding decrease in flux density. The curve produced by this interaction will deviate from the "virgin curve" produced during the initial magnetization.

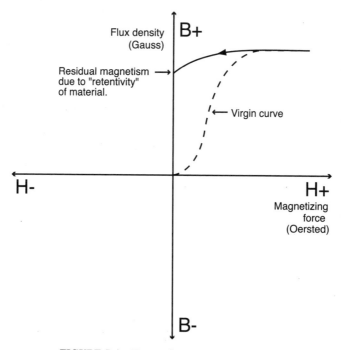

FIGURE 5-4 Hysteresis curve—residual magnetism.

As depicted by the dotted line in Figure 5-5, a point is eventually reached where the magnetizing force is no longer being applied, but there still is a positive flux density in the material.

This phenomenon is due to the "retentivity" of the material, which "retains" some of the flux density and is also referred to as a "remnant field" or "residual magnetism." The amount of residual magnetism will be determined by the retentivity of the material and the amount of magnetizing force initially applied. The retentivity value is highest among the hardened ferromagnetic materials.

Consider what has happened at the magnetic domain level. The domains were initially rotated from their random rest positions by the initial magnetizing force to a position of alignment. When the magnetizing force stops, some of the domains rotate back to a random orientation. Some of the domains, however, remain aligned, resulting in residual magnetism. These domains will remain aligned until a force applied in the opposite direction causes them to rotate back and will actually rotate some of the domains into alignment in the opposite direction.

In Figure 5-5, the effects of applying this opposing magnetizing force can be observed. As this negative force is increased (a force with opposite polarity), the flux density is reduced. The force required to reduce the net flux density to zero is called the "coercive force."

In Figure 5-6, the effects of increasing the opposite magnetizing force beyond the coercive force can be seen. The material exhibits a flux density flowing in the opposite di-

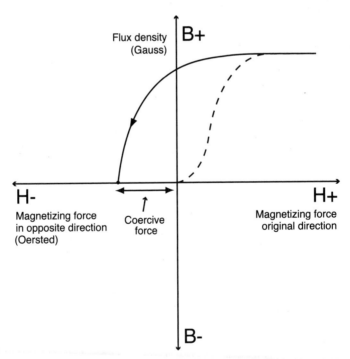

FIGURE 5-5 Hysteresis curve—coercive force.

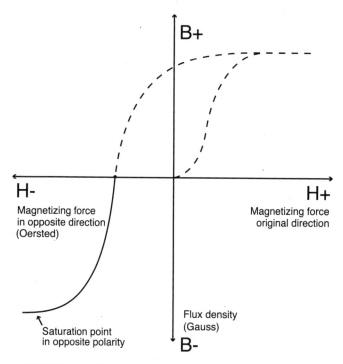

FIGURE 5-6 Hysteresis curve—saturation point with opposite polarity.

rection, which is expressed in the graph as flux density having a polarity opposite to the original. It is also measured in Gauss. The magnetic domains are now rotated such that their north and south poles are aligned 180° out of phase from the initial alignment. A condition is once more achieved whereby all of the domains become aligned in this manner, and saturation occurs in the opposite polarity when all of the domains are aligned.

If this "opposite" magnetizing force is now reduced back to zero, a reduction of the flux density will be observed, but again, some flux density will be present at the point when there is no magnetizing force applied (Figure 5-7).

To eliminate this field and bring the flux density to zero once again, magnetizing force in the original direction must be applied. Again, the amount of force required to achieve zero flux density is called "coercive force" (See Figure 5-8).

A further increase in the magnetizing force will produce an increase in flux density until saturation is once again achieved in the original direction, thereby completing the loop known as the "hysteresis loop" (Figure 5-9). The hysteresis loop can be used to display several of the material's properties. For example, permeability is indicated by the gradient of the virgin curve; reluctance can also be determined from this gradient; a steep curve indicates high permeability, whereas a shallow curve indicates high reluctance.

The residual magnetism is indicated by the flux density value at a magnetizing force of zero. It can be seen in Figure 5-9 that once the virgin curve has been used to produce saturation, the graph continues to follow the "S" shaped hysteresis loop and will never re-

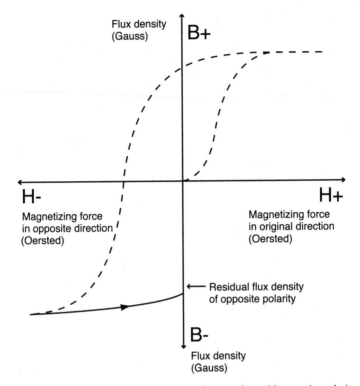

FIGURE 5-7 Hysteresis curve—residual magnetism with opposite polarity.

turns to zero, due to the residual magnetism. This condition will remain until the material is demagnetized.

Magnetization
In the creation of a hysteresis loop, the increase and decrease of magnetizing force was made possible by positioning a powerful permanent magnet closer to or farther away from the object being magnetized. The "negative magnetization" or magnetization in the opposite direction was accomplished by reversing the orientation of the powerful magnet, such that the opposite pole was closest to the object. Magnetism can also be produced by electrical means. If an electrical current flows through a conductor, a magnetic flux will be produced that flows around the conductor and also in the air surrounding that conductor. This flux flows in a circular direction at 90° to the direction of the electric current flow, as shown in Figure 5-10.

Electromagnetic Field Direction
By convention, the actual direction of the flux lines relative to the current flow can be illustrated if the conductor is held in one's left hand. Consider that the electron flow due to the electric current is in the same direction in which the thumb is pointing. The magnetic

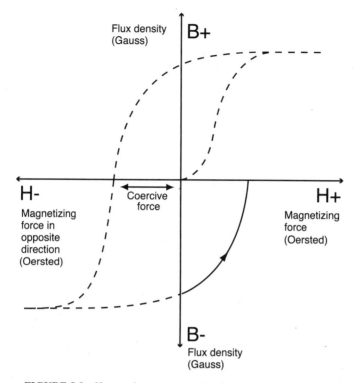

FIGURE 5-8 Hysteresis curve—coercive force with opposite polarity.

flux will flow around the conductor in the same direction in which the fingers are pointing (Figure 5-11).

This "left-hand rule" applies only if it is considered that the "electron flow" is from negative to positive. This is a relatively recent belief. The original "electric current flow," concept given by French physicist Ampere was that the current flowed from positive to negative. If this were the case, then the "right hand rule" would apply.

If the conductor were the actual ferromagnetic test object, concentrated circular flux lines would be produced in the object. The magnetization of a part in this way is referred to as "direct" magnetization. Current is passed directly through the part, creating the circular flux field.

If a non-magnetic conductor were formed into a "loop," the magnetic flux circulating in the air around and through the loop would be as shown in Figure 5-12. It can be seen that at the center of the loop, the flux is basically linear. This effect becomes more pronounced if the conductor is wound into several loops close together so as to form a coil. If the turns of the coil are close enough to each other, the individual flux fields from each loop will join together to create a field for the whole coil (see Figure 5-13).

When a ferromagnetic object is placed inside this coil, flux lines are induced into the object, resulting in a longitudinal flux field in the object in approximately the same direc-

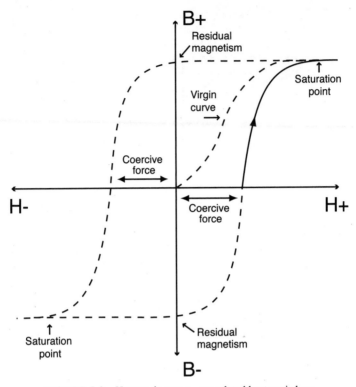

FIGURE 5-9 Hysteresis curve—completed hysteresis loop.

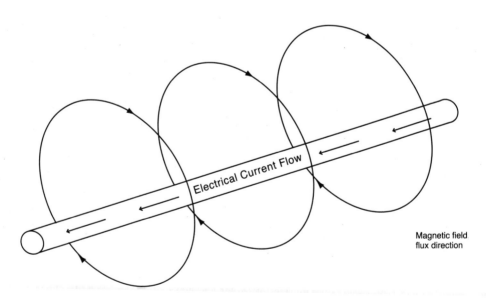

FIGURE 5-10 Flux lines around a current carrying conductor.

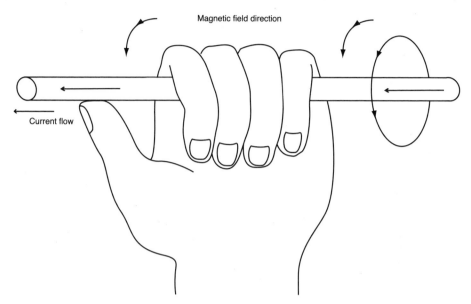

FIGURE 5-11   Flux line direction based on the left hand rule.

tion as the coil axis. Magnetization of a part within a coil is accomplished by magnetic induction and is referred to as "indirect" magnetization. It is therefore possible to produce both circular and longitudinal magnetic flux fields in the test object using electromagnetism.

Detection of Discontinuities

Distorted Fields
The lines of force in the internal field of a magnetized material will tend to distribute themselves evenly through the material, provided that the material is homogenous. The presence of a discontinuity presents an interruption to the field and an increase in reluctance. The lines of force prefer the path of least reluctance and will therefore redistribute themselves in the material by bending around the discontinuity. The field becomes "distorted" by the discontinuity.

Leakage Fields
As a discontinuity gets larger, the remaining metal path in the part becomes more restricted and the magnetic flux approaches saturation for that part of the material. Some of the magnetic lines of force then find that a path through air or across the discontinuity presents a lower reluctance than the remaining metal. As a result, some flux lines "break out" of the surface of the metal into the air. This is called a "leakage field." It is interesting to note that a leakage field may exist both at the near surface and also at a remote or hidden surface.

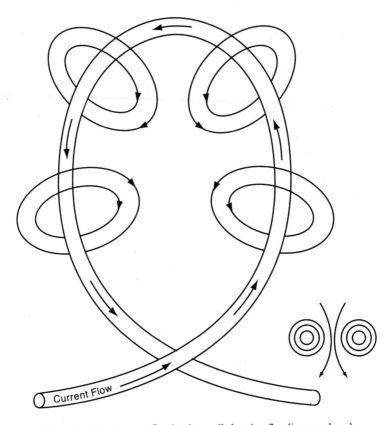

FIGURE 5-12 Current flowing in a coil showing flux lines produced.

In order to create a leakage field, the discontinuity must interrupt the lines of force in the material. A narrow discontinuity oriented parallel to the flux lines will not create a leakage field. In order to produce a leakage field, a discontinuity must interrupt the field usually considered to be within 45° to the perpendicular. It follows that in order to detect a discontinuity with any orientation, the part must be magnetized in at least two directions, 90° to one another.

Making the Leakage Field Visible
When a flux leakage field occurs, a north and south pole will be created at that location (see Figure 5-14). It has already been established that the maximum flux density will occur at the poles. Therefore, whenever a discontinuity disrupts the flux lines and flux leakage occurs, an area of high flux density will be produced. When ferromagnetic particles are applied to the surface of the test object, they are strongly attracted to this flux leakage area and will form an accumulation of particles, producing a visible indication of that discontinuity (see Figure 5-15). The creation of an indication relies on the flux lines being distorted sufficiently to produce flux leakage.

MAGNETIC PARTICLE TESTING   **5.15**

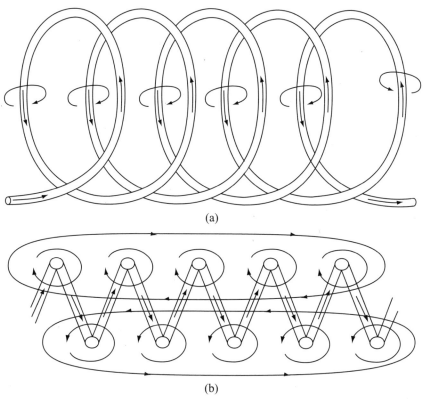

FIGURE 5-13 (a) Current in a multiturn coil and combined flux lines produced. (b) Plan view of a multiturn Coil.

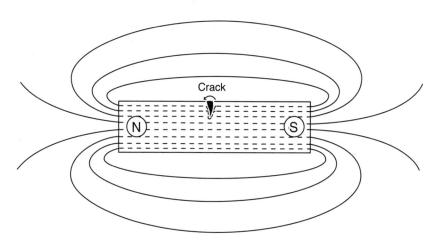

FIGURE 5-14 Flux distortion and leakage field at a crack.

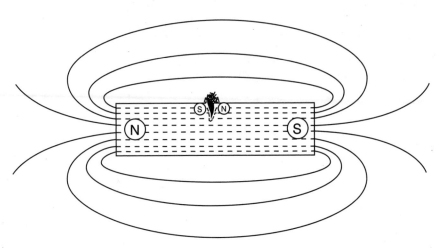

FIGURE 5-15 Particles applied.

Part Geometry
Some aspects of the part's geometry may also produce flux leakage. For example, threads, keyways, or other abrupt changes in section thickness will cause flux leakage and produce indications referred to as "nonrelevant" (see Figure 5-16).

Control of Magnetization
The dimensions of the discontinuity in relation to the component thickness, together with the strength of the applied magnetizing force, will determine the strength of the leakage field. When the component has a complex geometry, no one value of magnetizing force

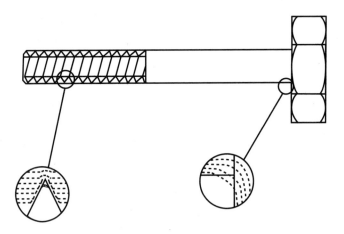

Flux leakage due to geometry at thread roots.

Flux leakage due to geometry at section changes.

FIGURE 5-16 Flux leakage at threads and geometric changes.

will be suitable if nonrelevant indications from section changes are to be avoided. It is therefore necessary to control the magnetizing force to suit the geometry of the part. With electromagnetism, there is far greater control than with permanent magnets. By varying the applied current, the magnetic flux density produced can be controlled to minimize nonrelevant indications while still complying with inspection requirements.

Types of Electrical Current

Direct Current (DC)
Current produced from a battery is constant in amplitude and direction and is called "direct current" or DC. A graphical representation of DC field is illustrated in Figure 5-17.

Altenating Current (AC)
Current provided from a conventional electrical outlet constantly varies in amplitude and also reverses in flow direction (polarity) many times a second. It begins at zero and flows in one direction, rising from zero to maximum flow, then decreases back to zero before reversing in flow direction, rising from zero to a maximum flow in the opposite direction before falling back to zero once again to complete a full cycle. The actual number of cycles per second is called "frequency." The frequency of AC power in the United States is 60 cycles per second or 60 "hertz" (1 hertz = 1 cycle per second). A graphical representation of this cycle is shown in Figure 5-18. This type of current is referred to as "alternating current" because it is continuously alternating in direction.

Half-Wave Rectification
This alternating current can be modified in several ways. Its frequency can be altered by using an oscillating circuit (although this is seldom done in MT) or the "negative" portion of the current flow can be removed to allow the current to flow only in one direction. This

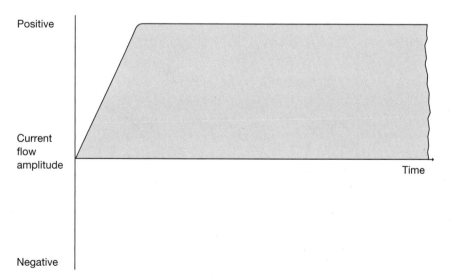

FIGURE 5-17   Graphical display of direct current waveform.

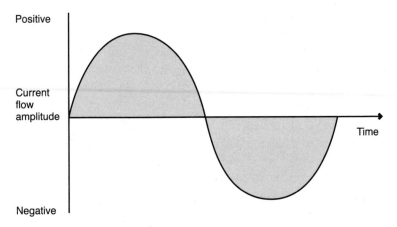

FIGURE 5-18 Graphical display of alternating current waveform.

is called "half-wave rectified" alternating current (HWAC) and its effects on the waveform can be seen in Figure 5-19. Obviously, if half of the electrical current is removed, this will reduce the magnetic flux produced.

Full-Wave Rectification
One other alternative is to reverse the polarity of the negative portion of the wave to make it also positive. This is called "full-wave rectified" (FWAC) alternating current and it causes little or no loss of the original AC wave's power (see Figure 5-20).

Electrical Power
The power developed in an electrical circuit is measured in watts (W) and is the product of the applied voltage (V) and the resulting current (I):

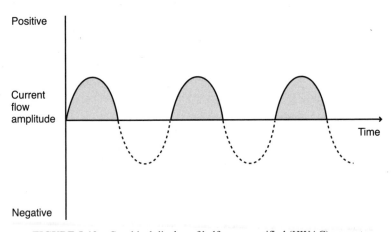

FIGURE 5-19 Graphical display of half-wave rectified (HWAC) current.

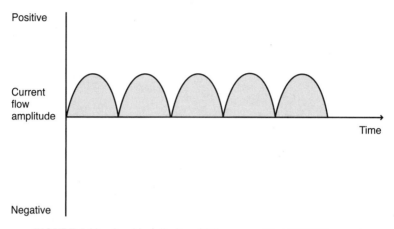

FIGURE 5-20   Graphical display of full-wave rectified (FWAC) current.

$$\text{Watts} = V \times I$$

One of the byproducts of the current developed is heat, which may damage metals if the current is applied for an extended period of time. At the same time, the flux density achieved with electromagnetism is proportional to the current flowing. These two factors need to be balanced when selecting the magnetizing parameters for MT. For safety reasons, most MT equipment operates at a low voltage (typically 3 volts), and so the current is the greater contributor to the power in the circuit.

The electrical current amplitude is measured in amperes (amps). With DC, the amplitude is constant; however, with any of the sinusoidal wave forms produced by AC, half-wave or full-wave rectified, the "peak" amperage produced at the maximum flow rate at the peaks of the waves will only apply for an instant. The average current includes the zeros, the peaks, and all points in between. The average value for an AC wave or full-wave rectified AC will be approximately 70% of the peak value. This value is referred to as the "root mean square" (RMS). It can be seen that the power developed at 1000 amp DC will be higher than at 1000 amp AC. With half-wave rectified AC, the difference becomes much more pronounced because half of the AC wave is not used; therefore, the actual output will only be approximately 35% of that produced by DC.

Care should be exercised when setting amperage values based on meter readings. Some meters indicate peak values and some will indicate RMS values.

With three-phase current, when the AC wave is one-third into its cycle, a second wave is generated; and when the second wave is one-third into its cycle, a third wave is generated. The result is three AC waves being generated simultaneously, with each one overlapping the others, thus providing more current flow and a smoother distribution. See Figure 5-21.

Early versions of MT equipment required banks of batteries to produce pure direct current. These were cumbersome and long charge cycles to produce relatively few "shots" before recharging again were necessary. Equipment with DC capability relies on modifying AC current into DC by full-wave rectification. This is further enhanced by using smoothing capacitors to slow down the rate of decay of the peaks and help to "fill in" the troughs. If three-phase current is used with full-wave rectification, this smoothing of

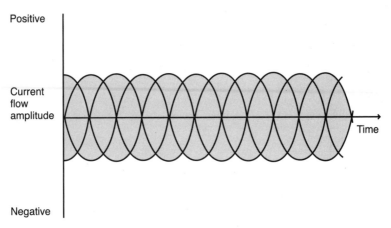

FIGURE 5-21 Graphical display of three phase wave AC.

the waveform and current flow is improved to a point where almost pure DC results (see Figure 5-22).

The Influence of AC and DC Currents
In addition to the amount of flux produced, the different currents used will affect the distribution of the flux in the part.

With DC, the current tends to use the entire cross section of the conductor, decreasing to zero at the center. The lines of magnetic force tend to spread through the material thickness and similarly decrease in density toward the center. With any of the AC wave forms, the current and lines of force tend to be confined to a region close to the surface of the material. The depth of penetration of both depends on the frequency of the AC. For 60 hertz this depth is roughly 3 mm (0.125″). This phenomenon is called the "skin effect." The higher the AC frequency, the shallower the depth of penetration.

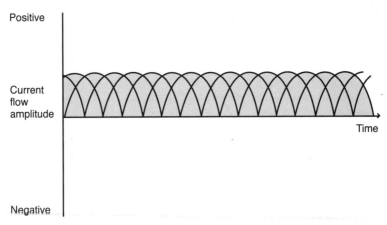

FIGURE 5-22 Graphical display of three phase full-wave rectified AC.

This difference in flux distribution can be used to an advantage. If it is desired to detect discontinuities slightly deeper in a part, then DC will result in a greater penetration of the flux. If the goal is to detect surface discontinuities, the concentration of the flux at the surface due to the skin effect will enhance detection. It is for this reason that DC is preferred for the detection of manufacturing induced discontinuities, which may occur anywhere in the part. The deeper the discontinuity is below the test surface, the broader and less distinct the indication becomes. To enhance the contrast it is necessary to use dry particles, since the particles are larger than those used in a suspension. The larger particles are also less mobile and more reluctant to gather at the region of the flux leakage, so that some form of external agitation may be required (blowing on the surface or tapping the component). Also, the amperage required to achieve sufficient flux density using DC will be higher than for AC and this will increase the heating effect. Before deciding to use DC current for deeper penetration, it may be worth considering whether another NDT method would be more suitable.

AC current is preferred for the detection of surface and near-surface discontinuities, particularly during in-service inspection, where discontinuities, by nature, tend to be surface-breaking (e.g., fatigue cracks). For a given current, the flux density near the surface is higher when using AC and this means that lower currents than those required for DC can be used. This, in turn, produces less heat. Another advantage is that an AC field, by its cyclic nature, tends to vibrate the particles, thereby increasing their movement toward the area of flux leakage. Most users regard magnetic particle testing as a method for the detection of surface and near-surface discontinuities only.

Detection Media

The detection media used for MT are basically similar to the ferromagnetic particles used in producing a magnetograph; however, they are of a higher level of refinement. Their size and shape are carefully controlled to produce a variety of different sizes and shapes. Different shapes are desired to meet different aspects of indication formation. Elongated particles will rotate to align with the flux lines, due to their increased length-to-diameter ratio, whereas rounded particles have greater mobility to move to the areas of flux leakage.

Different sizes are necessary. The smaller particles have the ability to accumulate at small discontinuities and produce a sharper indication. This is often useful in identifying the nature of the discontinuity. For example, a crack indication will have a jagged appearance. Larger particles have the ability to bridge across larger discontinuities and increase contrast.

Particles may be applied in a dry form or may be suspended in a liquid such as kerosene or a similar petroleum distillate. They may also be suspended in a water carrier containing suitable additives such as wetting agents, and antifoam liquids.

The particles used in the wet suspensions are usually iron oxides as opposed to iron filings, due to their lighter weight and greater ability to remain in suspension. The disadvantage of using iron oxides is the reduced permeability of the oxides compared to iron filings. This results in a lesser response to the weaker leakage fields produced by subsurface discontinuities. A distinct advantage of using suspended particles is their greater mobility in the liquid, which improves the detection of small surface discontinuities. Whenever the wet suspension is used, it is essential that periodic checks be made to determine the particle concentration in the liquid carrier. This is referred to as the "settling test" and will be discussed in detail at the end of Section IV.

The visual impact of indications is also assisted by the addition of a coating to the particles to provide a greater contrast with the test surface. This results in a significant increase in "seeability" or sensitivity. The coatings fall into two distinct groups, color contrast and fluorescent.

Color contrast coatings are available in several colors, including blue, red, gray, and black. See Figure 5-23 for examples of dry particles. The choice of color depends upon the color of the test surface and the degree of contrast to be provided. If suitable contrast cannot be achieved, another possibility is to coat the part with a specially formulated white contrast paint, which is quick drying and easily removed after inspection by using a solvent. This will create a good contrast between the black particles in a liquid carrier and the white background (see Figure 5-24.)

The fluorescent particles are viewed under black light illumination in a darkened viewing area. The fluorescent particles emit light when activated by the black light, providing excellent contrast against the dark background of the part due to the darkened viewing conditions. (See Section III for more information about black lights).

Demagnetization

Demagnetization of the test part is sometimes necessary before, during, and at the end of an inspection, for a variety of reasons. The principal reasons are described below.

Before and During the Test
If the test part has been left with a residual field in a different direction to the field about to be applied, the resulting field will be the vector sum of the residual magnetism and the

FIGURE 5-23 Examples of dry particles.

FIGURE 5-24 Weld with white contrast paint.

newly applied field. The direction of the resulting field will therefore not be precisely in the direction intended; therefore, sensitivity may be reduced. Demagnetization is recommended in this case.

After the Test
A residual field in a component being returned to service may have a detrimental effect on future operations of the part or nearby equipment. In an aircraft or ship, a compass may be affected. High-speed rotating parts may induce eddy currents in nearby structures, causing a braking effect on the part and heating of the surface. If the part is to be welded after inspection, the residual field may deflect the weld arc and make it difficult to control the weld deposit.

Demagnetization Process
In order to effectively demagnetize a part, two requirements must be met:

1. The polarity must be successively reversed
2. The field strength (flux density) must be decreased

This means that each successive current application will be in the opposite direction to the previously applied force and will be at a lower level. It should be below the saturation point, but must be above that required to produce a higher flux field than retained. This will produce successively lower residual fields at each application until an acceptably low field exists. It should be noted that although total demagnetization will not result, the field can be brought to an acceptable level.

These two requirements can be met using a reversing DC field; however, AC is preferred, due to its inherent reversing nature. The most satisfactory method of demagnetization using AC is to gradually reduce the applied current to zero or, more commonly, to pass the test part through an AC demagnetizer.

One way to completely demagnetize is by heating the part to a temperature above its curie point and allowing it to cool without any magnetizing force acting upon it. If this can be accomplished, the residual field will be totally removed and the part will be totally demagnetized. The curie points for steels range from 720 °C (1296 °F) to 800 °C (1440 °F). For most NDT applications however, this approach is not practical.

III. EQUIPMENT AND ACCESSORIES

Different MT equipment is available for production line and field-type applications. To facilitate the varied applications, several basic groups of test equipment are available. They are classified as "stationary," "mobile," or "portable" units.

Stationary Units

A stationary unit is referred to as a wet horizontal unit and usually has capabilities for producing longitudinal and circular fields. Some units will also be capable of demagnetizing the parts, although this is usually accomplished with a separate demagnetizer. Figure 5-25 illustrates a typical wet horizontal stationary unit. This unit has a fixed headstock

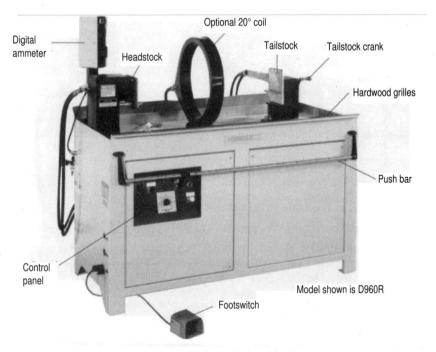

FIGURE 5-25 Wet horizontal unit.

and a sliding tailstock. The part is placed between these and the tailstock is adjusted to just slightly longer than the length of the part. When the tailstock is locked in this position, a foot-switch controlled, pneumatic headstock closes and grips the part tightly to permit the current to flow, thereby producing a circular magnetic field.

If hollow parts such as tubular products or ring-shaped parts are to be inspected, a central conductor technique can be used. This involves the use of a bar of high electrical conductivity, typically copper or aluminum, that is gripped between the headstock and tailstock. When current is passed through the conductor bar, a circular magnetic field will be induced in the component. It should be noted that the flux density in the part will be highest at the inner surface. Advantages of this technique are that several components may be inspected simultaneously and there is no risk of overheating the parts or arcing between the unit and the part(s) (see Figure 5-26), since this is an indirect magnetization technique.

For longitudinal magnetization techniques using coils (as in Figure 5-25), the part can either remain in this position if a centralized coil position is used or it can be placed on the lower inside surface of the coil. The ratio of coil inner diameter to component outer diameter, or cross-sectional area ratios, will determine which position should be used. This will vary based on specification requirements.

The amperage can be adjusted using a variable self-regulating current control. An adjustable timer controls the current duration. Demagnetization is accomplished using either the coil in the wet horizontal unit or a separate AC coil demagnetizer. The stationary units are very versatile in their range of capabilities; however, the main limitation is their inability to be taken to a field location. For this, "mobile" or "portable" equipment must be used.

FIGURE 5-26 Central conductor setup with a Keytos ring.

Mobile and Portable Units

The main difference between mobile and portable equipment is in their ability to be moved. A mobile unit must be transported to a location, then moved around on its own wheels or castors (see Figure 5-27). It can weigh up to 1000 lbs (500 Kg). A portable unit can usually be carried by one person, and will probably weigh no more than 50 lbs. (25 kg) (see Figure 5-28).

Despite their size difference, both of these types of units can accomplish similar inspections with the use of various accessories. The power outputs of these units vary considerably. As expected, the mobile unit will have a much higher output than the portable unit. Some of the mobile units can produce as much as 6000 amps, whereas the portable units are usually limited to a maximum of 1500 amps.

Common accessories include clamps and prods for producing circular magnetization. Great care must be exercised when using prods or clamps to ensure that good contact is maintained with the test specimen, otherwise electrical arcing can result. This can produce localized areas of surface damage and potential stress risers in the part.

Cable wrapped coils and rigid coils are used with the mobile and portable units for producing longitudinal magnetization. Both mobile and portable units usually have selectable AC or DC positions, which also enables demagnetization to be accomplished.

Electromagnetic Yoke

Even greater portability can be achieved by using an electromagnetic yoke, which is referred to as an AC yoke or contour probe (see Figure 5-29). This unit can be used with AC and is also available in a battery pack version, which further increases its portability by eliminating the need for an AC power source. Many yokes have articulating legs to facilitate various inspection area profiles. These yokes produce only longitudinal magnetization; repositioning is required to achieve flux line orientation in at least two 90° opposing directions.

FIGURE 5-27 Mobile MT unit.

FIGURE 5-28 Portable DC prod unit.

Permanent Magnets

Small permanent magnets can also be used to produce localized longitudinal magnetization, but there are some limitations to their use (see Figure 5-30). Due to their strong fields, particles tend to become attracted to the legs more readily than the test surface. At times, the use of these magnets can become unwieldly.

FIGURE 5-29 AC yoke.

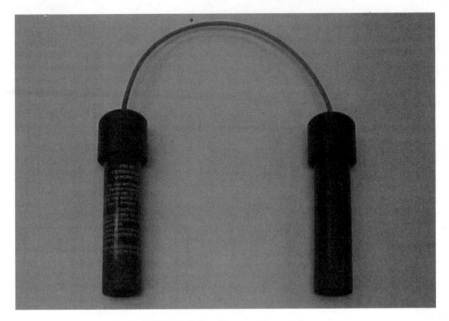

FIGURE 5-30 Permanent magnets.

Black Lights

The use of fluorescent particles requires a black light (see Figure 5-31). One type of blacklight contains a regulated current ballast transformer to limit the current drawn by the arc in the mercury vapor bulb. The light produced may also contain some white light and harmful UV radiation. It is therefore essential that a correct black light filter be used. This filter will allow the relatively harmless portion of the ultraviolet spectrum to pass through with a center wavelength of 365 nanometers (nm). The condition of the filter must be regularly checked to ensure that white light or harmful UV is not present.

Flux Direction Indicators

Flux direction can be determined by using one of several different indicators, as detailed below.

Pie Gauge

This is one of the most commonly used devices in MT. It consists of eight low-carbon steel segments, furnace brazed together, to form an octagonal flat plate that is then copper plated on one side to hide the joint lines. The gauge is placed on the test specimen during magnetization, with its copper face up. The particles are applied to this face and the orientation of the resultant field is displayed by the indications produced at the joint lines (see Figure 5-32).

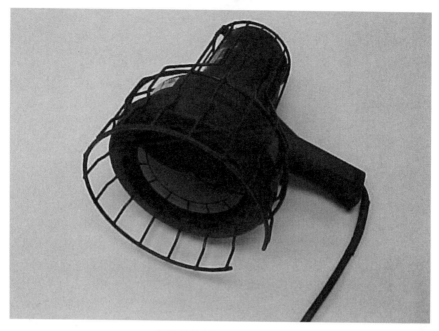

FIGURE 5-31 Black light.

Burmah Castrol Strips

These strips consist of a sandwich structure made up of a thin ferromagnetic foil with three fine slots running lengthwise sandwiched between two nonferromagnetic foils that conceal the slots. This is used in the same manner as the pie gauge. The slots in the strips will indicate the orientation of the flux lines (see Figure 5-33).

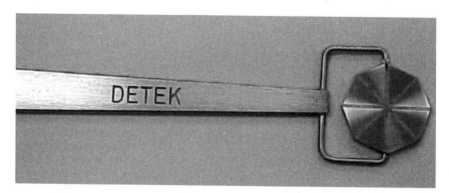

FIGURE 5-32 Pie gauge.

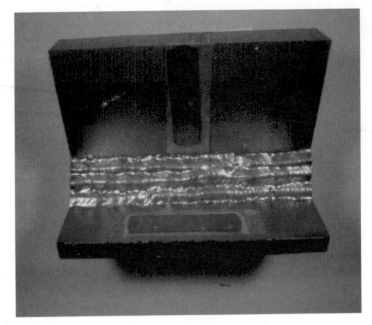

FIGURE 5-33 Burmah castrol strip.

Quantitative Quality Indicator (QQI)

This is a recent development of the foil concept and consists of a thin metallic foil, but instead of having slots in only one direction, the slots form circles or crosses, thereby making them able to indicate flux line direction without repositioning the indicators. The portion of the circle or the leg(s) of a cross determine which flaw orientation should be detected (see Figure 5-34).

AC Demagnetizing Coils

Separate coils are used when the demagnetization cannot be accomplished on the MT equipment. Many portable and mobile units have provisions for stepped reduction in current. The demagnetizing coil is often operated at a fixed current value and the reduction in field strength is achieved by passing the part(s) through the coil and beyond the coil along the central axis for a distance typically two to three times the length of the part(s).

This is effective for removing most of the residual fields because the residual vector of each cycle rapidly swings around to longitudinal. Thereafter, as the distance between the coil and the part increases, the field reduces until it is close to zero.

IV. TECHNIQUES

Magnetic particle testing techniques can be categorized into several key groups. They are:

MAGNETIC PARTICLE TESTING

FIGURE 5-34 Quantitative quality indicators.

1. Method of magnetization
2. Continuous or residual
3. Particle types

Magnetization Techniques

There are several factors that will dictate which magnetization technique is chosen for a particular test. The primary factor is the probable direction of anticipated discontinuities. The direction can usually be determined for fatigue cracks by studying the component loading and stress points, but in most other cases, the direction can be entirely unpredictable. Whether the direction is known or not, the component must inspected using two field directions perpendicular to one another, as discussed in Section II. This can be achieved using one or more of the magnetizing techniques described here.

The next factor in order of importance is the strength of the field to be applied, which will be discussed in Section V, Variables. The remaining factors to be considered include the end use of the part, the location of the part (in a factory, at a remote site, or in situ in a structure), and the susceptibility of the part to cracking due to heating or arc burns.

Magnetic Flow Techniques (Indirect)

The "indirect" or magnetic flow technique is defined as one in which the test part becomes part of the magnetic circuit by bridging the path between the poles of a permanent magnet or electromagnet. Figure 5-35(a) illustrates permanent magnets used in this way. The magnets can be turned 90°, as shown in Figure 5-35(b).

The AC yoke is an example of an electromagnet that is used in much the same way. They can also be used as magnetic flow devices, as shown in Figures 5-36(a) and 5-36(b). The main advantage of the indirect technique is that the risk of arc burning of critical components does not exist. Also, the use of permanent magnets or yokes can be very convenient for in-situ inspections in confined spaces or remote locations.

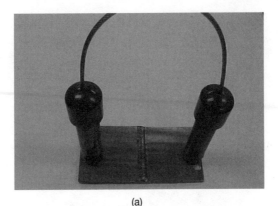

FIGURE 5-35 (a) Magnetic flow technique. (b) Magnetic flow technique at 90°.

Current Flow Techniques (Direct)

The "direct" technique is defined as one in which the magnetizing current flows through the part, thereby completing the electrical circuit. This is accomplished by placing the part between the "heads" of a stationary unit such as the one shown in Figure 5-25 or by using "prods," as shown in Figure 5-28.

The magnetic field formed with the direct technique is at right angles to the current direction of the current. In the case of a round bar held between the heads in a wet horizontal unit, the magnetic lines of force circulate around the bar as the current flows from one end to the other. This technique produces "circular magnetism." The flux line pattern in a flat plate when prods are used is illustrated in Figure 5-37.

One of the dangers in using the direct technique is that contact head or prod can cause a burn in the part if the high current is passing through a small contact area. For this reason, the contact faces on the heads should be flexible, and the tips of the prods should have a low melting point in order to spread the thermal load.

FIGURE 5-36 (a) Electromagnet. (b) Electromagnet at 90°.

Coil Technique

Figure 5-38 shows a cylindrical component inserted into the windings of a coil. This produces longitudinal magnetism in the part as the current flows through the coil and creates a longitudinal field through and around the coil. This is referred to as the "longitudinal magnetization" technique. It has the advantage of not causing thermal damage, since a current does not pass through the part (indirect). In wet horizontal units, it is usual to have a fixed coil, usually containing five turns; but with mobile and portable units, it is more usual to wrap the current-carrying cable around the part to form a coil.

Central Conductor Technique

"Central" conductor is sometimes a misnomer because the opening through which the nonmagnetic conductor is positioned is not always central. The technique is sometimes referred to as the "threader bar" technique to show the distinction. Figure 5-39 is a sketch of a tubular component being tested using this technique. Like the coil technique, it is also indirect because the current passes through the conductor and not the part. This technique produces circular magnetization and can be used to inspect the insider and outside surfaces of the part.

Continuous and Residual Techniques

There are four techniques that can be used to inspect parts using these methods of magnetization, they are:

1. Dry continuous
2. Dry residual
3. Wet continuous
4. Wet residual

There are advantages and limitations associated with each technique and careful selection must be made depending upon the application. The term "continuous" is used when the

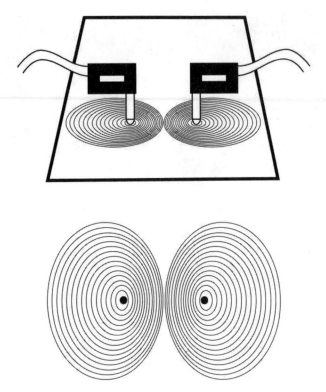

FIGURE 5-37 Sketch of flux lines—DC prod.

magnetic particles are applied while the current is still flowing. "Residual" is used when the material has sufficient retentivity to allow application of the magnetic particles after the current has ceased. "Dry" means that the magnetic particles are applied in fine particle form. "Wet" means that the particles are applied suspended in a liquid carrier. The particles may be suspended in kerosene or other similar petroleum distillates or may be suspended in water containing specially formulated additives. A brief description of the key parameters of each technique follows.

Dry Continuous
This technique uses dry particles that are applied while the magnetizing force is on. The particle application must cease before the current flow ceases. The use of dry particles is useful for detecting slightly subsurface discontinuities, since the particles have higher permeability compared to the particles in a wet suspension. This higher permeability is needed to detect the weaker leakage fields produced by slightly subsurface discontinuities. Applying the particles during the magnetization provides maximum sensitivity, since the flux density will always be at its maximum during the current flow.

One disadvantage of dry particles is their relatively poor mobility when used with DC current. Mobility can be improved by gently tapping the test part or by using half-wave rectified or AC current. The pulsed nature of the half-wave rectified and AC current creates a vibratory effect on the particles, causing them to "dance" toward the leak-

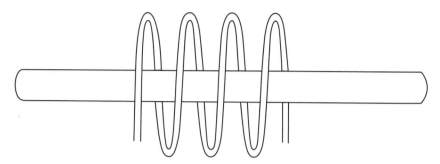

FIGURE 5-38 Coil technique.

age fields. This technique is used extensively for field inspections using prods or yokes. Care must be exercised when removing the excess particles from the test surface to minimize any effect on the particles held by discontinuities. In this respect, the current should also be flowing while the excess particles are being removed with low air pressure.

Dry Residual

This technique also uses dry particles, but differs from the previous technique in that the magnetizing current is applied for a brief period, which produces a residual magnetic field in the part (DC is preferred). The dry particles are applied after the magnetizing force has ceased. The retentivity of the material produces a residual magnetic field, which results in leakage fields at discontinuities open to the surface. This technique is suitable only with materials having high retentivity. This must be determined prior to selecting this technique.

There are further disadvantages when compared to the continuous techniques. The leakage fields produced will be weaker, thereby reducing the sensitivity. Small and subsurface discontinuities are difficult to detect. The lack of the magnetizing current's vibratory effect during particle application will also reduce sensitivity due to the decrease in particle mobility.

FIGURE 5-39 Central conductor/threader bar.

The main advantage of this technique is that multiple parts or batches can be magnetized simultaneously, then examined individually after particle applications. This technique would only be used when a lower sensitivity of high-retentivity parts is acceptable.

Wet Continuous
This technique uses particles suspended in a liquid carrier and is applied simultaneously with the magnetizing current. Compared to the dry particles, the suspended particles are generally of a lower permeability, which makes this technique less favorable for the detection of slightly subsurface discontinuities.

There are advantages and limitations of the different liquid carriers. Kerosene and petroleum distillates are more expensive and may produce some health and flammability problems, but they also help to lubricate the sliding mechanisms on the wet horizontal unit and do not constitute a corrosion source for the equipment or the parts being inspected. Water, on the other hand, is inexpensive, readily available, poses no health or flammability hazards, but is a source of corrosion. Using corrosion-inhibiting additives can reduce the corrosion risk, although it can never be completely eliminated.

The greatly improved mobility of the particles makes this technique very suitable for the detection of small surface discontinuities. Additionally, the suspension will adhere to complex shapes better than dry particles, due to the surface tension combating the effects of gravity. These advantages combined with the high flux densities of the continuous technique provide the maximum sensitivity for surface discontinuities. When using this technique, the suspension flow should cease before the current stops, otherwise indications may be washed away.

Wet Residual
This technique also uses suspended particles, but the suspension is applied after the magnetizing force has been stopped. As previously detailed, the residual method will be of a lower sensitivity than the continuous method, but has the advantage of increased inspection speed, due to multiple part or batch inspection possibilities.

Color and Fluorescence
Each of the above techniques may be further subdivided into color contrast or fluorescent particles techniques depending on the availability of equipment and materials, type of discontinuity, and sensitivity requirements. Some inspection requirements will mandate the use of fluorescent particles, due to the increased sensitivity provided by the higher contrast ratio.

Summary of Technique Choices

It is obvious that there are many options and choices to consider when developing the most suitable technique for the part to be inspected. In order to provide the optimum inspection approach, some of the key areas to be evaluated include:

- Material type
- Part configuration
- Dimensions
- Surface condition
- Type and location of discontinuities anticipated
- Code and specification requirements
- Customer-specific requirements

Once these are known, the choice of technique can be made. These key items should be considered:

- Type of current (AC, HWDC, FWDC)
- Wet suspension or dry particles
- Visible or fluorescent particles
- Direct or indirect magnetization
- Continuous or residual
- Stationary or portable equipment

The best combination can be determined through qualification by validation through the use of controlled test specimens.

Quality Control Considerations

In order to maintain consistency and high level of control over the MT process, it is highly recommended (and in many cases required) that certain checks be completed on the system. Probably the best ongoing quality check is to periodically inspect a controlled test specimen with known discontinuities. The test specimens should be carefully maintained and the periodic checks recorded.

A suspension concentration test, sometimes referred to as a "settling test," should be performed daily. This test is usually performed with a 100 mL ASTM pear-shaped centrifuge tube (see Figure 5-40). A sample of the suspension (after thorough agitation) is deposited directly into the tube from the hose dispenser and allowed to settle, typically for a period of 30 minutes. To assist in the settlement, the tube may be "run" through the demagnetizer. The purpose of this check is to assure that the proper concentration of particles is being maintained in the liquid carrier. The applicable specifications will detail the concentration range. Typical ranges are:

- For visible particles, 1.0 to 2.4 mL/100 mL
- For fluorescent particles, 0.1 to 0.4 mL/100 mL

A settling test tube in a stand is illustrated in Figure 5-40.

The Ketos ring seen in Figure 5-41 is a device made of tool steel and is designed to show the effectiveness of the MT field and the relative penetration based on the number of holes that display indications. Both wet suspension and dry particles are used, and general practice calls for the following:

	Number of Hole Indications	
Amperage (FWDC)	Wet Suspension	Dry Particles
1400	3	4
2500	5	6
3400	6	7

As with all the quality checks, it is essential that results of the ring test be recorded.

A final check for the presence of residual magnetism should always be made using a simple field indicator.

Other QC checks may include:

- Suspension contamination
- Ammeter accuracy

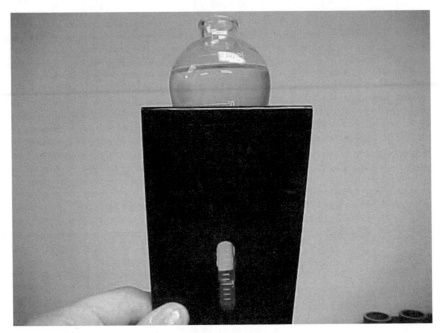

FIGURE 5-40 Settling test.

FIGURE 5-41 Ketos ring.

FIGURE 5-42 Field indicator.

- Timer performance
- Black light intensity
- White light intensity
- Ambient visible light
- Water in oil-based suspension liquid
- Fluorescent particle brilliance

The key to successful MT and control of the system is to periodically perform the appropriate checks and maintain a detailed record of them in a logbook.

V. VARIABLES

As mentioned in previous sections, there are many variables to consider when deciding upon the best inspection technique for a particular application. A listing of some of the considerations and options follows.

Considerations

Customer requirements—specifications, standards, or drawing requirements
Equipment availability—AC/DC, stationary, portable, black light, etc.
AC power availability—needed for all except permanent yokes or magnets
Part size requirements—e.g., will it fit into a wet horizontal unit?
Part location—will it require portable equipment?
Types of discontinuities anticipated—orientation and location
Retentivity of test material—must be high if residual technique is to be used

Options

AC or DC (HWDC or FWDC) Magnetizing Current
Use AC for surface discontinuities or those that would be expected during in-service inspection. Use DC for deeper penetration and possible detection of slightly subsurface discontinuities.

Continuous or Residual
Use the residual technique for inspection of highly retentive materials, especially where large batches are to be inspected. Use the continuous technique for all other applications, especially where higher sensitivity is required.

Wet or Dry Particle Application
Wet particles have better mobility, but generally lower permeability. Use wet for small surface discontinuities or service-induced discontinuities. Use dry for slightly subsurface discontinuity detection.

Color Contrast or Fluorescent Particles
Color contrast does not require a darkened viewing area or black light illumination, however, the sensitivity will be lower than that of fluorescent particles. Use color contrast for field inspections where highest sensitivity is not required. Use fluorescent where maximum sensitivity is essential.

Stationary, Mobile, or Portable Units
In a production or maintenance facility, the versatility of the wet horizontal unit makes it the ideal choice; however, many applications involving immovable or large test objects eliminate this possibility, resulting in the need for mobile or portable units.

Amperage to be Used
Will be determined based on specification and standard requirements and the test part dimensions.

Amperage Selection

Many international standards give tables of current values based on the type of current available in the magnetizing equipment, the sensitivity required, and the component geometry and material. In the absence of a specified standard, the following general "rules of thumb" have been widely used.

For direct circular magnetization of basically round sections use a range of 300–800 amps per inch diameter (12–32 amps/mm). This is usually 500 amps per inch (20 amps/mm) based on peak value for full-wave rectified current. Values above 500 amps/inch diameter (20 amps/mm) are only recommended for low-permeability materials or for slightly subsurface discontinuity detection.

Example 1. For a 1.5" diameter bar, 9" long, 1.5" × 500 amps per inch = 750 amps.

For central conductor circular magnetization of hollow round components with **conductor bar placed concentrically** with the component, the same current values stated above would apply, based on the outer diameter or inner diameter, depending on which surface is to be examined.

Example 2: Tubular component, 2" outside diameter, 1.5" inside diameter, using a 1" diameter central conductor (looking for OD discontinuities), 2" outside diameter × 500 amps per inch = 1000 amps.

MAGNETIC PARTICLE TESTING

For central conductor circular magnetization of hollow round parts, with the conductor bar placed against the inner surface of the component the same formula should be used for calculating the direct current amperage. However, instead of basing the calculation on the outer diameter, it should be based on the diameter of the central conductor plus two times the wall thickness of the part. The area that is considered to be effectively magnetized using this technique is a distance around the inner diameter equating to four times the central conductor diameter. The part should be rotated and reinspected as often as necessary to achieve a minimum of a 10% overlap between each examination area.

Example 3: Outside diameter 2″, inside diameter 1.5″, 1″ diameter central conductor, wall thickness 0.25″:

Central conductor = 1.0″

2 × wall thickness; 2 × 0.25″ = 0.5″

Part diameter = 1.5″

Part diameter × 500 amps per inch = 1.5″ × 500 amps per inch = 750 amps

For longitudinal magnetization using a rigid coil, when the part cross sectional area (cross sectional area is found by squaring the radius and multiplying by π) is **less than one tenth (10%) of the coil inner cross sectional area (low fill factor coil),** the number of "ampere turns" to be used can be found using one of the following formulas:

If part is placed against inside of coil, ampere turns
= 45,000 ÷ (part length) × (part diameter)

If part is placed centrally in the coil, ampere turns
= 43,000 × coil inner radius (6 × length/diameter ratio) − 5

Note 1. The length-to-diameter ratio must be in the range 2 (minimum) to 15 (maximum). Parts shorter than twice the diameter should not be inspected using the coil technique. Parts with a length greater than 15 times the diameter should be tested using a number of "shots" until the entire length is covered.

Note 2. The "ampere turns" value must then be divided by the number of turns in the coil to determine the amperage to be used.

Example 4. Inspecting a 1.5″ diameter round bar with a total length of 9″ using a five-turn coil with an internal diameter of 12 inches. Determine if the coil is a low fill factor by comparing area ratios:

Area of part = 0.75″ (radius) squared × π = 1.767 square inches

Area of coil = 6″ (radius) squared × π = 113 square inches

Ratio of areas = 1.767 divided by 113 = 1.56% (low fill factor)

Amperes to be used if the part positioned the side of coil:

$= 45{,}000 \div \dfrac{9}{1.5}$

$= 45{,}000 \div 6$

$= 7500$ ampere turns

Divide by 5 turns $= \dfrac{7500}{5} = 1500$ amps

Amperes to be used if the part is positioned centrally in coil:

$= \dfrac{(43{,}000 \times \text{coil radius})}{[6 \times (\text{length/diameter ratio})]} - 5$

$$= \frac{(43{,}000 \times 6)}{(6 \times 9''/1.5'') - 5}$$

$$= \frac{(258{,}000)}{(6 \times 6) - 5}$$

$$= \frac{(258{,}000)}{(36) - 5}$$

$$= \frac{258{,}000}{31}$$

= 8322 ampere turns

Divide by 5 turns = $\frac{8322}{5}$ = 1664 amps

For longitudinal magnetization using a coil (either rigid or wrapped cable), where the cross-sectional area of the part is at least one-half (50%) of the coil inner cross-sectional area (high fill factor coils), the number of amperes to be used can be calculated by using the following formula:

Ampere turns = 35,000 ÷ $\frac{\text{(part length)}}{\text{(part diameter)}}$ + 2

Again, the actual amperage setting is determined by dividing the "ampere turn" value by the number of turns.

Example 5. Inspecting a 1.5″ diameter round bar with a total length of 9″ using a five turn coil with an internal diameter of 2″. The fill factor is determined by dividing the coil area by the part area:

Area of coil = 1″ (radius) squared × π = 3.142″

Area of part = 0.75″ (radius) squared × π = 1.767″

Ratio of areas = 1.767 divided by 3.142 × 100 = 56.23% (high fill factor coil)

Ampere turns = 35,000 ÷ $\frac{(9)}{(1.5)}$ + 2

= 35,000 ÷ (6) + 2

= 35,000 ÷ 8

= 4375 ampere turns

Divide by five turns = $\frac{4375}{5}$ = 875 amps

For longitudinal magnetization using a coil where the part cross sectional area is between one tenth (10%) and one half (50%) of the coil inner cross sectional area (intermediate fill factor coils), the number of Amperes can be calculated by using the formula:

Ampere turns = [(Ampere turn value for high fill factor coil) × 10 − (coil cross-sectional area/part cross sectional area) − (8)] + [(ampere turn value for low fill factor coil) × (coil cross-sectional area/part cross-sectional area − 2)] ÷ 8

MAGNETIC PARTICLE TESTING

Example 6. Inspecting a 1.5" diameter round bar with a total length of 9" using a five turn coil with an inside diameter of 4", the fill factor ratio is determined by dividing the coil inner diameter by the part outer diameter:

Area of the coil = 2" (radius) squared × π = 12.566 square inches

Area of the part = 0.75" (radius) squared × π = 1.767 square inches

Ratio = 1.767 divided by 12.566 × 100 = 14.06% (intermediate fill factor coil)

From Example 5 (high fill factor coil), amperage for this part would be 875 amps. From Example 4 (low fill factor coil) ampere turns for this part would be 1500 amps. Insert the above values into the formula to give

$$875 \times \frac{(10 - 7.11)}{8} + \frac{1500 \times (7.11 - 2)}{8} = \frac{875 \times 2.89}{8} + \frac{1500 \times 5.11}{8}$$

$$\frac{2528.75}{8} + \frac{7665}{8} = 316 + 958 = 1274 \text{ amps}$$

Note: Amperages to be used are the calculated amperages ± 10%.

For circular magnetization using prods, the amperage to be used is usually determined by the prod spacing and part thickness. For part thicknesses up to $\frac{3}{4}''$ (19 mm), 90–115 amps per inch (3.5–4.5 amps per mm) of prod spacing should be used. For part thicknesses over $\frac{3}{4}''$ (19 mm), 100–125 amps per inch (4–5 amps per mm) of prod spacing should be used. The prod spacing will usually be between 2" (50 mm) and 8" (200 mm).

For longitudinal magnetization using a yoke. When using an AC yoke, the yoke should be proven to having a field strength capable of lifting a dead weight of at least 10 lbs (45 newtons) with a 2–4" (50–100 mm) spacing between the legs.

When using a DC yoke or a permanent magnet, it should be proven to have sufficient lifting capability to lift a dead weight of 40 lbs (135 newtons) with a 2–4" (50–100 mm) spacing between the legs. Alternatively, a weight of 50 lbs (25 newtons) with a leg spacing of 4–6" (100–150 mm) spacing can be used. It should be noted that there may be different weight requirements depending upon the applicable code or specification.

Note: Some codes require that the leg spacing for the lift test be the same as to be used during inspection.

Caution!

Although the above formulas will produce exact values for required amperages, it should be mentioned that the correct amperage requirements vary based on the applicable specification. Some specifications take into account the variations in flux density due to the differences in permeability between the different steels and also consider the type of current used and the type of meter used to measure the current.

There are specifications that recommend the use of a hand-held gauss meter (such as in Figure 5-43) held against the part surface to determine if sufficient flux density is being generated in the part. This practice does not take into consideration the permeability variations within the part. A part with a high permeability will concentrate more of the flux within itself, producing the high-density flux necessary for acceptable sensitivity. The flux density in the air surrounding the part will be low as a result of this. A low reading will therefore be produced on the gauss meter. If the part were to be replaced with a part with lower permeability, the flux density within the part would not be as high and might

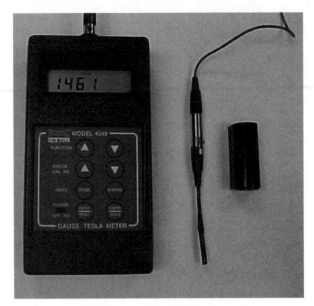

FIGURE 5-43 Hand-held gauss meter.

not be sufficient to produce reliable indications. However, the reading on the gauss meter would be higher due to the higher flux density in the air surrounding the part.

In summary, although code and specification requirements are very detailed, the practice of determining if sufficient flux is being generated is not an exact science. The ideal way to determine if sufficient flux is being generated in the part is to have an identical part with an actual discontinuity of the type, size, and location as those being sought. If this can be reliably found, then the probability of finding an identical or larger discontinuity in the test part is very high.

VI. EVALUATION OF TEST RESULTS AND REPORTING

The evaluation of test results is the most critical stage in the magnetic particle inspection process and greatly relies on the qualifications of the inspector. The ability of an inspector to observe an indication depends upon the correct application of the proper techniques, and using equipment and accessories that are functioning correctly. Another consideration is the inspector's ability to perform the test correctly. Competence should have previously been demonstrated by satisfying qualification requirements. The qualification process should also include a vision examination to prove visual acuity and ability to distinguish the colors associated with the MT technique to be used.

Classification of Indications

Once detected, the indications should be classified as either "false," "nonrelevant," or "relevant" before final evaluation.

False Indications

False indications can be produced by improper handling and do not relate to the part's condition or use. An example is "magnetic writing." This is typically produced by the formation of indications at local poles that are created when the part comes in contact with another magnetized part prior to or during inspection. This can be eliminated by demagnetization and repeating the inspection.

Magnetic writing is most likely to occur when using the residual method, through poor handling that allows the individual parts to touch. The continuous technique may require the demagnetization of parts before the next inspection to preclude the possibility of magnetized components touching. This type of false indication can be eliminated through careful handling.

Other sources of false indications may be caused through the use of excessively high magnetizing currents or inadequate precleaning of the parts to remove oil, grease, corrosion products, and other surface contaminants.

Nonrelevant Indications

These are the result of flux leakage due to geometrical or permeability changes in the part. Examples of geometric causes include splines, thread roots, gear teeth, keyways, or abrupt section changes (see Figure 5-16). A concern with these conditions is that they may also be stress risers and could be the origin for fatigue-induced cracks. These conditions are therefore some of the most critical; the possibility that one of these nonrelevant indications can conceal a crack must be considered. Other potential sources of nonrelevant indications include localized permeability changes in the part, which may be due to localized heat treatment or variations in hardness, and may also occur at the fusion zone of a weld.

Relevant Indications

These are produced by flux leakages due to discontinuities in the part. When these discontinuities are not in compliance with a code, they are classified as rejectable. If they meet the acceptance criteria they are considered to be acceptable discontinuities. Discontinuities that do not permit the part to be used for its original purpose or can potentially cause the part or fail are classified as defects.

Visual Appearance

Generally speaking, **surface** discontinuities will provide sharp distinct indications, which resemble very closely the shape and size of the discontinuity producing the leakage field (see Figure 5-44).

It is not feasible to describe every possible type of discontinuity and its appearance when using the MT method. However, a detailed description of discontinuities and their appearances and causes is contained in Chapter 2.

Subsurface discontinuities will produce weaker, diffuse, and broad indications, which become less well defined as their depth below the surface increases.

Depth Limitations

It is not possible to define at exactly what depth a discontinuity can be detected because of the many variables present. It has been claimed that subsurface discontinuities have been detected as deep as 0.240" (6 mm) from the surface.

Sensitivity varies greatly with depth below the surface. A surface crack will usually be easily detected providing the correct current and flux line direction are employed. It is

FIGURE 5-44   Indication of a quench crack in a saw blade.

reasonable to expect reliable detection of cracks on the order of at least 0.040" or (1 mm) in length.

Reporting

When the parts have been inspected and all indications evaluated, it will be necessary to prepare a report detailing the results of the test and, if applicable, the size, location, and orientation of discontinuities found. This report may vary considerably from company to company, but as a minimum, it should meet customer requirements and should typically include the following data:

1. Contract and customer
2. Inspection company
3. Date of inspection
4. Inspector's name and qualification and certification level
5. Report number or identification
6. Applicable codes, specifications, or procedures, including type and technique of inspection
7. Acceptance criteria
8. Component description, part number, and serial number

9. Flux line direction with respect to discontinuity orientation
10. Other identification details as requested in the contract; for example, batch or order number.
11. Material batch number (particles, liquid carrier, etc.)
12. Results of inspection, including recording of indications as detailed below
13. Signature or inspection stamp of inspector

A detailed, concise report will enable future evaluations by other inspectors. There are several ways of achieving this:

1. A descriptive written report including significant dimensions and indication location can be complied.
2. A photograph may be taken under the correct viewing conditions. Black light illumination in a darkened environment works fine if the correct exposure (usually several seconds) is used and a camera is mounted on a tripod. It may be necessary to vary the exposure time several times to ensure that the best lighting and exposure parameters are met.
3. A free-hand sketch may be used to supplement the data. Again, key dimensions must be included. Unfortunately, not everyone is a good artist, and the quality and usefulness of sketches can be questionable.
4. A piece of transparent tape may be used to lift the indication from the test surface if using dry particles. When peeled off, the tape will retain the shape and size of the indication through the adherence of the particles to its adhesive layer. The tape can then be applied to the report or other suitable background material to render the indication more visible.
5. Aerosol-based, strippable lacquer may be applied in several thin layers, allowing each layer to dry before applying the next, until sufficient thickness exists, allowing the solidified, flexible film to be peeled off the part.
6. Magnetic rubber inspection may be used to create a permanent record of indications. Magnetic rubber inspection involves using a two-part liquid rubber kit consisting of:
    a) Room temperature vulcanizing (RVT) rubber supplied in a liquid form. This liquid rubber also contains ferromagnetic powder.
    b) A catalyst, which when mixed will cause the mixture to solidify at a controlled rate.

   When mixed together, the rubber solution is poured onto the area of inspection and a magnetizing force applied during the "cure" time. The cure rate should be slow enough to allow the ferromagnetic powder to migrate to the flux leakage fields, but not so slow as to delay the inspection longer than necessary. Magnetic rubber materials with curing ranges from 5 minutes to 4 hours are available. When the rubber is solidified, it can be peeled off and the indications created by the flux leakage can be observed and retained as a record.

   The magnetic rubber inspection technique has uses beyond recording indications. It can also be used to inspect areas and surfaces that are inaccessible for standard MT inspection, such as inside blind fastener holes, particularly threaded holes. The rubber can be poured into the hole and a permanent magnet applied across the hole during the cure. After solidification of the rubber, the solid (but, flexible) rubber plug is removed from the hole and the indications from within the hole can be viewed on the outer surface of the plug. A small stick or other device can be cast into the plug to facilitate easy removal.

VII. APPLICATIONS

The magnetic particle test method is effective for the detection of surface and slightly subsurface discontinuities in ferromagnetic parts. It can be used as an inspection tool at all stages in the manufacture and end use of a product.

Stages of a Product's Life Cycle

Discontinuities can occur in any of the following stages of the product's life cycle and ican be classified accordingly.

Inherent
As the metal cools and solidifies to produce the original ingot, numerous discontinuities can be introduced. Inherent discontinuities that can be detected by magnetic particle testing include nonmetallic inclusions (sometimes referred to as stringers") and seams located at the surface.

Primary Processing
When the original material has become solid, it must be worked and formed to produce a rough shaped product. During these processes, the part is said to be in its "primary processing stage." Typical primary processes include, forging, casting, rolling (hot and cold), extrusion, and drawing. During these processing operations, discontinuities may be produced or existing discontinuities may be modified. Examples of primary processing discontinuities detectable by magnetic particle testing include:

- Forging bursts
- Forging laps
- Rolling laps, seams, and stringers
- Rolling seams
- Laminations (at the edges of plates and sheets)
- Casting shrinkage (at the surface)
- Casting inclusions (at the surface)
- Casting cold shuts (at the surface)
- Casting hot tears (at the surface)

Secondary Processing
After rough shaping the metal in the primary processing stage, it is further refined and shaped to produce finished products. This stage is referred to as "secondary processing" and can include such processes as machining, grinding, plating, and heat treatment. These secondary processes may also produce discontinuities or change the appearance of existing ones. Examples of secondary processing discontinuities detectable by magnetic particle testing include:

- Quench and heat cracks
- Grinding cracks (or checks)
- Machining tears
- Plating cracks

Welding

When parts are joined together by a welding process, numerous discontinuities may be created. Examples of the welding processes are found in Chapter 2. The following is a list of welding related discontinuities that may be detected by magnetic particle testing when at or close to the surface.

- Cracks (longitudinal, transverse, and crater)
- Lack of fusion
- Incomplete penetration (if accessible)
- Entrapped slag
- Inclusions
- Overlap or "cold" lap

Service

When the part is put into service and subjected to stresses and environments that may be detrimental to its structure, service discontinuities may result. Those that are detectable by magnetic particle testing include:

- Fatigue cracks
- Stress corrosion cracks
- Static failure cracks (overstressed structures)

Types of Components

Examples of materials, structures, and parts that may be inspected using the magnetic particle test method include:

- Ingots
- Billets
- Slabs
- Blooms
- Bar stock
- Sheet
- Rod
- Wire
- Castings
- Shafts (plain, threaded, or splined)
- Welds
- Bearings and bearing races
- Nuts and bolts
- Gears
- Cylinders
- Discs
- Forgings

- Tubular products
- Plate

Industrial Sectors

Magnetic particle testing has many applications throughout industry including but not limited to the following sectors:

- Petrochemical
- Construction
- Aircraft and aerospace
- Automotive
- Defense
- Nuclear
- Transportation
- Shipping (marine)

VIII. ADVANTAGES AND LIMITATIONS

Advantages

The following are advantages of magnetic particle testing as compared to alternative NDT methods.

1. Test results are virtually instantaneous, in that indications will form within one or two seconds of particle application. No developing or processing times are involved.
2. Permanent records of indications can be produced using photography, magnetic rubber, or transparent tape techniques.
3. MT can be applied "in-situ," without the need for an AC power supply, by using permanent magnets or battery-powered yokes.
4. Indications are easy to interpret.
5. The indications formed by the particles closely represent the shape and type of the discontinuity.
6. Training and experience requirements prior to becoming certified are significantly less stringent than for UT, RT, or ET, since MT is a relatively simple process.
7. MT equipment can be much less expensive than other NDT equipment. Depending on the degree of automation or scale of operation, it may also be more economical than many other NDT methods.
8. Virtually any size or shape of component can be inspected.
9. Inspections can be performed during all stages of manufacturing.
10. Test part surface preparation is less critical than with penetrant testing.
11. MT can be used to inspect through metallic and nonmetallic coatings or plating with some techniques. It should be noted, however, that a reduction in sensitivity will occur as the thickness of the coating increases. Maximum coating thickness should be

established through qualification tests or stipulated in customer specifications or code requirements.
12. There are no known personnel hazards associated with the process because the magnetic fields generated are of short duration; however, the usual electric shock, manual lifting, and chemical (petroleum distillate) precautions apply. Additionally, the parts may become heated during the process if high-amperage current is applied for an extended period.
13. Many parts can be inspected simultaneously if using the residual magnetism technique.
14. MT can be automated for certain production line applications.

Limitations

The following are limitations of magnetic particle testing as compared to other NDT methods.

1. It is only effective for the examination of ferromagnetic materials.
2. Discontinuity detection is limited to those at or near the surface.
3. Demagnetization may be required before, between, and after inspections.
4. Discontinuities will only be detected when their major axis interrupts the primary flux lines. This necessitates inspection in more than one direction to assure discontinuity detection regardless of orientation.
5. Some magnetic particle testing techniques may cause damage to the part as a result of arcing or localized overheating of the parts (for example, when using DC prods).
6. Paint and/or coating removal is necessary from localized areas on the part to facilitate good electrical contact when using direct magnetization techniques.
7. Uniform, predictable flux flow through the parts being tested may not be possible due to complex shapes.
8. Nonrelevant indications due to abrupt changes in component profile or local changes in material properties may make interpretation difficult.

IX. GLOSSARY OF KEY TERMS

Alternating current (AC)—Electric current that flows through a conductor in a back and forth manner at specific intervals. It provides the best sensitivity for the detection of discontinuities at the surface.
Background—The general appearance of the surface on which discontinuities are being sought.
Black light—Electromagnetic radiation in the 320 to 400 nanometer wavelength range, which is invisible to humans. Also see *Ultraviolet light*.
Burning—Local overheating of the component at the electrical contact area arising from high resistance and the production of an arc.
Carrier fluid or liquid—The fluid in which magnetic particles are suspended to facilitate their application.
Circumferential magnetization (circular magnetization)—Magnetization that establishes a flux around the periphery of a part.

Coil method—A method of magnetization in which part or all of the component is encircled by a current-carrying coil. (The use of this term is usually restricted to instances in which the component does not form part of a continuous magnetic circuit for the flux generated).

Current induction method—A method of magnetizing in which a circulating current is "induced" into a ring component by the influence of a fluctuating magnetic field that links the component. Also known as indirect magnetization.

Demagnetization—The process by which a part is returned substantially to the near-unmagnetized state.

Demagnetizing coil—A coil of wires carrying alternating current that is used for demagnetization.

Diffuse indications—Indications that are not clearly defined; e.g., indication of subsurface discontinuities.

Direct current (DC)—Electric current that flows through a conductor in only one direction at all times. DC from a battery source has been phased out in favor of "rectified" forms of AC for surface and subsurface discontinuity detection.

Dry particle technique—The application of magnetic particles in dry form (without the use of a liquid carrier).

Ferromagnetic—Having magnetic permeability, which can be considerably greater than that of air and can vary with flux density.

Fluorescence—The ability to absorb electromagnetic radiation with a wavelength outside the human visible spectrum (black light) and reemit electromagnetic radiation within the white light spectrum, usually in the yellow to green range.

Flux—Invisible lines of magnetic force.

Flux density (β)—Magnetic field strength per unit volume within a ferromagnetic test part; expressed in "gauss." (See also *Tesla*.)

Flux field penetration—The ability to establish and drive high-density magnetic lines of force into the test part.

Full-wave rectified current (FWDC)—Electric current that flows through a conductor in one direction only with an increased rate of pulsation surges and drops at specific intervals. FWDC is recommended for effective surface and subsurface discontinuity detection when using wet-suspension techniques.

Half-wave rectified current (HWDC)—Electric current that flows through a conductor in one direction only with pulsating surges and drops at specific intervals—hence the name half-wave. It is most effective for surface and subsurface discontinuity detection when using the dry particle technique due to the vibratory effect produced on the particles.

Hysteresis loop (and related terms)—See Section II.

Linear indication—Any indication having a length dimension at least three times greater than its width (as defined by some codes).

Longitudinal magnetization—Magnetization in which the flux lines are oriented in the part in a direction essentially parallel to its longitudinal axis.

Magnetic domains—Ferrous material atoms or molecules, normally represented as small bar magnets with north and south poles.

Magnetic particle inspection—A nondestructive test method that provides for the detection of linear, surface, and near-surface discontinuities in ferromagnetic test materials.

Magnetic poles—Those parts of a magnet that are the source of the external magnetic field.

Magnetic writing—Spurious indications arising from random local magnetization.

Magnetism—A form of energy directly associated with electrical current and characterized by fields of lines of force.

Magnetizing force (H)—Magnetic field strength per unit volume in air, measured in oersteds.

Mercury vapor lamp—A bulb used for producing ultraviolet radiation. (See also *ultraviolet light* and *black light*.)

Particle mobility—The ability to impart activity or motion to the magnetic particles applied to the test surface.

Permeability—The ease with which a material can be magnetized. The ability of a material to conduct magnetic lines of force.

Prods—Hand-held electrodes attached to cables that transmit the magnetizing current from the power supply to the part under examination.

Reluctance—The opposition of a material to conduct magnetic lines of force.

Residual magnetism—The magnetism remaining in a part after the magnetizing force has ceased.

Retentivity—The ability of a material to retain magnetism following magnetization.

Tesla—An expression of flux density. 1 tesla = 10^4 gauss or 1 weber/meter^2.

Ultraviolet light—Electromagnetic radiation in the 200–400 nanometer wavelength range. The 200–320 manometer wavelength range contains harmful UV radiation, which can damage the skin tissue or the eyes. It is essential that this be filtered out using the correct black light filter. (See *black light*.)

Weber—1 weber = 10^8 lines of force. Note: 1 meter^2 = 10,000 cm^2; therefore, there are 10,000 flux lines/cm^2 (1 gauss).

X. REFERENCES

Nondestructive Evaluation and Quality Control, Vol. 17. ASM International Handbook, 1992.

Metals Test Methods and Analytical Procedures, Vol. 03.03, *Nondestructive Testing.* American Society for Testing and Materials 1999 Annual Book of Standards.

Lovejoy, D. *Magnetic Particle Inspection—A Practical Guide.* Chapman and Hall, 1993.

CHAPTER 6
RADIOGRAPHIC TESTING

I. HISTORY AND DEVELOPMENT

The history of radiographic testing (RT) actually involves two beginnings. The first commenced with the discovery of x-rays by Wilhelm Conrad Roentgen in 1895 and the second with the announcement by Marie and Pierre Curie, in December of 1898, that they had demonstrated the existence of a new radioactive material called "radium."

The Discovery of x-Rays

One of the true giants in radiography is the man who discovered x-rays—Wilhelm Conrad Roentgen (Figure 6-1). Roentgen was born on March 27, 1845 in Lennep, Germany. His birth house is still in existence and is presently a magnificent museum containing many of the artifacts that he used in his early days of experimenting. It also includes many radiographic devices and interesting radiographs that had been taken over the years as a result of his discovery.

Roentgen was educated in Utrecht and Zurich and ultimately became a Professor of physics at Strasbourg in 1876, Giessen in 1879, and Wurzburg in 1888, where his famous discovery took place. He later became a Professor of physics in Munich in 1899.

There are various accounts as to how x-rays were actually discovered during those days in his experimental laboratory at the University of Wurzburg (Figure 6-2). Most accounts agree that the discovery took place on November 8, 1895, while he was working in his laboratory. He apparently had been working in a semidarkened laboratory room and was experimenting with a vacuum tube, referred to as a Crooke's tube, which he had covered with black photographic paper in order to better visualize the effects that the cathode ray tube produced. As he was experimenting that day, he observed that as the tube was energized, a cardboard coated with barium platinocyanide happened to be lying on a bench not too far from the tube. Even though no visible light escaped from the Crooke's tube because of the dark paper that encompassed it, Roentgen noted that the barium platinocyanide screen fluoresced. He also observed that as he moved the cardboard with the barium platinocyanide coating closer to the tube, it fluoresced with an even higher intensity, which proved that some unknown ray was emanating from the tube. Another account indicates that the cardboard contained a letter "S," which a student had painted on it with barium platinocyanide salts, and the glowing image was actually in the form of that letter "S." Regardless of which of these accounts is factual, it remains that this discovery that occurred on that November day in 1895 had a major impact, not only on industrial nondestructive testing, but also the world of medicine. As a result of this discovery, Roentgen began conducting further experimentations that involved taking radiographs of different objects in his laboratory, including a wooden box containing metal weights, a

FIGURE 6-1 Wilhelm Conrad Roentgen. (Courtesy Charles J. Hellier.)

lock on the laboratory door, and a double-barreled shotgun with different size pellets inside (Figure 6-3). He also x-rayed other materials, such as playing cards, a book, wood, and even some thin sheets of metal. He also observed while he was experimenting that as he placed his hand between the tube and the cardboard containing the fluorescent salts, a faint shadow outline of the parts of his fingers appeared. The actual structure of his bones within his living flesh was observed. Following this came the most historic demonstration of his new discovery, when his wife, who was apparently concerned about all the time he was spending away from home while at the laboratory, came for a visit. It was during this visit that Wilhelm Conrad Roentgen took an x-ray radiograph of his wife's hand (Figure 6-4).

Notice in that early radiograph, that the bones produce a lighter image, since they were higher in density than the surrounding flesh. History does not give any further details as to why Roentgen x-rayed his wife's hand rather than his own. Some suggest that even in those early days, Roentgen was aware of the biological effect that radiation may have had, but that is pure conjecture. In the first days of his discovery, he referred to this unknown ray as "X light," but many of his fellow scientists referred to them as "Roentgen rays." In fact, the expression "Roentgen rays" was quite popular into the early 1900s. In 1902, the term that Roentgen designated for this discovery, "x-rays," began to take hold.

Obviously, Roentgen was very involved with this remarkable finding and on December 28, 1895 (less than two months after first observing x-rays and making the x-ray of

(a)

(b)

FIGURE 6-2 University of Wurzburg. (a) Then and (b) now. Author is standing (and pointing) outside Roentgens' laboratory with Dr. Dietmar Henning. (Courtesy Charles J. Hellier.)

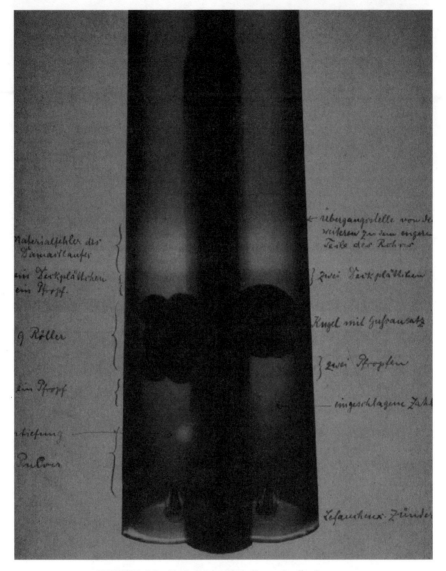

FIGURE 6-3 Early Roentgen radiograph of a shotgun.

his wife's hand) he presented his first official paper entitled, "On A New Kind of Rays" to the Wurzburg Physical Medical Society. He obviously did not celebrate New Year's Day because, on January 1, 1896, he mailed reprints of this paper together with copies of the x-ray of his wife's hand and other radiographs he had at taken to many of his colleagues. As a result of this, the very first news of this discovery appeared in the Sunday edition of the *Vienna Presse* on January 5, 1896. From there, it was wired to London and

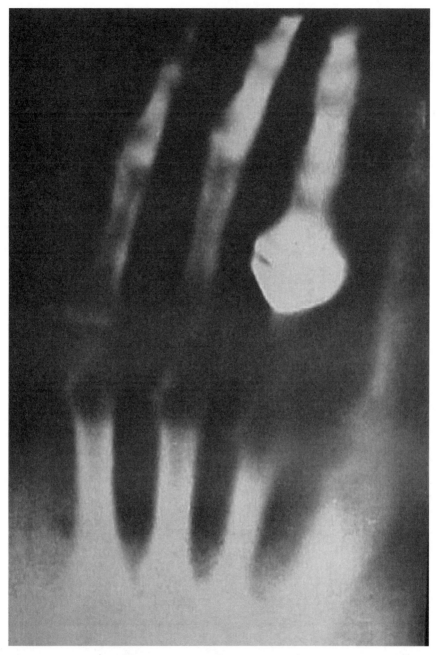

FIGURE 6-4 x-Ray of Mrs. Roentgen's hand.

then on to newspapers around the world. The earliest recorded newspaper article in the United States was in the *Milwaukee Journal* on January 7, 1896. Even in those days, before e-mail and satellite transmission of information, the good news traveled fast. In fact, the *Milwaukee Journal* account was published before it was announced in Roentgen's own hometown newspaper, *The Wurzburg General Anzeiger,* which finally reported the event on January 9, 1896. Roentgen made his first public presentation before the Wurzburg Physical Medical Society on January 23, 1896 and after the lecture, Roentgen took an x-ray of the hand of a famous anatomist named Kolliker who proposed that the new discovery be called Roentgen's Rays.

The first recorded use of x-rays in the United States was on February 8, 1896 when a young man by the name of Eddie McCoughey from Hanover, New Hampshire, was x-rayed for a broken wrist. That first American x-ray was taken by a Dr. Frost and his wife, who was the first head nurse at the Dartmouth College Medical Center.

One might wonder what this all has to do with nondestructive testing, since medical x-ray examinations cannot, theoretically, be considered nondestructive. But if it weren't for this early discovery, many of the early industrial x-ray experiments would have never occurred. In 1901, Roentgen was awarded the first Nobel Prize in science for his work with x-rays (Figure 6-5).

There is so much more to Roentgen's life and how science recognized him, but his desire was not to be in the limelight. He rejected all of the different commercial offers relating to his discovery. There were false reports that it was really his laboratory assistant who was responsible for first observing x-rays, and he had taken all the credit for the discovery. In his later years, Roentgen withdrew from public life. He died on February 10, 1923 of carcinoma of the rectum and was buried beside his wife in the family grave in Giessen. Another remarkable fact regarding Roentgen is that he refused to patent this dis-

FIGURE 6-5 Nobel Prize for Physics awarded to Roentgen in 1901. (Courtesy Charles J. Hellier.)

covery, so that the world could freely benefit from his work. At the time of his death, he was nearly bankrupt as a result of the inflation that followed World War I.

In the first 20 years that followed Roentgen's famous discovery, x-rays were primarily used in the medical arena by doctors who were developing medical x-ray techniques, looking inside the human body while the patient was still alive. In 1912, Laue, Knipping, and Frederick made the first successful experiment using the diffraction of x-rays. Almost immediately after this experiment, W. D. Coolidge of General Electric Co. invented the Coolidge tube, which allowed a much greater intensity of x-rays to be produced and permitted uniform control of the radiation. On the nondestructive testing scene, Dr. H. H. Lester began experimenting with different materials in the early 1920s at the Watertown Arsenal in Watertown, Massachusetts. This laboratory is illustrated in Chapter 1, Figure 1-7.

While some work continued, the importance of x-rays in NDT did not really become fully known, nor the technology widely utilized, until World War II. The importance of industrial radiography was reflected in the first name of the technical society now known as the American Society for Nondestructive Testing. It was founded in 1941 under the name of The American Industrial Radium and X-ray Society. The developments, triggered by World War II, resulted in significant innovation and invention with radiographic equipment that produced more intense beams and higher energies. Figure 6-6 depicts a "mobile" x-ray unit that was capable of producing energies of up to 250 kV. Certainly,

FIGURE 6-6 Early "mobile" 250 kV x-ray unit. (Courtesy Charles J. Hellier.)

the works of Horace Lester and William Coolidge were significant; however, other contributions were made by early scientists and engineers, such as Gerald Tenney, James Bly, Donald O'Connor, Edward Criscuolo, Daniel Polansky, Samuel Wenk, and many others too numerous to name.

The Discovery of Gamma Rays

There are many accounts about Marie and Pierre Curie (Figure 6-7) and there is no doubt that these two physicists led very interesting, unique lives. Marie Sklodowska, as she was known before she was married, was born in Warsaw on November 7, 1867. When she was twenty-four, she traveled to Paris to study mathematics and physics. While there, she met Pierre Curie, who was thirty-five years old, eight years older than Marie. He was an internationally known physicist and a very serious scientist who was dedicated to spending his life doing scientific work. They married in July of 1895 and began to work together on various scientific experiments.

One of their early observations was that thorium gave off the same rays as uranium. They also discovered that there were some reactions coming from materials containing bismuth and barium. When Marie took away a small amount of bismuth, a residue with a much greater activity remained. At the end of June 1898, the Curies had obtained a sufficient amount this substance and proved it was about three hundred times more active than uranium.

They also extracted a substance from pitchblende, similar to bismuth, which they claimed contained a metal never known before. It was suggested that this new, highly active material be called polonium after the country of Marie's origin. In those early days of experimenting, the term "radioactivity" was introduced. Finally, on December 26, 1898,

(a)   (b)

FIGURE 6-7 Pierre (a) and Marie Curie (b).

they produced evidence that there was a very active new material that reacted and behaved chemically, almost like pure barium. At that point, they suggested the name radium be used for this new element.

In order to provide proof that this element did exist, they would have to produce it in a sufficient amount and to determine such characteristics as its atomic weight. In order to do this, they would need a great amount of the pitchblende from which the radium could be extracted. They were able to obtain several tons of pitchblende and the intensive, laborious work of separating the radium from the pitchblende began. Obviously, it was very hard work and Marie Curie was quoted as saying, "Sometimes I had to spend a whole day stirring a boiling mass with a heavy iron rod nearly as big as myself. I would be broken with fatigue at day's end." She worked in a shed with an earth floor (see Figure 6-8), which had a glass roof and did not provide adequate protection against rain. In the summer, it was like a hothouse. And yet, history records that this is where they spent the best and happiest years of their lives. While they were conducting all these experiments, both Pierre and Marie were still teaching, so their days must have been very tiresome and long. Ultimately, from the several tons of pitchblende, they were able to isolate one decigram of almost pure radium fluoride and, at that time, determined the atomic weight of the radium to be 225. Marie presented this work in her doctoral thesis on June 25, 1903. As a result of the discovery of this new element, Marie Curie was awarded the Nobel Prize. She was the first woman to receive this award.

Working closely with radium, which is highly radioactive, and the gas radon, a byproduct, took its toll. Pierre Curie had scarred and inflamed fingers from holding small glass tubes containing radium salts and solutions. He actually conducted medical tests on himself by wrapping a sample of the radium salts on his arm for up to ten hours and then evaluated the results of the exposure, which was in the form of a burn, day by day. After

FIGURE 6-8 Laboratory where the Curies extracted radium from pitchblende. (Courtesy Charles J. Hellier.)

many days, a scar still remained and Pierre suggested that the radium might one day be used for the treatment of cancer. Fifty years after those days of experimentation, the presence of radioactivity was still observed in the laboratory where the Curies had worked so hard, requiring extensive clean-up.

Marie also had started to notice the effects of radiation exposure. Her fingers were cracked and scarred. Both Pierre and Marie were beginning to experience signs of fatigue. They obviously had no idea as to the effect of radiation on the human body at that time. Contrary to the expected, Pierre Curie did not die as a result of his exposure to radiation. He was killed when he was run over by a horse-drawn wagon in Paris on April 1906. This left Marie, who was only thirty-eight years old, alone with two daughters, aged nine and two at the time. Ultimately, Marie Curie would die of leukemia on July 4, 1934.

The work of the Curies marked the beginning of many other developments that would ultimately lead to the radioactive sources that are primarily used in industrial radiography today, namely Iridium 192 and Cobalt 60. The scientific community owes much to the Curies for their early sacrificial work, especially considering the damage that they personally suffered as a result of early exposure to radium.

Before the Curies began their work in isolating and identifying radium, the scientist Becquerel discovered that certain radiations were given off by uranium ore. He was responsible, in some respects, for directing and motivating the Curies to begin their work with radium. Another key scientist from England, Rutherford, was one of the scientists who identified the radiation emitted by certain radioactive elements. He also developed the theory that elements had isotopes and was responsible for identifying the existence of the neutron.

With the advent of nuclear reactors capable of generating high neutron intensities, the possibility of producing artificial isotopes became a reality. Iridium 191 and cobalt 59, both elements existing in nature, are quite stable. When exposed to thermal neutrons, the stable isotopes capture a thermal or slow neutron, becoming one unit heavier in mass. With the addition of this neutron, the iridium 191 becomes iridium 192 and cobalt 59 becomes cobalt 60. Both isotopes are unstable and therefore radioactive. More details about these two artificially produced radioactive isotopes will be presented in Section II. Other key developments in the area of radioactive isotopes involve the equipment used to encapsulate the radioactive isotopes and the projectors or "cameras" that store the isotopes, and from which they are operated.

II. THEORY AND PRINCIPLES

There are many excellent references that contain in-depth information regarding the principles of radiation and radiography. It is not the intent of this chapter to cover principles and theory in great depth but to provide the reader with an understanding of the basics, so that the entire process of producing radiographs with a radiation source will be comprehended. Whether the radiation is emitted from an x-ray tube or a gamma ray source, there are some essential components that apply to the process of radiographic testing.

The first component is the source of radiation. x-Rays emanate from an x-ray tube and gamma rays are emitted by a radioactive isotope. The second component is the test specimen that is to be examined. The third includes the development of the technique. The fourth involves the taking of the radiograph and the processing of the film. The final component, which is extremely critical, is the interpretation of the radiographic image.

Characteristics of Radiation

There are certain unique characteristics relative to radiation that must be understood in order to realize the physics and variables involved with producing a radiograph with the use of a radiation source. These characteristics apply to both x-radiation and gamma radiation. Recall that the only difference between x- and gamma radiation is their origin. x-Rays are produced by an x-ray tube and gamma rays come from a radioactive source that is disintegrating. The following characteristics apply to the radiation that will be used in the nondestructive examination of materials.

1. Radiation is Absorbed and Scattered by Material
There are four common absorption processes that influence the amount of radiation that passes through a part. The four absorption processes are:

1. Photoelectric effect
2. Rayleigh scattering
3. Compton scattering
4. Pair production

The *photoelectric effect* is that process in which a photon of low radiation energy (less than 500 kV) transfers all of its energy to an electron in some shell of the material atom. The energy may simply move an electron from one shell to another, or if there is energy above that required to interact with the orbital electron in the material, it will impart kinetic energy and the electron will be ejected from the atom.

Another name for *Rayleigh scattering* is "coherent scattering" and it is a direct interaction between the photon and the orbital electrons in the atom of the material. However, in this case the photon is deflected without any change or reduction of the kinetic energy of the photon or of the energy of the material atoms. In this process, there are no electrons released from the atom. It is estimated that Rayleigh scattering accounts for no more than about 20% of the total attenuation.

Compton scattering is a direct interaction between photons in the 0.1–3.0 MeV energy range and an orbital electron. In this case, when the electron is ejected from the material atom, only a portion of the kinetic energy of the photon is used. The photon then scatters in a different direction than the direction from which it came and actually emerges with a reduced energy and, therefore, a lower wavelength. Compton scattering varies with the atomic number of the material and varies, roughly, inversely with the energy of the photon.

Pair production is an attenuation process that results in the creation of two 0.51 MeV photons (as a result of annihilation of the electron-positive pair) of scattered radiation for each high-energy incident photon that is at or above an energy of 1.02 MeV. The two 0.51 MeV photons travel in different directions, causing the production of electromagnetic radiation through interaction with other material particles. Energies exceeding 1.02 MeV result in additional kinetic energy being applied to the pair of particles.

These four forms of attenuation or absorption are depicted in Figure 6-9.

Total absorption, therefore, is the combined sums of the four different types of absorption.

2. Radiation Penetrates
The variables relating to the penetration of the radiation can be expressed with the term "half-value layer." The half-value layer is defined as the thickness of a specific material that will reduce the radiation intensity to one-half of that entering the part. If the initial ra-

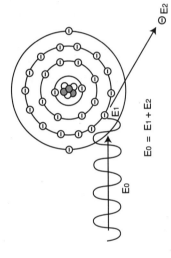

a) Photoelectric Effect

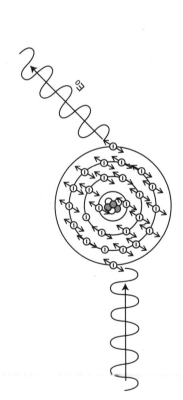

b) Rayleigh or Coherent Scattering

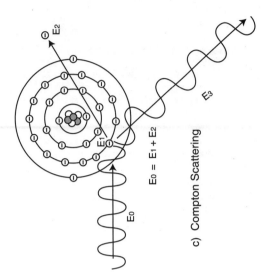

c) Compton Scattering

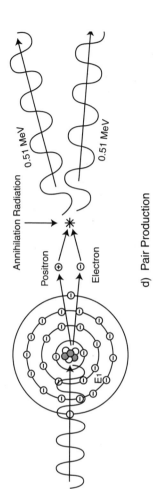

FIGURE 6-9 Forms of scatter.

diation intensity is 100 roentgens (100 R), a material that is exactly one half-value layer will reduce that 100 R to 50 R. Another half-value layer thickness will reduce that 50 R to 25 R, and so on. If this is carried forward, the radiation never reaches zero. Another term used is the "tenth-value layer." The same principle as the half-value layer applies, except that the thickness is somewhat greater and reduces the initial radiation to one-tenth on the opposite side. The factors involved with these half- and tenth-value layers are:

- *Energy.* The higher the energy, the thicker the half- or tenth-value layers. This is supported by the fact that the higher energy radiation produces shorter wavelength radiation, resulting in better penetration. The half- and tenth-value layers are especially useful for calculating the thickness of shielding when designing an enclosure or room for radiography. (See Table 6-1 for examples of half and tenth-value layers).
- *Material type.* The greater the material density, the thinner the half-value layer. Material that is low in density, such as aluminum or titanium, will allow more radiation to pass through, and there will be less absorption and scattering. Materials with higher density, such as steel and lead, provide a much greater chance of interaction because of their higher atomic number. With higher-density materials there will be more absorption and the half-value layer thickness will be considerably less than with the lower-density materials.
- *Thickness.* As mentioned, the half-value layer thickness is specific for a given material and energy. As the thickness of the material increases, the amount of absorption and scatter increases and the amount of radiation that passes through that thickness.

Figure 6-10 illustrates the half-value layer principle, using two different energies with the same test specimen. When there are discontinuities present in the material, the density through the cross section in that area of the test material will also vary, thereby affecting the amount of radiation that passes through the part. It is for this reason that radiography is so effective for the detection of discontinuities, especially those that result in the reduc-

TABLE 6-1 Half- and Tenth-Value Layers

	Half-value		Tenth-value		Half-value	
x-ray (kV)	Lead (inches)	Concrete (inches)	Lead (inches)	Concrete (inches)	x-ray	Steel (inches)
50	0.002	0.2	0.007	0.66	120 kV	0.10
70	0.007	0.5	0.023	1.65	150 kV	0.14
100	0.009	0.7	0.028	2.31	200 kV	0.20
125	0.011	0.8	0.035	2.64	250 kV	0.25
150	0.012	0.9	0.039	2.79	400 kV	0.35
200	0.020	1.0	0.065	3.30	1 MeV	0.60
250	0.032	1.1	0.104	3.63	2 MeV	0.80
300	0.059	1.2	0.195	3.96	4 MeV	1.00
400	0.087	1.3	0.286	4.29	6 MeV	1.15
1000	0.315	1.8	1.04	5.94	10 MeV	1.25
2000	0.393	2.45	1.299	8.60	16 MeV+	1.30
Isotopes						
Iridium 192	0.190	1.9	0.640	6.20		0.60
Cesium 137	0.250	2.1	0.840	7.10		0.68
Cobalt 60	0.490	2.6	1.620	8.60		0.87

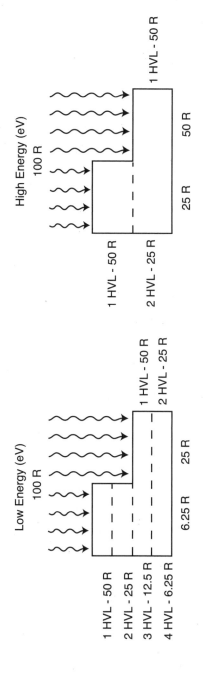

FIGURE 6-10 Half-value layers.

tion of material cross section. The densities of the different discontinuities vary greatly from the base material. If, for example, a weld or casting contains gas pores or other void-type discontinuities, the radiation will pass through these areas with relative ease, creating a higher exposure effect on the radiographic film, resulting in a darker region on the finished radiograph.

3. Radiation Travels in Straight Lines and at the Speed of Light
The speed of light is 186,300 miles per second or 299,800 km per second.

4. Radiation Exhibits Energy
On the electromagnetic spectrum, the wavelengths of x- and gamma rays are much shorter than that of visible light. Recall that as the radiation energy increases, shorter wavelengths are produced that provide greater penetration through the material being examined.

5. Radiation Ionizes Matter
Ionization is a change in the electrical nature or characteristics of matter. Electrons are displaced or knocked out of orbit, thereby changing the electrical balance. This, coincidentally, is what causes the greatest concern to humans. When radiation passes through living tissue, the cells are affected—electrically changed—and damage will result.

6. Radiation is Not Particulate
Radiation has no mass, and even though x- and gamma rays behave like particles, they are actually weightless. To describe the effect of radiation as it interacts with matter, the radiation is sometimes referred to as photons, which is another way of saying high-speed energy traveling at the speed of light that behaves like particles; but in fact, there are no particles in x- or gamma radiation.

7. Radiation Has No Electrical Charge
X- and gamma radiation are not affected by either strong electrical or magnetic fields.

8. x- and Gamma Radiation Cannot be Focused
If x- or gamma radiation is directed toward a glass lens, the radiation would not focus like light does. In fact, the radiation passing through that lens would be absorbed to a greater extent in the thicker portion of the lens and more radiation would pass through the thinner regions of the lens.

9. Humans Cannot Sense x- or Gamma Radiation
Radiation cannot be seen, tasted, or felt. It has no odor and if humans are subjected to radiation exposure, the effects are not realized for a period of time. In other words, the body goes through a latent period before the harmful effects of the radiation exposure are evident.

10. Radiation Causes Fluorescence in Some Materials
Some minerals and salts will fluoresce when subjected to radiation. Fluoroscopy utilizes a fluorescent screen that converts some of the radiation to light. When used in radiography, the light that is emitted from the fluorescent screen creates a much higher exposure effect on the film that is adjacent to it, thereby significantly reducing exposure.

11. Intensity of Radiation
Radiation is expressed in roentgens per hour (R/hr), roentgens per minute (R/min), or milliroentgens per hour (mR/hr). The intensity decreases with distance. This is a function of

the inverse square law, which states that the intensity of the radiation varies inversely with the square of the distance. As the distance from the radiation source is doubled, the intensity is decreased to one-fourth. This will be discussed in more detail in Section IV, Variables.

12. Radiation Produces Chemical Changes in Exposed Film
The film's emulsion is chemically changed through ionization as radiation interacts with it. If it weren't for this characteristic, it would not be possible to produce a radiographic image on film.

Principles of x-Radiography

X-rays are produced in a vacuum tube when high-speed electrons, which are negatively charged, are attracted by a positive potential in the anode and collide with a target material. The electrons are produced when a filament, usually tungsten, is heated to incandescence. The resulting electrons "boil" off and collide with the target material in the anode.

A higher current applied to the filament will result in a greater number of electrons and, therefore, higher radiation intensity. The energy of the radiation is a function of the applied voltage, or kilovoltage to the anode. This causes an increase in velocity of the electrons and, as higher voltage is applied, higher-energy radiation results. The energy of the radiation is its single most important characteristic. It is directly related to the wavelength and to the ability of that radiation to penetrate. The higher the energy, the shorter the wavelength, and the greater the penetration. The target in the anode in most x-ray tubes is typically made of tungsten. Tungsten has a high atomic number, which makes it an ideal material to interact with the high-speed electrons. It also has a high melting point, 5880 °F (3249 °C), which is essential since the target is subject to high temperatures as a result of electron impingement. Since almost all of the kinetic energy from the electrons is transformed or converted into heat, the anode must have the capability of dissipating that heat. In fact, 97–99% of the energy conversion results in the generation of heat and approximately 1–3% is converted to x-rays.

In summary, as the energy is increased, radiation with shorter wavelengths results in greater penetration. On the cathode side of the tube, which contains the filament and a focusing cup, the number of electrons increases with an increase of applied milliamperes, which results in an increase in intensity of the radiation. Figure 6-11a illustrates several typical x-ray tubes and a cross-sectional sketch of a tube can be found in Figure 6-11b.

The target size in the anode is very important, since this affects the image sharpness of the object being radiographed. This is directly related to the definition of the final image. The smaller the target or radiation source, the sharper the resultant radiographic image.

As mentioned above, the energy of the radiation is expressed in voltage or kilovoltage, with one electron volt being the amount of voltage that is necessary to energize and displace one electron. The common expression for x-ray energy is in units of kiloelectron volts (keV), but this is shorted to the kV. A kilovolt is equal to one thousand volts. The majority of industrial radiographic techniques use an energy range from about 100 kV to 400 kV. The intensity of the radiation is expressed in milliamperes (mA) (one milliampere = one one-thousandth of an ampere). In a general way, the energy of the radiation can also be described as the "quality" of the radiation and the intensity (in milliamperes) as the "quantity." Energies above the 400 kV range are considered to be high-energy x-radiation. Equipment producing high-energy x-rays include:

- X-ray tubes with a resonance-type transformer
- Betatrons

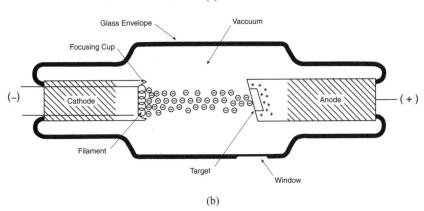

FIGURE 6-11 x-Ray tubes (a) and sketch of cross section sketch (b). (Courtesy Charles J. Hellier.)

- Linear accelerators
- Van de Graaff generators

The older of these high-energy x-ray units are the *resonant transformer* x-ray units, which were developed in the 1940s. Their energy range is from 250kV to 4 MeV. The x-ray tube is symmetrically located within the transformer. The electrons are accelerated to very high velocities by electrodes and focused by a coil as they enter an extension chamber. It is possible to penetrate up to 8″ of steel with this equipment.

Betatrons accelerate the electrons in a circular path by using an alternating magnetic field, and typically produce energies in the range of 10–30 MeV. The tube is doughnut-shaped and the electrons increase their velocity with each orbit until they reach a very high energy. They are eventually guided from their circular path to a target and x-rays are generated.

Linear accelerators produce high-velocity electrons by the use of radio frequency energy coupled to a waveguide and have energies from 5 up to about 25 MeV. These systems contains an electron gun, an accelerator, and a waveguide. The electrons are accelerated along the waveguide by an electrical field.

The *Van de Graaff generators,* also known as electrostatic generators, produce x-ray energies from about 500 kV to as high as 6 MeV. The systems consist of high-speed belts that travel about 4000 feet/minute and build up electrical charges that are carried to a high-voltage terminal. This results in a high electrostatic charge that is used to accelerate the electrons through the tube to the target.

Table 6-2 contains a list of the different energy levels produced by x-ray sources and approximate thickness penetration ranges that would be typical for steel.

Principles of Gamma Radiography

As mentioned earlier, the two most commonly used gamma ray sources in industrial radiography are iridium 192 and cobalt 60. Other sources that have been used to a limited extent in the industrial area are thulium 170 and cesium 137, and of course, radium 226. Table 6-3 contains a summary of these sources with characteristics and data regarding their half-life, energy, and penetration capabilities in steel.

TABLE 6-2 Radiation Energies and Practical Thickness Ranges

x-Ray energy	Thickness range in steel
100 kV	up to 0.100″
150 kV	up to 0.250″
200 kV	0.250″ to 0.500″
250 kV	0.330″ to 0.850″
300 kV	0.350″ to 2.00″
400 kV	0.500″ to 3.00″
1.0 meV	0.800″ to 4.0″
2.0 meV	1.00″ to 5.0″
6.0–10.0 meV	3.00″ to 16.0″
10.0–20.0 meV	10.0″ to 24.0″
15.0–30.0 meV	12.0″ to 30.0″
Sources	
Indium 192	0.500″ to 1.50″
Cesium 137	0.650″ to 2.0
Cobalt 60	1.0″ to 5.0″

TABLE 6-3 Gamma Ray Source Data

Source	Symbol	Atomic number	Atomic weight	Isotope	Half-life	Energy	Emissivity
Cesium	Cs	55	132.91	137	30 years	0.66 meV	4.2 R/C hr @ 1 ft
Cobalt	Co	27	58.9	60	5.3 years	1.17 meV 1.33 meV	14.5 R/C hr @ 1 ft
Iridium	Ir	77	192.2	192	75 days	0.61 meV 0.21 meV	5.9 R/C hr @ 1 ft
Radium	Ra	88	226	226	1,602 yrs.	2.2 meV 0.24 meV	9.0 R/C hr @ 1 ft

There are certain characteristics regarding radioactive isotopes that should be understood. The first is energy. Energy of a radioactive isotope is unique to that particular, specific radioactive material. Some of these radioactive isotopes actually give off a spectrum, or a number of unique energy bundles. For sources, energy never changes for a given radioactive isotope. Radioactive isotopes decay with time. With the decay comes reduction in activity or quantity. The radioactive half-life is the amount of time that it takes for a specific radioactive isotope to decay or be reduced to one half of its original activity. Refer to Table 6-3 for examples of isotope half-lives. Recall that with an x-ray source, the intensity or quantity of radiation is a function of the applied milliamperes. With a radioactive isotope, the activity, or the amount of radiation that is given off by a source, is expressed in curies or becquerels. One curie is equal to 3.7×10^{10} disintegrations per second and one becquerel is equal to one disintegration per second.

One way to understand disintegration is to visualize the atom of a radioactive isotope as having one neutron too many in the nucleus. In time, the nucleus literally breaks apart or disintegrates. In that act of disintegration, a bundle of radiation with a unique energy is given off. In addition, other by-product materials are created. When there are 37 billion of these disintegrations occurring in a unit time of one second, there is one curie of that radioactive isotope. Again, the half-life is the time it takes for the source activity to be reduced to one half of what it was originally. So, if an iridium 192 source is used when new and the activity is 100 curies, at the end of a half-life, which for iridium is approximately 75 days, the activity would be reduced to one-half, or 50 curies. If the source goes through another half-life, it will reduce to 25 curies; another half-life to 12½ curies; and so on. In other words, it does not matter how many half-lives the source goes through; it never reaches absolute zero. For cobalt 60, the half-life is 5.3 years. Its energy is higher than iridium. For iridium 192, there are seven principal energy packets produced during the disintegration process ranging from just over 0.2 to about 0.6 MeV (approximately 200 to 600 kV). With cobalt 60, there are two discrete gamma energies (sometimes expressed as photon energies), 1.17 and 1.33 MeV. With this higher energy, it is possible to examine materials that are somewhat thicker or denser. Another factor to consider when using radioactive isotopes is their specific activity. The specific activity of a source is expressed in curies per gram. The greater the number of curies for a given physical size, the higher the specific activity. The major benefit of a high specific activity source is the improved definition that will result from the high activity (curies) and small physical size.

III. RADIOGRAPHIC EQUIPMENT AND ACCESSORIES

x-Ray Equipment

In general, industrial x-ray equipment can be categorized by energy groups. Those x-ray units that produce energies up to about 125 kV are considered low energy; from 125 up to 400 kV, medium energy; and those systems that produce radiation energies above 400 kV are considered high energy. As mentioned above, the majority of examinations performed today are done in the medium energy range, from 125 kV up to about 400 kV. It is the most widely used for a broad range of applications and this chapter will focus primarily on that energy range.

When energy such as 300 kV is mentioned, it usually denotes the maximum energy output for that particular unit. For any given x-ray unit, the energy can usually be varied from a low energy up to the maximum energy for which the tube is rated. As discussed, x-rays are produced in a vacuum tube, by heating a filament that boils off negatively charged electrons. They are attracted to a positively charged anode, where there are very complex collisions and interactions that produce mainly heat, with only a small amount of the energy converted into x-rays.

Industrial x-ray systems come in different sizes, shapes, and duty cycles. There are x-ray units in cabinets that are ideal for radiographing smaller parts made of low-density material or with limited thickness. These cabinet units typically produce lower energies, usually up to about 150 kV, such as the one in Figure 6-12.

There are x-ray units that are portable, with gas filled tube heads. These are lightweight and are relatively easy to transport for field radiography. Many of the stationary units are self-rectified and continuous duty, such as the *constant potential units*, which can operate continuously for long periods of time. The gas filled tubes operate with partial duty cycles, such as 50%. This means that the x-ray tube can operate for a specific time with an equal amount of time permitted for cool-down before the next exposure. Figure 6-13 illustrates a 300 kV constant potential x-ray system.

Gamma Ray Equipment

As mentioned above, the first gamma ray source used in industrial radiography was radium. Figure 6-14 illustrates the "fish pole" technique in which the radium was transported inside a capsule on the end of a line that was attached to a fishing pole. The radiographer would place the "hook" on the end of the line through an eyelet on the capsule, then remove it from the lead storage pig and carry it to the area where the object to be radiographed was located, as illustrated.

The major component in a gamma ray system is the exposure device, such as the one depicted in Figure 6-15. These exposure devices have shields that contain an S-tube in which the iridium source is maintained inside a stainless steel capsule and is securely locked into place when not in use. The shield material is usually made from depleted uranium, unlike earlier gamma cameras that used shielding made of cast lead. Depleted uranium is slightly heavier than lead and has unusually good radiation shielding capabilities. A gamma ray set-up is illustrated in Figure 6-16, showing the camera with the source drive control, the source tube that is positioned in a source stand just above the pipe weld that will be radiographed. The radiographer "cranks" out the source to the extended position. The predetermined exposure time is allowed to elapse and then the source is cranked back into the camera.

FIGURE 6-12 x-Ray cabinet. (Courtesy Charles J. Hellier.)

Accessories
There are many different accessories that are used in conjunction with taking radiographs. These accessories include, but are not limited to:

- Film and cassettes
- Lead screens
- Lead numbers and letters
- Penetrameters (image quality indicators)

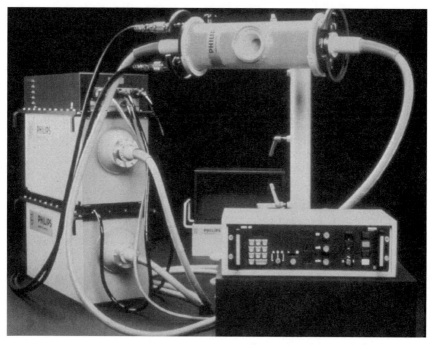

FIGURE 6-13 300 kV x-ray system. (Courtesy Charles J. Hellier.)

FIGURE 6-14 The "fishpole" technique. (Courtesy Charles J. Hellier.)

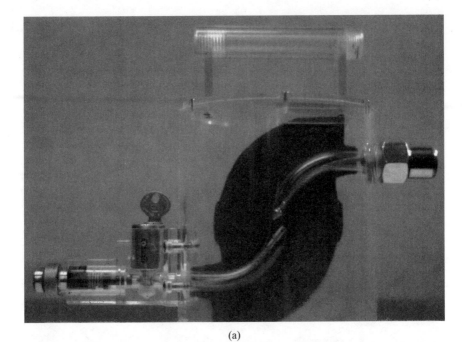

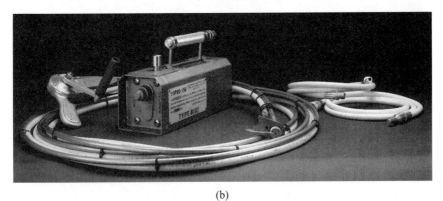

FIGURE 6-15 Gamma ray camera cutaway (a) and actual camera system (b). [(a) Courtesy Charles J. Hellier; (b) used with permission of Source Production and Equipment Company.]

- Densitometers
- High-intensity film illuminators
- Processing equipment (manual or automatic)
- Film hangers (for manual processing)
- Collimators (especially for use with radioactive sources)

FIGURE 6-16 Gamma ray exposure setup. (Courtesy Charles J. Hellier.)

IV. VARIABLES

Of all the nondestructive testing methods described in this book, radiography certainly has the most variables. These variables include:

1. Energy*
2. Exposure time
3. mA (x-ray) or curies (gamma ray)
4. Material type and density
5. Material thickness
6. Type of film
7. Screens used
8. Film processing (procedure development time and temperature, etc.)
9. Film density
10. Distance from the radiation source to the object
11. Distance from the object to the film
12. Physical size of the target (x-ray) or source (gamma ray)

---

*Recall that energy is a function of the applied kilovoltage for x-rays and can be varied up to the maximum for a specific x-ray tube. The energy is fixed with a specific gamma ray source. Also recall that energy is the most important variable, since it affects the contrast (to be discussed later), and determines the thickness of material that can be effectively examined. It also influences the amount of scatter, which affects the definition or sharpness of the image.

In order to control these variables so that the benefits can be maximized for each one, a technique chart should be used. Unfortunately, there are still many radiographs taken by the "trial and error" technique. Some radiographers will take three, four, or more exposures using different techniques, then, after processing the film, will decide which image looks best. The best way to produce a high-quality radiograph every time is through the use of exposure charts. An example of an exposure for a 270 kV x-ray machine is illustrated in Figure 6-17. Notice that all of the variables are addressed in this chart. The use of this chart is quite simple. It can be easily understood by following a few logical steps. (*Note:* This chart is for demonstraion purposes only and will not necessarily work for other x-ray machines of similar energy ratings.)

Step #1. Verify the material type (steel in this case) and find its thickness on the horizontal axis.

Step #2. Project a straight line vertically from that thickness up to the top of the technique chart and notice that there are a number of energies that can be used. If the material were one inch thick, for example, the energies that could be used would range from 160 kV to a maximum of 270 kV. It would not be advisable to use the lowest energy, since the exposure time would be unusually long. If, for example, 160 kV were to be used, the exposure time would be 50 minutes if a tube current of 10 mA were used (30 M or 30 thousand milliampere seconds divided by 10 mA = 3000 seconds divided by 60 = 50 minutes). This is done by projecting a line horizontally from the thickness/kV intersect point to the vertical. The other problem with this technique is that a greater amount of scatter will result due to the longer wavelengths that occur at the lower energies coupled with the long exposure time. Scatter reduces the clarity or sharpness of the image. If the highest energy were used (270 kV), the exposure time would be 22 seconds (220 milliampere seconds divided by 10 mA = 22 seconds). This would not be advisable, since high-energy techniques result in lower-contrast images. Contrast is the difference in density between two adjacent regions of the part, which are different in thickness or material density. So, it is not advisable to use the lowest or the highest kV based on the technique chart. The obvious choice would be to select energy somewhere in the middle. For this example, either 200 or 220 kV would be good choices.

Step #3. As explained in Step #2, once the energy is selected, a line is projected to the left vertical axis, where the exposure time and intensity are combined in units of milliampere seconds (mAs). Again, mA is the current applied to the filament that controls the quantity of radiation. Time, in this case in units of seconds, and mA are inversely proportional values. If, for example, a radiograph were taken with an exposure of 5 mA at 240 seconds (4 minutes), the exact same image would be produced if 10 mA were used for 120 seconds (2 minutes), a reduction in exposure time of 50%. In each case, there is an exposure time/intensity value of 1200 milliampere seconds (20 milliampere minutes). So, at the intersection of 1″ and the 200 kV curve, project over to the vertical axis, and the exposure time in milliampere seconds will be 1400 mAs (2 minutes 20 seconds), and at 220 kV, the exposure time would be about 650 mAs (1 minute 5 seconds). The exposure times were calculated using a tube current of 10 mA. Notice that with just an increase of 20 kV, there is virtually a 50% reduction in exposure time. The resultant image quality would be virtually the same using either of these kV settings.

A similar exposure chart for iridium 192 is illustrated in Figure 6-18. As with the x-ray technique chart, this is an example and may not provide exact results. Notice that there are no energy curves since energy is a fixed value for a given gamma ray source. In this technique chart, there are three different film types having different speeds. The way to establish the correct exposure time is similar to that of the x-ray technique chart. The first step (assuming the material is steel) is to project the material thickness vertically un-

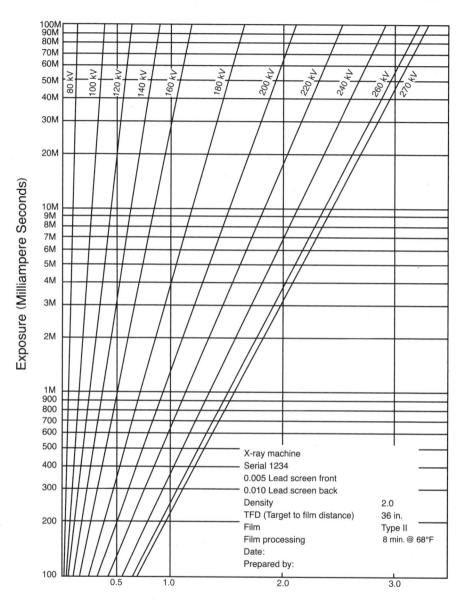

FIGURE 6-17 x-Ray technique chart.

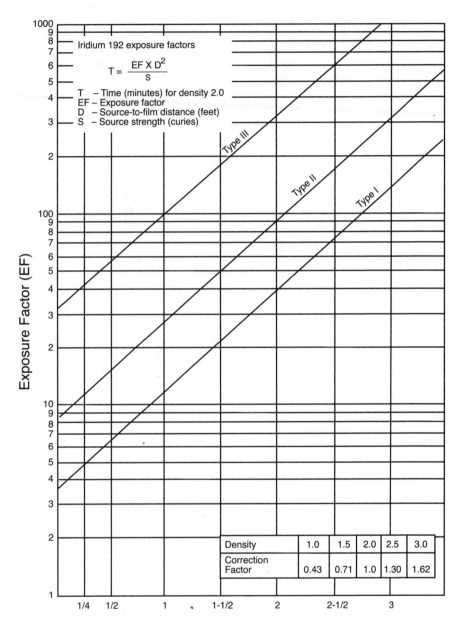

FIGURE 6-18 Gamma ray technique chart.

til it intersects with the film type being used. Then draw a line horizontally to the vertical axis. From that axis, the exposure factor (EF) is determined. Using the equation at the top of the chart and knowing the number of curies of iridium 192, an exposure time in minutes can be easily calculated.

Example:
Thickness = 1.5"
Type II film will be used
D (source to film distance) = 36" or 3'
S = 60 curies
T = exposure time in minutes (for a film density of 2.0)

Projecting vertically up the chart, the point of intersection with the Type II film is then projected horizontally. The EF (exposure factor) is 50.

$$T = \frac{50 \times (3')^2}{60}$$

$$T = \frac{450}{60}$$

$$T = 7.5 \text{ minutes}$$

If the material to be examined is other than steel, exposure times can be easily calculated by using the radiographic equivalent factors (REFs). Some REFs are illustrated in Table 6-4. (*Note:* It should be mentioned that these factors are approximate but will be sufficient to enable a useable technique to be established.) In order to obtain an equivalent thickness of a material other than steel, the REF should be multiplied by the material thickness. The relevant answer will be the equivalent thickness of steel. Once this thickness is known, the technique chart can be used as if the material were steel.

Example:
Material is aluminum
Thickness = 5.0"
Energy to used is 220 kV

TABLE 6-4 Radiographic Equivalence Factors

	x-rays (kV)							Gamma rays		
	50	100	150	220	400	1000	2000	Ir 192	CE-137	CO-60
Magnesium	0.6	0.6	0.05	0.08				0.22	0.22	0.22
Aluminum	1.0	1.0	0.12	0.18				0.34	0.34	0.34
Titanium		8.0	0.63	0.71	0.71	0.9	0.9	0.9	0.9	0.9
Steel	)	12.0	1.0	1.0	1.0	1.0	1.0	1.0	1.0	1.0
Copper		18.0	1.6	1.4	1.4	1.1	1.1	1.1	1.1	1.1
Zinc			1.4	1.3	1.3	1.1	1.0	1.1	1.0	1.0
Brass			1.4	1.3	1.3	1.2	1.2	1.1	1.1	1.0
Lead			14.0	12.0		5.0	2.5	4.0	3.2	2.3

Note: Aluminum is taken as the standard metal at 50 KV and 100 KV, and steel at the higher voltages and with gamma rays.

The REF obtained from the chart for aluminum at 220 kV is 0.18". So:

$$REF_{AL} \times \text{material thickness} = \text{equivalent thickness in steel}$$

$$0.18 \times 5.0" = 0.9"$$

Therefore, a technique for 0.9" of steel from the technique chart can be used for the 5.0" of aluminum.

In the event there is reason to use a film other than Type II, film characteristic curves (Figure 6-19) can help to determine the new exposure time. (*Note:* These characteristic curves are examples only. Film manufacturers can provide accurate curves that can be used to calculate precise correction values.) Notice that for gamma rays, the technique chart (Figure 6-13) has provisions for three different films.

Example:
If it is desired to obtain a radiograph with better definition, a change from the Type II film to a Type I would be appropriate. The 2.0 density (vertical axis) intersects with the Type II film at 1.9 (horizontal axis), which is the log relative exposure. Continuing on the 2.0 film density line over to the Type I film will result in a log relative exposure of approximately 2.55 or a difference of 0.65 in the log relative exposure. The inverse log of 0.65 is 4.5 (rounded off). The new exposure time would be calculated by multiplying the correction factor of 4.5 by the original exposure time used to achieve the 2.0 film density. If the original exposure was 3 minutes, the new exposure time required to achieve a film density of 3.0 would be 13 minutes, 30 seconds.

This film characteristic curve approach can also allow a given density to be changed to any desired density. For example, on the technique chart for x-rays, the resultant film density will be 2.0. For many codes and specifications, a film density of 2.0 is the minimum permitted. It would be much more desirable to produce a radiograph with a higher density. Higher densities produce better radiographic contrast. So, in order to achieve a film density that is different than the one stipulated on the technique chart, the film characteristic curves can be used.

Example:
Using a Type II film, a density of 3.0 is desired instead of the 2.0 as indicated on the technique chart (Figure 6-19). As in the previous example, the log relative exposure for a Type II film with a density of 2.0 is 1.9. The 3.0 density intersects the Type II film at approximately 2.1. The difference between the 2.1 and 1.9 log relative exposures is 0.2. The inverse log of 0.2 equals 1.6 (rounded off). If the original exposure time were calculated from the technique chart to be 3 minutes (for the 2.0 density), that time would be multiplied by the correction factor of 1.6, for a new exposure time of 4.8 minutes. If all other variables are the same as with the original exposure, a film density of 3.0 will be achieved if the new exposure time of 4 minutes, 48 seconds is used. It should be noted that correction factors are included on the technique chart for iridium 192.

In addition, a specific type of film with one density can be changed directly to another film type with a different density using the same approach.

Example:
If the Type II film and the film density as indicated on the x-ray technique chart is to be changed to a different film (Type I for example, with a density of 2.5), it can be accomplished as in the previous examples. As seen, the Type II film with the 2.0 density has a log relative exposure of 1.9. The Type I film with a 2.5 density has a log relative exposure value of approximately 2.65. The difference in log relative exposures is 0.75 and the inverse log is about 5.6. Multiplying the original exposure time by the correc-

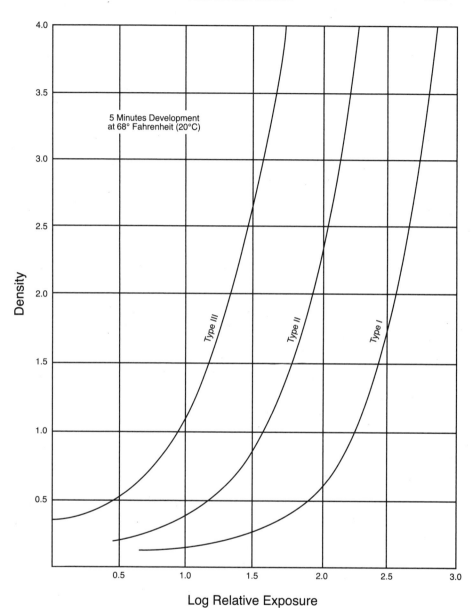

FIGURE 6-19 Film characteristic curves.

tion factor of 5.6 will give a suitable exposure time to achieve a film density of 2.5 with Type I film.

Another variable to be considered is the source to film distance (SFD). Occasionally, the source to film distance is also referred to as the target to film distance (TFD). The TFD generally applies when using an x-ray source and the SFD applies when radioactive isotopes are used. This distance is very critical and plays a major role in establishing the correct technique. Notice on the x-ray technique chart that the TFD is 36 inches. There may be occasions when the 36 inch TFD is not possible because of geometric limitations and a shorter distance may be necessary. On the other hand, the TFD may be increased, which would result in an image with better definition. There is a mathematical relationship between the exposure time and the distance. The technique chart, if used properly, will provide the correct exposure time at a distance of 36 inches. If a different distance is used, a simple equation that gives the relationship between time and distance should be used. The results are very precise and enable the radiographer to essentially use any TFD (x-ray) or SFD (gamma ray) that is appropriate for the object that is being radiographed. The equation for this calculation is:

$$\frac{T_1}{T_2} = \frac{D_1^2}{D_2^2}$$

where:
T_1 = original exposure time derived from the technique chart
T_2 = new exposure time
D_1 = original distance
D_2 = new distance

Example:
A new distance of 48″ is desired to achieve better definition. The original exposure time at 36″ was 3 minutes. The exposure time required at the new distance can be calculated as follows:

$$T_1 = 3 \text{ minutes}$$
$$D_1 = 36″ \, (3')$$
$$D_2 = 48″ \, (4')$$
$$T_2 = x$$

$$T_2 = 3 \text{ minutes} \times \frac{(4)^2}{(3)^2}$$

$$T_2 = 3 \text{ minutes} \times \frac{16}{9}$$

$$T_2 = 5.33 \text{ minutes or 5 minutes and 20 seconds}$$

The other variable that all radiographic personnel must be aware of is the inverse square law. This is important as it applies to technique development but also, more importantly, as it relates to safety. As mentioned, the intensity of radiation varies inversely proportional to the square of the distance and is expressed mathematically in the following equation:

$$\frac{I_1}{I_2} = \frac{D_2^2}{D_1^2}$$

where:
I_1 = intensity at distance D_1
I_2 = intensity at distance D_2
D_1 = initial distance
D_2 = new distance

Image Quality

By optimizing the variables that have been discussed, it is feasible to achieve a radiographic image with a high quality level. The goal of the radiographer should be to achieve the highest possible quality image of the object that is being radiographed. The quality of the radiographic technique is a established by comparison with the image of a "image quality indicator"(IQI), also referred to as a penetrameter, on the completed radiograph. There are two major types of penetrameters in use today: the shim and wire types. The "shim" or hole type contains a lead identification number at one end and three holes of different diameters. The thickness of the shim type penetrameter is based on a percentage of the material thickness that is being radiographed. The most common penetrameter thickness is 2% of that thickness. The three holes as seen in Figure 6-20a have different diameters, which are based on the penetrameter thickness.

The smallest hole located in the center is referred to as the 1T. The "T" in the penetrameter designation represents the penetrameter thickness. The 1T hole, therefore, will have a diameter equivalent to the thickness of the penetrameter, which again is usually 2% of the thickness of the material being radiographed. The hole diameter at the end opposite to the lead identification numbers is referred to as the 2T hole. Its diameter will be two times the penetrameter thickness. The largest hole is the 4T hole and that diameter will be four times the penetrameter thickness. The outline of the penetrameter and the image of the holes on the finished radiograph establish the quality level for that radiographic technique. It is unfortunate that the hole sizes are, at times, used as a comparison to estimate the size of discontinuities that are detected. This is not the intent of the penetrameter. For example, if the diameter of a 2T hole is a 0.020", a discontinuity of that size will not necessarily be detected. Penetrameter holes are very precisely manufactured into the penetrameter shim. They have sharp edges. Discontinuities are quite different from these holes. The penetrameter is used to establish the quality level of the radiographic technique and should not be compared to discontinuities. The typical designations for radiographic quality levels (RQL) using the shim type penetrameters will be a combination of two numbers and a letter. For example, the most common quality level is 2-2T. This designation means that a penetrameter whose shim is 2% of the material being radiographed, must display the 2T hole (which is two times the thickness of the penetrameter) clearly and distinctly on the radiograph. When the 2T hole is clearly discernible, the 4T hole, which is larger, should also be discernible. The 1T may, or may not, be clearly visible. If a 2-1T quality level is specified, the 1T hole must be clearly discernible in a 2% thick penetrameter shim. A 2-4T quality level will require the 4T-hole image to be discernable in the radiograph. Although the standard penetrameter shim thickness is typically about 2% of the material being radiographed, a 1% shim thickness may also be specified. This means that the penetrameter thickness is 1% of the material being radiographed. This designation is used when a higher level of quality is required in the radiographic technique. A listing of common shim type penetrameter designations is found in Table 6.5.

The second most commonly used image quality indicator is known as the "wire" penetrameter. The wire penetrameter has been in use for decades in Germany and other Euro-

(a)

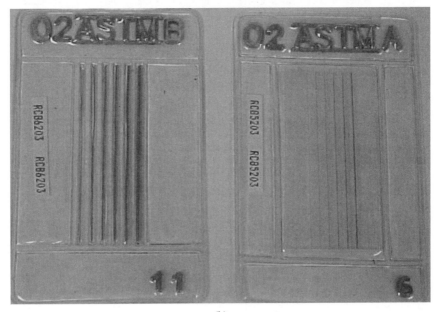

(b)

FIGURE 6-20 Hole-type penetrameter on a shim (a) and wire penetrameters (b). (Courtesy Charles J. Hellier.)

pean countries. In fact, the first wire penetrameter was developed in Germany and was known as the DIN (Deutsche Industrie-Norm) type penetrameter. Typical ASTM wire penetrameters are illustrated in Figure 6-20b. Notice that they contain a series of six wires encased in a clear plastic holder with lead identification numbers. Each of the six wires is of a slightly different diameter. There are four general wire sets described in Table 6-5.

Both shim and wire type IQIs must be placed on the source side of the object being radiographed whenever possible. There may be occasions when it is impossible to put the penetrameter on the source side due to geometric considerations or accessibility and, in those cases, some codes permit them to be placed on the film side, between the object being radiographed and the film. When a film side penetrameter is used, the required penetrameter designation is less than stipulated for the source side, since the quality level is somewhat easier to achieve, due to the closeness of the penetrameter to the film.

There are a number of other different designs and types of penetrameters; however, the shim and wire types are the most widely used. There is also a step-hole type that is used, primarily in France. There are codes today that permit the use of either the shim-type penetrameter or the wire-type penetrameter and their equivalence has been proven. Table 6-6 makes a comparison between the shim-type penetrameter and the wire-type penetrameter for the purpose of comparable quality levels.

The importance of achieving the highest possible quality level cannot be stressed enough. Those variables affecting quality level are listed in Table 6-7.

Note that the two major factors that influence sensitivity are contrast and definition. Contrast is the difference in density on two adjacent regions of the radiograph. Definition is the sharpness, or clarity, of the image as has been mentioned. Contrast is affected primarily by energy. Other factors, such as film type, differences in thickness of the specimen, density of the test object, and the degree of scatter radiation will also influence contrast. Placing a filter at the source of radiation also affects contrast. On the other hand, the sharpness of the object's image is mostly controlled by geometric considerations such as the size of the source of radiation, the source to object distance and the object to film distance. The type of screens that are used and their contact with the film will also affect sharpness. In fact, many codes stipulate that the unsharpness, or lack of definition in a radiographic technique, be calculated by using the unsharpness equation:

$$Ug = \frac{F \times t}{d}$$

where:
Ug = geometric unsharpness
F = the physical size of the target or source of radiation (*Note:* For some codes the physical size is defined as the maximum projected dimension, i.e., the diagonal.)
t = the distance from the source side of the test object (area of interest) to the film
d = the distance from the source to the op of the test object (area of interest)

At times, this equation may be used improperly. The unsharpness equation should be used to verify or to assure that the image sharpness meets the various code and specification requirements. Often, radiographers will use this equation to calculate the absolute minimum source to film distance that can be used and still meet the code required unsharpness values. In fact, in far too many cases the goal is to just "barely meet" the quality level requirements. It seems that the variables that directly reduce the quality of the image are those that some radiographers will employ to take the radiographs in the shortest practical time. For example, it should be obvious that when the source to film distance is greater, the image will be sharper, but this requires a longer exposure time. Many radiographers will use the absolute minimum source to film distance to reduce the exposure time. The

TABLE 6-5 Shim and Wire Type Penetrameter Data

Wire Penetrameter			
Set A		Set B	
Wire diameter in. (mm)	Wire ID No.	Wire diameter in. (mm)	Wire ID No.
0.0032 (0.08)[a]	1	0.010 (0.25)	6
0.004 (0.1)	2	0.013 (0.33)	7
0.005 (0.13)	3	0.016 (0.4)	8
0.0063 (0.16)	4	0.020 (0.51)	9
0.008 (0.2)	5	0.025 (0.64)	10
0.010 (0.25)	6	0.032 (0.81)	11
Set C		Set D	
Wire diameter in. (mm)	Wire ID No.	Wire diameter in. (mm)	Wire ID No.
0.032 (0.81)	11	0.10 (2.5)	16
0.040 (1.02)	12	0.126 (3.2)	17
0.050 (1.27)	13	0.160 (4.06)	18
0.063 (1.6)	14	0.20 (5.1)	19
0.080 (2.03)	15	0.25 (6.4)	20
0.100 (2.5)	16	0.32 (8)	21

Shim (Hole) Penetrameters				
Penetrameter designation	Penetrameter thickness	1T Hole diameter	2T Hole diameter	2T Hole diameter
5	0.005	0.010	0.020	0.040
7	0.0075	0.010	0.020	0.040
10	0.010	0.010	0.020	0.040
12	0.0125	0.0125	0.025	0.050
15	0.015	0.015	0.030	0.060
17	0.0175	0.0175	0.035	0.070
20	0.020	0.020	0.040	0.080
25	0.025	0.025	0.050	0.100
30	0.030	0.030	0.060	0.120
35	0.035	0.035	0.070	0.140
40	0.040	0.040	0.080	0.160
45	0.045	0.045	0.090	0.180
50	0.050	0.050	0.100	0.200
60	0.060	0.060	0.120	0.240
80	0.080	0.080	0.160	0.320
100	0.100	0.100	0.200	0.400

[a]The 0.0032 wire may be used to establish a special quality level if agreed upon between the purchaser and the supplier.

TABLE 6-6 Penetrameter Comparison

Shim (hole) type penetrameter designation	Wire diameter equivalence to hole diameter		
	1T	2T	4T
5	—	—	0.0006
6	—	0.004	—
8	0.0032	0.005	0.008
10	0.004	0.006	0.010
12	0.005	0.008	0.013
15	0.006	0.010	0.016
17	0.008	0.013	0.020
20	0.010	0.016	0.025
25	0.013	0.020	0.032
30	0.016	0.025	0.040
35	0.020	0.032	0.050
40	0.025	0.040	0.063
50	0.032	0.050	0.080
60	0.040	0.063	0.100
70	0.050	0.080	0.126
80	0.063	0.100	0.160
100	0.080	0.126	0.200

TABLE 6-7 Radiographic Quality Variables

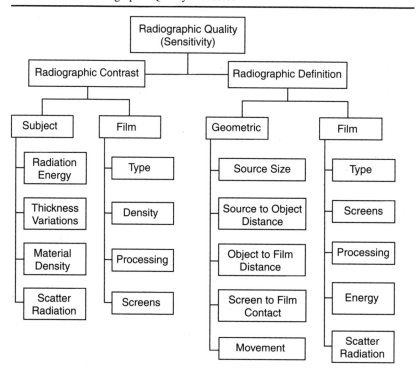

highest energy available with an x-ray unit should not be used because contrast is reduced. In too many cases, the higher energy is chosen and the resultant radiograph does not have that high-contrast image that makes interpretation much more meaningful. Radiographic film is another area where high quality is sacrificed for speed. Fast film, which results in images with noticeable graininess, is used in preference to the slow film in order to achieve the shortest exposure time. The slow, fine-grain film, which requires longer exposure times, produces a much sharper image with better definition. When all these factors are combined (a fast film with a short source to film distance, and the highest energy used), the resultant image will fall far short of the sensitivity goal that can be achieved. The goal should be to always use a technique that will provide the highest possible quality image.

Lead screens have a major effect on the quality level, primarily affecting definition. Lead screens serve two very important functions in radiography: they intensify the primary radiation and they absorb "soft" or secondary scatter radiation. Lead screens are typically 0.005 inches or 0.010 inches in thickness. Based on the energy of the radiation, either will be used. One lead screen will be placed on top and the other beneath the film, inside a light-tight cassette. The purpose of the top lead screen is to intensify the primary radiation that passes through the object. This intensification occurs as a result of the emission of secondary electrons from the interaction of the photons with the atoms in the lead. This intensification actually creates a sharper, denser image on the film because of this secondary emission. The lead screens also serve to absorb scatter radiation. These lead screens easily absorb scatter, which consists of long-wavelength, lower-energy radiation, whereas the primary radiation passes through with relative ease. The primary source of scatter radiation is from the object itself, and results when the radiation passes and interacts with the material atoms. Again, radiation will penetrate the object, be scattered, and be absorbed. With lead screens, the scattered radiation is absorbed to a great extent and the primary beam is intensified.

The lead screens under the film also improve the image quality by absorbing backscattered radiation, which comes from objects such as floors and tables, etc. underneath the film cassette. Lead screens are effective for both x-rays and gamma rays with energies above approximately 125 kV. When lower energies are used, the lead screens tend to absorb a portion of this lower-energy radiation rather than intensifying it. For gamma radiography using iridium 192, it is quite common to use 0.010 inch lead screens on the top and underneath the film.

Any type of contaminant or blemish on the lead screen surfaces, such as the lower screen in Figure 6-21, will cause corresponding images of that condition on the radiographic film, so it is extremely important that these lead screens be kept in excellent condition. They should be virtually mirror smooth and should be inspected periodically to assure that there are no conditions that would cause artifacts. It is recommended that a small, unique identification number be scratched into the lead at one of the corners. In the event that there is an artifact, it will be easy to see that number as an image on the film, facilitating the identification and removal of the faulty lead screen.

Other types of screens used in radiography are the fluorescent types. Fluorescent and fluorometallic screens result in a significant reduction in exposure time since the radiation causes visible light to be emitted that greatly increases the exposure effect on the film. The major problem with fluorescent screens is that light diffusion occurs, which reduces the definition of the image. It is however, an excellent way to radiograph test specimens that are either extremely thick or dense, where conventional radiography with lead screens would be impossible. Quality levels that would meet most code requirements are difficult to achieve with the use of fluorescent screens. This again is due to the light diffusion of the fluorescent material in the screens.

FIGURE 6-21 Lead screens (screen on bottom is unacceptable). (Courtesy Charles J. Hellier.)

V. TECHNIQUES AND PROCEDURES

The most effective technique used in radiography is one in which the radiation passes through a single thickness and the film is in contact with the surface opposite the source side. This technique is referred to as the "single wall exposure, single view technique" (Figure 6-22). The radiation passes through one wall of the object (a single thickness), and that thickness is evaluated or "viewed."

The second most common technique is referred as the "double wall exposure, single view technique." In this case, the radiation passes through two walls but only that area closest to the film is evaluated. This technique is depicted in Figure 6-23.

The third radiographic technique requires the radiation to pass through both walls of the object and both walls are evaluated. This technique is referred to as the "double wall exposure, double view technique" and is usually restricted to parts with small diameters, typically equal to or less than 3.5″. This technique is illustrated in Figure 6-24. In this technique, the source of radiation may be positioned directly over the area of interest, thus superimposing the top portion with the region directly under it. As an alternative, the source may be offset by an angle of approximately 15° in order to observe both the top and bottom walls. When the source is offset, the technique is often referred to as the "ellipse" or "elliptical" technique.

In the single wall exposure, single view technique, the penetrameter is placed on the source side of the object (the surface opposite the film). In the double wall exposure, single view technique, the penetrameter may be on the source side of the object or the film

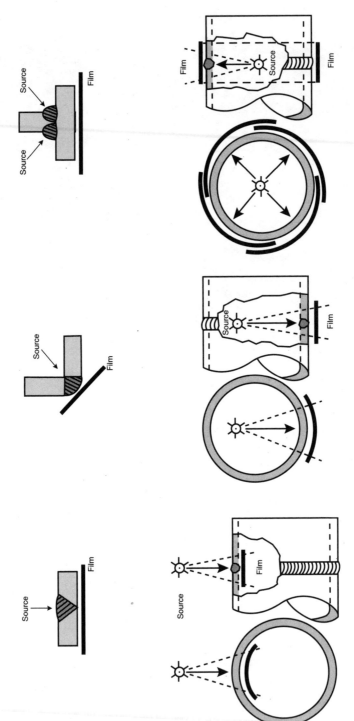

FIGURE 6-22 Single wall exposure, single view technique.

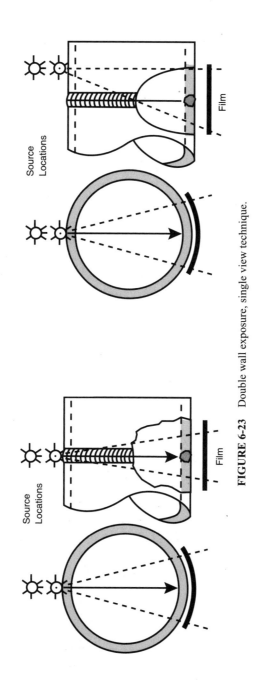

FIGURE 6-23 Double wall exposure, single view technique.

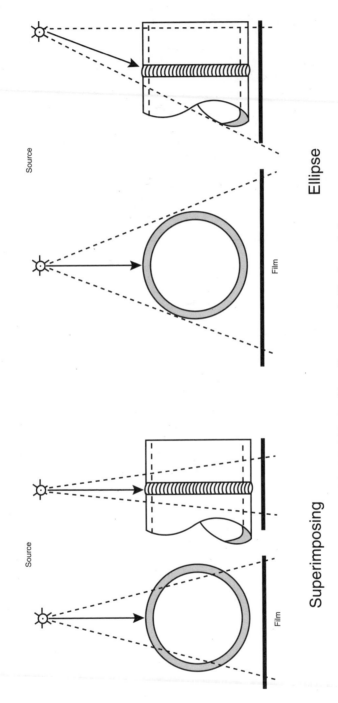

FIGURE 6-24 Double wall exposure, double view technique.

side, depending upon accessibility. The double wall exposure, double view technique usually permits the penetrameter to be placed on the source side.

In all of these techniques, the goal is to achieve the highest possible quality level. The penetrameter placement should be as close to the area of interest as possible. If the object being radiographed is a weld, the penetrameter should be as close as possible to the weld, without interfering with the weld image. In the case of castings, it may be necessary to place the penetrameter in the area of interest, depending upon the size and shape of the casting. In other cases, the technique may permit an object of similar radiographic density and thickness to be placed adjacent to the test object. This will serve as a location for the penetrameter to be placed. If the weld has reinforcement, it may be necessary to use a shim of radiographically similar material upon which the penetrameter is placed, as illustrated in Figure 6-20a. The purpose of the shim is to provide the penetrameter with a through thickness that will closely approximate the weld thickness, including the reinforcement. This is especially important when the codes and specifications require that the density range in the area of interest have a density within a certain percentage of the density through the penetrameter image. More details on this will be found in the section on density (page 6.47).

With all of these techniques, it is necessary that there be some form of "shooting sketch" that indicates the key aspects of the technique, such as the source location, film location, penetrameter placement and test part position. This serves as a great aid to the radiographic film interpreter who may not be totally familiar with the radiographic technique used.

One other important aspect of radiography has to do with the source placement. In far too many cases, the radiographic technique is designed to show the best possible image of the penetrameter. It seems that the goal in many cases, is to produce a good "picture" of the penetrameter and the essential hole, whereas the goal should be to display the best images of discontinuities, especially those that may not be oriented in a direction favorable to the radiation source. Radiography is extremely sensitive to the orientation of tight planar discontinuities. If a tight planar discontinuity is expected to be at an angle to the source of radiation, it will be difficult if not impossible to detect. These aspects of discontinuity orientation should be taken into consideration when developing the technique. Since the purpose of radiography is to detect discontinuities, the nature, location, and orientation of the expected discontinuities should always be a major factor in establishing the technique.

Procedure

A typical procedure is outlined below to provide guidance, especially when radiography of a specific object is being attempted for the first time. The recommended steps in a radiographic procedure are:

1. Understand the codes, specifications, and customer requirements thoroughly
2. Develop a technique based on the thickness and type of material
3. Prepare a shooting sketch
4. In the darkroom, carefully place the radiographic film in the cassette with the proper lead screens
5. Place the film under the area of interest
6. Ensure that the correct source to film distance is being employed
7. Place the appropriate station markers and identification numbers in the area of interest to assure easy correlation with a discontinuity if one is detected

8. Set up the exposure parameters
9. Make the exposure
10. In the darkroom, unload and process the film
11. Evaluate the film for artifacts
12. Evaluate the film for compliance
13. Complete a report and store the film

Radiographic Film
Radiographic film is, by far, the most important permanent record in nondestructive testing. Standard industrial radiographic film contains two emulsions, one on each side of the film base. There are seven layers that make up the cross section of the film. The base, which occupies the vast majority of the thickness, is made from cellulose, triacetate, or polyester. Both sides of this base have a very thin layer that contains an adhesive, and adhering to the adhesive are the layers of emulsion. The emulsion contains many small silver halide crystals, which are suspended in gelatin. This is where the chemical changes take place when the emulsion is exposed to radiation or light. On the two outer surfaces, there is a very thin layer of hardened gelatin, which is designed to protect the emulsion. The emulsion is very sensitive to handling, chemicals, abrasion, and various levels of light exposure, so it essential that the film be treated with extreme care.

There are four general classifications of industrial radiographic film. *Class I* is described as extra-fine grain, low speed, with very high contrast capabilities. This film is generally used for lower-density materials and can be used with or without lead screens. *Class II* is a fine-grain, medium-speed, high-contrast film that is also used for the lower-density materials with low- and medium-energy radiation. This film classification tends to be more widely used than the Class I since it provides very good definition, has fine grain, and is slightly higher in film speed than Class I. It can also be used with or without lead screens. *Class III* is a high-speed film, and therefore requires shorter exposure times. It is typically used for x-rays or gamma rays with higher energies, and can be used with or without lead screens. It is considered a medium-contrast film with high graininess. The *Special Class* includes films with very high definition and very fine grain. There are other classes of wide lattitude films that are not used as frequently as the four groups above.

Film Processing

After the radiograph has been taken, the film is ready for processing. A typical manual film processing system is illustrated in Figure 6-25. The first step in manual processing is to carefully place the film onto a stainless steel hanger, which is designed for manual processing, or carefully place it into the feed-in end of the automatic processor (Figure 6-26). The first step that the film will undergo is development. Developers are alkaline solutions that change the latent or chemically stored image in the radiographic emulsion into a visible image, resulting in various shades of gray or black, depending upon the amount of exposure. Agents in the developer reduce the exposed silver halide to metallic silver. There is also an accelerator, which speeds up the development; a preservative, which prevents the oxidation of the developer; and a restrainer, which helps to control and reduce the chemical fog that can occur during the processing of the film. It is fairly common to use a development time of 5 to 8 minutes at a temperature of 68 °F (20 °C). Some film manufacturers recommend 4 to 7 minutes at 68 °F. If the temperature is higher than 68 °F, a shorter development time will be appropriate. On the other hand, if the temperature of the developer is lower than 68 °F, a longer development time will be necessary. Just as it is

FIGURE 6-25 Manual film processing tanks. (Courtesy Charles J. Hellier.)

important to follow a procedure for taking the radiograph, it is also important to follow a consistent procedure for the processing of the film. A standard time and temperature for the development should be adopted and remain consistent throughout the radiographic operation.

When the radiographic film is being processed manually, it is necessary to agitate the film while it is in the developer solution. This provides a more uniform development and prevents the formation of air bubbles on the film surface. If the film is not properly agitated, the chemical activity in the development will be inconsistent and a condition called "streaking" may result, characterized by variations in film density.

The next step is to arrest or stop the development. This can be done in one of two ways. The film can be taken out of the developer and placed into a water bath for several minutes, or it can be placed in an acidic solution called stop bath. The stop bath is an excellent way to immediately and consistently arrest the development activity. The acid

FIGURE 6-26 Automatic processor. (Courtesy Charles J. Hellier.)

neutralizes the alkaline developer and protects the acidity of the fixer, which is the next step.

The fixer serves two major functions. First, it clears out the unexposed silver halide crystals remaining in the film and, second, it fixes or hardens the image. The fixer also influences the archival properties of the film. Many times, shortcuts are taken in the fixing process and the film is taken out before it is totally fixed or hardened. In this case, the film may still contain some unexposed silver halide crystals and they will continue to develop on their own. This could ultimately result in the image of the film having a yellowish or brownish appearance. It is good practice to keep the film in the fixer for about 15 minutes, although some film manufacturers recommend at least double the clearing time. Clearing time is the time that it takes for the remaining silver halide crystals to be removed from the emulsion, clearing up the image.

After fixing, the film goes into a water rinse for a period of time, typically 30 minutes. The water should be continually changed in order to remove any remaining traces of the developer or the fixer. If the film is removed from the water rinse prematurely, shortened archival properties can result and the film can eventually turn yellow or brown. At the end of the water rinse time, it is recommended that the film be placed in water containing a wetting agent. This uniformly wets or coats the film so that droplets of water will not form, causing artifacts when the film is dried. It also facilitates consistent drying of the film.

The final step in processing is to dry the film, which is normally done in a warm air recirculating drier designed for this purpose.

The entire film processing procedure (apart from the drying operation) is accomplished under safelight conditions in a darkroom. The darkroom safelights should be tested periodically for film safety. An unexposed film with a coin or some other opaque object on the emulsion should be placed on the worktable. After a period of time that exceeds the longest time that films would be exposed to the safelight, the object should be removed and the film processed. There should be no change in the density of the film in the area of the opaque object. If an image of the object appears on the film, it indicates that the safelight is not "safe" and should be changed.

With automatic processors, the dry-to-dry time is about ten minutes, compared to over an hour for manual processing. This is because the developer solution is more concentrated and the development temperature is higher, which speeds up the entire process. The automatic processor is the optimal way to process film, since it results in consistent images with fewer film artifacts compared to manual processing. Automatic processors are very expensive and there should be a sufficient quantity of radiographs to be processed in order to make it cost effective. The key to successful film processing with automatic processors is maintenance. The automatic processor must be cleaned often. The roller system must be removed and carefully maintained, otherwise there will be many problems. Even the most exceptional technique capable of producing the highest quality radiographic images can be rendered meaningless because of problems in processing.

Summary

The first and most important consideration in radiography is the handling of the film. In far too many cases, film artifacts are the cause for the rejection of the final radiograph. Handling the film properly in the darkroom will help to minimize this. The darkroom should have a "wet" side and a "dry" side. The dry side is the area where the film is stored and handled, i.e., loaded into and removed from cassettes. The wet side is where the processor is located, whether it is manual or automatic. The darkroom should be hospital-clean at all times. A darkroom that is dirty and untidy is usually a darkroom where film will be scrapped as a result of artifacts.

The five steps in the processing of the film are:

1. Developing
2. Stop bath
3. Fixing
4. Washing
5. Drying

The precautions that should be taken include:

1. Proper control of the development temperature and time
2. Periodic maintenance of the developer and fixer solutions
3. Agitation in the manual system during the development step
4. Maintaining a safelight condition in the darkroom
5. Most important, cleanliness

Density

In radiography, film density is defined as the quantitative measure of film blackening as a result of exposure and processing. It can be expressed mathematically as follows:

$$D = \log \frac{I_0}{I_t}$$

where:
D = density
I_0 = light incident on the film
I_t = light intensity transmitted through the film

If a film is processed and it has not been exposed to light or radiation, it will be virtually clear. A film that has zero density is totally clear and has a film density reading of zero. If a film is exposed and the resultant film density is one, the amount of light that passes through the film is 10% of the incident light. For a film density of 2.0, only 1% of the incident light passes through. A film density of 3.0 permits 0.1% of the incident light to pass through, a film density of 4.0, 0.01%, and so on. Most specifications require that the film density in the area of interest be between 2.0 and 4.0. There are other specifications that differ slightly, but generally, the density range of 2.0 to 4.0 is quite common. Film density is measured with a densitometer (see Figure 6-27).

It is essential that densitometers be calibrated prior to each use to assure that the readings are accurate. In the past, film density strips were used to compare known densities to densities in the area of interest. This required a great deal of conjecture and, even though still employed today, the readings are not nearly as accurate as the readings obtained with the densitometer.

As mentioned earlier, the base density of a radiograph is established by taking a density reading through the penetrameter. The density range in the area of interest must not

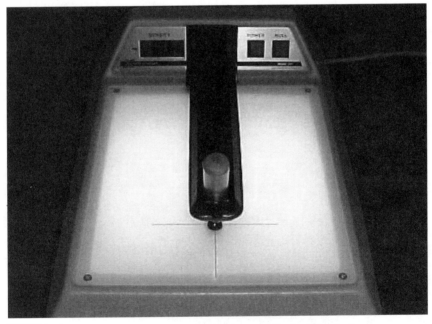

FIGURE 6-27 Densitometer. (Courtesy Charles J. Hellier.)

vary greater or less than a percentage of that base density. It is not unusual to have a density range requirement for the area of interest of +30% to −15% of the base density established through the penetrameter. This, in fact, is good radiographic practice. The quality level of the radiographic technique is established by the penetrameter image and it has been determined that that quality level will also apply to a defined density range. Table 6-8 shows density readings and the +30/−15% range. Notice that for a base density reading of 3.0 (i.e., the density through the penetrameter), the maximum density that is considered acceptable in the area of interest would be 3.9 and the minimum density would be 2.55. Most codes and specifications require that even with the application of this range, densities greater than a specified maximum (usually 4.0) and a minimum (usually 2.0) not be exceeded. The density readings in the area of interest should be taken at random in that area. Density readings should not be taken through an area containing discontinuity. It should be remembered that the density is a measure of establishing that the correct exposure has been employed. The density readings for both the penetrameter and the area of interest should be entered on the radiographic report.

TABLE 6-8 Film Densities

Density reading through penetrameter	+30%	−15%
1.5	1.95	1.28
1.6	2.08	1.36
1.7	2.21	1.45
1.8	2.34	1.53
1.9	2.47	1.62
2.0	2.60	1.70
2.1	2.73	1.79
2.2	2.86	1.87
2.3	2.99	1.96
2.4	3.12	2.04
2.5	3.25	2.13
2.6	3.38	2.21
2.7	3.51	2.30
2.8	3.64	2.38
2.9	3.77	2.47
3.0	3.90	2.55
3.1	4.03	2.64
3.2	4.16	2.72
3.3	4.29	2.81
3.4	4.42	2.89
3.5	4.55	2.98
3.6	4.68	3.06
3.7	4.81	3.15
3.8	4.94	3.23
3.9	5.07	3.32
4.0	5.20	3.40

Note: Typical acceptable density range for the penetrameter and area of interest is 2.0–4.0. Some specifications may require different ranges.

VI. RADIOGRAPHIC EVALUATION

The final step in the radiographic process, and by far the most important, is the evaluation of the radiograph. At times, radiographic interpretation has been referred to as a mix of art and science. Science involves all of the physics and principles involved with producing the radiographic image. Interpretation of the image involves art, to an extent. Interpretation of radiographs cannot be taught in a classroom, although that can provide a good foundation. It requires hours of reviewing and understanding the different types of images and the various conditions that are prevalent in industrial radiography. There are individuals who have been evaluating radiographs for decades who will admit that they are still learning how to interpret with each additional radiograph.

After the film density readings are completed, the next step in evaluating radiographs is to take an overall look at the appearance of the radiographic image, as well as the condition of the film itself. This initial look will generally indicate the overall quality that has been achieved. After that, the quality level should be confirmed by observing the image of the IQI and assuring that the essential hole in the shim-type penetrameter, or wire in the wire-type penetrameter, is clearly and discernibly displayed.

The radiographic interpreter should always wear cloth gloves (preferably cotton) when evaluating radiographs. This prevents fingerprints, smears, and smudges from being placed on the surface of the radiograph. After the quality level has been established, it is then appropriate to review the radiographic image for artifacts. There are a large number of artifacts that can appear in a radiograph. Artifacts, or false indications, are classified based on when they were formed, i.e., prior to processing, during processing, or after processing.

FIGURE 6-28 Interpretation of a radiograph with a high-intensity illuminator. (Courtesy Charles J. Hellier.)

The high-intensity illuminator (Figure 6-28) designed for viewing the radiographs should be in good working condition. Even before the radiograph is initially reviewed, the illuminator should be checked to make sure that the high-intensity light bulbs are working property and that the viewing area is clean and free from any scratches or artifacts that could be superimposed on the image of the radiograph. The illuminator should be properly designed to dissipate the heat that is created when the high-intensity lights are in operation. There should also be an intensity control to allow the interpreter to increase or decrease the brightness of the light source. Is essential that any light not coming through the radiograph be masked. The light intensity given off from these high-intensity illuminators is very intense and can cause severe discomfort to the eye, so masks are very strongly recommended to assure that the light coming through the film is the only light reaching the eye. Most high-intensity illuminators are operated with a foot switch, which permits the interpreter to turn the light on and off while keeping both hands free to handle the film. This is another area where caution is warranted. The light switch should not be activated until the film is in place on the viewing area of the viewer and the light should be switched off prior to its removal.

There seems to be a lack of agreement as to whether magnification should be employed during the evaluation of radiographs. Magnifiers are, in fact, encouraged when they can assist in the proper detection and identification of the different discontinuities. When magnification exceeds a certain power, it can become quite useless, since the grains in the film are also magnified and the images of the discontinuity tend to be diffused.

Radiographs should be viewed in a darkened area. It is advisable to allow the eyes to become used to the dimly lit conditions prior to radiographic viewing. Recommendations vary regarding the length of dark adaptation time. Experience has shown that periods from one to as high as 15 minutes may be necessary to properly distinguish the density changes within the image. The issues to consider are those such as:

a) The minimum density changes sought. Small or tight discontinuities may appear as very slight changes in the density of the radiographic image.

b) If the interpreter has recently been in bright sunlight, longer times may be required for dark adaptation.

For critical interpretation, it may be necessary to mask the lighter parts of the image where the low-density (lighter) areas allow the brighter transmitted light to reduce the interpreter's perception of changes in density.

Evaluation for Artifacts

Film Artifacts or False Indications that are Caused Prior to Processing
These include:

- Lead screen marks
- Static marks
- Film scratches
- Exposure to light
- Fog due to exposure to low levels of light or aging
- Finger marks
- Pressure marks
- Crimp marks

Artifacts that are Caused during the Processing of the Film
These include:

- Pressure marks (from rollers in an automatic processor)
- Chemical and delay streaks
- Pi lines (in automatic processors)
- Chemical spots
- Dirt

Artifacts Caused after Processing
These are primarily in the form of scratches and fingerprints. Fingerprints will be minimized with the use of cotton gloves.

Evaluation for Discontinuities

After the radiograph is evaluated for artifacts, it is finally time to evaluate for discontinuities. Discontinuity conditions that are normally found in welds include those in the following subsections, listed in order of severity.

Cracks
There are many different types of cracks that are classified by their orientation and location. They will always appear as dark, irregular, linear indications in a radiograph and are the most serious of all discontinuities. They are generally tight and not always detectable by radiography unless their orientation is somewhat in the same plane as the direction of the radiation. The most common types of cracks are those that are oriented longitudinally (along the length of a weld). Others may occur as star-shaped patterns, usually at the end of a weld pass. Some will occasionally be transverse to the length of a weld. Cracks can also appear at the toe of the weld or at the edge of the root pass. Underbead cracks will typically be angular, extending from the fusion zone.

Lack of Fusion
This serious discontinuity results from an absence of metallurgical fusion, either between a weld pass and the base material (weld edge prep) or between two successive weld passes. Lack of fusion is usually very narrow, linear, and tends to be straighter than the crack. At its extremities, the condition "feathers" down to an unusually sharp edge. In many cases, the image of lack of sidewall or lack of fusion appears somewhat straight along one edge and slightly wavy on the other.

Incomplete Penetration
This discontinuity is an absence of weld metal or an area of "nonfusion" in the root pass of the weld. Its appearance is very straight, dark, linear, and usually "crisp" in sharpness. It can be short but usually has significant length.

Inclusions (Dense and Less Dense)
Inclusions are basically materials that have been entrapped in the weld that do not belong there. They will have a variety of shapes and dimensions ranging from short and isolated to linear and numerous. The lighter-density inclusions will result in a darker image on the radiograph and the more dense inclusions, such as tungsten, as a lighter image.

Porosity

When gas is trapped in a weld metal, the void-type condition created is referred to as gas or porosity. Porosity comes in different shapes (globular, tailed, elongated) and distributions (linearly aligned, clustered, isolated, scattered). Porosity will always appear darker, since they are gas filled, and are the easiest of all weld discontinuities to detect.

Geometric Conditions

There are also geometric conditions that can occur in welds that are observable in a radiograph and should be further addressed by visual examination and dimensional checks. When possible, these conditions should be measured with mechanical gauges. These geometric conditions include the following.

Concavity—a concave condition at the root of a weld resulting in a thinner region through the weld cross section. The image of concavity in a radiograph is usually easy to identify since it is a broad indication with a gradual change of density as compared to the abrupt change associated with incomplete penetration.

Convexity—usually considered the opposite of concavity since this condition is a thicker protrusion at the root pass. It is caused by excessive deposit of the root pass and is sometimes referred to as excessive penetration. In extreme cases, the weld metal will "burn through" and may form droplets that are referred to as "icicles." The concern is the abrupt change in contour and the possibility of flow constriction in a pipe due to the protrusion of additional weld metal.

Undercut—a depression that occurs at the edge of the weld where it has fused into the base metal at the outer or inner surface. It is like a slight "valley" that continues along the length of the weld and varies in depth. The depth is the measurable dimension that determines its seriousness.

Underfill—sometimes interpreted as undercut since it can appear in the same region of the weld. In fact, it is a different condition that occurs when the weld groove is not completely filled with weld metal. Also, unlike the undercut, the contour of the bottom of this condition is more V-shaped or notch-like as compared to the generally broad gradual configuration of the undercut.

Overreinforcement—this condition results from an excessive depositing of weld metal on the outer surface of the weld. The concern is the geometric change that may create a stress riser condition at the outside surface where stresses normally tend to be higher by design.

Casting Discontinuities

Discontinuities that occur in castings that can be detected radiographically include the following, listed in order of their severity.

Hot tears and cracks—both serious ruptures or fissures that typically occur in an isolated zone due to the high stresses that build up during the cooling of the casting. Hot tears usually form during initial cooling and cracks later, usually at or near room temperature. On a radiograph, both conditions appear linear and branch-like and are most likely to be in or near an area of thickness change, where the different rates of cooling cause stresses to build up.

Shrinkage—usually in the form of a zone of minute fissures as a result of stresses during cooling. Shrinkage comes in various shapes. Sponge shrinkage has a "spongy" appearance and can be isolated, scattered, or significant in size and density variations. Microshrinkage is feathery in appearance and the change in density is often quite minor.

Slag and sand inclusions—the entrapment of inclusion materials and sand cause these conditions, which will have irregular shapes and variations in density due to the nature of the included matter.

Gas voids and porosity—unlike the inclusions, gas voids and porosity are more uniform, typically globular and dark in appearance. In fact, these discontinuities just look like voids, are normally easy to detect (they are not subject to alignment limitations like cracks), and readily recognizable.

Cold shuts—very tight discontinuities that occur when a surface that has begun to solidify comes in contact with other molten metal as the casting is in the process of being poured. There is usually a thin film of oxide present that prevents total metallurgical fusion. It is a very difficult discontinuity to detect with radiography due to its tight condition and angular orientation.

Geometric Conditions

As with welds, there are also geometric conditions in castings that can be observed radiographically. These geometric conditions include the following.

Misrun—this condition is actually an absence of metal due to the inadequate filling of the casting mold. It is easily detected by a simple visual test.

Unfused chaplets—metal supports that are strategically placed in the casting mold for support of the mold walls. They are designed to be consumed when the molten metal comes into contact with them, and when this does not happen, the circular shape of the chaplet will be apparent in the radiograph.

Note. A compendium at the end of this chapter contains radiographs showing examples of false indications, and weld and casting discontinuities.

Responsibility of the Interpreter

In order to effectively evaluate radiographs, it is essential that the interpreter be thoroughly familiar with the parts, dimensions, and material, in order to properly judge whether it is acceptable or not. The interpreter should be thoroughly familiar with the technique that was used to produce the radiograph, how the film was processed, the codes and standards that apply, and acceptance criteria.

If there is a rejectable discontinuity, it is extremely important that the condition be suitably identified as to its exact location. For this purpose, it is good practice for the radiographer to prepare a transparent skin that is overlaid on the radiographic image; then, with a wax pencil, the identification numbers are marked on the skin, as well as the outline and location of the discontinuity. This permits an accurate marking of the discontinuity on the actual part and facilitates precise repair. After repair, the region that originally contained the discontinuity must be reexamined to assure that the discontinuity has been completely removed.

It is essential that the entire evaluation process be entered on a radiographic report form. Report forms differ in format with each individual company. It would be extremely beneficial to the radiographic industry if a consistent, generic report form could be adopted. To this end, sample radiographic report forms for welds and castings are offered (see Figure 6-29).

VII. APPLICATIONS

Although the majority of applications in radiographic testing appear to involve welds and castings, it has been effectively applied to many other product forms spanning a

RADIOGRAPHIC TESTING   **6.55**

Customer _____ RT Report # _____ Date _____
Part ID/ Location _____
Weld ID/Location _____   **TECHNIQUE DATA**
               PART DATA   IQI# _____ ☐ s/s ☐ f/s   Screen Type _____
Base Mat'l Thickness _____ Shim Thickness _____ Front Thickness _____
Weld thickness _____ Film Mfr. & Type _____ Inner Thickness _____
Material Type _____ Film Size _____ Back Thickness _____
Welding Process _____ # of Film /Cassette _____
Weld Status _____ Radiation Type: ☐ Isotope _____ ____ ci. ☐ X-Ray _____ kV
Type of Weld Joint _____ Exposure: _____ ci-min / _____ mam
Diameter or Weld Length _____ f = _____ t = _____ d = _____ Ug = _____
RT Procedure _____ Rev. _____ Film Processing: ☐ Manual ☐ Automatic
Acceptance Criteria _____

SHOOTING SKETCH
(Show location and orientation of radiation source, part, film, IQI & location markers)

INTERPRETATION DATA

EXPOSURE: ☐ Single wall ☐ Double wall   VIEW: ☐ Single wall ☐ Double wall

Part #	Weld #	Location #	Acc	Rej	Disc. Code	Quality	Density	Artifiacts	Remarks

POR - Porosity   **DISCONTINUITY CODES**   EUC - External Undercut
SI - Slag Inclusion   SP - Spatter   DT - Drop Through   IUC - Internal Undercut
ESI - Elong. Slag Inclusion AS - Arc Strike   W - Tungsten Inclusion   UI - Unconsumed Insert
IP - Incomplete Penetration CX. - Convexity   CR(L) - Longitudinal Crack SO - Surface Oxidation
IF - Incomplete Fusion   CV. - Concavity   CR(T) - Transverse Crack SURF - Surface
BT - Burn Through   HiLo - High Low   CR(C) - Crater Crack   UF - Underfill

Interpreted by: _____ NDT Level _____ Date _____

Customer: _____   Date _____

(a)

FIGURE 6-29 Radiographic report form for (a) welds. (*Continues*)

6.56 CHAPTER SIX

Customer _____ RT Report # _____ Page ___ of ___ Date _____
Part ID/ Location _____
Casting ID/Location _____ TECHNIQUE DATA
        PART DATA IQI# _____ ☐ s/s ☐ f/s Screen Type _____
Mat'l Thickness _____ Shim Thickness _____ Front Thickness _____
Material Type _____ Film Mfr. & Type _____ Inner Thickness _____
RT Procedure _____ Rev. _____ Film Size _____ Back Thickness _____
Acceptance Criteria _____ # of Film /Cassette _____
        Radiation Type: ☐ Isotope _____ ___ ci. ☐ X-Ray ____ kV
        Exposure: _____ ci-min / _____ mam
        $f =$ _____ $t =$ _____ $d =$ _____ $U_g =$ _____
        Film Processing: ☐ Manual ☐ Automatic

SHOOTING SKETCH

(Show location and orientation of radiation source, part, film, IQI & location markers)

INTERPRETATION DATA

EXPOSURE: ☐ Single wall ☐ Double wall   VIEW: ☐ Single wall ☐ Double wall

Part #	Casting #	Location #	Acc	Rej	Disc. Code	Quality	Density	Artifacts	Remarks

DISCONTINUITY CODES

CR - Crack	SI - Sand Inclusion	SH - Shrinkage	UC - Unfused Chaplet
MP - Microporosity	INC - Inclusion	DR - Dross	UCL - Unfused Chill
POR - Porosity	DI - Dense Inclusion	HT - Hot Tear	CS - Core Shift
WP - Worm Hole Porosity	SS - Sponge Shrinkage	MR - Misrun	SURF - Surface
BH - Blow Hole	MS - Micro Shrinkage	COL - Cold Shut	SEG - Segregation

Interpreted by: _____ NDT Level _____ Date _____

Customer: _____ Date _____

(b)

FIGURE 6-29 (*Continued*) Radiographic report form for (b) castings.

wide variety of industries. The major industries include but are not limited to the following.

1. *Power Generation.* A wide range of weld configurations, thicknesses, and materials are used at power plants. Radiographic testing requires radiation sources that are capable of providing energies from the low end of the spectrum to the highest commercially available. Pressure vessels are complex structures that require high-quality initial examinations. The associated piping systems, with their many weld joint geometries, provide continual challenges for the radiographer. Cast valve bodies are usually radiographed at the foundry. In-service examinations offer even greater challenges. There are other conditions the radiographer must deal with, such as accessibility, the environment (temperature; in the case of nuclear power plants, radiation, air quality, and circulation, etc.), the transporting and positioning of equipment, the concern for the security of radiation areas, and the weather when examinations are to be conducted outdoors. There are other non-conventional applications for RT in power plants. There are ongoing concerns regarding the degradation of structures and components as a result of corrosion, vibration, and wear. Many times, the areas are covered with insulation or are very difficult to reach. There are portable RT units that provide for the examination of these conditions and provide contour data through the insulation. When there are similar concerns for the internal conditions of valves, or for the determination of the position of the mechanical components, a higher-energy source may be necessary.

2. *Aerospace.* There are countless uses for RT in the aerospace industry. The critical components in the aircraft engine itself call for thousands of radiographs. The airframe and other related structures depend greatly on radiography for initial and in-service inspection. Water ingress in honeycomb cells can cause problems due to freezing when the aircraft is airborne. Radiographic techniques can effectively detect this condition in most cases.

3. *Petrochemical.* Radiography is of great importance in the initial construction of petrochemical plants. Considering the many service conditions under which this type of complex operates, establishing a baseline through radiography is of extreme importance. Just as with power plants, in-service inspection is necessary and RT plays a key role. The detection of areas of degradation, corrosion, erosion, wear, and other conditions that develop through the extended operation should be detected and corrected before an unplanned shutdown becomes necessary. Other than the cost of initial construction, unplanned outages are the most costly events in this industry. It should be noted that this also applies to most other major industries.

4. *Nonmetals.* This category includes a wide variety of materials that are inspected by radiography, including plastics, rubber, propellants, ceramics, graphite, concrete, explosives, and many more. Perhaps the electronic industry, because there are some metals involved, also belongs in this industry segment.

5. *Medicine.* There are numerous applications in medicine that do not involve the examination of living creatures. On occasion, foreign objects or contaminants have inadvertently found their way into various drugs and medicines during processing. These elements most usually have densities that are quite different from the products into which they have become included. This makes their examination an ideal application for RT. A prerequisite of course, must dictate that the product will not be affected by exposure to radiation. When a large quantity of product is involved, real-time radiography may be the most cost-effective approach. This provides for a two-dimensional image to be displayed on a monitor almost immediately, as a result of the conversion of the radiation into an electronic or optical signal.

6. *Law Enforcement and Security.* Radiography is especially adaptable and useful for the detection of hidden contraband and weapons and for the examination of explosive de-

vices such as shells and projectiles. Sealed boxes, envelopes, and other packages have been found to contain devices that were potentially hazardous if they had not been detected. When passing through the security area in an airport, carry-on articles are subjected to real-time radiographic inspection.

7. *Food.* Various food products are examined with some form of RT for the purpose of detecting foreign objects that may have been introduced during processing. In some cases, food items may be inspected for content, distribution of additives (such as nuts, candy, etc. in ice cream), or for quantity. Again, the fastest approach would include fluoroscopy or some other real-time technique.

8. *Objects of Art or Historic Value.* This industry is perhaps the most unusual and most interesting for radiographers. Famous paintings have been radiographed (using low-energy, high-contrast techniques) to disclose hidden paintings or images beneath their outer layer. Mummies, usually of some very prominent ruler or an individual with some historic significance, have been inspected to evaluate the condition of the remains. Statues, vases, old pottery, figureheads from historic seagoing vessels, and the famous Liberty Bell have all been subjected to radiographic inspections. Entire jet engines, cars, and complicated assemblies have been examined, many times with a single exposure.

Summary

The uses of radiography are limited only by the principles inherent in the process and the human mind. There are those that would say that the world of radiography has reached its peak. It is easy to dispute such a narrow opinion when considering the many recent developments and innovations, including industrial computed tomography, real-time radiography, flash (high-speed) imaging, in-motion radiography, microfocus radiography, digital imaging, and the photostimulable luminescence (PSL) process, also known as storage phosphor plates, for capturing radiographic images.

The future of radiography will continue to offer new and exciting techniques that will expand its world of applications and opportunities. Certainly, Wilhelm Conrad Roentgen and Marie and Pierre Curie would have been proud.

VIII. ADVANTAGES AND LIMITATIONS OF RADIOGRAPHY

Advantages

Radiographic testing has many advantages, some of the most significant of which are listed below. It:

1. Provides an extremely accurate and permanent record
2. Is very versatile and can be used to examine many shapes and sizes
3. Is quite sensitive, assuming the discontinuity causes a reasonable reduction of cross section thickness
4. Permits discontinuity characterization
5. Is widely used and time-proven
6. Is a volumetric NDT method

Limitations

1. There are safety hazards with the use of radiation devices
2. RT has thickness limitations, based on material density and energy used
3. RT can be time-consuming
4. RT is very costly in initial equipment and expendable materials
5. It is also very dependent on discontinuity orientation
6. RT requires extensive experience and training of the personnel taking the radiographs and during the interpretation

Safety Considerations

Radiation ionizes matter through the ejection of electrons from their orbit, thereby changing the electrical characteristics of the atoms. It is this process that results in biological effects to humans. Living tissue will be damaged to an extent determined by the dose received. Absorption of radiation by a human is expressed in REM (roentgen equivalent man). It is the product of roentgens or milliroentgens (mR) and the Quality Factor (QF) of the radiation type. Since x- and gamma rays have a QF value of 1, the exposure as measured in Roentgens will be the dose received in units of REM. The age-old rules of protection against radiation exposure involve time, distance, and shielding.

Time. The shorter the exposure, the better. Every effort should be made to keep personnel exposure time as close to zero as possible. When performing radiography in the field, zero exposure is virtually impossible. So every attempt should be made to minimize the time of exposure to the body to as short a time as possible.

Distance. The further an individual is from a radiation source, the less the exposure. Recall the inverse square law and how radiation intensity varies inversely proportional to the distance. If there is an intensity of 200 mR/hour at a certain distance from a radiation source, simply doubling the distance will decrease the intensity by a factor of one-fourth, or 50 mR. Doubling the distance again will reduce the intensity to 12.5 mR, and so on. Using the inverse square law will permit the radiographer to calculate the approximate exposure at any distance.

Shielding. The greater the thickness and density of the shielding material, the less the radiation exposure. This is a function of the half- and tenth-value layers, as discussed in Section II. Lead, of course, makes an excellent shielding material because of its high density.

To summarize, the exposure to radiation should be kept as short as possible, the distance from the radiation as far as reasonable, and the shielding as thick as practical.

There are many excellent references regarding the biological effects of radiation exposure. Unfortunately, there is not a great deal of data to support the effects of radiation on humans. The limited data from the end of the Second World War and some of the more notable accidents involving radiation have been the primary source for the exposure effects still referred to today. The ALARA (as low as reasonably achievable) concept is the most appropriate way to treat the issue of radiation exposure. If every person working with radiation took every precaution and made every effort to minimize exposures, needless damage would not occur to nearly the same extent as it does.

Since humans cannot sense radiation, it is essential, and in many industries a requirement, that personnel-monitoring devices be used. Such devices include:

1. *Film badges* are small plastic holders that contain a special film, which are worn during the time working in or near a radiation area. The film is quite sensitive and can de-

tect as low as several mR of low-energy x-rays and up to as high as 2000 R. They usually contain small metallic filters that enable the type and energy range of the radiation to be identified. After a period of time, they are sent to a laboratory for processing and reading. This is the main source of data that are entered on a lifetime radiographic exposure history report that should be maintained for all radiation workers.
2. *Ionization chambers* are sometimes referred to as pocket dosimeters. They are small, lightweight, and are capable of being read directly from a scale inside the chamber.
3. *Thermoluminescent dosimeters,* referred to as TLDs, are widely used due to their rapid readout response time and broad range, which is quite linear. They can be reused and are excellent under field conditions.
4. *Photoluminescent glasses* are useful down to about 1 R. They build up fluorescent centers when exposed to radiation and emit visible light when viewed with ultraviolet light. The intensity of the light is proportional to the exposure of radiation.

In addition, area monitors should always be employed whenever radiographic activities are in process. Some of the monitors, or survey meters commonly used include:

- Ionization chambers
- Geiger—Mueller (G–M) counters
- Scintillation instruments

(Note: Two survey meters are better than one in the event of a failure.)

Radiation should never be taken for granted. One cannot be too cautious when dealing with this potentially hazardous form of electromagnetic energy. And yet, it can be safely and effectively used in radiography by following the rules and being familiar with the regulations. Personnel who do not respect radiation and try to do things their way cause most accidents and overexposures. Radiation does deserve respect!

IX. COMPENDIUM OF RADIOGRAPHS

Examples (Figures 6.30–6.44) of a number of conditions (artifacts) and discontinuities are included in the following pages as a guide. It must be stressed that discontinuities are unique and no two are alike. There are many other types that are not covered in this chapter. A thorough knowledge of the materials, how they were processed, the radiographic techniques employed, and an understanding of the variables involved in radiographic interpretation are all essential to effective evaluation.

Figures 6-30–6-33 illustrate artifacts; Figures 6-34–6-42 illustrate weld discontinuities; and Figures 6-43 and 6-44 illustrate casting discontinuities. (*Note.* For an example of Radiographs showing other casting discontinuities, refer to Chapter 2: hot tear—Figure 2-5b, sand and slag inclusions—Figure 2-20, and an additional example of shrinkage, Figure 2-5a.)

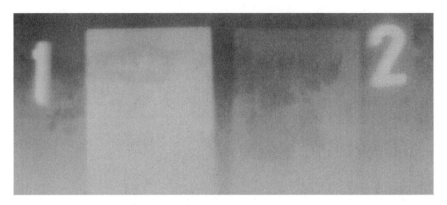

FIGURE 6-30 Chemical stains. (Courtesy Charles J. Hellier.)

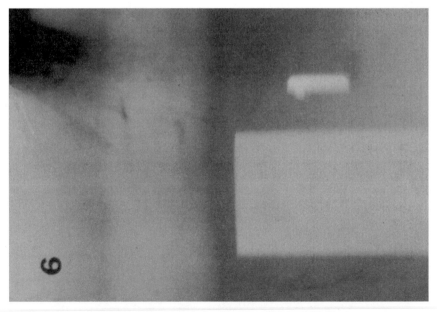

FIGURE 6-31 Light leaks. (Courtesy Charles J. Hellier.)

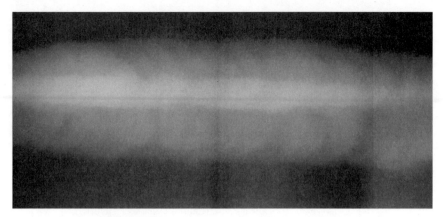

FIGURE 6-32 Pi lines. (Courtesy Charles J. Hellier.)

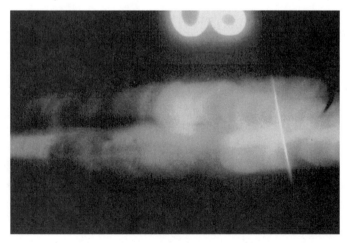

FIGURE 6-33 Sliver of paper (located between the lead intensifying screen and the film). (Courtesy Charles J. Hellier.)

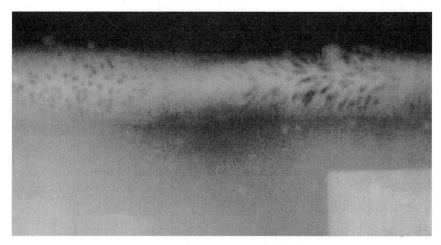

FIGURE 6-34 Porosity. (Courtesy of Quality Consulting Company, Inc.)

FIGURE 6-35 Slag inclusions. (Courtesy Charles J. Hellier.)

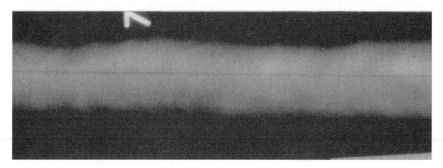

FIGURE 6-36 Incomplete penetration. (Courtesy Charles J. Hellier.)

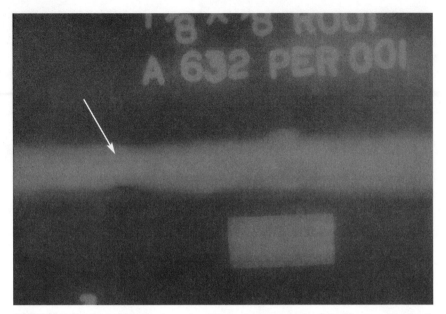

FIGURE 6-37 Lack of fusion (arrow) and undercut (at edge of weld). (Courtesy Charles J. Hellier.)

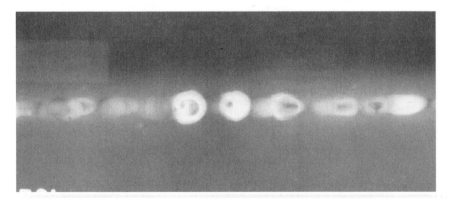

FIGURE 6-38 Icicles and burnthrough. (Courtesy Charles J. Hellier.)

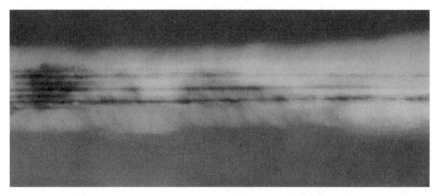

FIGURE 6-39 Slugging (weld rods placed in weld groove). (Courtesy Charles J. Hellier.)

FIGURE 6-40 Tungsten inclusion (Courtesy of Quality Consulting Company, Inc.)

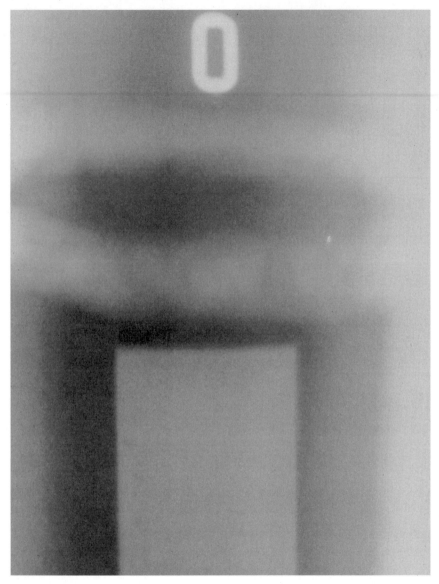

FIGURE 6-41 Tungsten inclusion (double wall double view ellipse technique). (Courtesy Charles J. Hellier.)

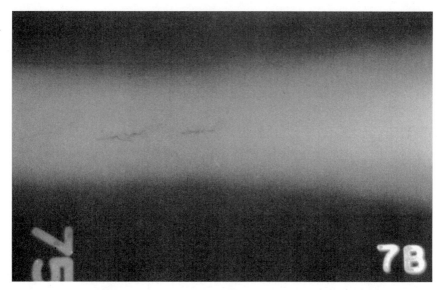

FIGURE 6-42 Cracks. (Courtesy Charles J. Hellier.)

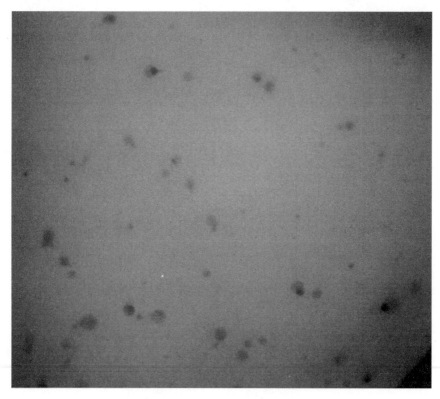

FIGURE 6-43 Gas. (Courtesy Charles J. Hellier.)

FIGURE 6-44 Shrinkage. (Courtesy Charles J. Hellier.)

X. GLOSSARY

Absorption—The process whereby photons of radiation are reduced in number or energy as they pass through matter.

Angstrom unit (Å)—A unit of length that is used to express the wavelength of electromagnetic radiation (i.e., light, x-rays, gamma rays). One angstrom unit is equal to 0.1 nanometers (1 nm = 10^{-9} m).

anode—The positive electrode of a discharge tube. In an x-ray tube, the anode contains the target.

artifact—A false indication on a radiograph arising from, but not limited to, faulty manufacture of the film, storage, handling, exposure, or processing.

cassette—A light-tight packet or container for holding radiographic film during exposure; may include intensifying or conversion screens.

collimator—A device made of radiation absorbent material intended for defining the direction and angular divergence of the radiation beam.

composite viewing—The viewing of two or more superimposed radiographs from a multiple film exposure. The film may be the same or different speeds.

constant potential x-ray unit—An x-ray system that operates on full-wave rectified current.

contrast sensitivity—A measure of the minimum percentage change in an object that produces a perceptible density/brightness change in the radiographic image.

definition, image—The sharpness of delineation of an image in a radiograph. Generally used qualitatively.

densitometer—A device for measuring the density of radiographic film.

density, radiographic—The quantitative measure of film blackening when light is transmitted.

duty cycle—The usable time of a device versus the time that it has to rest when operating "continuously."

equivalent penetrameter sensitivity—That thickness of penetrameter, expressed as a percentage of the part thickness radiographed, in which a 2T hole would be visible under the same radiographic conditions.

exposure table—A summary of values of radiographic exposures suitable for the different thicknesses of a specific material.

film contrast—A qualitative expression of the slope or steepness of the characteristic curve of a film; that property of a photographic material that is related to the magnitude of the density difference resulting from a given exposure difference.

film speed—A numerical value expressing the response of a radiographic image film to the energy of penetrating radiation under specified conditions.

filter—Uniform layer of material, usually of higher atomic number than the specimen, placed between the radiation source and the film for the purpose of preferentially absorbing the lower-energy radiation.

focal spot—In x-ray tubes, the area of the anode that emits x-rays when bombarded with electrons.

fog—A general term used to denote any increase in optical density of a processed photographic emulsion caused by anything other than direct action of the image-forming radiation.

gamma radiography—A technique of producing radiographs using a radioactive source emitting gamma rays.

geometric unsharpness—The penumbral shadow in a radiographic image that is dependent upon 1) the radiation source dimensions, 2) the source to object distance, and 3) the object to film distance.

graininess—The visual impression of irregularity of silver deposit in a processed film.

half-life—The time required for a radioactive isotope to decay to one-half of its original activity.

half-value layer (HVL)—The thickness of a material required to reduce the intensity of a beam of incident radiation to one-half its original intensity.

image quality indicator (IQI)—A device or combination of devices whose demonstrated image or images provide visual or quantitative data, or both, to determine the radiographic quality and sensitivity. Also known as penetrameter. (*Note:* it is not intended for use in judging size or establishing acceptance limits of discontinuities.)

intensifying screen—A material that converts a part of the radiation into light or electrons and that, when in contact with a recording medium during exposure, improves the quality of the radiograph or reduces the exposure time required to produce an image, or both.

IQI sensitivity—The minimum discernible image and the designated hole in the shim-type or the designated wire image in the wire-type image quality indicator.

isotope—One of a group of atoms that have the same atomic number (same chemical characteristics) but have a different mass number. The nuclei have the same numbers of protons but different number of neutrons, resulting in differing values of atomic mass.

Kinetic energy—The energy of a body with respect to its motion.

latent image—A chemical change in the film emulsion as a result of a condition produced and persisting in the image receptor by exposure to radiation. It can to be converted into a visible image by processing.

location marker—A number or letter made of lead or other high-density material that is placed on an object to provide traceability between a specific area on the radiograph and the part.

material density—A material's mass per unit volume.

milliamperes (mA)—The current applied to the filament in the cathode portion of an x-ray tube that controls the intensity of the x-rays.

object to film distance—The distance between the surface of the source side of the object and the film.

photon—A "packet" of electromagnetic radiation.

primary radiation—Radiation coming directly from the source.

radiographic contrast—The difference in density between an image and an adjacent area on a radiograph.

radiographic quality—A qualitative term used to describe the capability of a radiograph to show changes in the area under examination.

radiographic sensitivity—A general or qualitative term referring to the size of the smallest change in a test part that can be displayed on a radiograph.

secondary radiation—Radiation emitted by any substance as the result of radiation exposure by the primary source.

source—A machine or radioactive material that emits penetrating radiation.

source to film distance—The distance between the radiation-producing area of the source and the film.

step wedge comparison film—A strip of processed film carrying a stepwise array of increasing photographic density for comparison to radiographs for the purpose of estimating density.

subject contrast—The ratio (or the logarithm of the ratio) of the radiation intensities transmitted by selected portions of the test specimen.

target—That part of the anode of an x-ray tube that emits x-rays as a result of impingement of electrons from the filament.

transmitted film density—The density of radiographic film determined by measuring the transmitted light, usually through the use of a densitometer.

tube current—The current, measured in milliamperes, passing from the cathode to the anode during the operation of an x-ray tube.

XI. BIBILIOGRAPHY

"Radiographic Inspection." From *ASM Handbook,* Vol. 17, *Nondestructive Testing and Quality Control,* 2nd Edition, May 1992. ASM International, XXXX.

Radiography and Radiation Testing. *Nondestructive Testing Handbook,* Vol. 3, 2nd Ed. American Society for Nondestructive Testing, XXXX. 1985.

Charles J. Hellier and Samuel A. Wenk. "Radiographic Interpretation." NDT Handbook, Section Eight. American Society for Nondestructive Testing, XXXX. 1984.

CHAPTER 7
ULTRASONIC TESTING

I. HISTORY

The use of sound energy to determine the integrity of solid objects is probably as old as mankind's ability to manufacture objects in pottery and metal. The English language has many words and phrases that illustrate the acceptance of this fact and hint at the way in which sound was used in the past to test for integrity. Expressions such as "the ring of truth" or "sound as a bell" are commonplace in everyday speech to indicate quality, honesty, or good health.

Both phrases allude to the fact that a sharp tap on a solid object will set up a vibration at the natural frequency of the object; that is how a bell or any percussion instrument makes music. Any major disruption of the homogeneity of the object will distort that natural frequency and indicate that there is a problem. The instinct to tap an object is so ingrained into human nature that it probably accounts for all those people who, when viewing a prospective second-hand car purchase, unconsciously kick a tire!

The problem with this simple approach to testing an object is that it takes a relatively large imperfection to cause a significant change in sound for the human ear to detect. This is because the determining factor is the wavelength of the sound wave encountering an imperfection in relation to the size of the imperfection. Wavelength, in turn, depends on the speed of sound in the object and the frequency, or pitch, of the sound wave. Within the human range of audible sound frequencies, the wavelength is rather large in most metals. It wasn't until the ability to generate and detect sound waves at much higher frequencies existed that smaller discontinuities could be detected in metals.

The first steps toward this ability were taken in the 1870s with the publication of Lord Rayleigh's work on sound, "The Theory of Sound." This work explained the nature and properties of sound waves in solids, liquids, and gasses, which led to the development of the techniques that are currently in use in nondestructive testing.

The means for generating and detecting sound waves at frequencies above the audible range followed shortly after with the discovery of the piezoelectric effect by the Curie brothers and Lippmann. In 1880, the Curies found that an electrical potential could be generated by applying mechanical pressure to plates cut in a particular fashion from certain crystals. The following year, Lippmann discovered that the reverse was true and that the application of an electrical signal to these plates caused a mechanical distortion. Naturally occurring crystals of quartz, tourmaline, and Rochelle salt were among those materials displaying the piezoelectric effect.

Over the years, there have been many uses made of this effect, from crystal microphones and gramophone pickups to spark generators for cigarette lighters and, of course, ultrasonic transducers for NDT. However, growth in use of piezoelectricity was slow.

As early as 1912, following the Titanic disaster, it was suggested that sound waves could be used to detect icebergs at sea, an idea that received further stimulation during World War I for the detection of submarines. The pulse echo system developed for this application gave rise to peacetime uses between the two world wars in the fields of hydrographic surveys charting the ocean depths and fishing, where echo sonar was used to detect shoals of fish.

S. Y. Sokolov, in Russia, was the first to suggest using ultrasonic waves to detect discontinuities in metals. In 1929, he described some experiments in which he generated ultrasonic waves in metals including cast iron and steel samples, which were subsequently sectioned. In 1935, he described his design for piezoelectric transducers for generating and detecting ultrasound, including a method of coupling sound to the metal. His method worked through a transmission technique using continuous waves with quartz transducers and mercury as a couplant. An alternating current generator was used to drive the quartz crystal transmitter and vibrations reaching the receiver caused an alternating signal that could be measured.

Much work was carried out using this technique between the wars, particularly in Russia and Germany. However there was a problem with continuous wave testing and much of the experimental work concentrated on trying to find a solution. The principle of the continuous wave technique was that energy transmitted through the sample would generate a signal of particular amplitude when no obstruction was present. Some of the sound energy would be obstructed by a discontinuity so that the resulting received signal would be weaker.

This principle was fine as long as there was only enough energy to complete one full transition of the sample, as was the often the case with large castings, where grain size rapidly attenuated the sound. If, however, there was sufficient energy to set up multiple reflections of the sound within the sample, the reflected waves joined in with later sound waves to produce confusion. If the reflections were in phase with the continuous wave, there was an apparent increase in transmitted energy. On the other hand, if the reflections were out of phase, the signal became weaker, giving a false indication. The same effect could be observed in narrow samples, where the beam could reflect from the side walls. This reflected beam path, being longer than the straight-through path, could also create constructive or destructive interference with the main signal.

This problem limited the development and use of ultrasonic flaw detection until World War II, when several workers on either side of the combating nations adopted the technique used in echo sonar known as pulse echo. In this system, short pulses of sound are transmitted at regular intervals, the transmitted pulses and the resulting echoes being displayed on a cathode ray tube (CRT). The interval between transmitted pulses is arranged to be sufficient to allow all internal reflections to die away before the next pulse is started, thus avoiding the interference effects previously encountered. Both the transmitter and receiver crystals can be positioned on the same surface. The sweep time of the CRT can be arranged to display one or more transit periods, and the position of the initial pulse and subsequent echoes can be used to determine the depth of the various reflecting surfaces, including discontinuities.

The pulse echo system was devised around 1942. In Britain, the development was attributed to D. O. Sproule, and in the United States to F. Firestone. After the war, the two approaches were compared, together with developments that had taken place in Germany. The main difference was that Sproule had used separate transmitter and receiver elements, whereas Firestone used the transmitter element as a receiver during its quiescent periods. Since there were merits in both approaches under various circumstances, developments since World War II have tended to use either single or dual element transducers.

At this stage of the development of ultrasonic flaw detection, only "straight beam" compression wave techniques, suitable for detecting reflectors parallel to the scanning surface, were in use. Attempts to angle the beam to reflect from surfaces at other orientations gave confusing results because of the existence of mode conversion in solids. With only a small angulation of the beam at the test surface, both compression waves and shear waves were generated in the object being tested. These traveled at different angles of refraction and at different velocities in the sample, making interpretation of the displayed echoes difficult.

In 1947, Sproule described a transducer design that would generate only shear waves in the sample, and this advance allowed the development of ultrasonic techniques for many other types of discontinuities in the welding, aerospace, and foundry industries. Rapid increase in the use and application of ultrasonics followed this development.

During the next 20 years, much of the development of ultrasonic techniques, as opposed to instrumentation, centered on the accurate sizing of the reflectors detected by the beam. Various approaches using beam parameters were tried, with varying degrees of success. Some were intended to estimate actual size, some were intended to assess the minimum theoretical reflector surface area, and some were intended to provide a common "go, no-go" reporting standard. In Britain, the intensity drop technique developed by C. Abrahams used the plotted –6dB or –20dB edges of the sound beam to plot the longitudinal and vertical extents of the reflector. In Germany, Krautkramer developed the DGS (distance gain amplitude) system to compare reflector amplitude against known-diameter circular reflectors (flat-bottomed holes) at defined depths for defined transducers. In the United States, signals were compared and reported against a DAC (distance amplitude correction) curve to produce a common reproducible reporting standard.

Each of the above techniques proved to be unsatisfactory for the application of fracture mechanics to determine fitness for purpose. Only the intensity drop technique claimed to assess the critical "through-thickness" dimension and this technique was heavily dependent on operator skill and reflector profile; repeatability of results was poor.

Sproule, in the 1950s, had described the tip diffraction signal originating at the tip of a discontinuity and defined its amplitude as being 30dB smaller than a corner reflector at the same depth. Whitford, in the 1960s, developed an alternative sizing system to the intensity drop system that he called the "maximum amplitude" system, in which the last maximum, defining the edges of the reflector, was the tip diffracted shear wave echo. In theory, this diffraction signal gave a much more precise location of the limits of the reflector. However, the signal from the diffracted shear wave is weak and difficult for the ultrasonic practitioner to positively identify.

It was Silk, in 1977, who first used the time of flight diffraction (TOFD) technique to display the top and bottom edges of discontinuities in a way that would allow greater accuracy of through-thickness measurement. The method employs angled compression-wave transducers, located on the same surface, to both transmit and receive sound. Lateral wave and tip diffraction signals allow accurate triangulation of the top and bottom edges of the reflector. Recent advances in instrumentation allowing real-time and postinspection analysis of results using computer technology have increased the number of users exploiting this technique.

If physicists have advanced the understanding of the theory and developed more and more ways of using ultrasonics, engineers have been no less productive in improving the instrumentation over the last 50 years. Early ultrasonic flaw detectors used vacuum tubes (valves), needed generated electricity, and were heavy (see Figure 7-1). Using quartz crystals, signal amplitude was poor and resolution very poor. After a shaky start, semiconductor technology has produced flaw detectors that are light, very portable, and together with synthetic crystal materials offers performance that is greatly enhanced. Much of this had been achieved by the mid 1970s.

During the 1980s and 1990s, microchips have been incorporated into the flaw detector, allowing the operator to store calibration parameters and signal traces. This, in turn, allows off-line analysis and reevaluation at a later date. Digital technology and the use of LCD display panels instead of CRTs during the 1990s has further reduced the size and weight of the flaw detectors.

FIGURE 7-1 Early ultrasonic instrument called the Supersonic Reflectoscope.

II. THEORY AND PRINCIPLES

Nature of Sound Waves

Sound waves are simply vibrations of the particles making up a solid, liquid, or gas. As an energy form they are therefore an example of mechanical energy, and it follows that, since there must be *something* to vibrate, sound waves cannot exist in a vacuum.

The only human sense that can detect sound waves is hearing, and that sense is restricted to a relatively narrow range of vibration frequencies called "the audible range". It follows that there will be vibration frequencies that are so low or so high that they cannot be detected by the human ear.

The unit of frequency is the hertz, abbreviated as Hz, defined as "one cycle of vibration per second." Sounds below approximately 16 Hz are below the limit of human hearing and are called "subsonic vibrations," and sounds above approximately 20,000 Hz are too high to be heard and are called "ultrasonic vibrations." Between those two values, in the audible range, it is more common to use the term "pitch" to refer to frequency; a high-pitched sound means high audible frequency, and low-pitched means low audible frequency. A piano key pitched at "middle C" is at a frequency of 260 Hz.

Abbreviations are used for high frequencies; 1000 Hz is shortened to 1 KHz (one kilohertz), 1,000,000 Hz becomes 1 MHz (one megahertz), and a billion cycles per second becomes 1 GHz (one gigahertz). In ultrasonic flaw detection, most testing is carried out in the MHz range (0.5 MHz to 25 MHz).

It is fortunate that there are devices called "transducers" that will change sound waves into electrical energy that can be displayed as visual signals on a cathode ray tube (CRT) or liquid crystal display (LCD) screen. This allows all sounds, including those outside the audible range, to be detected and studied. A transducer is defined as a device that will change one form of energy into another, and vice versa. Materials exhibiting the piezoelectric effect are commonly used to both generate and detect sound waves.

Vibration and Periodic Motion

A vibration is an example of periodic motion, a term that suggests that the body or particle concerned is undergoing some repetitive change of position with time. Another example is a pendulum swinging back and forth at a steady rate or frequency. To study the essential requirements for a vibration, consider Figure 7-2. The weight, W, is suspended from a beam by a spring. At rest, two equal and opposite forces are acting on the weight—gravity (G), acting downwards, is opposed by the tension (T) in the spring. The weight is said to be in a state of equilibrium. If the weight is lifted, slackening the spring, then released, gravity will try to restore the weight to its original position. If the weight is pulled down, the tension in the spring will increase; and when the weight is released, this extra tension will try to restore the weight to its original position.

This arrangement provides all the essentials to sustain a vibration. First, there must be something (the weight) to move, and second, there must be a restoring force that will try to counteract that movement or displacement (in this case gravity and the spring).

Imagine that the weight is pulled down from its normal rest position A in Figure 7-3 to position B, and then released. The extra stretch in the spring will exert a force on W so that it will begin to accelerate back to position A. As it moves, the stretch on the spring decreases until, at position A, both force G and force T are equal again. Since the weight has been accelerating all this time, it has now reached its maximum speed.

Any mass in motion possesses inertia, and this inertia will carry the weight on past the equilibrium position A. But, of course, as soon as it passes A, the spring slackens so that T becomes less than G. In other words, gravity starts to slow the weight down. Eventually, the weight comes to rest at a new position C. Now gravity outweighs the tension in the spring, trying to accelerate W back to position A. On reaching A, inertia ensures that the weight overshoots again and the whole train of events starts again.

If a pen is fixed to the weight and allowed to write against a piece of chart paper pulled

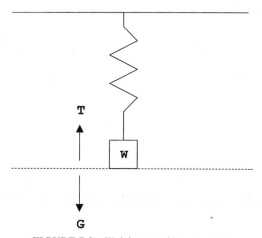

FIGURE 7-2 Weight on spring extended.

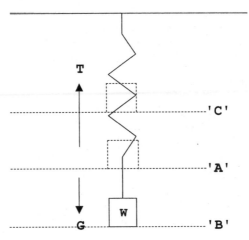

FIGURE 7-3 Weight on spring extended 2.

at a steady speed past the pen, a trace of the movement of the weight with time, as it bobs up and down, will be drawn, as shown in Figure 7-4. This trace is typical of all periodic motion and faithfully shows what has happened to the weight.

The steeper the line traced out, the faster is the movement of the weight. At the maximum displacement up or down, the trace is flat, showing that the weight has stopped briefly. As the weight passes through the normal rest position each time, the line is steepest, showing maximum speed. On the trace, t_1 represents the weight, traveling upwards and passing position A. The next trace position, t_2, shows the weight traveling down but again passing A, and t_3 again shows the weight traveling upward at position A.

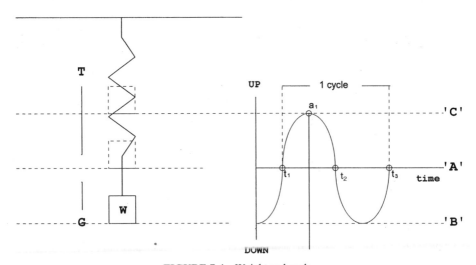

FIGURE 7-4 Weight and cycle.

Between t_1 and t_3 the weight has traveled up to its top limit, down to its bottom limit, and back to where it started. The trace between t_1 and t_3 has drawn "one cycle" of the motion of the weight in a set period. If the weight is allowed to carry on bouncing up and down until it eventually comes back to rest, the trace will only go on drawing repeats of that motion, each cycle occupying the same time span. The number of cycles completed in one second has already been defined as the frequency of the vibration.

The maximum displacement of the weight from its normal rest position is shown on the trace as a_1 and is known as the amplitude of the vibration. With sound vibrations, frequency is perceived as the pitch of the sound, whereas amplitude is the loudness of the sound.

Sound Vibrations

For sound waves in solids, liquids, and gases, the vibrating bodies are the particles making up the substance, and the restoring forces are the elastic bonds holding the substance together. The particles can be imagined to be joined together by springs. If one particle moves toward its neighbor, the spring gets squashed and tends to push the invader back "home." Similarly, if it moves away from its neighbor, the spring gets stretched and the particle is pulled back into place.

Audible sound is an example of a vibration mode called a "compression wave." It travels from the source by a succession of shunting actions from one particle to the next. Each particle vibrates at the frequency of the sound, oscillating to and fro by a distance that is the amplitude or loudness of the sound. As each particle oscillates, it squashes the "spring" to the next neighbor and starts the neighbor oscillating. As the oscillation passes from one particle to the next, and the next, and so on, the sound wave is said to travel or "propagate" through the material. Note that individual particles do not migrate to another place; they only oscillate about a mean position.

Modes of Propagation

The type, or mode, of sound wave propagation described above (compression wave) can exist in solids, liquids, or gases. Other modes of vibration can exist, but only in solids. The various ways in which sound can propagate are usually described in terms of the direction of particle motion in relation to the direction in which the sound wave travels. Compression waves can be defined on this basis as waves in which the particle motion is in the same plane as the direction of propagation.

All three media have forces that bind the particles together to resist squashing or pulling apart (compression or tension). In solids, this is provided by the modulus of elasticity, known as "Young's modulus of elasticity." The pressure of an entrapped gas as it is squashed rises to oppose the squashing force, and the pressure drops if the volume is increased, the partial vacuum applying the restoring force (see Figure 7-5).

Solids, unlike liquids and gasses, also have rigidity that is a resistance to shear loads. It is the rigidity that has to be overcome when snapping a stick, for instance. The name for this resistance to shear loads in solids is called "the modulus of rigidity," and it allows sound to propagate in a different way under certain circumstances. This new mode of propagation is known as a shear wave and is defined as a wave in which the particle motion is at right angles to the direction of propagation.

If a shear wave is set up so that it just skims along the surface of a solid, it again changes mode to one, which is contour following with a peculiar particle motion. This contour following wave is called a surface wave and is defined as a wave in which the particle motion is elliptical, with the major axis of the ellipse perpendicular to the direction of propagation.

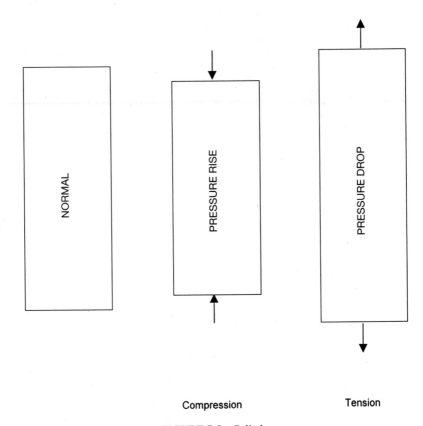

FIGURE 7-5 Cylinders.

Lamb waves, like surface waves, propagate parallel to the test surface and have a particle motion that is elliptical. They occur when the thickness of the test material is only a few wavelengths at the test frequency and where the test piece is of uniform thickness. In other words, they can exist best in plate, tube, and wire.

Finally, there is a special type of compression wave that skims along the surface rather like a surface wave and is called a creeping or lateral wave. Its use is described under TOFD techniques.

The four main modes of propagation are the compression wave, the shear wave, and the surface wave. Each of these has an alternative name that is sometimes used. These alternative names are:

Compression waves are sometimes called longitudinal waves

Shear waves are sometimes called transverse waves

Surface waves are sometimes called Rayleigh waves

Lamb waves are sometimes called plate waves

Properties of Sound Waves

Velocity

Sound travels at different speeds through different materials. This is noticeable when, for example, a railroad worker is observed from a distance striking a rail with a hammer. Since the speed of light is much faster than that of sound, the observer first sees the hammer strike the rail. If the observer is standing close to the rail, he or she next hears the sound of the blow coming from the rail. Finally, the airborne sound resulting from the blow is heard.

This shows us that the speed of sound in the rail is faster than the speed of sound in air. It is true that sound travels faster in liquids than in gasses and faster in metals than in liquids. However, it is also true that sound travels at different speeds in different metals. There is a distinct speed of sound for each material. In ultrasonics, this is called the velocity of sound for that material. This being so, it would be useful to have an understanding of the reasons for the difference.

Factors Affecting Velocity

The two main factors affecting velocity are the density and the elasticity of the material. To grasp a logical explanation for this, imagine that the molecules of any material are balls whose weight is analogous to the material density. So for lead, the balls would be heavy; they would be lighter for aluminum, and featherweight for air. Also imagine that these balls are joined together by springs representing the elasticity, or strength, of the material. For steel, the springs would be strong; they would be weaker for lead, and very weak for air. With these two concepts in mind, the scene is set.

The speed with which sound propagates through a material depends on how quickly one ball can get its neighbor to take up the vibration; in other words, to pass on the message. To get its neighbor moving, it has to overcome the inertia of that neighbor. Suppose two balls of a given weight are joined by strong springs and two more of equal weight are joined by weak springs. If the first balls in each pair are moved simultaneously, the ball facing the strong spring will quickly build up enough force to overcome the inertia of its neighbor. On the other hand, the ball facing the weak spring will have to move further and thus take longer to build up the equivalent force. From this, the logic tells us that it is reasonable to expect materials with a high value for Young's modulus of elasticity to have a high velocity of sound, as is the case, for example, with steel.

Consider two more pairs of balls. This time the springs are all the same strength but one pair of balls is light and the other heavy. If, again, the first balls in each pair are moved simultaneously, the lightweight pair quickly exchange messages but this takes longer in the heavy pair. Again, the general rule is indicated: the higher the density of a material, the lower the velocity of sound. Lead, for example, has a lower velocity than steel.

Density and elasticity are the dominant factors affecting velocity, but there is another one that plays a relatively minor, but nonetheless significant, role, and it is called Poisson's ratio. It is easy to see that when an elastic band is stretched it also gets thinner. The more it is stretched, the thinner it gets. Poisson's ratio relates the thinning to the stretching and can be calculated by dividing the change in diameter of the elastic band by the change in length.

The velocity of the compression wave for a given material can be calculated from the equation

$$V_c = \sqrt{\frac{E}{\rho} \cdot \frac{1-\sigma}{(1+\sigma)(1-2\sigma)}}$$

where
V_c = compression wave velocity
E = Young's modulus of elasticity
ρ = material density
σ = Poisson's ratio

Shear waves are able to exist in solids and they do not travel at the same velocity as the compression wave in a given material. This is because it is the modulus of rigidity, rather than Young's modulus, that dictates the velocity, and the modulus of rigidity is lower than the modulus of elasticity. This means that the shear wave velocity is always slower than the compression wave velocity in a material. As a rule of thumb, the shear wave velocity is roughly half the compression wave velocity. The velocity can be calculated from

$$V_s = \sqrt{\frac{E}{\rho} \cdot \frac{1}{2(1+\sigma)}} \quad \text{or alternatively} \quad V_s = \sqrt{\frac{G}{\rho}}$$

where
V_s = shear wave velocity
G = modulus of rigidity
ρ = material density
σ = Poisson's ratio

Surface (Rayleigh) waves also have their own particular velocity, which is generally taken to be approximately 90% of the shear wave velocity.

Although the velocity for each of these modes of propagation can be calculated, it requires a precise knowledge of all the parameters, and these are not usually available to the ultrasonic practitioner. Parameters such as density and strength vary with alloying, heat treatment, casting, rolling, and forging processes, all of which make it difficult to know that the correct values are being used. Instead, it is more normal to carry out a routine called "calibration" during the setting up procedure, for an ultrasonic inspection. In the calibration procedure the flaw detector timebase is adjusted to give a convenient scale against a calibration sample of known thickness and made of the same material as the work to be tested.

Wavelength

The distinction between the oscillating motion of the particles making up a solid, liquid, or gas and the velocity of the sound moving through the substance has already been made. As the particles are completing each cycle of their vibration, the sound wave is moving on in the direction of propagation at the characteristic velocity for that material. It follows that during the time taken to complete one cycle of vibration, the sound wave will move a certain distance depending on the velocity in that material. For gasses with low velocities, that distance is small compared to the distance in metals, which have high velocities. This distance for a given material and sound vibration frequency is called the wavelength.

Wavelength is given the Greek symbol λ (lambda) and for any material and sound frequency, can be calculated from the equation

$$\lambda = \frac{V}{f}$$

where
λ = wavelength

V = Velocity
f = frequency

Example 1. Calculate the wavelength of a 5 MHz compression wave in steel, given that the velocity of sound in mild steel is 5960 meters per second (m/sec).

$$\lambda = \frac{V}{f}$$

$$\therefore \lambda = \frac{5960}{5,000,000} \text{ meters (m)}$$

$$\therefore \lambda = 0.00192 \text{ m}$$

It would be better to express such a small distance in millimeters (mm) by multiplying the answer by 1000:

$$\lambda = 0.00192 \times 1000$$

$$\lambda = 1.192 \text{ mm}$$

At ultrasonic frequencies, the wavelength of sound in metals is relatively short and so it is usual to express the wavelength in millimeters. This is done at the start of the calculation by changing the velocity from meters to millimeters per second by multiplying by 1000.

Example 2. Calculate the wavelength of a 5 MHz compression wave in aluminum, given that the velocity is $0.252''/\mu$ sec (6400 m/sec).

$$\lambda = \frac{V}{f}$$

$$\lambda = \frac{6,400 \times 1000}{5,000,000}$$

$$\lambda = 1.28 \text{ mm}$$

Wavelength is useful in many ways in ultrasonic flaw detection. In the first place, the smallest reflector that can be detected must have a major dimension of at least half a wavelength at the test frequency. If the critical size of the discontinuity that must be detected is known, the knowledge helps with selection of an appropriate test frequency. Wavelength is also used in the calculation of the sound beam shape and the near-field distance. The significance of these will be discussed later.

Reflection

The boundary between one medium and another—for instance, steel to air at the far side of a steel plate—is called an "interface." At an interface, a proportion of the sound may be transmitted to the next medium and the remainder reflected back to the first medium. In the case of a steel to air interface, almost all the energy reflects and virtually none goes into the air. If the steel is under water, so that there is a steel to water interface, 88% of the energy is reflected and 12% is transmitted into the water. The proportions that will be reflected or transmitted depend upon the properties of the materials on either side of the interface.

In order to understand this phenomenon, consider again the balls and springs. This time, consider a train of tightly packed heavy balls joined by strong springs, representing steel. The steel train leads to a train of widely spaced light balls attached to weak springs, representing the air. If a compression wave is initiated at the start of the steel train, the message is passed on along the train, with each ball limited in its oscillation by its neighbor until the last ball in the steel train. It finds more space and lighter neighbors and is able to move a greater distance from its mean position in the direction of the air. In doing this, it stretches the spring joining it to the last but one steel ball until the spring tension arrests its motion. The last steel ball now starts to accelerate back into position and is so fast when it gets there that it overshoots, crashes into the last but one ball and starts a compression wave going back the other way. This is a reflection.

The two big differences between air and steel are density and elasticity, and these are the factors that decide how much energy is reflected and how much is transmitted at the interface. Each material is given a factor that is used to calculate reflectivity at an interface. This factor is called the "acoustic impedance" and given the symbol Z. Acoustic impedance is the product of density and velocity for that material. Stated mathematically

$$Z = \rho \times V$$

where
Z = acoustic impedance
ρ = material density
V = material velocity

To calculate the percentage of energy reflected at an interface between any two materials, the following formula is used:

$$\text{Reflected energy} = \left(\frac{Z_1 - Z_2}{Z_1 + Z_2}\right)^2 \times 100\%$$

Where Z_1 & Z_2 are the acoustic impedance of the materials on either side of the interface.

Example 3. Calculate the percentage energy reflected at a steel to water interface, given that the acoustic impedance for steel is 46.7 and that for water is 1.48.

$$\text{Reflected energy} = \left(\frac{Z_1 - Z_2}{Z_1 + Z_2}\right)^2 \times 100\%$$

$$\text{Reflected energy} = \left(\frac{46.7 - 1.48}{46.7 + 1.48}\right)^2 \times 100\%$$

$$\text{Reflected energy} = \left(\frac{45.22}{48.18}\right)^2 \times 100\%$$

Reflected energy = $(0.93856)^2 \times 100\%$

Reflected energy = $0.8809 \times 100\%$

Reflected energy = 88.09%

Example 4. Calculate the percentage energy reflected at a steel to air interface, given that the acoustic impedance for steel is 46.7 and that for air is 0.0004.

$$\text{Reflected energy} = \left(\frac{Z_1 - Z_2}{Z_1 + Z_2}\right)^2 \times 100\%$$

$$\text{Reflected energy} = \left(\frac{46.7 - 0.0004}{46.7 + 0.0004}\right)^2 \times 100\%$$

$$\text{Reflected energy} = \left(\frac{46.6996}{46.7004}\right)^2 \times 100\%$$

Reflected energy = $(0.99998)^2 \times 100\%$

Reflected energy = $1.0000 \times 100\%$ *approx*

Reflected energy = 100%

When a beam of sound traveling through a metal sample encounters a discontinuity such as a crack, lamination, void or nonmetallic inclusion there is an interface. On one side is the sound metal and on the other the discontinuity. At this interface, some energy will be reflected and some transmitted. If the discontinuity side of the interface is air then the reflection is total; but even for a nonmetallic inclusion, most of the energy will be reflected. This property of sound waves allows for the detection of discontinuities in materials.

Couplant

The property of reflection can also be a problem because if a transducer is simply placed on a part there must be an air gap, however small. But a solid to air interface creates 100% reflection, so the sound goes straight back into the transducer without transmitting into the metal! To overcome this problem there has to be some way to exclude the air using a medium that will match the acoustic impedance of the transducer to the metal. Since this medium must also allow the transducer to be scanned over the surface of the metal it needs to be a liquid, grease, or paste. Such a substance is called a "couplant."

There are many suitable substances that can be used as couplants, the main criteria being the best possible match and no adverse chemical reaction between the couplant and the metal. Most couplants only allow limited matching because liquids in general have a low acoustic impedance. In immersion testing, the couplant is usually water, which only allows about 12% of the energy into the steel, and, of course, only 12% of any echoes to pass back across the interface and back to the receiving transducer. Most couplants permit between 10% and 15% sound transmission. The best of these is glycerin at around 15%. Commonly used couplants are:

- Water
- Kerosene
- Oil
- Grease
- Wallpaper paste
- Glycerin
- Special gels designed for the purpose

Refraction and Mode Conversion

So far only sound entering the metal perpendicular to the surface has been discussed. When the sound is introduced at an angle to the surface called the "angle of incidence," several things may happen, depending on the actual angle of incidence.

Figure 7-6 shows a beam of sound traveling toward an interface at an angle of incidence $i°$ to the perpendicular that is usually called the "normal." The velocity of sound in Medium 1 is V_1 and in Medium 2, on the other side of the interface, the velocity is V_2.

Assume for this example that V_1 is slower than V_2, as would be the case if Medium 1 had been water and Medium 2 steel. As the beam travels toward the interface, the whole beam is moving at the same speed until the left-hand edge first strikes the interface. The moment the edge of the beam reaches Medium 2, it speeds up. But the sound still in Medium 1 stays at the old speed. Gradually, as the entire wave-front sweeps the interface, it speeds up until at last the right hand edge passes across the interface and the entire beam travels on in Medium 2 at the new speed.

During this transition, the beam slues around to a new angle in Medium 2 called the "angle of refraction." It is a bit like somebody passing through a doorway, catching their pocket on the door handle, and being diverted by the braking action on that side. The reason for this refraction is the velocity difference on either side of the interface. Snell's law allows the new angle to be calculated if the two velocities and the angle of incidence are known. Snell's law states that the sine of the angle of incidence divided by the velocity in Medium 1 equals the sine of the angle of refraction divided by the velocity in Medium 2. Stated mathematically:

$$\frac{\sin i°}{V_1} = \frac{\sin R°}{V_2}$$

where
 $\sin i°$ = The sine of the angle of incidence
 $\sin R°$ = The sine of the angle of refraction
  V_1 = The velocity in Medium 1
  V_2 = The velocity in Medium 2

Refraction refers to the transmitted portion of the sound energy at the interface; the proportion of energy reflected is the same as before but the reflected energy leaves the interface at an angle of reflection equal to the angle of incidence, as shown in Figure 7-7. This diagram shows the angle of incidence (i°), angle of reflection (r°), and angle of refraction (R°).

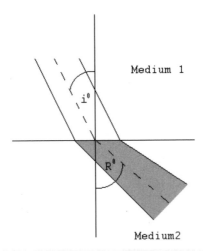

FIGURE 7-6 Refraction.

Example 5. Calculate the angle of refraction in steel for an incident angle in water of 10° given that the compression wave velocity of sound in water is 1480 m/sec and in steel 5960 m/sec.

$$\frac{\sin 10°}{1480} = \frac{\sin R°}{5960}$$

$$\frac{5960 \times \sin 10°}{1480} = \sin R°$$

$$\frac{5960 \times 0.1736}{1480} = \sin R°$$

$$0.6993 = \sin R°$$

$$R° = 44.37°$$

The relationship between velocity and refraction can be seen in the above example, the compression wave velocity of sound in steel is roughly four times that in water, and the refracted compression wave angle is roughly four times the incident angle.

Mode Conversion

As the beam of sound is introduced at an angle of incidence to a solid, another phenomenon begins to arise, and that is mode conversion. Although the incident beam is a compression wave, a refracted shear wave begins to develop in the solid as the sound crosses the interface, in addition to the refracted compression wave. For small angles of incidence, the amplitude of the shear wave is small and can be ignored, but as the angle of incidence increases, so does the amplitude of the shear wave. Eventually, both the shear wave and the compression wave are about equal in amplitude. Snell's law shows us that the two modes will not refract through the same angle because the velocity of the shear wave is less than the compression wave.

Example 6. Calculate the angle of refraction of the shear wave in steel for an incident compression wave of 10° in water, given that the shear wave velocity in steel is 3240 m/sec.

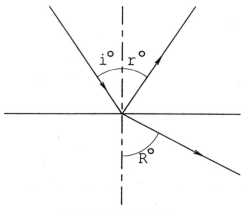

FIGURE 7-7 Snell's law.

$$\frac{\sin i°}{V_1} = \frac{\sin R°}{V_2}$$

$$\frac{\sin 10°}{1480} = \frac{\sin R°}{3240}$$

$$\frac{3240 \times \sin 10°}{1480} = \sin R°$$

$$\frac{3240 \times 0.1736}{1480} = \sin R°$$

$$0.3801 = \sin R°$$

$$R° = 22.34°$$

The problem for ultrasonic flaw detection is immediately obvious—two beams traveling at different speeds and in different directions spells chaos! This was a problem that beset the early practitioners until Sproule came up with a solution in 1947. What he did was to get rid of one of the beams. He did this by increasing the angle of incidence until the refracted compression wave refracted to 90°. Any further increase in incident angle leaves only a refracted shear wave. The compression wave is said to have undergone total internal reflection in Medium 1.

The angle of incidence giving a 90° refracted angle for the compression wave is called the "first critical angle." The first critical angle for a water to steel interface is about 15°, and for Plexiglas (Lucite or Perspex) to steel the first critical angle is about 28°. Above the first critical angle of incidence, only a shear wave remains. By suitable choice of an incident angle above the first critical angle, a shear wave beam of any desired angle can be achieved. For immersion testing, the transducer is simply tilted through the calculated angle of incidence in water. For manual scanning, the transducer is mounted on a Plexiglas wedge angled to the desired incident angle.

If the incident angle is increased more and more beyond the first critical angle, eventually the shear wave will also be refracted to 90°. The angle of incidence to achieve this is called the "second critical angle." At the second critical angle, the shear wave undergoes another mode conversion; this time it becomes a surface (Rayleigh) wave, which is the contour-following wave. Any increase in angle of incidence beyond the second critical angle leaves no sound in Medium 2 at all; there is total internal reflection in Medium 1. For water to steel, the second critical angle is about 27°, and for Plexiglas to steel about 58°. Figure 7-8 shows the angle of refraction in steel obtained for increasing angles of incidence in Plexiglas at a Plexiglas to steel interface, and Figure 7-9 shows the same for a water to steel interface.

Reflective Mode Conversion

Mode conversion also takes place inside a solid when an ultrasonic beam strikes a reflector at an angle of incidence other than perpendicular. Figure 7-10 shows a compression wave, C, striking a steel to air interface at an angle of incidence, i°, to the normal. The reflected compression wave, C_r, is at a reflected angle r° equal to the angle of incidence. However, there is also a mode converted shear wave, S, at an angle derived from Snell's Law, s°. In ultrasonic flaw detection, this mode conversion can cause confusion, depending on the relative amplitudes of the reflected compression wave, and the mode converted shear wave.

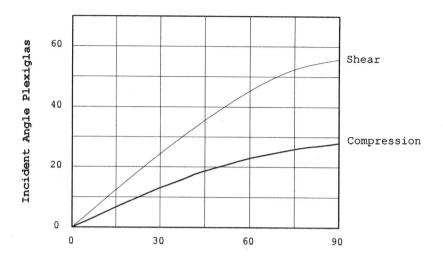

FIGURE 7-8.

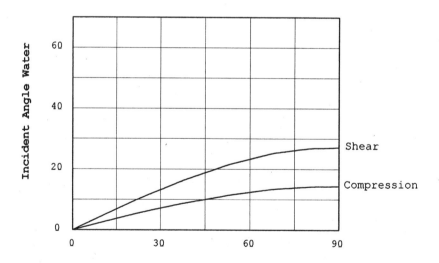

FIGURE 7-9.

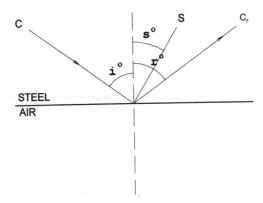

FIGURE 7-10 Incident and reflected angles.

Figure 7-11 shows the relative amplitudes of the reflected compression wave and the mode conversion shear wave for increasing angles of incidence, of the compression wave at a steel to air interface. It can be seen that at low angles of incidence the shear wave is weak and can be ignored. At an angle of incidence of about 25°, the reflected compression wave and the shear wave are at the same amplitude, and at about 70°, the reflected compression wave is very weak whereas the shear wave is still very strong.

Figure 7-12 shows the angle of the mode-converted shear wave with respect to the normal for increasing angles of incidence for the compression wave at the steel to air interface. These two graphs show the strength and direction of the mode conversion. The practitioner may encounter these circumstances when carrying out a compression wave (straight beam) test if a reflecting surface is not parallel to the scanning surface.

Mode conversion can also take place when a shear wave meets a reflecting surface, as shown in Figure 7-13. In this case, the reflected shear wave is again at the same angle as the angle of incidence, but the mode-converted compression wave is at an angle $\alpha°$, which can be calculated from Snell's law.

Figure 7-14 shows the amplitudes of the reflected shear wave and the mode-converted compression wave relative to the amplitude of the incident shear wave.

Figure 7-15 shows the angle of the mode-converted compression wave. It can be seen from the graph that when a shear wave is incident to a reflecting surface at about 30°, less than 10% of the shear wave is reflected, but the mode-converted compression wave amplitude rises steeply, and is far greater than the reflected shear wave. Figure 7-15a shows that for this incident angle of shear wave, the mode-converted compression wave will be at an angle of about 65° to the normal. This situation will occur if a 60° angle beam transducer is chosen to examine a weld with a vertical fusion face, such as an electron-beam weld, or the root face of a double "V" weld preparation. As shown in Figure 7-15b, the shear wave will meet any vertical nonfusion at an incident angle of 30°, and the strong mode-converted wave will reach the transducer. The path taken by this wave and its velocity will cause confusion if the practitioner is unaware of the problem. Obviously, a 60° beam angle is a wrong choice in those circumstances.

Beam Characteristics
Many of the illustrations so far used have treated the sound as if it were a single ray, but in fact, the sound propagates as a beam. Within the beam, intensity or amplitude of the

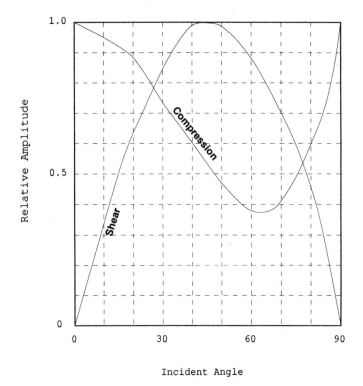

FIGURE 7-11 Incident angle graph.

sound energy varies. The following paragraphs deal with those variations and with the shape of the beam. For convenience, the beam is divided into two distinct zones called the "near field" and the "far field." In these two fields, different mechanisms are at work to vary the sound intensity. The word used to describe what effectively is a gradual loss of sound energy is "attenuation." Attenuation is the combined effect of a number of parameters:

- Interference and diffraction effects
- Interference Absorption (friction and heat)
- Interference Scatter
- Interference Beam spread

Interference and Diffraction Effects. Huygens developed a convenient way of looking at wave energy propagating from a source. He said that a point source was rather like dropping a stone into a pond; the disturbance moves out as an expanding circle on the pond, but from a sound source the circle becomes an expanding sphere—it moves out in all directions. A sphere is a three dimensional object that is difficult to portray on a sheet of paper, so for this exercise a circle will have to do.

Figure 7-16 shows a point source surrounded by concentric circles representing successive pressure waves of sound frozen in time a short time after the sound starts. The

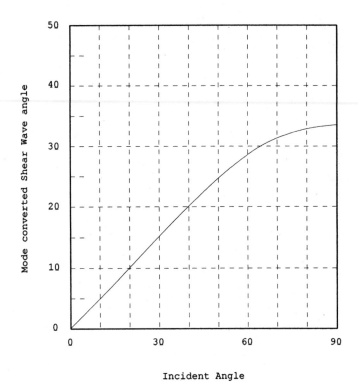

FIGURE 7-12 Incident angle graph 2.

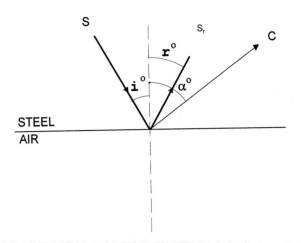

FIGURE 7-13 Incident and reflected angles 2.

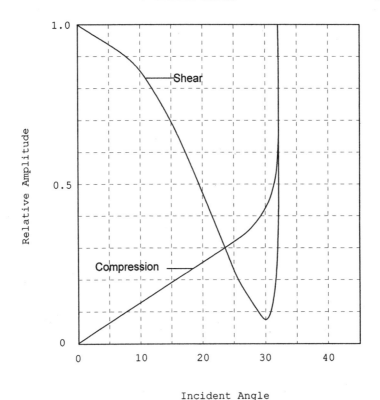

FIGURE 7-14 Incident angle graph 3.

spaces between the circles represent the rarefaction part of each cycle of sound. If the pattern had been frozen a little later in time, the outer circle would have been bigger in diameter. The space between each circle represents one wavelength of the sound in the material. But an ultrasonic transducer is not a point source, it has a diameter and a surface area, all of which is active. This is called a "finite source" and Huygens said that this could be considered as being made up of an infinite number of point sources. Figure 7-17 shows a finite source with a few of these infinite point sources frozen a short time after the vibration has been started. It can be seen that the wave fronts from the point sources combine to make a united wave front as the "beam" propagates from the source. But notice how a little bit of the sound is lost around the edges of the source; it is said to "diffract" around the edges; this is one of the energy losses in the near field.

The next source of loss needs a little more explanation. Figure 7-18 shows a finite source again, but this time only the point sources in the center and at the edges are shown, for simplicity. In front of the source is a point "P," which is waiting for sound to arrive. In the diagram, the first pressure wave from the middle of the source has already arrived, but sound from the edges has some way to go yet. "P" is given a gentle nudge in the direction of propagation by the sound from the middle of the source. Figure 7-19 shows the situation a short time later, when the first pressure wave from the edges of the source arrive at "P."

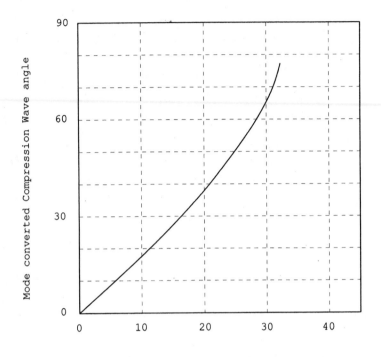
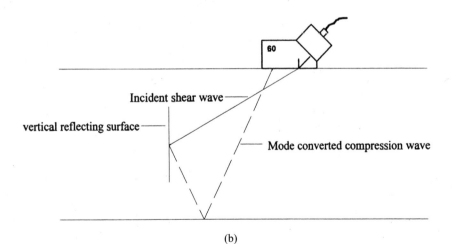

FIGURE 7-15 (a) Incident angle graph 4. (b) Mode converted beam.

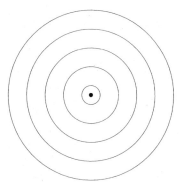

FIGURE 7-16 Point source.

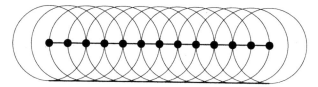

FIGURE 7-17 Wavefront.

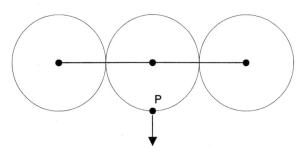

FIGURE 7-18 Point source 2.

The result of these two nudges from either side is again to move "P" in the direction of propagation. But that is without considering what is arriving from the center of the source. In the illustration, the third pressure wave is arriving from the center, resulting in an extra-large nudge for "P," representing the combined nudges from the edges and middle. These three simultaneous nudges are called "constructive interference" because the end effect is a local increase in sound intensity. It has happened because 'P' is an exact number of wavelengths from both the center and the edges of the source for the frequency of the sound wave. A change in frequency or a shift in the position of "P" might result in the sound from the center and from the edges arriving at "P" out of phase, as shown in Figure 7-20.

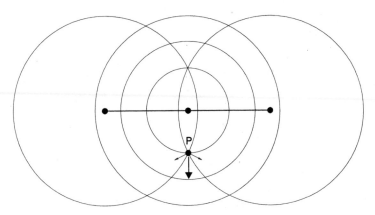

FIGURE 7-19 Point source 3.

In this figure, the first pressure waves from the edges have arrived at 'P' as a rarefaction arrives from the center. Two forces are pushing and pulling at "P." This is called "destructive interference" and leads to a local reduction in sound energy. For an absolutely pure frequency continuous wave sound, the destruction could be total; in other words, there would be no sound at all at "P." The pulsed, broadband transducers used in ultrasonic flaw detection never quite cancel out.

The reason for this destructive interference is the difference in path length from 'P' to the center of the source and to the edges compared to the wavelength. This being so, eventually there will be a distance for "P" in front of the source where the path difference becomes significantly less than a wavelength (see Figure 7-21) and the interference effects cease. This distance is the end of the "near field." The near field distance, *NF*, can be calculated from

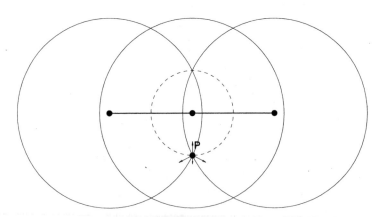

FIGURE 7-20 Point source 4.

$$NF = \frac{D^2}{4 \times \lambda}$$

Alternatively, where the wavelength is unknown

$$NF = \frac{D^2 \times f}{4 \times V}$$

where:
D = Transducer diameter
F = Frequency
V = Velocity

Example 7. Calculate the near field distance in steel for a 5 MHz compression wave when using a transducer that is 10 mm in diameter.

$$NF = \frac{D^2 \times f}{4 \times V}$$

$$NF = \frac{100 \times 5{,}000{,}000}{4 \times 5{,}960{,}000}$$

$$NF = \frac{5{,}000{,}000}{23{,}840}$$

$$NF = 20.97 \text{ mm}$$

Absorption

Sound propagates through the vibration of particles of a solid, liquid, or gas and the movement of those particles causes friction and absorbs some of the energy. The rate at which energy is absorbed depends on the material through which the sound is passing and the frequency of the sound. In general, the higher the frequency, the greater the absorption; or put another way, the lower the sound frequency, the further it penetrates into the material.

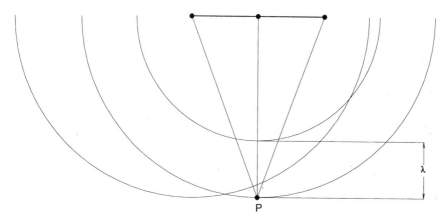

FIGURE 7-21 Point source 5.

Scatter

Sound waves will reflect from interfaces within the material being tested, and grain boundaries in solids are interfaces that may be randomly orientated to the beam. This causes some of the sound to reflect in random directions or "scatter." Very fine-grained material causes very little scatter but coarse-grained material causes considerable scatter. Scattered energy that does not reach the receiver transducer is "lost" energy. Scattered energy that does reach the receiver is worse! It creates small signals across the timebase. This condition is called "noise," "grass," or "hash" and it tends to mask signals from discontinuities.

Both absorption and scatter exist as sources of lost energy in both the near field and the far field. Beam spread is the remaining cause of energy loss affecting the far field.

Beam Spread

In the near field, the beam is taken to be roughly cylindrical and the same diameter as the transducer crystal. Beyond the near field, in what is called the "far field," the beam spreads out like a cone. The angle of the cone, as shown in Figure 7-22, can be calculated from

$$\sin\frac{\theta}{2} = \frac{1.22\,\lambda}{D}$$

Where:
$\theta/2$ = Half angle of beam spread
λ = Wavelength
D = Transducer crystal diameter

The above equation includes the constant 1.22. This calculates the beam spread to the absolute limit of the beam where sound ceases to exist. This is not a practical limit for the ultrasonic practitioner because if sound doesn't exist, it can't be detected or measured. In practice, it is more usual to replace the constant 1.22 with either 0.56 or 1.08. The 0.56 value predicts the limits of the beam where the sound has dropped to one half of the intensity at the beam center. The 1.08 value defines the limits where the sound is one tenth of that at the beam center.

Note: The constants used above (e.g., 0.56, 1.08, and 1.22) are commonly used for calculation of theoretical beam shapes. If the shape of the beam is required for discontinuity sizing purposes, it is more practical to plot the beam shape using a special calibration block, rather than calculate the beam spread (see Intensity Drop Technique on page 7.92).

Example 8. Calculate the beam spread angle for a 5 MHz compression wave in steel when using a 10 mm diameter transducer (λ = 1.192 mm).

$$\sin\frac{\theta}{2} = \frac{1.22\,\lambda}{D}$$

$$\sin\frac{\theta}{2} = \frac{1.22 \times 1.192}{10}$$

$$\sin\frac{\theta}{2} = \frac{1.45424}{10}$$

$$\sin\frac{\theta}{2} = 0.145424$$

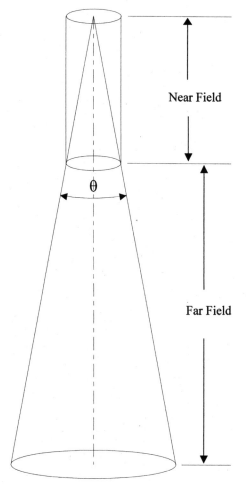

FIGURE 7-22 Near and far fields.

$$\sin\frac{\theta}{2} = 8\cdot36°$$

$$\therefore \theta = 16\cdot72°$$

Figure 7-22 shows the overall beam shape, including the near-field portion.

Figure 7-23 shows the way in which amplitude changes along the beam center. In the near field, there are fluctuations in amplitude because of the interference effects. The last maximum amplitude marks the end of the near field and the beginning of the far field. This is called the Y_0 point. In the far field, the intensity can be seen to decay exponentially. From a practical point of view, Figure 7-23 implies that it is unreliable to use amplitude as an acceptance criterion for flaws detected in the near field. In some ap-

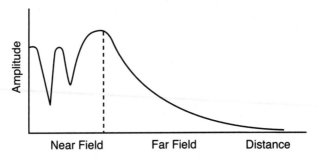

FIGURE 7-23 Near and far field plots.

plications, this problem is overcome by using a delay line (a Plexiglas column) between the transducer and the work piece so that the near field is contained within the delay line.

The Decibel System

In order to compare data, some form of measurement is necessary; e.g., for length, the standard used is either inches or millimeters. Because electric power is proportional to the square of the voltage produced, it can be said that the voltages produced at the transducer are relative to the sound intensity. When measuring sound intensities, the unit of measurement is the bel. The bel is named for Alexander Graham Bell (1847–1922), inventor of the telephone.

The bel being a large unit of measurement, it needs to be broken down into smaller units. These units are called decibels or dB. "Deci" is a prefix that is borrowed from the Latin. It means "one tenth," so a decibel is one tenth of a Bel. The decibel is a unit used to express the intensity of sound energy. It is equal to twenty times the common logarithm of the ratio of the pressure produced by the sound energy, to a reference pressure. In other words, it is used to express the ratio of the magnitudes of two reflections (signals), each having different magnitudes, equal to twenty times the common logarithm of the voltage or current ratio. In practical terms, if there are two signals on the screen and the difference between these signals needs be known, it can be calculated. Alternatively, if a signal of known amplitude were to be reduced by a certain percentage, this reduction in gain can be calculated. For example, if a signal height of 100% is to be reduced to 10%, the reduction in gain can be calculated.

Of course, this can be measured by using the gain control or attenuator on an ultrasonic instrument, assuming that the instrument is linear in the vertical axis (see subsections on linearity in Section IV).

To calculate the difference between two signal amplitudes, the following formula is used:

$$dB = 20 \times \log\left(\frac{A_1}{A_2}\right)$$

where
A_1 = the first percent signal height
A_2 = the second percent signal height

TABLE 7-1 Amplitude Ratios

dB	Ratio
3	1.41:1
6	2.00:1
9	2.82:1
12	3.98:1
14	5.01:1
20	10.00:1

Example. Two signals are noted on the screen. The first has an amplitude of 80% full screen height (FSH) and the second is noted to be 40% FSH. Calculate the difference in dB between the two.

$$dB = 20 \times \log\left(\frac{80}{40}\right)$$

$$\log 2 = 0.3010$$

$$20 \times 0.3010 = 6.02 \text{ dB}$$

The 80% signal is 6 dB greater than the 40% signal. It can be seen from this calculation that +6 dB is twice the amplitude.

Conversely, if the 80 and the 40 were inverted, that is, if the 40 were placed over the 80 and divided, the result would be 0.5. The log of 0.5 is the same as the log of 2; however, it calculates to negative or −.3010. If this is multiplied by 20, it can be noted that −6 dB is one half the amplitude

This means that if the signal amplitude were to increase by 6 dB, the signal height would double. If the signal amplitude were decreased by 6 dB, the signal height would be half of its original amplitude.

The above formula applied to a first signal amplitude of 100% and second signal amplitude of 10% will result in a difference of 20 dB (see Table 7-1).

Having noted the significance of the pressure (amplitude) differences of reflections, it follows that the same rationale applies to the cross section of the ultrasonic energy "beam." If the beam intensity were shown as a slice across the beam section, it would appear as in Figure 7-24. Information regarding the actual pressure differences across the beam section is very important, particularly when using transducer movement to investigate or evaluate the characteristics of a discontinuity (see Section IV).

III. EQUIPMENT FOR ULTRASONIC APPLICATIONS

As with computers, the technology concerning ultrasonic equipment and systems is becoming somewhat transitory. Ultrasonic systems are either battery operated portable units, multicomponent laboratory ultrasonic systems, or something in between. Whether they are based on modern digital technology or the fast disappearing analog original, "systems" (often defined as instrument plus transducer and cable) basically comprise the following components. The appropriate controls are shown in the ellipses in Figure 7-25.

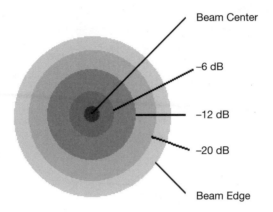

FIGURE 7-24 Beam slice.

1. Transducer
2. Pulser (clock)
3. Receiver/amplifier
4. Display (screen)

To understand how a typical ultrasonic system operates, it is necessary to view one cycle of events, or one pulse. The sequence is as follows.

1. The *clock* signals the *pulser* to provide a short, high-voltage pulse to the *transducer* while simultaneously supplying a voltage to the *time-base trigger* module.

2. The *time-base trigger* starts the "spot" in the CRT on its journey across the screen.

3. The voltage pulse reaches the transducer and is converted into mechanical vibrations (see "piezoelectricity"), which enter the test piece. These vibrations (energy) now travel along their "sound path" through the test piece. All this time, the spot is moving horizontally across the CRT.

4. The energy in the test piece now reflects off the interface (back wall) back toward the transducer, where it is reconverted into a voltage. (The reconverted voltage is a fraction of its original value.)

5. This voltage is now received and amplified by the *receiver/amplifier*.

6. The amplified voltage is sent to the "vertical (Y axis) plates" (top and bottom) in the CRT. At this time, the upper Y axis plate attracts the spot upward. This motion produces the "signal" on the screen that signifies the time that the energy has taken to make the round trip through the test piece, from the moment the energy leaves the transducer until it is received by the transducer. The spot is set to start its trip at the time the energy enters the test piece. This is manually adjusted by using the *delay* or *zero* control. This step is particularly necessary when using a Plexiglas delay line (see Glossary of Terms).

7. The same "packet" of returning energy has by this time reflected down off the test piece's top interface and now makes a second trip down through the test piece. (The spot continues its horizontal journey across the screen.) The energy reflects once more off the back wall interface and returns again to be received and amplified. The amplifier once again sends the voltage to the Y axis plates. The spot is again drawn up toward the upper Y axis plate, this time at a "later" position on the time base. This is a "repeat signal" (multi-

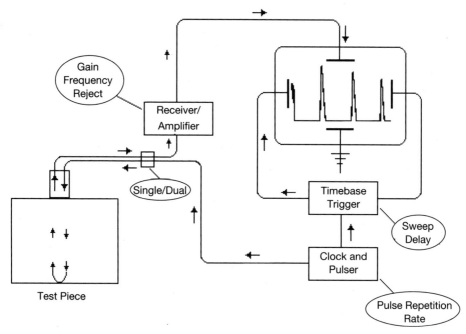

FIGURE 7-25 Block diagram, analog.

ple) that is lower in amplitude because of factors such as attenuation and other losses. The spot is then released to continue its journey across the screen, and the above sequence repeats again and again until the energy in the test piece has been attenuated. The display will show multiple repeat signals, as many as are available in the calibrated time base and according to the amount of amplification (gain) selected. For example, if the screen is calibrated for 5 inches of steel and the test piece is 1 inch thick, there will be five signals on the screen, representing five "round trips" of 1 inch thickness. (The energy has, in fact, traveled two inches each trip, one forward and one back, but it is displayed as a series of 1 inch trips on the screen.)

8. The clock now sets off the pulser a second time and the next pulse is produced. The complete scenario is repeated over again, n number of times per second. The n number of pulses per second is referred to as the *pulse repetition frequency* (PRF) or the *pulse repetition rate* (PRR).

Digital Instrumentation

In principle, the above also applies to digital flaw detection instrument (see Figure 7-26).

The Controls and Their Functions
Instrumentation varies by manufacturer; however, there are three controls that are common to most ultrasonic flaw detection equipment. These controls are sweep (range), delay, and gain.

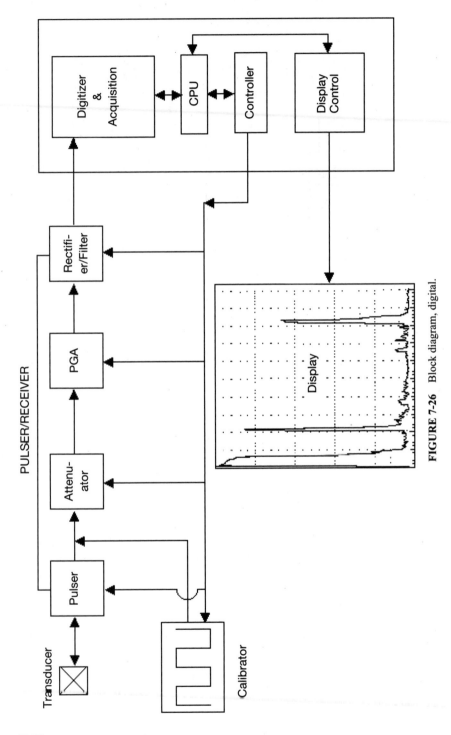

FIGURE 7-26 Block diagram, digital.

Sweep (Range—Coarse and Fine). These controls usually exist on analog instruments. On digital units, the controls are usually combined into one, designated as "range." The function of this control is to adjust the speed of the screen spot in order to accommodate displays of varying sound paths. The spot will move slower when the screen is to represent a long sound path, e.g., displaying a time base for a long shaft of steel (slow time base). Conversely, the spot will move very fast across the screen when displaying a few millimeters of steel across the full screen (fast time base).

Delay (Zero). This control delays the spot from beginning its journey across the screen. For example, when using a dual transducer or a transducer with a Plexiglas wedge (standoff or delay line), the start time of the spot has to be delayed for an instant to allow the sound to travel through the Plexiglas wedge before it enters the test piece. The display need only show the area from the top of the test piece and onward. Displaying the wedge on the screen is unnecessary and also confusing. When calibrating the system, test piece material zero should appear at screen zero. The delay (zero) control will be used to accomplish this.

Gain (Attenuator). Because everything is relative to something else, and to facilitate the ability to make decisions about the nature of the signals that are observed on the CRT when conducting an ultrasonic examination, certain comparisons need to be made. The meaning of a signal on the screen is rather limited unless it is related to something tangible such as another signal from a known reflector.

If the signal from a reflector in a test piece is compared with a signal from a known reference reflector, theoretically (all things being equal), the size of the reflectors can be compared. [Note: signal amplitude in itself does not necessarily indicate reflector (discontinuity) size.]

If accept or reject decisions are made based on signal amplitude, consideration of a signal that saturates the screen may be impossible. That is because it exceeds 100% full screen height (FSH). Since the signal is above the viewable screen, it is impossible to make any comparisons unless the signal height is adjusted so that the top can be seen. If the instrument gain were to be adjusted so as to lower this signal until it becomes the same height as our reference signal, the actual reference signal height may be reduced so much that it cannot be quantified; or, worst case, it may be so low in amplitude as not to be visible at all, so signal amplitudes can not be compared.

The solution to this problem is to employ a "volume" control, much the same as those found in stereo systems. This device is known as either a "calibrated gain control" or, in some cases, an "attenuator." The difference between the two devices is their functionality. This is discussed below. If a numeric value could be attributed to the amount that the signal amplitude (gain) is adjusted, numeric signal height comparisons could be conducted.

Gain controls and attenuator controls operate using similar circuitry. The difference to the user is that when using an attenuator, increasing the attenuation results in diminished signal height. When using the calibrated gain control, increasing the gain level increases the signal height. The differences in signal height and variations in gain need to be known in order to obtain accurate data.

Other may vary between instruments by respective manufacturers; however, the following features are generally common to most:

Single and Dual Transducer Selection. This switch isolates the transmitter side of the circuit from the amplifier. It allows the reception of voltage generated solely by the receiver side of a dual transducer or from a transducer used as a receiver in a "pitch catch" or through transmission mode. In the case of a single transducer, the first voltage to reach the amplifier is from the pulser and transducer (in combination). This produces the signal on the left-hand side of the screen that is often referred to as the "initial pulse" or "main bang." It is obvious that the initial pulse "blinds" the display for an area under the transducer (front surface) in the test piece. Reflectors close to the test surface will not be resolved because they occur within the time of the initial pulse. Using the "dual" setting with a dual transducer essentially eliminates the "initial pulse" signal from the left hand side of the screen, thus increasing the available test time close to the front interface (top) of the test piece.

Frequency Selection. Transducers operate at a predetermined nominal frequency based on their thickness. The transducer oscillates at its resonant frequency, but it also produces other frequencies, some higher and some lower than the nominal center frequency. It is sometimes necessary to filter out undesirable frequencies as they can produce low-level noise. This can adversely affect (reduce) the "signal-to-noise" ratio. It is important that the signal from a reflector be visible above the background noise caused by material grain and other factors such as instrument circuit noise. To this end, some ultrasonic instrumentation is designed so that individual frequencies are user-selectable. Other instrumentation is designed so as to accept a range of frequencies that are not user-selectable. These are classified as having either narrow band receivers or broad band receivers. The narrow band receiver is usually user-selectable. In this case, the user selects the frequency nearest to that of the transducer being used. The effect is that the receiver processes only the frequency selected, within a certain "bandwidth"; e.g., a 5 MHz selector may be receptive to energy from 4 MHz to 6 MHz, depending on the design specifications. This circuitry will filter out frequencies outside this bandwidth. Another name for this type of circuitry is called a "bandpass filter." There are other types of filters. Those that allow frequencies higher than a certain value to be processed are called "high-pass filters." Conversely, filters that blank out frequencies above a certain value are called "low-pass filters." When conducting a test on very grainy material, low-frequency energy is used to help overcome the grain noise. In this case, it is advantageous to use a low-pass filter to reject the scattered higher-frequency energy. This usually helps to increase the signal-to-noise ratio and provide superior data.

Gates

Electronic gates are used to produce some action based on a signal being present in the gate. A gate is a device that is inserted into the time base at a user-selected location, as shown in Figure 7-27. It is usually seen as an extra electronic line on the time base. A "positive gate" will enable any signal interrupting this gate to cause a voltage to be sent to a selected apparatus, for example, an audible or visible alarm. Conversely, if used in a negative direction, the absence of a signal in the gate can cause a similar action to take place. For instance, the back-echo signal can be gated. If there is some diversion causing the back-echo to disappear, this occurrence could signal an alarm of some kind. The alarm may also be in the form of a paint gun used to mark a discontinuity area on the item being examined, or a pen recorder used to record the event on a strip chart or an X–Y recorder. Depending on the circuitry, the signal can cause the gate to deliver a voltage that is proportional to the signal amplitude. The gate can be used to produce a numeric display of the signal's horizontal position or percentage amplitude on the screen. The gate threshold can be selected so that it functions at a predetermined signal amplitude. The user, dependent upon the application, also determines the gate position and width. Modern instrumentation uses the gated signal to provide information that is used to calculate and numerically display flaw depth or, in the case of an angle beam transducer, the location distance of a reflector in front of the transducer by programmed trigonometry. Gated signals are also used to produce a "C scan" image (see Figure 7-31). Generally, during a manual ultrasonic examination, it is not necessary to use a flaw gate, although it can also be used to electronically mark a position or amplitude on the time base.

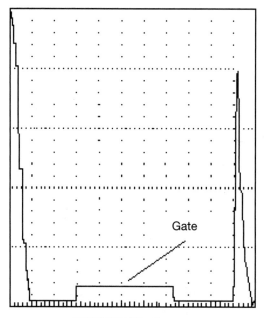

FIGURE 7-27 Gate.

Reject. This control is used to dismiss low-level "noise" on the screen. The effect of reject is visual. The principle of reject can best be described by the exercise of holding a ruler horizontally along the bottom of the screen and raising it until the noise level has been masked by the ruler. (It can be imagined that this practice would affect the instrument's vertical linearity; however, modern circuitry overcomes this problem). Reject can on rare occasions be a useful tool when conducting thickness measurements. [Note. The effect of reject is to electronically suppress the low-level signals on the time base. As a result of this, low amplitude signals from discontinuities may not be observed.

Storage Memory
Digital instrumentation usually provides the facility to store calibrations or waveforms. This is a very useful function. A multitude of calibrations can be stored for retrieval at any time. Waveforms (screen dumps) are also stored and can usually be uploaded to a computer for inclusion in subsequent printed reports.

Displays
There are a few different ways that the ultrasonic information can be displayed. Typically, "A scan" presentations are viewed with conventional ultrasonic flaw detection equipment. There are other displays to discuss. Figure 7-28 shows a test piece containing two reflectors with an "A scan" presentation. (Note that the presentation has been rotated in the figure to show the reflectors on the screen relative to their actual positions in the test piece).

A scan displays can be "rectified" or "unrectified," as illustrated below Figure 7-29. The unrectified trace has both positive and negative deflections. Other display options can be seen in Figures 7-30 and 7-31.

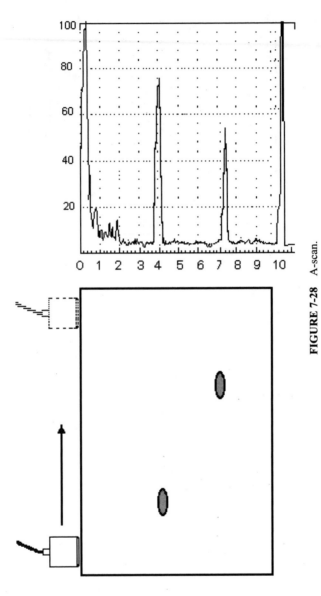

FIGURE 7-28 A-scan.

7.36

"B scan" displays show a "slice" through a section of the test piece. In other words, a cross-sectional view. The signals show as bright spots or lines on the screen. In the display shown in Figure 7-30b, the spot is synchronized with the search unit as it is moved across the test piece (at the same speed as the search unit). Note how the back wall signal drops out as the internal reflectors "shadow" the sound energy from the back wall as the transducer passes over them. Figure 7-30a illustrates an ultrasonic test B-scan display of a plastic specimen with various back surface variations.

A "C scan" is a "map" type of display. It is a "plan" view of the test piece (see Figure 7-31). The first C scan recordings were produced with external recorders that were activated by a signal entering an electronic gate set up on the time base, generally between the top and bottom surfaces of the test piece where reflectors were expected to occur. A pen device "wrote" on the recording paper in the areas that were activated by the signal in the gate. Technology today allows the recording of these images digitally and displays them in different colors on a monitor. The transducer is moved back and forth and indexed so as to scan the entire test piece as can be noted in Figure 7-31. (This pattern is known as a "raster" scan.)

The B scan and C scan techniques are well suited to applications where a permanent record is required. It is common to digitally store A, B, and C scan data.

Transducers

The transducer is the actual "front end" of the system. It is analogous to the microphone in a public address system. If the public address system is of the best quality and a poor microphone is used, the sound will be only as good as the microphone. The same principle can be applied to an ultrasonic system. The use of an inferior quality transducer on the best system can result in deficient data.

Other terms used to describe the transducer are "probe," "search unit," and "test head." The word "transducer" is from the Latin "transducere," which means to lead

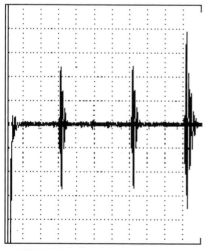
Unrectified "A" Scan Display

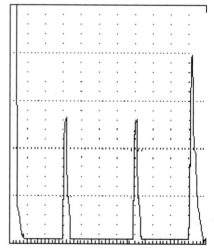
Rectified "A" Scan Display

FIGURE 7-29.

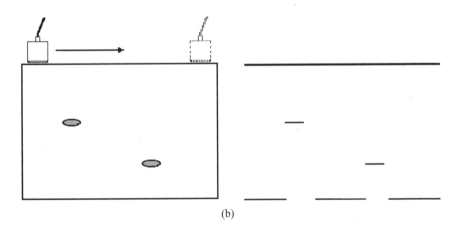

FIGURE 7-30 (a) B-scan instrument. (b) B-scan.

across or transfer. The function of the transducer is to transfer electrical energy to mechanical energy and vice versa.

As far back as the year 1880, the Curie brothers, Pierre and Jacques-Paul, discovered that when sectioned in specific planes certain crystal materials would generate a voltage when distorted. This is called "piezoelectricity"—electricity due to pressure. The opposite effect is also valid; i.e., if a voltage is applied to the crystal material, it will distort.

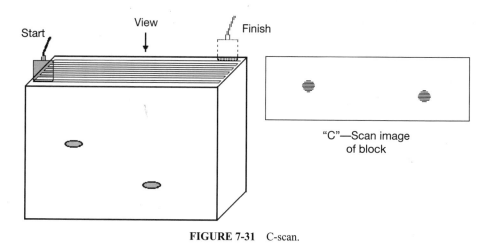

FIGURE 7-31 C-scan.

Lippman documented this about a year later. Quartz crystal is a prime example of this type of crystal. Other naturally occurring piezoelectric materials exist, such as tourmaline and Rochelle salt. These crystals were used in the early days of ultrasonic testing until polycrystalline ceramic materials—materials that do not exhibit piezoelectric properties in their original state—were developed to perform this function. Some of the more common polycrystalline materials used in transducers are lead zirconate titanate (PZT) and lead metaniobate (PMN). The material is mixed in the form of a slurry, poured into a mold, then dried under pressure. It is then sliced to the required thickness. This is the thickness at which the element will resonates at its designed frequency. (Materials resonate according to their formulation and thickness. For example, a 5 MHz transducer element from PZT may be a different thickness than its counterpart made from PMN). The slices are then placed on a lapping table and precision lapped to the final thickness. The next step is to coat the element with a very thin layer of conductive material, usually silver. This is sometimes electrostatically applied. At this stage, the element is not yet active. It comprises many microscopic piezoelectric elements that are randomly oriented. These have to be aligned or polarized in order for the element to be useful for the purpose of generating and receiving ultrasonic energy. This is referred to as the "poling" process. Electrodes are attached to the faces and the element is immersed in a bath of oil. The oil is heated to a temperature that is around the "Curie" temperature of the specific material also referred to as the "critical" temperature). A high polarizing voltage is applied to the element. The element is then allowed to cool while the field is active. Once cool, the voltage is eliminated and the element is now polarized. This means that the element will have + (positive) and − (negative) polarity. For longitudinal wave generation, the elements are polarized so that the element deformation is as shown in Figure 7-32a. Shear waves can be generated when the elements are polarized to deform as shown in Figure 7-32b.

Note 1. Transducers used for angle beam shear wave applications are usually longitudinal. The shear wave component is generated upon the energy traversing two different material velocities at predetermined angles (see Refraction and Mode Conversion on page 7.13).

Note 2. Heating a transducer above its Curie temperature will allow the microscopic elements to depolarize. This will remove the piezoelectric properties. When conducting

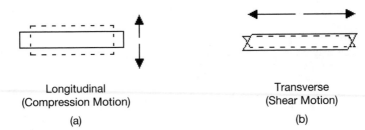

FIGURE 7-32 Elements.

high-temperature ultrasonic examinations, precautions are necessary so that the operational temperature does not approach the Curie temperature of the transducer. *Approximate* Curie temperatures of the more common element types are as noted in Table 7-2.

The primary advantage of polycrystalline ceramic transducers is their relative efficiency. With some of these materials, the efficiency can be as much as 60 to 70 times more efficient than their quartz equivalent.

Elements polarized as shown in Figure 7-32a will function with a "piston" type motion. The element "grows" taller and thinner, and as the voltage is removed from the element, it collapses and becomes "shorter and fatter." This motion continues until it has stopped ringing. This is analogous to striking a bell with a hammer just once. The bell rings and takes a while to stop ringing or oscillating. The transducer operates in much the same manner as the bell. A voltage from the pulser strikes it and it rings. This ringing has to be arrested or "damped" quickly. Superior resolution characteristics require that the transducer exhibit the minimum number of oscillations possible. If the oscillation is not damped, signals from reflectors that are relatively close together in a test piece will appear on the screen as one combined (reflector) signal instead of two separate signals. The ringing of the element has to be shortened in order to improve resolution of reflectors, whether they are close to the surface (near-surface resolution) or close together in time (spatial resolution). Refer to the example of the bell. If the bell is struck many times and in quick succession, it is difficult to establish the number of "strikes" on the bell. If a hand is placed on the bell while it is being struck, the vibrations are quickly damped after each strike. The number of times that the bell is struck may now be individually counted. This principle applies to the transducer. This quality is accomplished by placing "damping" or "backing" material onto the rear of the element, similar to placing a hand on the bell.

Transducer technology may vary between manufacturers, each having their own zealously guarded methods of damping transducers. The bond between the backing material and the element is of primary importance, as is the acoustic impedance match between the element and the backing material. Ideally, the transducer should have a short pulse length

TABLE 7-2 Approximate Curie Temperatures

Element type	Curie temperature, °C
Barium titenate	120
Lead zirconate titenate	190–350
Lead metaniobate	400
Quartz	575

and a high-energy output. These two features usually contradict each other (see Bandwidth). Return to the example of the bell. If there is a hand placed on the bell, it will not ring as loudly as when it is rung without the arresting hand. There is obviously a compromise between heavy damping and high-energy output. It boils down to a matter of selecting the right tool for the job to be done. Damping is accomplished by using a substance that is acoustically matched to the element material and that is also dense in structure. Usually, a tungsten and epoxy mixture is used. Another requirement of a damping material is that it attenuate the energy fast. It is extremely undesirable to have signals from energy bouncing around in the damping material. Designs of the backing material include the shaping of the backing and inclusion of scattered particles of reflective material that scatters the energy within the backing piece.

Regardless of the technique used, the transducer performance should match its useful purpose, that is, the introduction of ultrasonic energy into the part being examined. Applications that require the detection of reflectors that exist close together in the part being examined will require the use of a transducer that has a short pulse. It is preferable that the transducer is acoustically matched to its application. The transfer of energy from the transducer to the test piece or to the wedge material (in the case of angle beam units) has to be optimized (see Acoustic Impedance). The transducers used as "wedge drivers" should have a front "face," usually made from an epoxy, that has a good impedance match to the Plexiglas wedge to which it is attached. Transducers intended for use in direct contact applications such as conventional zero degree units, have front faces that are made from material that is acoustically close to the test piece, usually aluminum or steel. The material used for this application is generally aluminum oxide. Not only is the impedance match to the element and a steel or aluminum test piece reasonably good, but the material is also by nature extremely hard and therefore exceptionally wear resistant. It therefore goes without saying that the use of a wedge driver for a direct contact test is not recommended. Not only will the face wear very quickly, but from the acoustic standpoint the transfer of the energy will not be optimum.

The layer thickness of the facing material is extremely important. It should be equal to one-quarter wavelength in the facing material. This arrangement considerably enhances the emitted energy. See Figure 7-33, an example of single element transducer construction, and Figure 7-34 for examples of an angle beam unit. Note that the damping material is made at some angle so as to redirect the internal energy away from the element.

Piezocomposite Transducers
Primarily used in the medical field, the idea of piezocomposite transducers was considered as early as the mid-1970s. The technology necessary to manufacture this type of transducer was not readily available at that time and solutions to problems such as the facility for slicing the crystal accurately in a production mode were not possible at that time. Piezocomposite transducers function in a similar manner to regular ceramic transducers. The primary difference is that the piezocomposite ceramic is made up of small-sectioned pieces of ceramic material (PZT, PMN, etc.). See Figure 7-35.

Initially, the ceramic is sliced into squares. The small areas between the squares are filled with epoxy and the transducer is lapped to the required thickness, silvered, and polarized in a similar manner to the regular transducer elements. The difference to the user is in the performance. Because of the damping material (epoxy) around each square, the transducer exhibits superior bandwidth and therefore superior resolution. There is little or no need to apply a backing; therefore, the efficiency is greater than the conventional damped ceramic transducer. Because of the lack of backing, the height profile of these units can be reduced for accessibility to small areas. These transducers are particularly useful when testing grainy materials having increased signal-to-noise ratios.

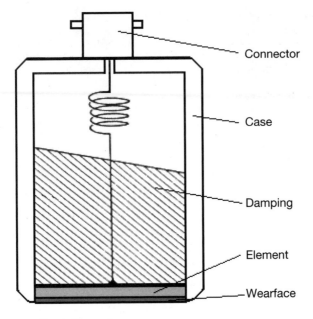

FIGURE 7-33 Single transducer.

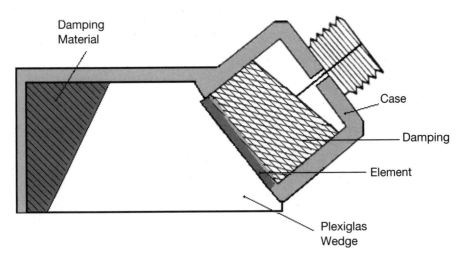

FIGURE 7-34 Angle beam transducer.

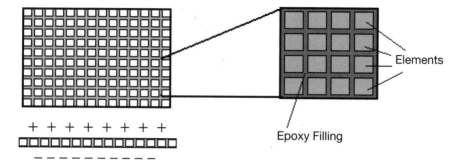

FIGURE 7-35 Piezo-composite element.

PVDF (*P*oly*v*inyli*d*ene *F*luoride)

Ceramics are not the only materials that can produce the piezoelectric effect. Piezoelectric polymer sensors have been available in the nondestructive testing industry for some time. In the 1960s, it was discovered that whalebone exhibited a very weak piezoelectric effect. This inspired the search for other materials that could exhibit piezoelectric activity. Relatively high activity was found to exist in polarized polyvinylidene fluoride. The advantages of this material are:

1. PVDF has an acoustic impedance close to that of water, so that it allows efficient transfer of energy.
2. PVDF has a large frequency range and therefore a very broad bandwidth. Because of this, it exhibits very favorable spatial and near-surface resolution compared with conventional piezoelectric element material.
3. The PVDF material is flexible and can be shaped for beam focusing. It is therefore ideal for high-resolution immersion applications.

    The disadvantages are:

1. PVDF is relatively low in power compared with conventional piezoelectric transducers and, in most cases, requires additional amplification.
2. It cannot be used in contact applications.

Noncontact Methods

EMAT (*E*lectro*M*agnetic *A*coustic *T*ransducer) (see Figure 7-36) technology is an alternative method of generating and receiving ultrasonic energy. These are transducers that are made up of coils that are placed in close proximity to the test piece. The coils produce a magnetic field that interacts with the metal, producing a deformation in the surface of the material. This deformation produces the wave of ultrasonic energy. The advantages of these units are:

1. There is no need for a couplant. An EMAT is a noncontact transducer.
2. They lend themselves to applications that normally have limitations, such as the examination of high-temperature components. Because this type of transducer depends on

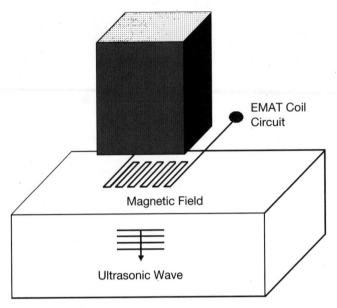

FIGURE 7-36 EMAT transducer.

the induction of a field, the transducer has to work in close proximity to the work surface. The strength of the magnetic field is reduced as the distance between the transducer and the component surface increases.
3. The gap between the transducer and the work face need not be composed of air. Examination of components that have been coated with some protective layer is possible. It is the front surface of the component material that actually generates the ultrasonic energy.
4. Focusing of the beam is also possible, as is steering the beam at various angles.
5. Horizontally polarized shear wave energy can be produced. The polarity is important in that horizontally polarized shear waves do not mode convert when striking surfaces that are parallel to the direction of polarization. This has certain advantages, particularly when examining austenitic welds and other materials with dendritic grain structure, e.g., certain cast stainless steels.

The disadvantages include:

1. Low efficiency compared with piezoelectric transducers.
2. Relatively large transducer size.
3. Producing ultrasonic energy in nonconductive material is only possible if a conductive layer is applied to the surface.

Laser-Generated Ultrasound

The use of laser technology for the generation of ultrasonic energy has been known since the latter part of the 1970s. Acoustic propagation is accomplished by briefly heating or "ablating" the surface of the test material. This brief heating of the surface causes the generation of thermal expansion on the surface of the material, which in turn results in the formation of a wave front that travels through the material. The technology generally employs two separate lasers for this application—one to ablate the surface and produce the wave, and a second (a laser interferometer) to detect the movement of the surface due to the disturbance from the reflected wave. The technology is commercially available and has been developed for (but not limited to) applications such as the inspection of composite materials in the aircraft industry. Reception of the energy is also possible with a conventional transducer coupled to the surface.

The advantages are:

1. When used with laser interferometry for reception, there is no need for couplant.
2. The laser can be located remote from the test surface (10 inches or more in certain instances).

Bandwidth

Transducer bandwidth can best be described in ultrasonic terms as the spectrum or range of frequencies that occur when pulsing the transducer. To make this more meaningful for comparison purposes, the bandwidth is measured within a given amplitude range. This is usually at either the –3 dB or –6 dB levels from maximum amplitude (see Figure 7-37).

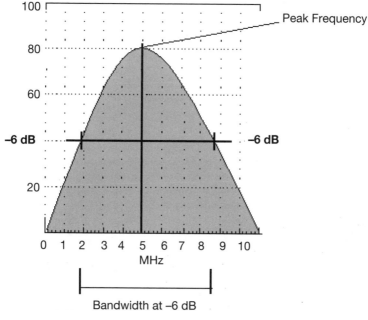

FIGURE 7-37 Spectrum.

This is a standard method of comparing the transducer's "quality factor" or "Q" that is varied by the mechanical damping placed on the element. The higher the Q, the greater the sensitivity. The lower the Q, the greater the resolution characteristics. A short-pulsed transducer provides a broader bandwidth but less energy, therefore a "low Q." See Figure 7-38.

Dual Transducers

Transducers that incorporate separate elements for transmission and reception of the ultrasonic energy are referred to as "dual" transducers. Designs vary, but a typical dual transducer construction is as shown in Figure 7-39. As can be seen, the elements are mounted on separate delay lines that are separated by an acoustically opaque (nontransmitting) material such as cork. The receiver element must not be able to receive energy directly from the transmitter element. In other words, they must not indulge in "cross-talk." The primary purpose of this type of transducer is to enhance the near-surface resolution capability of the ultrasonic system. The dual transducer can be designed to resolve reflectors that are very close to the scanning surface. To this end, they are designed with a certain "roof angle" or "squint angle." This produces energy in the test piece that is refracted in the direction of the receiver side of the transducer. It is obvious that the greater the roof angle, the better the resolution close to the scanning surface. It must be realized that the detection efficiency further out in the part will suffer. Again, it is a matter of selecting the correct transducer for the application. The dual transducer can be used in angle beam applications. In this case, the same advantages are present. The sound energy intersection area can be predetermined to enhance a specific area in the test piece or component under test. It is incorrect to refer to this type of transducer as a "focused" transducer, but the technique does limit the received reflected energy to the zone where the beam from the transmitter intersects the hypothetical area of the receiver beam. Note that there is no energy from the receiver; however, because the angle of incidence is equal to the angle of reflection, and because the receiver is "aimed" at the area of the transmitted beam, planar reflectors in the beam path will reflect

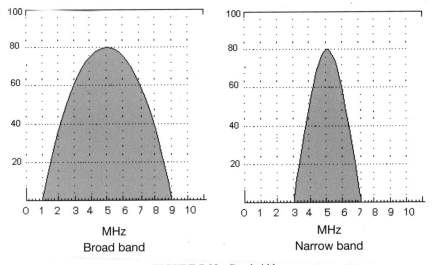

FIGURE 7-38 Bandwidth.

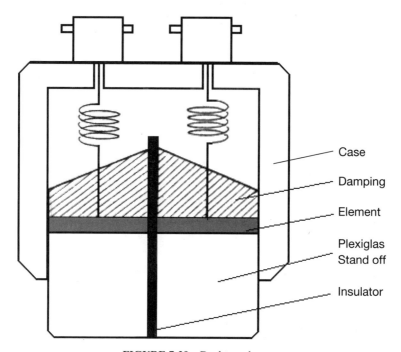

FIGURE 7-39 Dual transducer.

the energy towards the receiver. The energy that is detected by the receiver will be from the intersecting zone of the transmitter (see Figure 7-40).

Focused Transducers
The principles of optics apply to sonics. Focusing light is achieved with the use of lenses. Focusing sound energy is accomplished in a similar manner. Immersion testing, in particular, lends itself to the use of focused transducers (see Figure 7-41).

There are essentially three approaches that can be used to accomplish the task of focusing the beam:

1. Shaping the actual transducer element.
2. Attaching a concave lens to the transducer face.
3. Inserting a biconvex lens into the path of the sound energy, similar to focusing the light from the sun using a magnifying glass.

The sound energy can be focused so that it concentrates the energy at a particular depth or area within the test piece. This improves the detection of small reflectors at a predetermined depth. This is due to the small beam dimension at the focal point relative to the size of the reflector.

Contact angle beam transducer units that focus the energy have also been produced. This is accomplished by contouring the wedge under the transducer to a specific radius, dependent upon the desired focal length (see Figure 7-42). This radius is filled with a fluid having a velocity different from the wedge material, thus providing a lens to the ener-

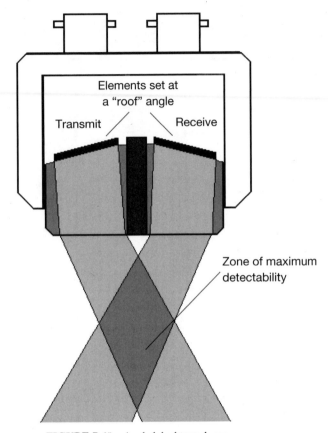

FIGURE 7-40 Angled dual transducer.

gy. This will produce a refracted (converging) beam in the wedge that refracts again to a focal point after entry into the test piece.

Phased Array Transducers
These transducers incorporate elements that are arranged in certain patterns for the purpose of dynamically focusing or steering the energy. Sequentially pulsing the elements, using a combination of elements in the array, and timing the pulses used to excite these elements, produces a beam focused at variable depths in the test material (see Figure 7-43). Multiple wave fronts are combined to form a beam of a particular shape. Element configurations can be circular or rectangular, depending on the desired beam shape and direction of energy propagation. Linear array units are more commonly used in the medical field for imaging than in the industrial sector.

Bubblers and Squirters
Immersion testing is not limited to placing the test material or component into a tank of water. An immersion test can be simulated by a device called a "bubbler" or a "squirter."

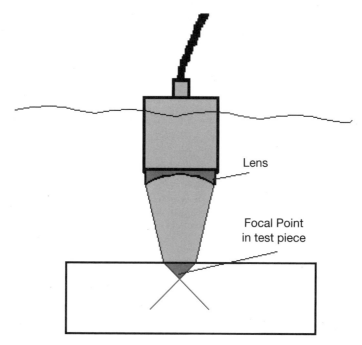

FIGURE 7-41 Focused transducer.

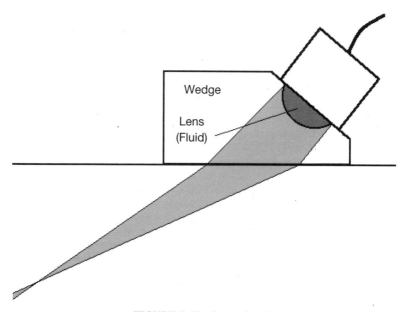

FIGURE 7-42 Focused wedge.

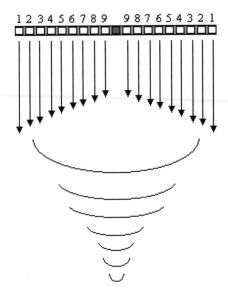

FIGURE 7-43 Phased array.

A bubbler is a device wherein the transducer is located within a housing that includes a fluid reservoir. The fluid is continuously pumped into the housing. The transducer is placed slightly above the level of the holder. When the unit is placed on the test surface, or more often, the test surface is placed over the unit, the housing fills with fluid and provides a film of couplant to the front of the transducer (see Figure 7-44).

A squirter (see Figure 7-45) comprises a holder with a nozzle. The transducer is located inside of this device and the smooth jet of water that exits the nozzle carries the sound energy to the surface of the test component. This device is generally incorporated into automated test systems such as those for production line inspection of pipe, or inspection of aircraft components where immersion of the component is impractical. Through-transmission techniques are best suited for the squirter. In the pulse-echo mode, extraneous signals may appear at the surface of the test component due to the water striking the surface of the test part. The size of the ultrasonic beam can be controlled by varying the size of the nozzle orifice. (Note that the example in Figure 7-45 does not have a smooth water jet.)

Wheel Transducers
Immersion testing is found in many forms. The wheel transducer is another variation. Transducer(s) are located in a holder(s) within a sealed wheel that has been filled with fluid. The fluid can be water, thin oil, or other suitable materials. The transducers can be placed at strategic angles within the wheel so as to provide the desired refracted test angles. This device is very useful for examination of moving components where surface couplant is to be limited. They are commonly used in the railroad industry to test carriage wheels and railroad track on the fly. The wheels usually require a fine spray of water to couple the tire to the test component. There are special wheel transducers that do not need couplant on the tire surface (dry coupled). The wheel tire material can be made so that it is virtually invisible to the ultrasound by matching the tire characteristics to the fluid in the wheel.

FIGURE 7-44 Bubbler. (Courtesy of Matec Instruments.)

Transducer Problems

As mentioned previously, the ultrasonic test begins with the transducer. It is therefore of utmost importance that the transducer is performing as required. Problems with the transducer need to be identified prior to any test. It is advisable to document the performance of each transducer when purchased so that its performance can be monitored during its lifetime. Items such as pulse width, amplitude, and angle (if appropriate) should be quantified. Photographs of the spectrum and wave-form should be acquired. These parameters should be checked on a regular basis and the performance verified. It is advisable to verify the parameters on the same system each time it is used. Variables such as pulsers, receivers, and cables, can exhibit anomalies that may be erroneously attributed to the transducer alone. Possibilities of changes in performance include, but are not limited to:

1. Pulse length. This can increase for the following reasons:
    a) The backing or damping material becomes detached from the transducer.
    b) The wear face becomes detached from the transducer.
    c) The element or wear-face is cracked.
    d) The tuning coil is detached.
2. Low sensitivity. This can occur for a number of reasons such as, but not limited to:
    a) Deterioration of or damage to the transducer element material.
    b) Detachment of the transducer face.
    c) Detachment of the tuning coil.
3. Beam skew (apparent) due to partial detachment of the transducer face.

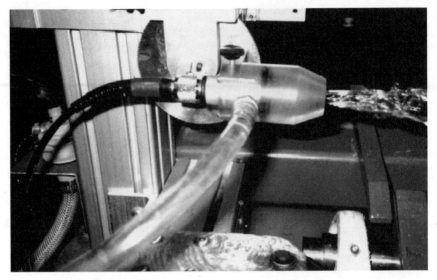

FIGURE 7-45 Squirter. (Courtesy of Matec Instruments.)

4. Faulty connections. Damaged or loose wires in the transducer housing.
5. In the case of angle beam transducers, wearing down of the Plexiglas wedge may cause the refracted angle to change.

In conclusion, the transducer is a precision component in the ultrasonic system. Treated with due care, it will provide many years of service.

IV. TECHNIQUES

The ways in which sound waves propagate through materials and are attenuated, reflected, or transmitted dictate the different ultrasonic methods or techniques used to detect the many types of discontinuities that can exist in materials. These techniques fall into two main categories, one called "pulse-echo," and the other called "through-transmission." Any of the techniques that may be used requires *calibration* of the ultrasonic system so that the time base can have some meaning in terms of material thickness. This is also true for through-transmission techniques where depth information is not available, since the practitioner needs to be sure that an adequate time base is available to show the transmitted energy.

Calibration Techniques

The process of calibration needs to be just as disciplined as the inspection technique. In fact, the whole inspection relies on the calibration process. Some of the following calibration techniques are concerned with measuring and documenting the characteristics of the transducer and flaw detector.

Transducer Characteristics

Compression Wave Transducers

Resolution. The ultrasonic pulse consists of a few cycles of sound energy at the test frequency. Therefore, the pulse occupies some space in time, or distance, within the material. The amount of space occupied is called the "pulse width." Physically, it is the number of cycles in the pulse multiplied by the wavelength of that frequency in the test material, or, mathematically:

$$W = n \times \lambda$$

Where
n = The number of cycles in the pulse
λ = Wavelength

Pulse width is important because while one pulse is being processed, another closely following pulse cannot be properly processed. This means that two reflectors that only have a small separation in depth will only show as a single, messy signal on the trace. Because the two signals overlap, the time difference between them cannot be measured. The ability of an ultrasonic system to discriminate between two reflectors that are close together is called "resolution."

Resolution for a given transducer can be determined in one of several ways:

1. By calculation of the pulse width and expressing that in terms of distance occupied in the test material. For example, the wavelength of a 5 MHz compression wave in steel is 0.047" (1.192 mm). If the transducer in question has two-and-a-half cycles in the pulse, then the pulse width for the transducer is 0.047" × 2.5, which comes to 0.117" (2.98 mm). Signals from surfaces closer together than this will begin to merge.

2. By measuring the pulse width on the calibrated trace. If the timebase has been calibrated for the test material, any signal positioned clear of any other can be used to measure pulse width. Figure 7-46 shows a trace calibrated to 25 mm full scale. The signal in the middle of the trace starts at 11 mm and ends at 14 mm. The pulse width is 14 − 11 = 3 mm.

3. By checking the resolution on a suitable calibration block. Such a block is the IIW V1 block, described in the "Calibration Blocks" section. The transducer is placed as shown in Figure 7-47, opposite the machined slot.

If the transducer is capable of resolving the distances, three distinct signals should be shown on the trace (see Figure 7-48). The block has been designed so that the distance between the first and second reflecting surfaces is 1 μsec at steel velocity. For the round trip, the time difference between the first and second signals is 2 μsec. A disadvantage of the resolution check that can be performed on the IIW V1 block is that it is confined to straight beam transducers. An alternative approach is illustrated in Figure 7-49, which shows a pair of holes drilled on a common axis from opposite sides of the block. The two holes are of different diameters and meet in the center of the block. A sound beam, at any angle, aimed at the region where the two holes meet, will reflect from both holes. If the two echoes are resolved on the trace, the transducer can be said to have sufficient resolution for that spacing. If the holes are 0.5" (12 mm) and 0.25" (6 mm) in diameter, respectively, the time difference between the two is equivalent to 0.125" (3 mm). The advantage of this approach is that it is applicable to both straight beam and angle beam transducers.

Dead Zone—Initial Pulse (Main Bang). When using single crystal transducers, the initial pulse and ringing of the crystal are connected to the receiver circuit and displayed

7.54 CHAPTER SEVEN

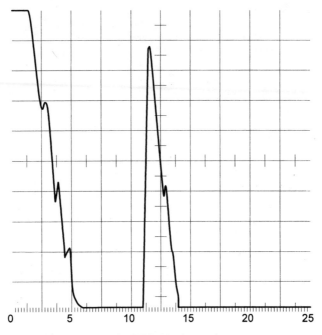

FIGURE 7-46 A-scan 3.

on the timebase. This signal occupies the early part of the trace, and indicates the "dead zone" in the material. While the crystal is transmitting, the trace cannot display received signals clearly in this region. For this reason, it is usual to use dual crystal transducers for components less than 0.5″ (12.5 mm) in thickness. The dead zone can be measured on a calibrated time base by noting the point at which the initial pulse dies away. It should be remembered that the initial pulse, and hence the dead zone, will increase as the gain is increased. So it is necessary to measure the dead zone at the working gain for the inspection.

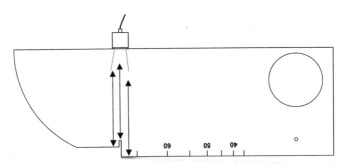

FIGURE 7-47 Resolution on IIW block.

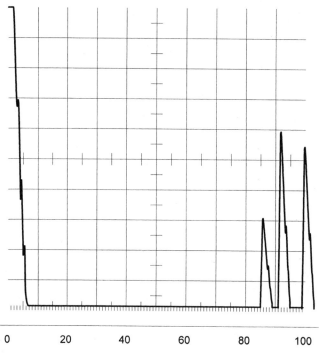

FIGURE 7-48 A-scan resolution.

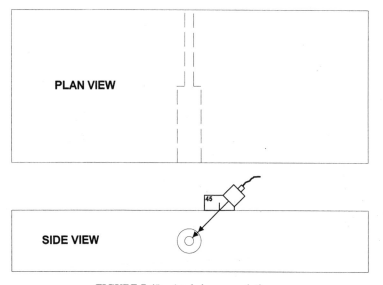

FIGURE 7-49 Angle beam resolution.

Angle Beam Shear Wave Transducers

Probe Index (Exit Point). With angle beam transducers for contact scanning, the crystal is mounted on a Plexiglas incidence wedge in order to generate the desired refracted angle in the component. The center of the sound beam emerges from the wedge at a position called the "probe index" or "beam exit point." This position is engraved on the side of the transducer wedge and is the position from which horizontal distance (HD) is measured. With use, the wedge may become worn and the actual exit point can move. The new probe index must be measured and marked if accuracy is to be preserved.

The IIW V1 block allows several angle beam calibration checks to be carried out. One of these checks is to determine the true probe index. Figure 7-50 shows a 45° shear wave angle beam transducer positioned on the IIW—V1 block so that the beam center is aimed at the 100 mm radius. The intersection of the narrow slot with the scanning surface marks the center of the 100 mm radius. As the transducer is moved forward or back past the slot, the echo from the 100 mm radius will be seen to rise and fall. When it reaches maximum amplitude, the beam center is aligned with the radius center, and is perpendicular to the tangent to the radius. The slot marking the center of the radius is now aligned with the true probe index. If this differs from the marked index, a new probe index must be marked on the wedge or the discrepancy measured and recorded so that allowance can be made during subsequent operations.

Beam Angle. The beam angle of an angle beam transducer can be measured on the V1 block once the true probe index has been established. The large Plexiglas-filled hole in the V1 block provides the target for beam angle measurement. From the center of this hole, angles have been projected to the edges of the block and the intersection engraved with the angle every five degrees from 35° to 70°.

When an angle beam transducer is scanned toward the hole, as shown in Figure 7-51, the signal maximizes when the beam is aiming straight at the center of the hole. As this occurs, the probe index is aligned with the true beam angle mark on the edge of the block. The actual beam angle should be checked against the nominal angle; any variation must be noted and the actual angle should be used when conducting discontinuity plotting and sizing techniques.

Beam Spread Diagram. Some discontinuity sizing techniques make use of beam spread diagrams to identify the edges of a discontinuity. For this, a practical beam spread diagram must be drawn for the probe being used. To obtain an accurate beam spread diagram, the following information is required:

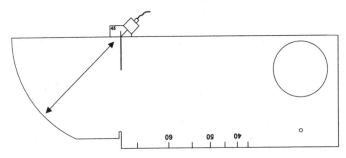

FIGURE 7-50 Angle beam on IIW block.

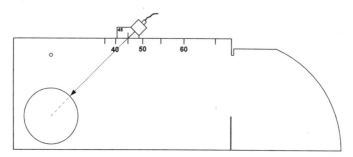

FIGURE 7-51   Angle Beam on IIW block 2.

1. Beam angle of the probe in question
2. The beam limits to be used (–6 dB or –20 dB)
3. The exact probe index (exit point) for the probe in question

Beam angle and probe index are defined in Calibration Techniques above. The choice of 20 dB drop, or 6 dB drop will be dictated by the relevant inspection code or standard. Once the above information has been obtained, the next step is to prepare the plotting aid shown in Figure 7-52.

The card shown in Figure 7-52 has a horizontal scale representing the horizontal distance along the scanning surface from the probe index to a reference mark (the center line of a weld, for instance). The vertical scale represents the thickness of the component. From the intersection of these two scales, the beam path scale is drawn in at the measured beam angle for the probe (not the nominal angle marked on the probe).

The next step is to measure the beam spread at various depths so that a beam spread diagram can be constructed. This is done using a beam calibration block such as the IOW block (see Calibration Blocks). These blocks contain a series of side drilled holes 1/16″ (1.5 mm) diameter and at various depths below the surfaces of the block. On the card shown in Figure 7-53, lines are drawn parallel to the horizontal scale at the depths of the holes that are to be used for the beam spread diagram. It is only necessary to plot the beam spread to the maximum range that will be used on the time base. In specimen thick-

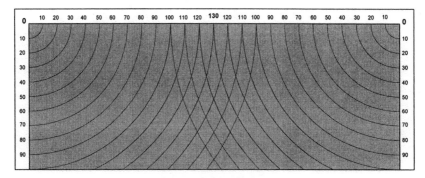

FIGURE 7-52   Plot card.

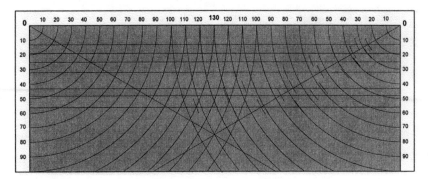

FIGURE 7-53 Plot card.

ness terms, this should be at least twice specimen thickness if discontinuities at full skip distance are to be plotted.

If, for example, the component being tested is 25 mm thick, then parallel lines would be drawn at IOW block hole depths of 13, 19, 25, 32, 43, 48, and 56 mm, as shown in the diagram. Each of the holes would then be scanned from the appropriate surface. The echo from the hole is maximized by moving the probe along the scanning surface, as shown in Figure 7-54. When the signal reaches its maximum amplitude, the beam center is pointing at the center of the hole. The signal height is then adjusted to a convenient value, usually 80% of full screen.

The gain setting in dB is noted and the gain reduced by the appropriate intensity drop (–6 dB or –20 dB). The new signal height is noted and marked on the screen with a wax pencil before returning the signal to the original (80%) height. The beam path length is

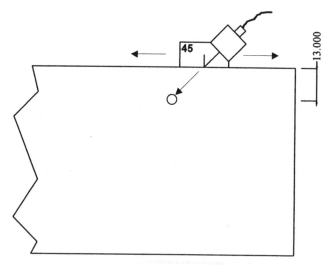

FIGURE 7-54 Depth hole.

taken from the time base and recorded in a table for that depth of hole and beam center position.

The transducer is then scanned forward until the signal amplitude has decreased to the wax pencil line (i.e., by the required intensity drop). The time base range is again noted and recorded as the beam *back-edge* position.

The last step is to move the transducer back to locate the maximum amplitude and then scan back until the signal again reaches the wax pencil line. Note this time base range and record it as the beam *front-edge* position.

The three recorded time base ranges are then transferred to the plotting card by drawing arcs centered on the card zero. The arcs cut the parallel line for the hole depth in question and the beam center arc also cuts the beam path scale. These arcs are shown in Figure 7-53. The above steps are repeated for each of the seven holes to be used in the beam spread diagram. Once all the arcs have been plotted, the edges of the beam can be drawn as shown in Figure 7-55 and the beam spread diagram is complete.

System Checks and Calibrations

The techniques discussed so far have concentrated on the transducer, but in the overall preparation for an inspection, the combination of transducer and instrument must be addressed. Many of the checks rely on the practitioner being certain of the position of a signal with respect to the transducer or scan surface, which, in turn, relies on time base calibration.

Time Base Calibration—Compression Waves

Time base calibration is concerned with establishing a known depth scale across the trace for the material to be inspected. This requires a calibration block made of the same material as the intended work piece. The IIW V1 block is a calibration block for steel. It has reflecting surfaces that allow direct calibration of the time base for 25, 50, 100, and 200 mm. By using multiples of those values, other timebase ranges can be established (see Calibration Blocks).

The objective for the time base calibration is that "zero" on the trace represents the scanning surface and "full scale" represents a known depth or thickness. Between zero and full scale, the graticule should represent a linear change of depth. There is a major problem facing the practitioner, because the point on the trace that represents entry of the

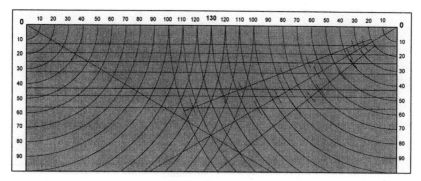

FIGURE 7-55 Plot card.

7.60 CHAPTER SEVEN

beam into the surface is not obvious. In the case of a single crystal transducer, sound starts as the applied pulse voltage is released. The start point is therefore somewhere in the "initial pulse," where the signal is saturated.

Of course, the calibration block is of known thickness, so all that needs to be done is to identify the first echo from that thickness on the block and position it at "full scale," right? Wrong! The zero has to be correct too, or the time base will be a little too long or too short. Somehow, there has to be a pattern display with two signals that are exactly a known distance apart, and whose start points can be identified.

When looking at a multiple echo pattern as sound bounces up and down inside the material, the one certain fact is that between one back wall reflection and the next, the sound has traveled through the specimen thickness. So, for instance, the space from the beginning of the first back reflection to the beginning of the second (first multiple of the back reflection) is directly related to the wall thickness. Figure 7-56 shows a single crystal

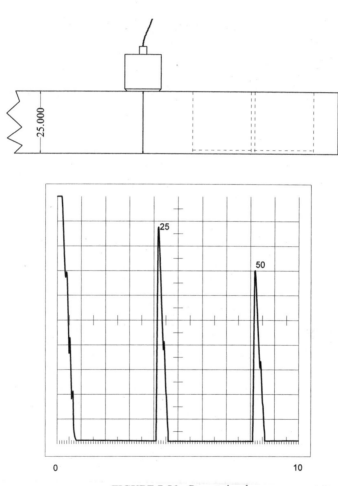

FIGURE 7-56 Repeat signals.

transducer positioned on the V1 block for the 25 mm range. To the left of the trace (time base) is the initial pulse and to the right are the back reflection and first multiple signals. The two signals are spaced apart by one round trip in the block.

To exploit this known distance, the left-hand edge of the first back reflection is moved using the "delay" control until it coincides with "zero" on the graticule. Then, using the "depth" or "range" control, the left-hand edge of the first multiple back-reflection signal is moved to coincide with "10" on the graticule. The position of the first back-reflection signal is checked and adjusted, using the "delay" control if necessary, and the first multiple signal checked and adjusted using the "depth" control. These adjustments are made until the left-hand edges of the two signals are correctly aligned with the graticule, as shown in Figure 7-57.

At this stage, the time base is calibrated for a range of 25 mm, but zero is 25 mm, and ten is 50 mm (i.e., the back reflection and first multiple echoes). The requirement was that zero means scanning surface (0 mm). However, all is not lost! Lock the depth control, then use delay to place the first back reflection echo at 10, and the job is done! Zero is now 0 mm and 10 is 25 mm.

The above technique, using delay and depth (range) controls to adjust the back reflection and first multiple echoes to fit the graticule, can be used for any of the direct ranges on the V1 block. Because there is a delay due to the Plexiglas column, and because the

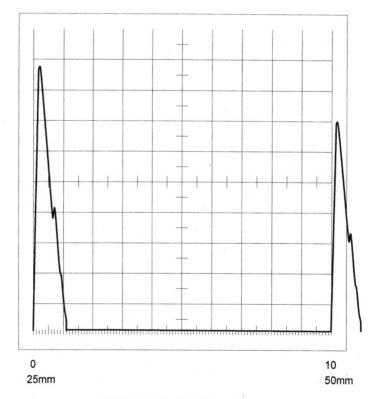

0
25mm

10
50mm

FIGURE 7-57 Signal from 50 mm.

transmitter is separated from the receiver, there is no way in which the beam entry point can be identified with a dual crystal transducer. The only way to calibrate is to use the above procedure.

Time Base Calibration—Shear Wave Transducers

Calibration of the timebase for a known time base range is also necessary in most angle beam inspections. This calibration can be carried out on the V1 block for 100 mm beam path range, and on the DIN 54 122 (*V*2) block for 25 mm and 50 mm. The procedure is similar to the one described above (see Figures 7-58 and 7-59).

The narrow slot that marks the center of the 100 mm radius also serves another purpose. It makes a corner reflector at the scanning surface to enable the first return echo to reflect back to the radius to give a multiple echo pattern. The calibration of the time base can therefore be set from the first back echo to the second. This is achieved by using the delay control to bring the first back echo to zero on the time base. The range control is then used to position the second back echo to 10 on the time base. The time base now represents 100 mm of the material, but is set from 100–200 mm. To correct this setting to read 0–100 mm, the delay control is used by adjusting the control until the first back echo is at 10.

Calibration of Amplifier (Test Sensitivity)

Just as it was necessary to calibrate the time base in order to correctly position reflectors, so it is necessary to calibrate the amplifier to a known amount of gain. This is necessary to make sure that enough gain is used to detect critical discontinuities, that the gain is not so high that spurious indications appear, and to ensure repeatability between inspections.

The simplest form of setting gain is to choose a known target in a calibration block, and use the gain to set the amplitude of the echo from the target to a predetermined height. The target could be the back wall, a side-drilled hole, a flat-bottomed hole (FBH) in the calibration block, or even a keyway or oilway in the component itself. The

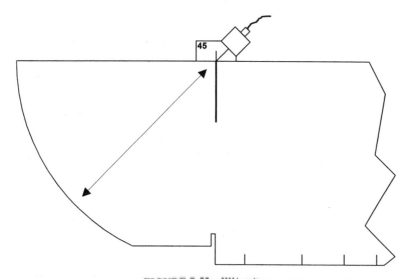

FIGURE 7-58 IIW radius.

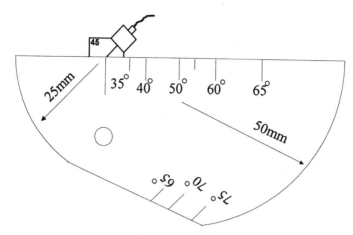

FIGURE 7-59 V2 block.

main objective is that the actual target used is documented so that anyone can repeat the test.

Distance and Area Amplitude Blocks

Since the amplitude of an echo can be a function of the area and depth of the reflector, targets of known area and depth can be used to set test sensitivity. There are calibration blocks that are designed to perform this function. Essentially, flat-bottomed holes are used as targets. They range in size from 1/64" (0.4 mm) to 8/64" (3.2 mm) diameter in 1/64" steps. Sets of blocks in those sizes are available in a range of scanning depths.

Sensitivity is set by selecting an appropriate FBH size and scanning depth in a block of the same material as the component to be tested. The transducer is placed on the block and scanned to obtain the maximum echo amplitude from the FBH. The gain is adjusted to bring this echo to a predetermined height. This is known as "reference sensitivity" or "reference gain." Reporting levels are generally compared to this signal amplitude.

Gain as a Function of Beam Path Distance

In the Section II, it was shown that the intensity of sound energy decreases with distance along the beam due to attenuation. This means that simple signal amplitude references, like those just described, do not apply the same sensitivity across the entire time base. There are however, techniques to overcome this problem.

Distance Amplitude Correction (DAC)

Imagine listening to the radio while sitting close to the speaker. If the distance from the speaker to the ear increases, the volume will decrease relative to the distance from the speaker to the ear. This is because the sound is being attenuated in the air. If it were necessary to hear the sound at the same level, regardless of the distance from the ear to the speaker, a remote volume control could be used and the volume (amplitude) could be gradually increased relative to the distance from the source. This would enable the sound from the speaker to be heard at the same level when near the speaker or far away (this is distance amplitude correction). By electronically maintaining the same signal amplitude to the ear as the distance from the speaker is increased or decreased, the volume to the ear has remained the same.

As in the above scenario, when the distance between the transducer and a reflector increases, the amount of energy reaching and returning from the reflector to the transducer decreases. This principle is made apparent by observing the amplitude of the reflected signal from a given size reflector placed at different distances from the transducer. The apparent amount of signal reduction will be dependent on, but not limited to, such factors as material attenuation, transducer frequency, and transducer size.

Ultrasonic systems can include circuitry that electronically increases the gain according to the distance of the reflector, taking into consideration the amount of attenuation in the test material and the other factors mentioned above. This is also known as "swept gain." It is effectively a controlled increase in gain along the time base (sweep). For example, a signal from a ¼" diameter flat reflector placed at a depth of one inch with its signal amplitude set at 80% can now electronically appear at the same amplitude on the time base if the reflector was at perhaps two, four, five, ten, or more inches, away from the transducer. This type of presentation is necessary when factors such as attenuation affect gating conditions in automated test systems. The variable of material attenuation is reduced or eliminated in this way.

A curve can also be plotted to graphically express the reduction in sound energy as it moves through the test material. This is known as a "DAC" curve see Figure 7-60. Certain specifications require that a DAC curve be generated for a specific examination. This is sometimes needed to determine an acceptance or rejection level based on the amount of energy that is returned to the transducer by a reflector (see note below). In this instance, a DAC curve is generated by observing and noting the amplitudes of signals from known sized reflectors, e.g., flat bottomed holes or side drilled holes at various depths in the test piece. The signal amplitudes are plotted on the screen as the reflectors are interrogated at the varying distances (depths) from the transducer. A line is drawn to connect the signal amplitude points along the time base.

Note. There are many variables to be considered when determining the size of a reflec-

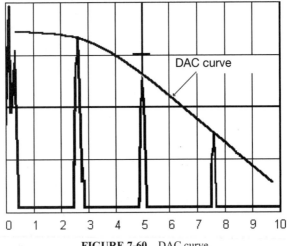

FIGURE 7-60 DAC curve.

tor. Estimating or measuring reflector size based solely on amplitude is not a very practical or accurate method.

Reasonable size comparisons using flat bottomed holes drilled perpendicular to the scanning surface can be made under controlled conditions. In terms of quantifying the amount of energy that returns to the transducer from a reflector, a number of variables need to be considered, e.g., the size, shape, and orientation of this reflector.

Transfer Corrections
When reference amplitudes are established using calibration or reference blocks, there may be differences between the calibration block and the component that affect the sensitivity. These may be due to differences in surface condition, material thickness, or attenuation. The differences can be compensated for; this process is known as "transfer correction" (transferring from calibration block to component). The following example is for angle beam transducers, but the same process also applies to straight beam transducers.

Transfer Correction Technique. The transducers are placed on the calibration block and the received signal from one full skip distance is maximized. The signal amplitude is set at 80% FSH (full screen height). The beam path length along the timebase is noted.

The transducer is moved to locate the echo from the second full skip. With a wax pencil, the signal amplitude and the beam path length at its time base location are noted.

By dividing the dB difference by the beam path length, the energy losses can be calculated in dB per unit distance (inches or mm). This value should be noted.

The transducers are then placed on the part and the above procedure is repeated.

The calculation is performed as above and the dB difference per unit distance is noted.

The difference is subtracted in order to determine the amount of change that needs to be made to the reference sensitivity for the examination of the part.

By performing the above procedure, compensation for attenuation, differences in curvature, surface condition, and beam spread losses will be taken into account.

Distance, Gain, Size (DGS) Technique
The DGS system was developed by Krautkramer in Germany as a method of standardizing inspection and to provide assistance with acceptance and rejection decisions. The principles are based on the known beam characteristics of each transducer. The way in which specific reflectors will respond to the beam can be predicted. DGS plots curves for two types of reflectors. The first is a total beam reflector, effectively back wall reflectors (i.e., bigger than the beam) at increasing depths. This forms one boundary of the DGS diagram. Inside the back wall curve is the second type—a series of curves for reflectors that are smaller than the beam. These reflectors are FBH targets; each curve represents one diameter of FBH at increasing depths (see Figure 7-61).

Curves are calculated and produced for each transducer type and are available in data sheet form and as transparent accessories to clip in front of the screen. The calibration procedure uses the echo from a full back wall as the setting up target. This is adjusted using the gain control to a specific height on the CRT (40% FSH, for example). The next step is to refer to the "size" curve for the FBH diameter specified. Read down the "distance" line to the distance equivalent to the component thickness and note the dB difference between the "back wall" curve and the "size" curve at that distance. Go back to the flaw detector and increase the gain by that dB difference. At the new gain setting, the reference FBH reflector at the back wall depth, would give a 40% FSH signal. At any other depth, the signal from the reference FBH would follow the size curve for the depth concerned.

In angle beam shear wave applications, there is not likely to be a back wall echo. Instead, the echo from the 100 mm radius on the IIW V1 block is used. Because this is not the back wall echo from the component, a transfer correction has to be made. This entails

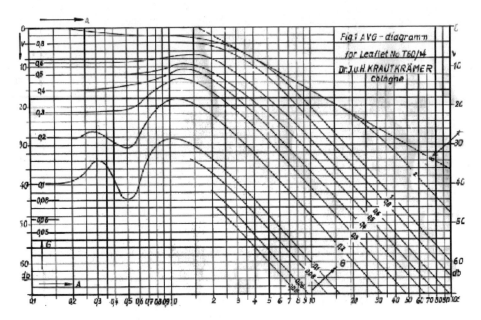

FIGURE 7-61 DGS diagram. (Courtesy of Krautkramer Branson.)

measuring the sound intensity through the V1 block and through the component, expressing that as a dB difference, and adding that difference to the reference echo as well as adding the size difference.

The greatest attribute of the DGS system is probably the fact that it allows the determination of what would be the smallest discontinuity that could be detected with that transducer at that depth. Since a FBH reflector is a near-perfect reflector, any real discontinuity would give a weaker response. Therefore, a lot of time can be saved by *not* looking for a discontinuity that can't be found.

System Checks

Various checks and calibrations are available for testing the entire system, i.e., transducer, cable, and flaw detector. The simplest is the check for overall system gain, using the large Plexiglas insert in the V1 block. The transducer is placed on the Plexiglas and the gain increased until 2 mm of noise (grass) appears or full gain is reached. The number of back wall echoes that appear on the screen is taken as a measure of the system gain for that transducer, frequency, and flaw detector.

Other checks need to be carried out from time to time to ensure that the performance of the system has not deteriorated. These include linearity of the amplifier (vertical linearity) and linearity of the time base (horizontal linearity).

Vertical Linearity. It has been established that signal amplitude has some meaning and that signal amplitudes are comparable in a measurable denomination (dB). In order to compare amplitude data, it is necessary that the measuring device, or comparator, be "linear"; in this case, that the measured signal amplitude is accurate within a given tolerance. Knowing this, there is confirmation that the measured signal amplitudes displayed on the screen are meaningful for the intended purpose. This provides information for the comparison of reflectors.

The procedure for verification of vertical linearity is one that demonstrates the instrument's ability to maintain an amplitude ratio between two signals, throughout the instrument's "linear range," while varying the gain. This practice is referred to as "screen height linearity verification." The basic procedure is as follows:

1. Two signals from either one or two reflectors are established at 80% FSH and at 40% FSH.
2. The gain is reduced until the 80% signal is positioned at 70% FSH. At this point, the lower signal should have an amplitude of 35%.
3. This procedure is repeated in steps that reduce the higher signal amplitude by 10%. The lower signal amplitude (if the display is linear) should remain at 50% of the higher signal amplitude throughout the vertical range. The specification to which this evaluation is being conducted may include acceptance criteria.

Note. There are several standards that provide the methodology for conducting these checks.

Amplitude Control Linearity. Verification that that the gain control is linear with respect to the signals noted on the screen is also an essential requirement. This procedure is known as "amplitude control linearity verification" and is as follows:

1. A signal from a reflector is set at an amplitude of 80% FSH. The gain is reduced by –6 dB on the control. The signal should now have an amplitude of 40% FSH.
2. If the signal amplitude is set to 10% FSH and 20 dB of gain is inserted, the amplitude should become 100% FSH.
3. Other values throughout the range of interest are selected and verified.

To summarize: screen height linearity checks are conducted to verify that the display is linear throughout its useable vertical range. Amplitude control linearity checks verify that the gain control on the instrument is in fact adjusting the signal amplitude by the selected number of dB.

Horizontal Linearity. Horizontal linearity is a measure of the instrument's ability to display time (distance) in a linear fashion across the screen. This means that in any given material, repeat signals from parallel surfaces will be represented on the screen as equidistant signals along the time base. For example, a ruler is calibrated to measure in inches or millimeters. If the ruler is not divided into some known denomination, it cannot be used as a measuring device. The same principle applies to the ultrasonic instrument. It is necessary to know that if the horizontal position of the signal on the screen indicates that a reflector is, for example, indicating 6 inches of steel thickness under the transducer, this measured distance is actually 6 inches.

With ultrasonic instruments, the ability to calibrate the circuitry (sweep) to display either inches, millimeters, or even microseconds is possible. The instrument screen displays a combination of the speed of sound in the test material and the distance that the sound has traveled. To do this, a "calibration standard" is necessary. This is referred to as "horizontal linearity." The instrument is calibrated and verified by using a standard having a known thickness. With the transducer placed on this standard, the instrument's controls (sweep and delay) are set to display signals in multiples of this thickness, which coincide with selected equal increments on the screen. If the signals can be aligned with these equally spaced increments, the instrument can be considered to have achieved horizontal linearity.

Inspection Techniques

Pulse-Echo Techniques

Contact Scanning Using Compression Waves. Compression wave pulse-echo techniques usually employ, either a single or dual crystal transducer directing ultrasonic energy perpendicular or near perpendicular to the scanning surface. These techniques are often known as "straight beam testing" techniques. There are some special techniques using compression waves at steeper angles used, for instance, in detecting cracks in ferrous materials under stainless steel cladding and for time of flight diffraction (TOFD) testing, but these will be covered separately.

In the standard compression wave techniques, reflections from the back wall and discontinuities are used to assess the suitability of a component for service. In order to obtain a reflection, it is necessary for the reflector to be orientated so that part of its surface is parallel to the scanning surface, in other words, normal to the beam. Laminar discontinuities and volumetric discontinuities like gas pores and nonmetallic inclusions are all suitably orientated. Discontinuities that are angled to the scanning surface may either not reflect at all or may reflect the sound away from the transducer.

Figure 7-62a shows a single crystal compression wave transducer set up for thickness measurement of a metal part, and Figure 7-62b shows the corresponding ultrasonic A-scan trace in which the time base has been calibrated for 25 mm full scale.

The initial pulse appears at zero on the left of the trace and the back reflection signal appears three-quarters along the time base, indicating a sample thickness of 18.75 mm. Thickness gauging is one of the simplest examples of compression wave testing. Notice that the initial pulse occupies almost a quarter of the time base so that 6 mm of metal path are obscured. This obstructed area is known as the "dead zone."

Figure 7-63a shows a dual element transducer set up for thickness measurement on a sample that is 4 mm thick. Figure 7-63b shows the trace for this sample with the time base again calibrated for 25 mm. Notice that the selection of "dual" operation of the flaw detector isolates the transmitter from the receiver circuit, so there is no initial pulse and, therefore, no dead zone. The first back reflection signal (also called "back wall echo" or "BWE") shows at 4 mm on the time base. Notice also that multiples of the BWE appear at 8, 12, 16, 20, and 24 mm on the time base.

One way in which the reading accuracy can be improved is to take a reading from a multiple and divide the result by the number of passes corresponding to that multiple. Take, for example, the reading at 24 mm (which is the sixth signal). Divide 24 by 6 and the answer is 4 mm. However, suppose the actual thickness was 4.15 mm. It would be difficult to read that accurately on the first back reflection signal, but the sixth signal would have been judged at 24.9. This number divided by 6 equals 4.15 mm. In practice, this sort of accuracy could only be expected on samples with very smooth scanning and back wall surfaces. For thickness measurement in the field on corroded surfaces, errors of up to ±0.5 mm are typical. Figure 7-63c illustrates an actual thickness gage being used to read the thickness of a plate that is corroded.

Figure 7-64a shows a single element transducer set up to detect laminations in steel plate 20 mm thick. The lamination is smaller than the beam. Notice in Figure 7-64b that the position of the lamination echo occurs on the screen at 11 mm below the scanning surface and the back reflection echo indicates a 20 mm thickness. The back reflection echo is reduced in amplitude because part of the beam is reflected by the lamination. If the lamination had been bigger than the beam, there would be no back reflection echo. If, on the other hand, the lamination had been smaller, the signal from the lamination would have been smaller in amplitude and the back reflection echo bigger. It might have been necessary to increase the equipment gain to see the lamination at all. In the extreme case, the

ULTRASONIC TESTING 7.69

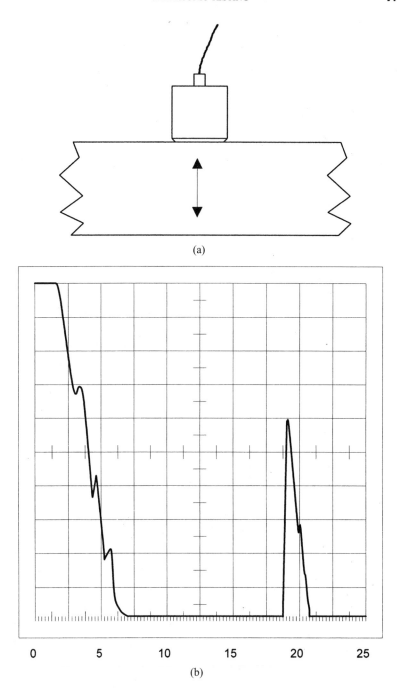

FIGURE 7-62 Thickness measurement.

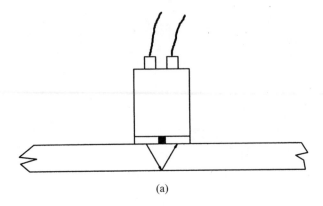

(a)

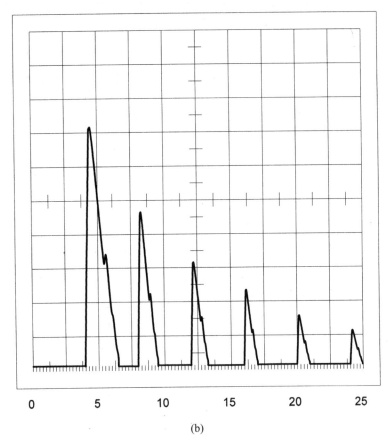

(b)

FIGURE 7-63 (a) and (b) Thickness measurement.

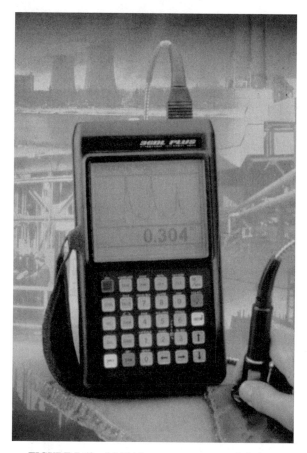

FIGURE 7-63 (c) Thickness gauge on corroded plate.

lamination might be so small that it could not be detected at the test frequency or gain used. Therefore, the detection of suitably orientated discontinuities is dependent on the size of the discontinuity, the test frequency, and the gain used. Higher frequencies can detect smaller reflectors due to their shorter wavelength.

It must be remembered that attenuation of the ultrasonic beam also has an effect on detection. As the energy penetrates deeper into the material, it weakens. Eventually, the beam is too weak to allow small echoes to get back to the receiver. The higher the test frequency, the greater the attenuation and the less penetration that can be achieved. The material and its grain structure also affect attenuation. The practitioner must balance the conflicting requirements of discontinuity size to be detected, material properties, and ultrasonic beam properties in the choice of transducer and test frequency.

Figures 7-65a and b show two discontinuities unfavorably oriented to the sound beam. The inclined discontinuity in Figure 7-65a is reflecting the energy away from the transducer, but also obscuring the back wall. The result would be no signals visible on the display, but there would be a reduction in the back reflection. In Figure 7-65b, the vertical

7.72 CHAPTER SEVEN

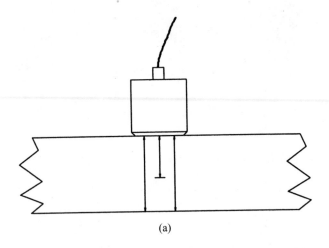

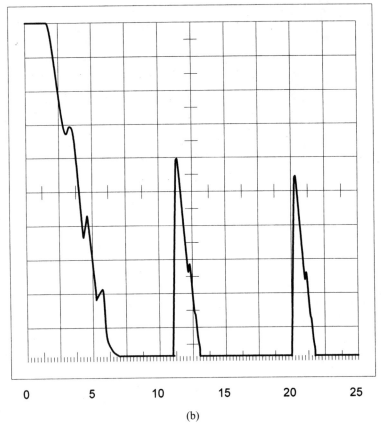

FIGURE 7-64 Perpendicular reflector.

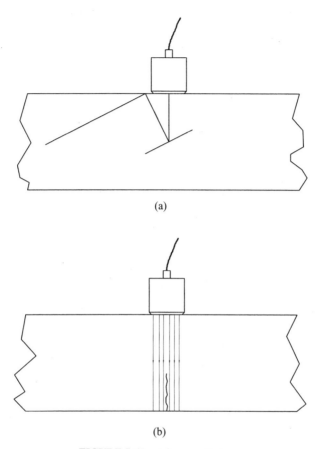

FIGURE 7-65 Adverse reflectors.

crack allows the sound to pass on either side without reflecting but would give a normal back wall echo. The possible orientation of the discontinuity must also be considered in devising a test procedure.

Finally, a test technique for the detection of laminations in thin plate is illustrated in Figures 7-66a, b, and c. The technique is called the "multiple echo" technique for reasons that are obvious from Figures 7-66b and c. The timebase has been calibrated for 50 mm for a sample 3 mm thick. With the transducer in position 1 (sound material), the multiple echo pattern stretches to 30 mm (14 signals) as shown in Figure 7-66b. With the transducer in position 2, over the lamination, the multiple echo pattern only stretches to 15 mm, as can be noted in Figure 7-66c. This is because the echoes are closer together; so close, in fact, that they interfere with each other, leaving no clear time base in between echoes.

Contact Scanning Using Angle Beam Shear Waves. If the possible orientation of any discontinuity is considered to be unfavorable to a beam perpendicular to the scanning surface, it will be necessary to tilt the beam to an appropriate angle to ensure that the beam strikes the discontinuity as near perpendicular as possible. For small angles (up to about

7.74 CHAPTER SEVEN

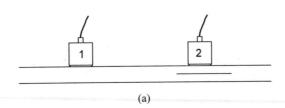

(a)

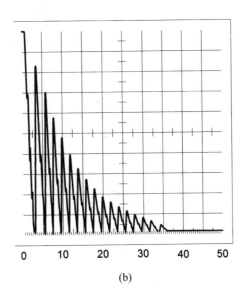

(b)

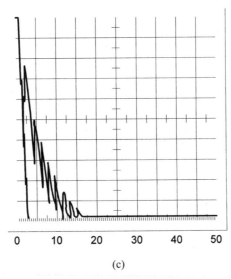

(c)

FIGURE 7-66 Lamination detection.

10° in the test material), compression waves may be used. However, for larger angles, mode conversion to shear wave energy makes the use of compression waves alone impossible. It therefore becomes necessary to increase the incident angle beyond the first critical angle, leaving only a shear wave in the part.

The lowest practical angle for testing with a shear wave alone is about 35° refracted shear wave angle. This does not mean that testing at angles between 10° and 35° is impossible. However, if an angle in this range needs to be used, the practitioner must consider carefully the geometry of the part. The next decision is whether to use the compression wave or the simultaneous shear wave, depending on what happens to the unwanted mode. Regular off the shelf transducers are either straight beam compression wave or shear wave angle transducers of 35° to 70°.

The common or "preferred" angles available in ultrasonics for shear wave testing are 45°, 60°, and 70°, although other angles can be made to order. The angles marked on a shear wave transducer are for steel, unless followed by an identifying letter for other materials. For instance, "45°Al" would denote a 45° shear wave angle transducer for aluminum.

Half Skip and Full Skip Techniques

The simplest angle beam test is one that looks only for vertical surface-breaking discontinuities originating at the scanning surface or back wall. Figure 7-67a shows a 45° shear wave transducer positioned on a 20 mm thick plate. The beam reflects at the bottom and then top surfaces at an angle equal to the angle of incidence of the shear wave beam, 45° in this case. Note that with no discontinuity present, no echo energy returns to the transducer.

Figure 7-67b shows the transducer aimed at a slot breaking the bottom surface. This slot will produce an echo arriving back at the transducer at a fixed time that is twice the beam path distance A–B divided by the shear wave velocity. The trace is shown at Figure 7-67c; the time base is calibrated for 100 mm return trip time, and the beam path distance is shown as 28 mm.

In Figure 7-67d, the distance A–B along the scanning surface is called the "half skip" distance and for a 45° probe this is equal to the specimen thickness. Position C is called the half skip position. Any discontinuity breaking the bottom surface of this part will produce a signal at 28 mm.

Figure 7-67d shows the transducer positioned to reflect from a top surface-breaking slot. The distance A–D is called the "full skip" distance and a full skip signal would appear at 56 mm on the time base, as shown in Figure 7-67e. If the transducer is scanned along a 20 mm plate containing top and bottom surface-breaking discontinuities, only three signal patterns are possible:

1. No signal representing sound material
2. A signal at 28 mm representing a bottom corner reflector
3. A signal at 56 mm representing a top corner reflector

For the practitioner, interpretation of results is relatively simple since there are only two screen locations on which to concentrate. The technique is commonly used to detect fatigue cracks during the in-service inspection of critical components.

Beam Path Distance Techniques

Of course, not all discontinuities occur at the top or bottom surfaces. In welds or castings, for example, planar and volumetric discontinuities may occur anywhere within the volume of the part. In order to detect, correctly assess, and position such discontinuities, it is necessary to determine the distance along the beam path at which the reflection occurs. This distance together with the known beam direction and angle allows the position of the discontinuity to be plotted.

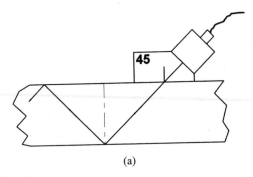

(a)

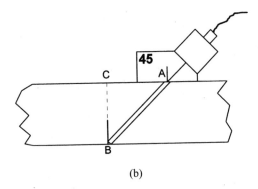

(b)

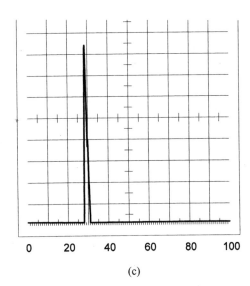

(c)

FIGURE 7-67 Discontinuity detection.

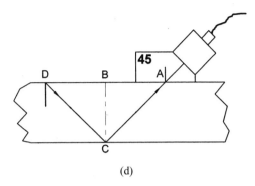

(d)

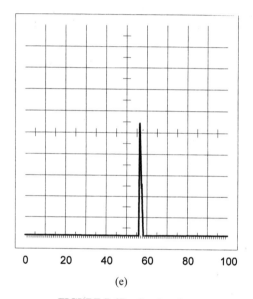

(e)

FIGURE 7-67 *Continued.*

Volumetric discontinuities, such as gas pores or slag inclusions in welds, are not very sensitive to beam angle. Approached from almost any direction, there is likely to be a facet of the discontinuity that will reflect back to the transducer. On the other hand, planar discontinuities, such as lack of side wall fusion in welds and angular cracks, are very sensitive to beam angle. The practitioner must be aware of the types of discontinuities that might occur during fabrication and service in a part to be inspected.

To be able to measure beam path distance, the time base must be calibrated for the shear wave velocity in the material to be inspected. This process of calibration requires suitable calibration blocks. Such calibration blocks should be capable of being used for a variety of beam angles. An example of a calibration block (IIW) is shown in Figure 7-68.

Figures 7-69a–f illustrate how this technique can be used to locate a region of lack of

7.78 CHAPTER SEVEN

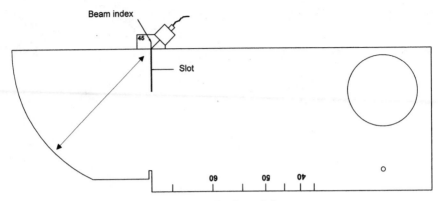

FIGURE 7-68 Beam index.

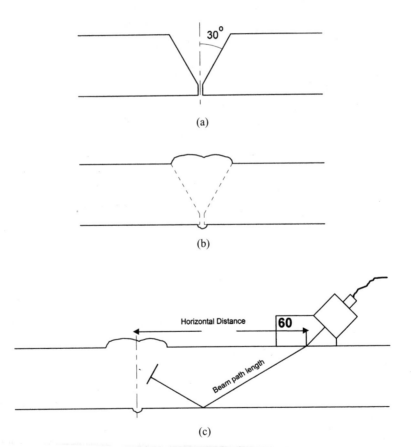

FIGURE 7-69 Discontinuity detection in weld.

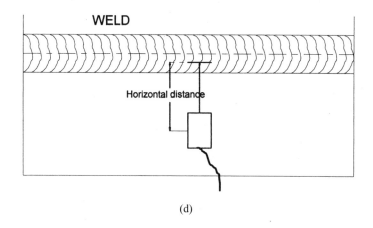

(d)

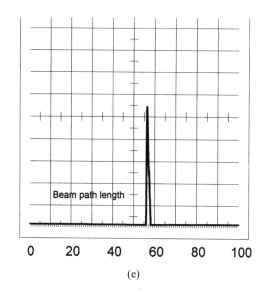

(e)

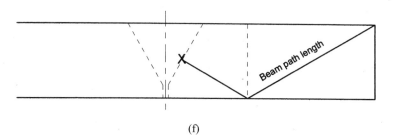

(f)

FIGURE 7-69 *Continued.*

side wall fusion in a weld 20 mm thick. Figure 7-69a shows the prepared edges before welding. Figure 7-69b shows the completed weld, with dotted lines to show the original prepared edges. In Figure 7-69c, a 60° angled shear wave transducer is shown positioned to obtain a maximum echo from the side wall discontinuity. The angle has been chosen to ensure that the beam meets the original prepared face at 90° for maximum sensitivity. Figure 7-69d is a plan view of the same set-up, showing that the center line of the weld has been marked as a reference for measurements. Figure 7-69e shows the trace as the echo maximizes.

The practitioner, once satisfied that the echo is maximized, then makes two measurements; one from the time base, the other on the scanning surface using a ruler. The first measurement is the time base range giving the beam path length to the discontinuity. The second measurement is the horizontal distance from the beam index to the weld center line. These two measurements, together with knowledge of the beam angle, are enough to accurately plot the position of the reflecting point.

Figure 7-69f shows the process for plotting this position. A scale drawing of the weld is made. The center line of the weld is drawn and another perpendicular line through the parent metal at the measured horizontal distance from the weld center line. From the intersection of this perpendicular with the scanning surface, a beam path line is drawn at 60° to the perpendicular to meet the bottom surface. A reflection line from this intersection is then drawn up toward the top surface. The position of the discontinuity is plotted by measuring along the beam path line a distance equal to the measured time base range. The position is marked on the scale drawing ("X" in Figure 7-69f). In practice, dedicated plotting aids are used to make the process easier.

Rod and Pipe Techniques

Figure 7-70 shows an angle beam transducer placed on the circumference of a metal bar in order to detect longitudinal discontinuities along the outer diameter (OD). The beam

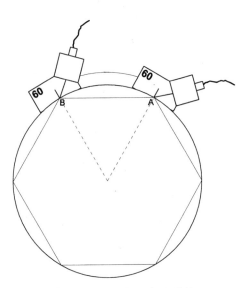

FIGURE 7-70 Angle beam on solid bar.

path length at which an echo from such a discontinuity will appear on the timebase can be calculated from

$$\text{beam path} = OD \times \cos \theta$$

where θ = probe angle.

If the time base is calibrated to a suitable range, the entire circumference can be examined by scanning from A to B in Figure 7-70.

A similar set up is shown in Figure 7-71a, but this time the probe is placed on a pipe. The angle at which the beam strikes the bore of the pipe depends on the transducer angle and the ratio of the inside diameter (ID) to the OD. The beam path length A–B to the ID can be calculated from

$$\text{beam path} = \frac{t}{\cos \theta}$$

where
t = pipe wall thickness
θ = probe angle

When inspecting pipe for OD or ID surface breaking discontinuities it is often easier to calibrate the time base using a reference block similar to the one shown in Figure 7-71b; the drilled hole provides both ID and OD reflecting points. If the pipe wall is very thick, the beam center never reaches the bore of the pipe. This is illustrated in Figure 7-72.

For any given beam angle the maximum wall thickness for a given pipe OD can be calculated from

$$t = \frac{OD \times (1 - \sin \theta)}{2}$$

where
t = maximum wall thickness
θ = Probe angle

From this equation, it is possible to derive another, which will allow the calculation of optimum probe angle to examine the bore of any given pipe. The equation becomes

$$\sin \theta = 1 - \left(\frac{2t}{OD} \right)$$

For convenience, Table 7-3 gives the maximum wall thickness that can be examined with common standard probe angles for a range of typical pipe diameters.

Multiple Transducer Techniques

Through-Transmission. Ultrasonic tests using through-transmission predate the pulse-echo technique and were used in the initial experimentation procedures. By definition, through-transmission implies that two transducers are placed on opposite sides of a test piece facing each other. Whereas with dual transducers only single-sided access is necessary, through-transmission techniques require access to two sides of a component. The test material is either moved through or rotated within the sound field and scanned. It is a valuable technique when used for velocity measurements and in the characterization of material by comparison with a standard.

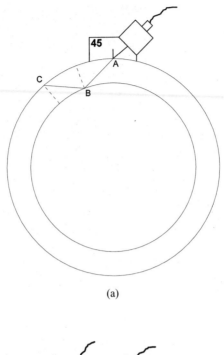

(a)

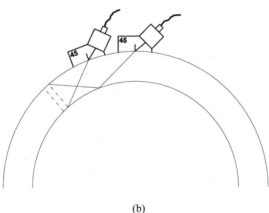

(b)

FIGURE 7-71 Single and tandem transducers.

Certain applications lend themselves to this technique. These are usually situations in which the material has high attenuation and pulse-echo techniques are not suitable. Composite material used in the aircraft industry is one example. Other applications include production plate scanning, where banks of transducers are used with water columns (squirters) in the through-transmission mode to inspect for laminations. Other situations exist where discontinuities are close to the surface and can only be detected by observing the loss of transmitted energy.

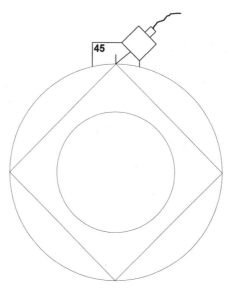

FIGURE 7-72 Angle beam on hollow bar.

Advantages of through-transmission are:

1. Discontinuities in highly attenuative materials may be detected by noting a reduction in the received energy.
2. The complete section can be tested. There are no initial pulse constraints.

The disadvantages are numerous. Through-transmission testing relies on energy reduction as it is "shadowed by" a material discontinuity. Energy reduction can be caused by factors other than the anomalies sought. The possibilities are:

1. Loss of couplant on either side
2. Misalignment of transducers
3. Change in material attenuation
4. There is no positive indication of a reflector and therefore discontinuity depth information is not possible.
5. Variations in surface finish (roughness)

TABLE 7-3 Approximate Maximum Wall Thickness

Pipe OD, inch (mm)	Probe angle 35°	Probe angle 45°	Probe angle 60°
4" (100)	0.825" (21)	0.57" (14.5)	0.250" (6.5)
6" (150)	1.26" (32)	0.87" (22)	0.375" (10)
8" (200)	1.67" (42)	1.14" (29)	0.5" (13)
10" (250)	2.1" (53)	1.44" (36.5)	0.65" (16.5)
12" (300)	2.5" (64)	1.75" (44)	0.8" (20)

Tandem Techniques. These are common terms for techniques using two or more transducers arranged so that reflections of the transmitted pulse are detected by a receiver (or receivers) positioned at their predicted exit points. As an example, Figure 7-73a shows a vertical discontinuity that does not break either the top or bottom surface of a metal plate. Such a discontinuity will not reflect sound back to the transmission point. This means that a single angle beam transducer cannot be used. Instead, the arrangement of two probes, one to transmit and one to receive, is used, as can be noted in Figure 7-73b. This is known as the "tandem" technique.

The distance A–B between the probes is dependent on the aiming point of the transmitter. The illustration shows an aiming point in the middle of the plate. If the aiming point were to be nearer the top surface, then the probes would need to be further apart. If the aiming point were to be lower, then the probes would have to be closer together.

Lamb Wave Techniques

Lamb waves, like Rayleigh waves, propagate parallel to the test surface and have a particle motion that is elliptical. They occur when the thickness of the test material is only a few wavelengths at the test frequency and where the test piece is of uniform thickness. In other words, they can exist best in plate, tube, and wire.

The Lamb wave affects the entire thickness of the test material in such a way that it flexes. Figure 7-74 illustrates a type of Lamb wave where the crests of the wave on the near and far surfaces coincide. These are called symmetrical Lamb waves. Figure 7-75 shows another type of Lamb wave where the crest on one side coincides with a trough on the other. These are called asymmetrical Lamb waves.

These waves are generated at incident angles that depend on the test frequency and material thickness. These parameters also determine the number of modes of Lamb wave that can exist in the test material. In order to generate a Lamb wave, the velocity at which

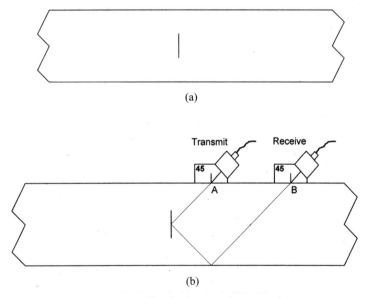

FIGURE 7-73 Tandem transducers on plate.

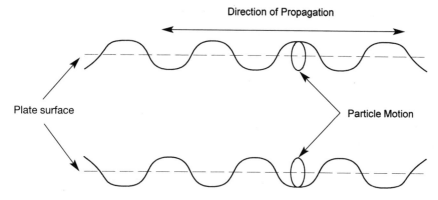

FIGURE 7-74 Lamb wave, symmetrical.

the incident compression wave in the Perspex (Plexiglas) sweeps along the interface must coincide with the velocity of the Lamb wave in the material. This is achieved by adjusting the angle of incidence $i°$. This velocity can be calculated from

$$V = \frac{V_c}{\sin i°}$$

where
V = the velocity of the incident wave front along the test surface
V_c = the incident compression wave velocity in Perspex (Plexiglas)
$i°$ = the angle of incidence in the Perspex

Figure 7-76 illustrates the above formula.

Techniques using Lamb waves usually involve transmit and receive probes facing each other on the test surface. Lamb waves can travel several meters in steel, so they can be used for rapid scanning of plate, tube and, wire.

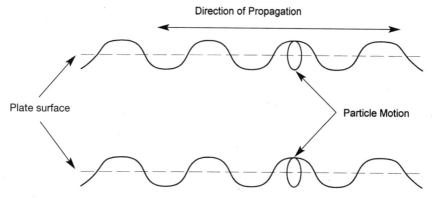

FIGURE 7-75 Lamb wave, asymmetrical.

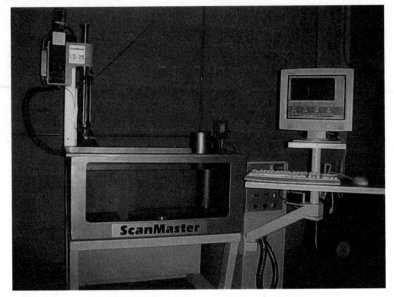

FIGURE 7-76 Multipurpose immersion system.

Immersion Techniques

Contact scanning techniques are mostly used for scanning small areas of components for in-service inspections and for scanning large components that cannot be moved to an immersion set-up. Immersion techniques are far quicker and more convenient for scanning large areas of plates, pipes, and wrought products during manufacture. The techniques lend themselves to automated scanning and recording systems (see Figure 7-76).

There are several advantages of automated immersion testing over manual scanning techniques. These include:

1. Consistent coupling conditions and scanning capability
2. Variable beam angles and beam focusing
3. Interface gating and contour following is possible
4. Improved near-surface resolution and use of higher frequencies

Most contact scanning techniques can be used in immersion testing. In addition, the same transducer can be used for both compression wave and angle beam techniques. To generate any angle of shear wave, the transducer is simply tilted to the appropriate incident angle derived from Snell's law.

Disadvantages include:

1. Setting up is more complicated
2. The part must be compatible with water
3. Air bubble formation interferes with the test

Figure 7-77 illustrates a typical immersion set-up for compression wave scanning. The water path distance (w) is important because there will be multiple echoes from the top surface of the part (entry surface). The spacing between these multiple echoes must be longer than the time taken by the transmitted sound to travel the back wall and re-

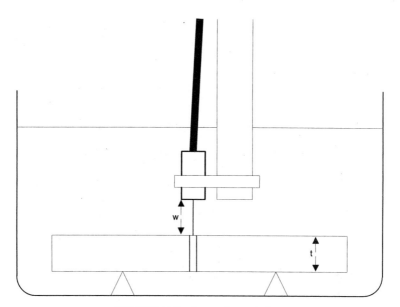

FIGURE 7-77 Immersion technique.

turn. As a rough guide, for steel and aluminum the minimum water path should be about one-quarter of the material thickness. Since sound in water travels at about a quarter of the speed it does in those metals, the back reflection (back wall echo) will appear just before the water multiple. Figure 7-78 shows a typical trace for the above immersion set-up. In this trace, no delay has been used. By using the delay control to get rid of the water path, and calibrating the time base for a suitable range, the trace shown in Figure 7-79 can be obtained.

This trace can be compared to that shown in Figure 7-64b in the contact scanning techniques above. Note that the entry surface echo in Figure 7-79 occupies less space than the initial pulse shown in Figure 7-64b and therefore offers better near-surface resolution.

Figure 7-80 shows the same transducer tilted to an angle of incidence to generate a shear wave beam at 45° to the normal. It can be seen that other refracted angles can be generated by simply changing the incident angle.

Normalization

Normalization is the name given to the process of ensuring that the transducer is actually perpendicular to the entry surface before any other calibration procedure is commenced. The name given to the probe holder in immersion testing systems is the "manipulator." The manipulator allows rotation of the probe in two directions perpendicular to one another. During the initial set-up for an immersion test, the entry surface signal is maximized by careful manipulation of the probe in each axis. As the entry surface echo maximizes in an axis, the rotation is stopped in that axis. When a maximum response has been achieved in both axes, the probe is perpendicular to the entry surface and the transducer has been "normalized."

The normalized position would be used for the compression wave techniques and as a starting point from which to set the incident angle for shear wave techniques.

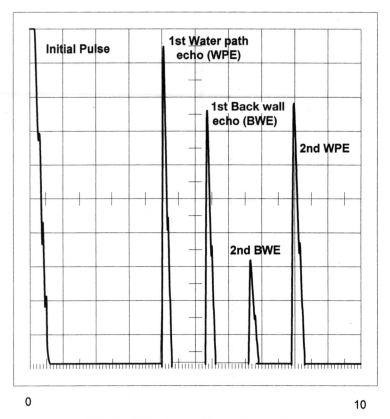

FIGURE 7-78 A-scan of immersion technique.

Interface Triggering (Gating)

Interface triggering or "gating" is often used when the entry surface, because of the component geometry, is not at a fixed distance from the probe face. For example, the stepped surface shown in Figure 7-81 presents three different water path distances (surfaces A, B, and C). Interface triggering starts the sweep as the first entry signal arrives. It acts as an automatic delay control that always ensures that zero on the trace represents the entry surface of the test piece regardless of water path. This is particularly useful, for example, if a round component is being scanned with a spiral scan pattern as the component is rotated past the scanning head. If the component is not truly round, or is being rotated eccentrically, without interface triggering, the whole signal pattern would be moving left and right. Interpretation of the trace visually or automatically under those circumstances would be difficult.

Offset Method for Generating Shear Waves

When carrying out immersion testing of tubular components for longitudinal discontinuities such as the one shown in Figure 7-82, it is not necessary to tilt the probe in order to generate shear waves of a given angle. The appropriate angle of incidence can be

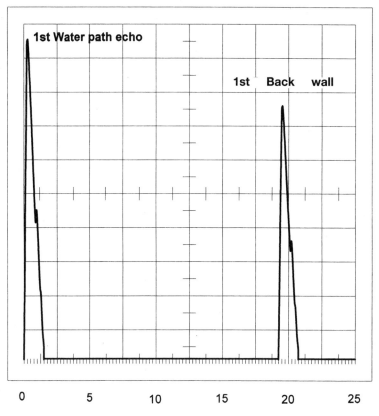

FIGURE 7-79 Delayed immersion A-scan.

achieved by offsetting the probe from the center line by an appropriate amount. The amount of offset for a given required incident angle can be calculated using the following equation:

$$\text{offset} = \frac{D}{2} \times \sin i°$$

where
D = component OD
$i°$ = required angle of incidence

Example. Calculate the offset required to generate a 45° shear wave in a 50 mm diameter steel rod. Given that the shear wave velocity in steel is 3240 m/sec, and the velocity of sound in water is 1480 m/sec, from Snell's law, the required angle of incidence is

$$\sin i° = \frac{1480 \times \sin 45°}{3240}$$

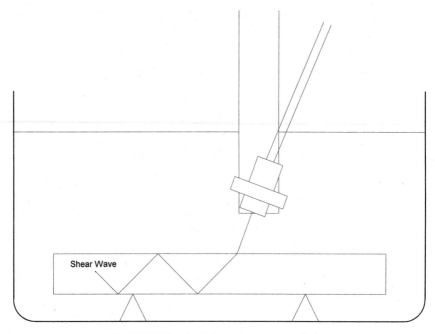

FIGURE 7-80 Angle beam immersion technique.

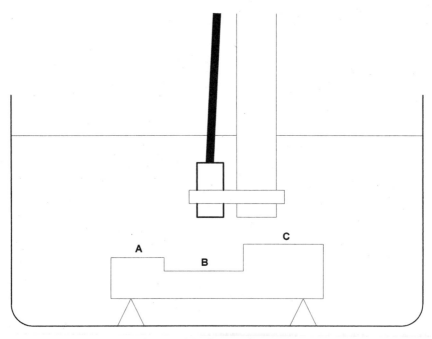

FIGURE 7-81 Contoured component.

7.90

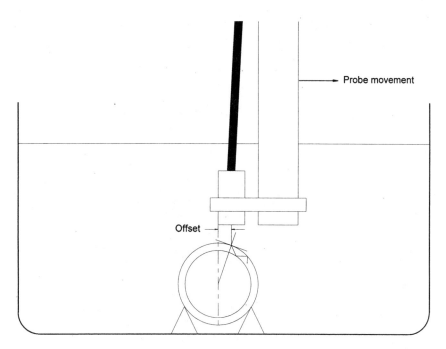

FIGURE 7-82 Offset transducer.

$$\sin i° = 0.3230$$
$$i° = 18.84°$$

Putting the calculated angle into the formula

$$\textit{offset} = \frac{D}{2} \times \sin i°$$

$$\textit{offset} = \frac{50}{2} \times \sin 18.84°$$

$$\textit{offset} = 25 \times 0.3229$$

$$\textit{offset} = 8.03 \text{ mm}$$

Discontinuity Sizing Techniques

The operating procedure for carrying out the inspection of a specific component must include instructions for setting test parameters that will ensure detection of a given minimum reflector. This can mean that discontinuities will be detected that do not render the component unfit for service. Each discontinuity detected must, therefore, be evaluated against the applicable inspection code or standard. In most cases, the acceptance criteria are set in terms of the length or vertical extent of the discontinuity. In order to correctly evaluate the results, it will be necessary to derive these critical dimensions from the ultrasonic information. Techniques for doing this are known as "sizing techniques."

The techniques available to the practitioner for assessing the severity of discontinuities fall into two categories. First, there are those procedures that quickly "screen" the echo on the trace with reference to an amplitude standard. Examples are the area/amplitude reference standards, the DAC curves, and the DGS curves. In most codes, if the signal exceeds a given proportion of the reference amplitude, the code calls for further investigation (see DAC, page 7.63).

The second category is those procedures that attempt to measure or plot the through-thickness dimension and length of the discontinuity. The techniques include intensity drop, maximum amplitude, and time of flight diffraction (TOFD).

Intensity Drop Technique. The intensity drop technique uses the beam spread to determine the edges of the reflector. Figure 7-83 shows a 20 dB beam spread diagram on a card to which a transparent cursor has been fitted. The cursor has a horizontal line scribed to coincide with the horizontal scale on the card, and it also represents the scanning surface of the component. A vertical line is scribed to denote the reference point (e.g., weld center line). In this example, a weld is being examined and the weld profile has been drawn in wax pencil on the cursor. Note that a mirror image of the weld has also been drawn to allow full skip ranges to be plotted.

Sizing Procedure—Through-Thickness Dimension

Figure 7-84 shows the plan view of a weld with the transducer at a position where a maximum echo height has been obtained from a discontinuity. At this point, the echo height is adjusted to the reference level (i.e., 80%) and the horizontal distance (HD1) measured from the probe index to the weld center line. The corresponding time base range (TB1) is also noted.

The transducer is scanned forward until the signal drops to the intensity drop wax pencil line. At this position, the bottom edge of the plotted beam will be reaching the top of the discontinuity. The new horizontal distance (HD2) and timebase range (TB2) are recorded.

Last, the transducer is scanned backward, through the maximum again and back until

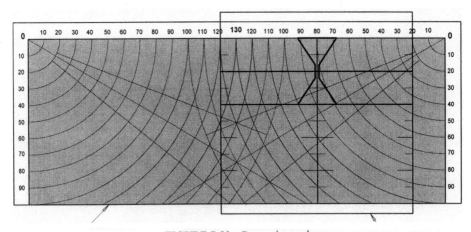

FIGURE 7-83 Beam plot card.

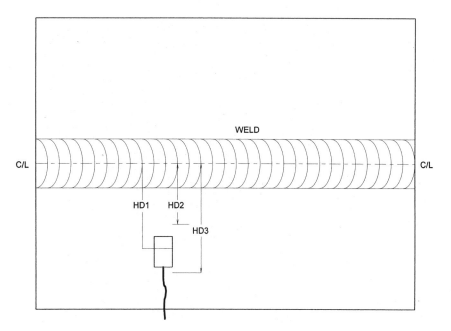

FIGURE 7-84 Weld center line.

the signal again drops to the wax pencil line at HD3. The distance and range (TB3) are recorded. This position represents the top edge of the beam just striking the bottom of the discontinuity.

There are now three sets of data for the through-thickness dimension of the reflector. These are the probe position (HD) and time base range (TB) for top, middle, and bottom of the discontinuity. These data are now transferred to the cursor of the plotting aid.

Figures 7-85a, b, and c show the steps for plotting the discontinuity. In Figure 7-85a, the cursor has been set with the weld center line positioned at a distance HD1 on the card's horizontal scale. Going down the beam path scale, a cross has been marked with a grease pencil at the time base range for the middle of the discontinuity (TB1). Note that in this case the distance TB1 has put the cross in the mirror image on the cursor. This means that the reflection has come from beyond the half-skip position when the beam is on its way up to the full-skip position. Because the weld profile has been sketched onto the cursor it is immediately obvious that this discontinuity is close to the weld side wall.

In Figure 7-85b, the cursor is shown to be repositioned at HD2 and a new cross marked at range TB2. Note that this cross is marked where the bottom edge of the beam on the beam spread diagram intersects with the beam path scale. The new cross marks the top of the discontinuity. Note that it also lies near to the side wall.

The bottom of the discontinuity is plotted as shown in Figure 7-85c. The distance HD3 and range TB3 place a cross where the top edge of the beam intersects the beam path scale. Again, the cross falls close to the side wall. The three crosses can now be joined to show the extent of the discontinuity in the cross-section of the weld. This can be measured and evaluated against the relevant acceptance standard.

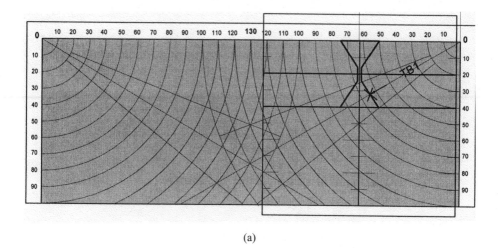

(a)

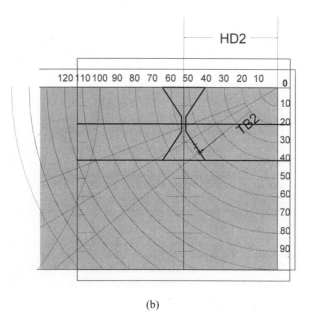

(b)

FIGURE 7-85 Plotting the discontinuity.

Sizing Procedure—Discontinuity Length

Using the intensity drop method to measure the length of the discontinuity is much simpler than the procedure for the through-thickness dimension. All the measurements and marks are made on the test surface. Figure 7-86 shows the transducer positions for the measurement.

Position 1 is where the transducer detects a maximum echo. The position is marked

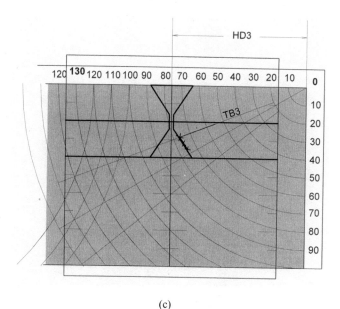

(c)

FIGURE 7-85 *Continued.*

with grease pencil. The transducer is scanned sideways until the echo has dropped by 6 dB and the position marked on the surface (position 2). The transducer is then scanned through position 1 and on to position 3, where the signal has again dropped by 6 dB. Position 3 is marked on the surface. The distance "L" between positions 2 and 3 is the length of the discontinuity. Care should be taken to ensure that the –6 dB point is the "end" of the discontinuity.

Maximum Amplitude System
The maximum amplitude system (MAS) uses the same plotting aid as the intensity drop system. However, with MAS only the beam center is used to plot the extent of the discontinuity. The method is particularly suitable for sizing discontinuities that are multifaceted, such as cracks or slag inclusions. Figure 7-87 shows an angle beam transducer aimed at an internal crack-like target.

Across the beam, various facets of the target are oriented perpendicular to that part of the beam. The result is that several echoes get back to the receiver in a cluster over a short time period. This gives a signal pattern similar to the one shown in Figure 7-88. The signal from each facet of the discontinuity reaches a maximum as the beam center moves onto it. If the transducer is moved, forward or back, from the position on the test surface at which the overall signal pattern is maximized, the signal pattern as a whole decays. So far everything is the same as for the intensity drop method. However, if the dynamic signal pattern is studied in detail during this decay, signals from the individual facets will be seen to rise and maximize as the beam center reaches that area of the discontinuity. As the transducer is scanned right across the discontinuity region, each facet in turn will grow, maximize and decay. Each maximum represents a point on the face of the discontinuity.

The sizing procedure starts in the same way as the procedure for intensity drop—by

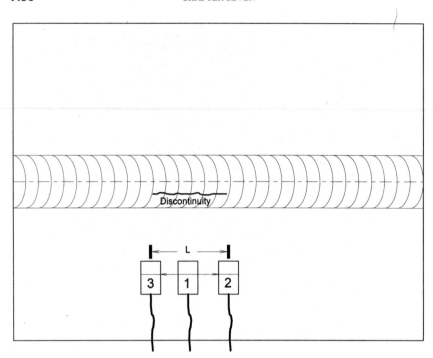

FIGURE 7-86 Plan view of discontinuity.

maximizing the overall signal from the discontinuity. The horizontal distance to the weld center line and the time base range are recorded as before. As the transducer is scanned forward from this position, the signal pattern is scrutinized to detect the rise in signal of a facet echo while the main pattern is falling. As each facet echo maximizes, the HD and TB distances are recorded until the last facet has started to fall and the overall signal is lost.

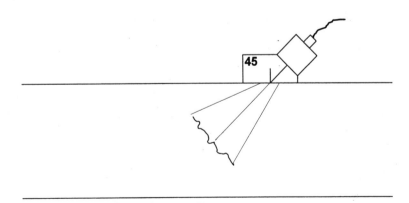

FIGURE 7-87 Sectional view of discontinuity.

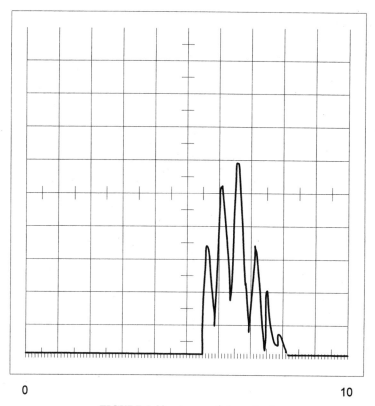

FIGURE 7-88 A-scan of discontinuity.

The process is repeated on the backward scan from the overall maximum position until the last maximum in that direction has been recorded. The coordinates of all the recorded points can then be transferred to the cursor of the plotting aid.

The set-up for the plotter is the same as that shown in Figure 7-83, except that only the beam path scale is used to plot the points. There will be more points to plot for each discontinuity than for the intensity drop method. Joining these points on the cursor traces the shape and plane of the discontinuity after the fashion of "connect the dots" picture books for children.

During the scanning procedure, as each maximum is being detected, the gain can be increased to bring in the weaker signals from the smallest facets. If this is done, the last maximum detected is likely to be a tip diffraction signal from the end of the discontinuity. This signal was described by Sproule to be –30 dB below the amplitude of a corner reflector at the same depth. In practice, this can be difficult to identify, but if it does appear, it very accurately pinpoints the extremes of the discontinuity.

Tip Diffraction Techniques

The tip diffraction techniques follow on from the "last maximum" in the maximum amplitude technique. If the extremities of the discontinuity can be accurately plotted, it follows that the discontinuity can be "sized." Since the diffraction signal originates at a "tip," it would appear to be an ideal feature to exploit. There are two basic approaches to tip diffraction sizing:

1. Using a standard angle beam transducer to carry out an additional scan specifically to size discontinuities that have already been detected
2. Using the time of flight diffraction (TOFD) technique in which two angled compression wave transducers are used to both detect and size in one pass

Using a Standard Angle Beam Transducer. When a beam of sound strikes the tip of a reflector, the reflected energy "diffracts" at the tip to form a spherical wave, as shown in Figure 7-89. The spherical wave will eventually reach a large part of the component surface. This means that the diffracted energy can be detected over a much larger area than the main beam, which is very directional. The diffraction energy is greatest as the beam center encounters the tip. The energy as this point is reached is typically 30 dB less (i.e., –30 dB) than a corner reflector at the same depth.

This means that the tip diffraction signal is weak and may only be two or three times the noise level (i.e., the signal to noise ratio is typically around 2:1–3:1). These small signals can be difficult to identify on the screen. Nevertheless, the signal can frequently be identified, and in these cases it becomes a useful sizing tool.

The procedure is similar to the maximum amplitude method. Once the discontinuity has been identified and positioned within the component, the transducer is arranged to obtain a corner reflection at a depth close to the discontinuity depth. The signal from the corner reflector is adjusted to a reference height (typically 30% FSH). The gain is then increased by 30 dB (+30 dB) and the transducer moved to the discontinuity region. As the transducer interrogates the discontinuity, the screen is studied for a signal that rises to the reference level as the main echo signal decays. As this signal reaches its maximum, the coordinates are noted and the tip position plotted as before.

It is often considered preferable to use an unrectified (RF) display to carry out tip diffraction techniques. This is because there is a phase reversal between diffraction signals originating at the top tip and those originating at the bottom tip of a discontinuity. With small discontinuities, it is sometimes possible to identify top and bottom of the discontinuity at the same time.

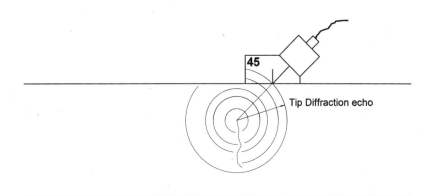

FIGURE 7-89 Tip diffraction echo.

Time of Flight Diffraction (TOFD) Technique. The TOFD technique, first used by M. G. Silk in 1977, uses tip diffraction to identify the top, bottom, and ends of a discontinuity in one pass. Silk chose to use an angled compression wave for the TOFD technique rather than a shear wave, for two reasons. First, the tip diffraction signal is stronger than a shear wave diffraction signal, and second, a lateral wave is produced that can be used to measure the horizontal distance between the transmitter and receiver.

The tip diffraction signal is generated at the tip of the discontinuity—effectively a "point" source. According to Huygens, a point source produces a spherical beam. Figure 7-90 shows both the lateral wave and a diffraction beam from the tip of a reflector.

Figure 7-91 shows a typical TOFD transducer set-up on a component with a vertical discontinuity. There are four sound paths from the transmitter to the receiver. Path "A" is the lateral wave path traveling just below the surface. Path "B" is the tip diffraction path from the top of the discontinuity. Path "C" is the tip diffraction path from the bottom of the discontinuity, and path "D" is the back wall echo path.

Figure 7-92 shows a typical unrectified trace for the four signals. Note that the phase relationships A and C are in opposite phase to B and D. The important difference to note is between B and C—the top and bottom diffraction signals are in opposite phase. This phase difference allows the practitioner to identify those points.

Assuming that the diffracting tip is centered between the two transducers, the depth of the tip below the surface can be calculated from

$$depth = \sqrt{\left(\frac{BPL}{2}\right)^2 - \left(\frac{HD}{2}\right)^2}$$

where

BPL = beam path length for the signal in question
HD = beam path length for the lateral wave

The distance measurements taken from the ultrasonic trace must be made from the same part of each waveform. In the example trace shown in Figure 7-92, the largest half-cycle would be selected. For signals A and C, this is negative, and for signal B, positive. Advances in computer technology have made it possible to carry out all the calculations

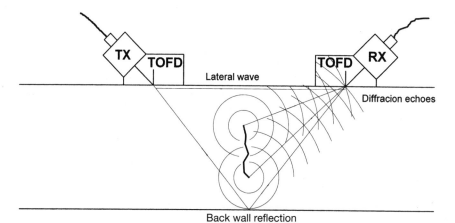

FIGURE 7-90   TOFD.

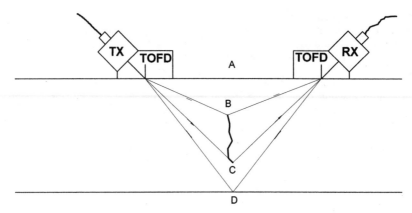

FIGURE 7-91 TOFD.

and for plotting to be handled automatically and stored for subsequent evaluation. The method that has been chosen to display this TOFD data presents the information in a special "B-scan" form that is easy to assimilate. The way in which the positive and negative half cycles are displayed needs explaining.

An echo arriving at the receiver is a pulse of a certain pulse width and amplitude. In conventional B-scan displays, this pulse is displayed as a bright spot whose diameter is proportional to the pulse width and whose brightness is proportional to the signal amplitude. In some ways, it is like a broad pencil tip that can be used to draw pictures in light or bold broad strokes. The pulse is really a short burst of a few cycles of alternating waveform. In the TOFD system, the waveform is depicted in grayscale, with positive half-cycles tending toward white, and negative half-cycles tending toward black (see Figure 7-93). This allows particular half cycles to be identified for measurement purposes, and phase changes to be recognized for determination of top or bottom echo.

Figure 7-93 shows a typical computer screen for a TOFD inspection. The image shows details of the component (in this case, a weld) as well as the TOFD B-scan image and an A-scan trace. In this image, left to right represents the component thickness, and the vertical dimension represents scan length.

The A-scan trace shown corresponds to a slice through the weld at the location indi-

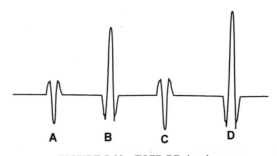

FIGURE 7-92 TOFD RF signals.

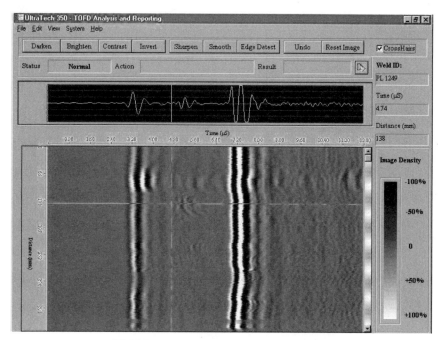

FIGURE 7-93 B-scan presentation of TOFD.

cated by the cross hairs of the cursor. The striped band on the left of the TOFD image represents the lateral wave, and the bold striped band to the right of the image represents the back wall signal. The difference in boldness is due to the different signal amplitudes. Following the horizontal cross hairs and about halfway between the lateral wave and back wall "stripes," a series of faint horseshoe-shaped stripes can be seen. These are diffraction signals from a small discontinuity. The A-scan trace shows the signal clearly.

In this example, the discontinuity has a very small dimension in the through-thickness dimension, but close study of the A-scan shows a small phase shift in the last half-cycle of the discontinuity signal. This tells the practitioner that the distance from top to bottom of the discontinuity is about the same as the pulse length for this particular discontinuity.

A much bolder indication can be seen toward the top of the lateral wave line, suggesting a discontinuity at, or just below, the surface. In Figure 7-94, the cursor has been moved to this location. The lateral wave signal can be seen to be longer and stronger than at the previous location. The fact that the wave shape stays in phase suggests that the diffraction echo, which is extending the signal, has the same phase as the lateral wave. In other words, it is a bottom tip signal. However, it is not possible in this case to see where the lateral wave ends and the bottom tip begins, and so it is not possible to say how deep the discontinuity extends below the surface. The TOFD method is limited in its ability to size near-surface discontinuities when the depth is similar to the pulse length.

The transducers used in these TOFD techniques are angled compression wave transducers (refracted longitudinal wave). The common angles used are 60° and 70°, al-

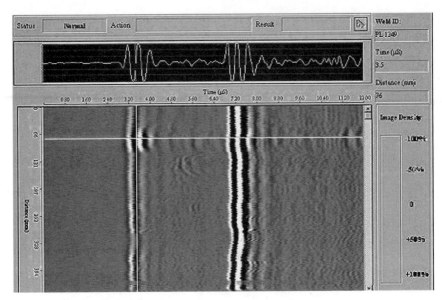

FIGURE 7-94 B-scan presentation of TOFD.

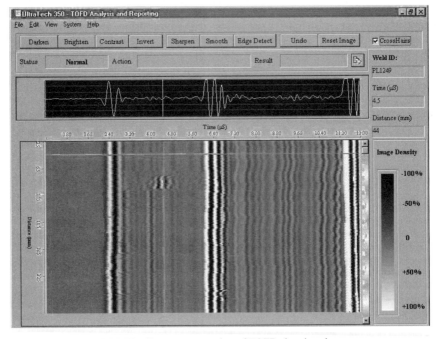

FIGURE 7-95 B-scan presentation of TOFD showing shear wave.

though other angles may be used if the component thickness makes it necessary. The design and construction of the transducer is important in order to promote a good lateral wave. Previous theory has suggested that a shear wave should also exist in the component, and this is true—it does. Figure 7-95 shows a little more of the trace for the above example. On the extreme right of both the A-scan and TOFD B-scan, the shear wave can be seen. Since it arrives well after the other signals, it does not present a problem in this application.

V. VARIABLES

There are a number of variables that can influence an ultrasonic examination. Because of the many different materials tested, the component configurations, and the anomalies that can occur during an ultrasonic inspection, it is impossible to list or predict all of these variables. The following subsections deal with some of the more common variables that may have an affect on the results of a test.

Temperature and Velocity

Temperature is an important factor when considering ultrasonic examination. Temperature affects the velocity of sound in most materials to a greater or lesser degree depending on the material. Water, for instance, undergoes velocity changes with temperature. The velocity of water at 68°F (20°C) is approximately 1480 meters per second, whereas at 86°F (30°C) the velocity is approximately 1570 meters per second. At the other end of the scale, water at approximately 34°F (1°C) is approximately 1414 meters per second. Velocity changes in metals are not as dramatic as with water, but changes do occur. When conducting examinations, consideration should be given to temperature variations between the calibration standard and the component. Some specifications require that the temperature difference of the calibration standard and the component be within 20°F (–7°C). Of greater influence than the component is the temperature in the Plexiglas wedge or delay line. Temperature variations can cause beam angle changes and/or alter the apparent delay on the time base. When conducting examinations at elevated temperatures, special high-temperature transducers and couplants are usually necessary. Calibration blocks need to be heated to a temperature similar to that of the component to be examined prior to calibration, and the transducers should be allowed to warm up to the examination temperature before using them to calibrate the system. Temperature variations will also result in dimensional changes to the part, which must be taken into consideration when calibrating and testing material.

Attenuation

In addition to the change in velocity, temperature can also affect the amount of attenuation in some materials. Apparent changes in attenuation can be indicative of changes in the material structure. Hardened steel, for example, may exhibit less attenuation than its untreated counterpart. Sudden changes that are noted in signal amplitude can yield valuable data about the component being examined (see Hydrogen Embrittlement in Section VII). Microporosity in the material may have the same affect on the apparent attenuation. Material that exhibits high attenuation may require examination using lower test frequencies due to the probable larger grain size.

Frequency and Grain Size

For details regarding this variable, see Scatter in Section II, page 7.26.

Resolution

Pulse length can affect the resolution characteristics of the system. Refer to Resolution, page 7.53.

Surface Conditions

An important variable is that of surface condition. The differences in surface finish can result in large variations in the results of an examination. Paint or other coatings can have similar effects. Tightly adhering coatings on the surface generally allow good transfer of the energy. Loose or flaking coatings are undesirable and should be removed prior to conducting the examination. When calibrating the equipment for reference sensitivity on critical applications, it is essential to evaluate the component for any energy losses due to surface condition and apparent attenuation variations. The procedure for this is fairly simple and is performed by using two transducers in a "pitch–catch" technique. See Transfer Correction Technique in Section IV (page 7.65) for details of this procedure.

Diameter Changes

Changes in the diameter of the test surface can result in changes in test sensitivity. The effective transducer size is limited to its contact area. It is highly desirable to perform the system calibration on a surface with a diameter similar to the one being scanned on the component under test. Compensation for diameter changes should be made by either adding or subtracting gain after calibration, prior to examination of the component.

Contact Pressure and Couplant

The amount of couplant used and the contact pressure on the transducer can create differences in signal amplitude. Too little couplant will leave the surface dry and therefore create an air boundary between the transducer and the component surface. Excessive pressure can squeeze the couplant from under the transducer.

Dendritic Structures

Dendrites are branch-like grains that exist in certain metal structures and can cause problems, particularly in stainless steel welds. These dendrites form in the direction of heat dissipation during the welding process. A single grain can grow from one weld pass to the next, leaving elongated grains that can effectively redirect the sound energy. Indications that are plotted to originate from, e.g., the fusion line may actually originate from the root of the weld if the energy has been redirected. Special procedures including the use of refracted compression wave transducers may be necessary to reduce this effect.

Gain

The use of excessive gain can exaggerate otherwise insignificant indications. This emphasizes the need for precise calibration.

Other Factors

Other factors such as transducer frequency, diameter, and angle can affect the examination results. It is necessary to follow a qualified procedure when carrying out an examination, particularly when repeatability is an issue. Ultrasonic examination requires careful consideration of all the variables. The practitioner needs to be mindful of these variables and others that may present themselves. Attentiveness, awareness, and the ability to recognize anomalies are important.

VI. EVALUATION OF TEST RESULTS

In order that the terminology used in this section be understood, the flow chart in Figure 7-96 should be considered.

When NDT is specified, the following information must be provided:

1. The component description
2. The test method
3. The specification for the test

The specification for the test is most important. If there is no agreed upon definition regarding the type, size, and quantity of discontinuities, meaningful accept/reject determination will not be possible. The specification used should consider the component's "fit-

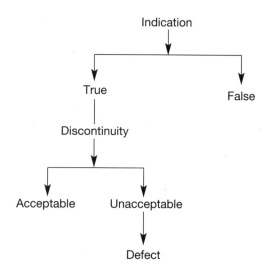

FIGURE 7-96 Procedure flow chart.

ness for purpose." This means that the component should not have any discontinuities that would lead to its failure. This determination should be made by the designer of the specific component, who will use "fracture mechanics" to make this judgment. Evaluation of any indication should be made in consideration of the code or specification referenced in the purchase order or contract.

Reporting

Unless a report is generated at the conclusion of an examination, there will be no record of it having been conducted. Examination repeatability is important, especially where the condition of a component is to be monitored periodically. Test reports should be developed that require certain information. A situation may arise where specific test reports are unavailable and the examiner may have to improvise.

Always ensure that as a minimum the following details are recorded clearly and concisely.

Identification

    Date, Time, and Place of examination

    Examiner's name and certification level

    Component examined and its serial number if applicable

    Procedure, specification, or standard to which the test was performed. Note any revision number, changes, or deviations from the procedure.

Equipment

    Instrument used, including serial number

    Transducers used, including frequency, serial number, and angles

    Calibration standards or reference block(s)

    Couplant used—include batch number where appropriate

    Calibration

    Time base (range and units per division)

    Calibration sensitivity and DAC curve (if used)

    Scanning sensitivity

    Reference level for recording

    Examination details

    Area scanned

    Limitations and interferences

    Percentage complete coverage

Results

    Indications noted—percent DAC (as appropriate), location of indications

    Classification of discontinuities (if required)

Scale drawing or plot of indications

Comments—surface condition, temperature, etc

Note. Reports should be concise and accurate. The person reading the report may not be very familiar with UT. A drawing or sketch is easier to follow than a lengthy description.

VII. APPLICATIONS

Product forms commonly subjected to ultrasonic inspection can be wrought, cast, welded, composite, or other materials. There are two main applications for nondestructive testing (NDT) in general. These are production-type tests of either raw materials and/or new components, and inspections that take place at a time when the component is in service. There is virtually an infinite list of areas where ultrasonic testing can be applied. It is, however, prudent to remember that the shape and orientation of a discontinuity does not always lend itself to ultrasonic inspection.

Economic Factors

The application of almost any method of nondestructive testing is dependent on factors such as the nature of the material, accessibility to critical areas, component geometry, discontinuities that are sought, and cost factors. From an economic standpoint, the cost of dismantling a system for NDT plus the cost of the NDT itself has to be considered against the cost of replacement of the component or system. The amount and method of inspection necessary to ensure an acceptable product or system as well as the probability of discontinuity detection has to be evaluated. For example, it is certainly possible to scan a complete tank floor and map the corrosion ultrasonically; however, the cost of performing this may exceed the cost of replacing the tank floor. In this case, there are other less time-consuming and therefore more economical means of accomplishing this task; in this instance, magnetic flux leakage.

A general description of some applications for UT follows.

Thickness measurement

Probably the most common UT application is the measurement of component thickness. The "one side only access" condition for ultrasonic thickness measurement is very desirable. Thickness measurement of tubes and pipes using inside diameter (ID) access can be conducted with either contact transducers or immersion-type systems. The contact systems usually incorporate a device in which the transducer(s) is spring loaded against the tube wall. In the case of immersion systems, either static or revolving transducer(s) scan the tube wall for thinning or corrosion. The data output is either digital or in the form of a "B scan" profile. Thickness measurements are either tabulated or can be determined from the "B scan" type data. Hand-held thickness gauges are usually in abundance at many plant maintenance departments. They are relatively simple to use, very portable, relatively accurate, and also generally inexpensive.

Discontinuity Detection

A discontinuity is an imperfection that may render the component in which it exists unfit for service. It is the time of the formation of the discontinuity that determines the inspection category into which it falls; i.e., whether it is production or service related. Components that are presented for inspection prior to being placed in service are said to be subjected to preservice inspection (PSI). Those components or systems that are to be inspected during service or during plant "shut down season" are said to be subjected to in-service inspection (ISI).

When considering PSI-type discontinuities, cognizant personnel should bear in mind that certain acceptable discontinuities can mask those that may occur in service. An example of this would be a lamination in a pipe near the weld joint. This may cause an "acoustic shadow" that will reflect the sound energy away from a weld discontinuity. Design personnel should be aware of the inspection requirements of their designed components and should also be knowledgeable of the technology of NDT to the extent that it may influence their design philosophies.

Elevated Temperature

In situations where the component is "on line" and shutting down the system is costly, ultrasonic testing can be extremely cost effective; for instance, in cases where a pressure vessel is working to temperature and pressure. Transducers can be designed for either thickness measurement or discontinuity detection at elevated temperatures.

Hydrogen Embrittlement

Hydrogen embrittlement occurs in steel during the solidification process. This is usually due to moisture entrapment and manifests itself as extremely small ruptures close to the grain boundaries. These discontinuities, by nature, are extremely small and usually cannot be detected by most other nondestructive methods. With ultrasonics, the effect of hydrogen embrittlement is to scatter the sound energy. This appears as increased attenuation. The ultrasonic methodology includes the comparison and measurement of apparent attenuation differences in the component. Attenuation is calculated in dB per unit length and measurements are made at the suspected sites. These measurements are compared with a standard section of material that does not contain discontinuities. This process should be conducted with care. Attenuation measurements are made by observing signal amplitude; therefore, transducer contact is critical. Variations in contact or contact area can cause changes in signal amplitude.

Bond Testing

Because there is reflection at an interface, and since the amount of reflection varies with the material on either side of the interface, bond integrity can be assessed with ultrasonic techniques. In its simplest form, the increase in signal height at a bond line will usually indicate a lack of bonding. This is due to a change in the acoustic impedance between the two (or more) layers. A change in impedance also causes a change in the phase of the signal. Special equipment is available that is designed specifically to detect these changes.

Multiple bond layers can be interrogated from one surface. An example of this is multiple layer "skin" on aircraft bodies.

Inaccessible Components

Components that are inaccessible may be inspected remotely using ultrasonic techniques by attaching transducers to rods or manipulators that can be inserted into an opening so that they reach the component to be examined. Examples of these conditions could be high-radiation areas or the rotor shaft within the gearbox of a helicopter, where the component cannot seen without dismantling the gearbox.

Fluid Level Measurement

An ideal method of measuring the level of a fluid in a vessel, pipe, or container is with ultrasonics. Sound passing through the liquid will be transmitted to the opposite side of the container, and the level of the fluid can be established by monitoring the presence or absence of a signal from the opposite side of the vessel. (The opposite side must of course be parallel to the scanning side.) Alternatively, the height of fluid can be monitored by placing a transducer on the bottom of the container and monitoring the signal from the fluid to air interface.

Stress Analyses

The monitoring or measurement of component stress, either residual or induced, can be accomplished with ultrasonic techniques. This is because the sound velocity changes in the material with mechanical strain. Because these changes are relatively small, instrumentation designed for this purpose incorporates extremely accurate velocity measurement timing devices. Stress measurement can be made either along the surface of a component with precision surface wave transducers, or through a section, usually with the use of transducers that oscillate in the "Y" axis and produce shear wave energy in the material. Because shear wave energy is polarized, the direction of stress can be determined by rotating the transducer and observing velocity changes in anisotropic material.

Ultrasonic Extensiometers

High-technology fasteners that have precision tensioning specifications can be tensioned using ultrasonics to measure this parameter. A transducer is sometimes located within a socket wrench and the extension of the bolt being torqued is measured. This is a more accurate tension-measuring device than a conventional torque wrench, as it is not affected by thread friction.

Liquid Flow Rate

Usually used in medical applications, this technique is used industrially when liquid flow rates in pipes must be monitored. This technique usually incorporates a through-transmission configuration where the sound frequency is monitored. Variations in the flow velocity produce changes in the frequency of the sound due to the "Doppler effect."

Other Applications

New applications involving ultrasonic technology evolve almost on a daily basis. With computer technology, the application horizon broadens and is often limited only by the imagination.

VIII. ADVANTAGES AND LIMITATIONS

There is no nondestructive test method that is a panacea. Each method has its advantages and limitations. It is a matter of selecting the test method that offers the most effective approach to solving the examination problem. When determining whether ultrasonics is the most appropriate test method, consideration should be given to the following:

1. Part and geometry to be examined
2. Material type
3. Material thickness
4. Material process—cast, wrought, etc.
5. Type of discontinuities to be detected
6. Minimum discontinuity size to be detected
7. Location of the discontinuities—surface-breaking or internal
8. Orientation of discontinuities (very important when selecting a test technique)
9. Accessibility to areas of interest
10. Surface conditions
11. Type of examination record required

Ultrasonic inspection is ideal for locating small, tight discontinuities assuming the following:

1. The sound energy can be projected at some angle that will respond favorably to the orientation of the reflector.
2. The relationship between the size of the discontinuity and the material's grain structure allows for an acceptable signal to noise ratio.
3. The surface condition is suitable for scanning. A poor scanning surface will not only require a more viscous couplant but possibly the use of a lower test frequency. This may not provide the necessary resolution for the test.

The advantages of ultrasonic examination are as follows:

1. Inspection can be accomplished from one surface
2. Small discontinuities can be detected
3. Considerable control over test variables
4. Varieties of techniques are available using diverse wave modes
5. High-temperature examination is possible with the correct equipment

6. Examination of thick or long parts
7. Inspection of buried parts, e.g., shafts in captivated bearing houses
8. Accurate sizing techniques for surface-breaking and internal discontinuities is possible
9. Discontinuity depth information
10. Surface and subsurface discontinuities can be detected
11. High speed scanning is possible with electronic signal gating and alarm system
12. "Go/No-Go" testing of production components
13. Test repeatability
14. Equipment is light and portable
15. Area evacuation of personnel is not necessary
16. Special licenses are not required as with radiation sources
17. Minimum number of consumables

Some of the limitations of ultrasonic examination are as follows:

1. Discontinuities that are oriented parallel with the beam energy will usually not be detected. Orientation of the discontinuity (reflector) is the most important factor in detecting discontinuities.
2. Discontinuities that are similar to or smaller than the material's grain structure may not be detected.
3. Thin sections may present resolution problems or require the implementation of special techniques.
4. Uneven scanning surfaces can reduce the effectiveness of the test.
5. Signals can be misinterpreted. This includes spurious signals from mode conversion or beam redirection, etc.
6. In general, this method requires a high level of skill and training.
7. Permanent record of the examination results is not typical. The records are limited to physical documentation rather than an actual reproduction of the test, e.g., as is possible with radiography.

IX. GLOSSARY OF TERMS

A-scan—A method of data presentation on an ultrasonic display utilizing a horizontal baseline, that indicates distance, and a vertical deflection from the baseline, that indicates amplitude.

A-Scan presentation—A method of data presentation utilizing a horizontal baseline to indicate distance, or time, and a vertical deflection from the baseline to indicate amplitude.

Amplitude—The vertical height of a signal, usually base to peak, when indicated by an A-scan presentation.

Angle beam—A term used to describe an angle of incidence or refraction other than normal to the surface of the test object, as in angle beam examination, angle beam search unit, angle beam longitudinal waves, and angle beam shear waves.

Area amplitude response curve—A curve showing the relationship between different areas of reflection in an material and their respective amplitudes of ultrasonic response.

Attenuation—A factor that describes the decrease in ultrasound intensity or pressure with distance. Normally expressed in decibels per unit length.

B-scan presentation—A means of ultrasonic data presentation that displays a cross section of the specimen, indicating the approximate length (as detected per scan) of reflectors and their relative positions.

Back reflection—An indication, observed on the display screen of a UT instrument, that represents the reflection from the back surface of a reference block or test specimen.

Back echo—See *back reflection*.

Back surface—The surface of a reference block or specimen that is opposite the entrant surface.

Beam spread—A divergence of the ultrasonic beam as it travels through a medium.

Bubbler—A device using a liquid stream to couple a transducer to the test piece.

C-scan—An ultrasonic data presentation that provides a plan view of the test object and discontinuities.

Collimator—A device for controlling the size and direction of the ultrasonic beam.

Contact testing—A technique in which the transducer contacts directly with the test part through a thin layer of couplant.

Couplant—A substance, usually a liquid, used between the transducer unit and test surface to permit or improve transmission of ultrasonic energy.

Critical angle—The incident angle of the ultrasonic beam beyond which a specific refracted wave no longer exists.

DAC—Distance amplitude correction. Electronic change of amplification to provide equal amplitude from equal reflectors at different depths. Also known as swept gain, time corrected gain, time variable gain, etc.

DAC curve—A curve (usually drawn on the screen) derived from equal reflectors at different depths.

Damping, search unit—Lmiting the duration of a signal from a search unit subject to a pulsed input by electrically or mechanically decreasing the amplitude of successive cycles.

dB control—A control that adjusts the amplitude of the display signal in decibel (dB) units.

Dead zone—The distance in the material from the surface of the test specimen to the depth at which a reflector can first be resolved under specified conditions. It is determined by the characteristics of the search unit, the ultrasonic instrumentation, and the test object.

Decibel (dB)—Logarithmic expression of a ratio of two amplitudes or intensities. (UT) $dB = 20 \log_{10}$ (amplitude ratio).

Delay line—A column of material such as Plexiglas that is attached to the front of a transducer. It behaves similarly to a water path and allows the initial pulse to be shifted off the scree. This often improves "near surface resolution."

Delay sweep—An A-scan or B-scan presentation in which an initial part of the time scale is not displayed.

Discontinuity—A lack of continuity or cohesion; an intentional or unintentional interruption in the physical structure or configuration of a material or component.

Distance amplitude, compensation (electronic)—The compensation or change in receiver amplification necessary to provide equal amplitude on the display of an ultrasonic instrument for reflectors of equal area that are located at different depths in the material.

Distance amplitude, response curve—See *DAC*. A curve showing the relationship between the different distances and the amplitudes of an ultrasonic response from targets of equal size in an ultrasonic transmitting medium.

Distance linearity range—The range of horizontal deflection in which a constant relationship exists between the incremental horizontal displacement of vertical indications on the A-scan presentation and the incremental time required for reflected sound to pass through a known length in a uniform transmission medium.

Doppler effect—The change in frequency of a sound wave due to movement of the reflector. Movement toward or away from the sound will result in a change in frequency (e.g., the tone of a train whistle changing as the train passes).

Dual search unit—A search unit containing two elements, one a transmitter, the other a receiver.

Dynamic range—The ratio of maximum to minimum reflective areas that can be distinguished on the display at a constant gain setting.

Entrant surface—The surface of the material through which the ultrasonic waves are initially transmitted.

Far field—The zone of the beam (beginning at the Y_0 point) where equal reflectors give exponentially decreasing amplitudes with increasing distance.

Flaw—A discontinuity in a material or component that is unintentional.

Flaw characterization—The process of quantifying the size, shape, orientation, location, growth, or other properties of a flaw based on NDT response.

Frequency (examination)—The number of cycles per second (Hz).

Frequency, pulse repetition—The number of times per second that a search unit is excited by the pulser to produce a pulse for ultrasonic imaging. This is also called pulse repetition rate or pulse repetition frequency (PRF).

Gate—An electronic means of selecting a segment of the time range for monitoring, triggering an alarm, or further processing.

Immersion testing—An ultrasonic examination technique in which the search unit and the test part are submerged (at least locally) in a fluid, usually water.

Impedance, acoustic—A mathematical quantity used in computation of reflection characteristics at boundaries. It is the product of wave velocity and material density (ρc).

Indication—A response or evidence of a response disclosed through an NDT that requires further evaluation to determine its full and true significance.

Initial pulse—The response of the ultrasonic system display to the transmitter pulse (sometimes called "main bang").

Lamb wave—A specific mode of propagation in which the two parallel boundary surfaces of the material under examination (such as a thin plate or wall of a tube) establish the mode of propagation. The Lamb wave can be generated only at particular values of frequency, angle of incidence, and material thickness. The velocity of the wave is dependent on the mode of propagation and the product of the material thickness and the examination frequency.

Linearity, amplitude—A measure of the proportionality of the amplitude of the signal input to the receiver and the amplitude of the signal appearing on the display of the ultrasonic instrument or on an auxiliary display.

Linearity, time or distance—A measure of the proportionality of the signals appearing on the time or distance axis of the display and the input signals to the receiver from a calibrated time generator or from multiple echoes from a plate or material of known thickness.

Longitudinal wave—A wave in which the particle motion of the material is essentially in the same direction as the wave propagation. (also called compressional wave).

Metal path—See *Sound path*

Mode—The type of ultrasonic wave propagating in the material as characterized by the particle motion (e.g., longitudinal, transverse, etc.)

Mode conversion—Phenomenon by which an ultrasonic wave that is propagating in one mode refracts at an interface to form ultrasonic wave(s) of other modes.

Multiple back reflections—Successive signals from the back surface of the material under examination.

Near field—The region of the ultrasonic beam adjacent to the transducer having complex beam profiles and intensity variations. Also known as the Fresnel zone.

Noise—Any undesired signal (electrical or acoustic) that tends to interferes with the interpretation or processing of the desired signals.

Normal incidence (also see *Straight beam*)—A condition in which the axis of the ultrasonic beam is perpendicular to the entrant surface of the part being examined.

Penetration depth—The maximum depth in a material from which usable ultrasonic information can be obtained and measured.

Probe—See *Search unit*.

Pulse-echo technique—An examination method in which the presence and position of a reflector are indicated by the echo amplitude and time.

Pulse length—A measure of the duration of a signal as expressed in time or number of cycles.

Range—The maximum distance that is presented on a display.

Rayleigh wave—An ultrasonic surface wave in which the particle motion is elliptical and the effective penetration is approximately one wavelength.

Reference block—A block of material that includes reflectors. It is used both as a measurement scale and as a means of providing an ultrasonic reflection of known characteristics.

Reflector—An interface at which an ultrasonic beam encounters a change in acoustic impedance and at which at least part of the sound is reflected.

Reject, suppression—A control for minimizing or eliminating low-amplitude signals (electrical or material noise) so that true signals are emphasized.

Relevant indication—An indication caused by a discontinuity that requires evaluation.

Scanning—The movement of a transducer relative to the test part in order to examine a volume of the material.

Search unit—An electroacoustic device used to transmit and/or receive ultrasonic energy. The device generally comprises a piezoelectric element, backing, wearface and/or wedge. Sometimes known as a "probe" or "transducer."

Sensitivity—A measure of the smallest reflector that produces a discernible signal on the display of an ultrasonic system.

Shear wave—wave motion in which the particle motion is perpendicular to the direction of propagation (transverse wave).

Sound path—The path of the sound energy from the time that it leaves the transducer and reflects back to the transducer.

Skip distance—In angle beam testing, the distance along the test surface from sound entrant point to the point at which the sound returns to the same surface. It can be considered the top surface distance of a complete "vee" path of sound in the test material.

Transducer—A piezoelectric element used to produce ultrasonic vibrations.

Through-transmission technique—A technique in which ultrasonic waves are transmitted by one search unit and received by another at the opposite surface of the material being examined.

Vee path—The angle beam path in materials starting at the search-unit examination surface, through the material to the reflecting surface, continuing to the examination surface in front of the search unit, and reflecting back along the same path to the search unit. The path is usually shaped like the letter V.

Water path—The distance from the transducer to the test surface in immersion or water column testing.

Wedge—In angle beam examination by the contact method, a device used to direct ultrasonic energy into the material at an angle.

Wheel search unit—An ultrasonic device incorporating one or more transducers mounted inside a liquid-filled flexible tire. The beam is coupled to the test surface through the rolling contact area of the tire. Also known as a "wheel probe" or "roller search unit."

X. REFERENCES

J. C. Drury, *Ultrasonic Flaw Detection for Technicians,* Ninth Edition, OIS plc, Stockton-on-Tees, Cleveland, U.K., 1997.

Nondestructive Evaluation and Quality Control, ASM International Metals Handbook, Ninth Edition, 1989.

CHAPTER 8
EDDY CURRENT TESTING

8.1 HISTORY AND DEVELOPMENT

Eddy current testing is one of the oldest nondestructive testing (NDT) methods. However, it wasn't until the last few decades of the twentieth century that the eddy current method started to reach its true potential in the marketplace. One reason for this is that general purpose, user-friendly eddy current instruments are a relatively recent phenomenon. Whereas portable ultrasonic instruments offering considerable versatility have been available since the 1960s, comparable eddy current portables only became available in the 1980s. In addition, it is only recently that eddy current theory became widely understood by NDT professionals. The early 1980s, in particular, produced excellent explanatory material that made eddy current theory understandable to persons without advanced technical backgrounds. Modern microprocessor-based instruments, plus the availability of high-quality operator training, ensure the continued growth of this versatile, high-performance NDT method.

8.1.1 Significant Discoveries about Electromagnetism

Development of the eddy current method was based on certain discoveries made during the early nineteenth century about the relationship between electricity and magnetism. In fact, the relevant electromagnetic principles were discovered in the same sequence in which they occur during an eddy current test.

In 1820, Hans Christian Oersted, a Dane, discovered electromagnetism—the fact that an electrical current flowing through a conductor causes a magnetic field to develop around that conductor. Oerstead discovered electromagnetism accidentally. While demonstrating that heat is developed when an electric current passes through a wire, Oerstead observed that the needle of a magnetic compass deflected perpendicular to the wire while the current was passing through it. Electromagnetism is the principle on which eddy current coils operate. Whereas Oersted was using direct current developed from a battery voltage when he discovered electromagnetism, an eddy current instrument employs alternating electric current flowing through the test coil in order to develop an alternating magnetic field around the coil.

In 1831, an Englishman, Michael Faraday, discovered electromagnetic induction—the fact that relative motion between a magnetic field and a conductor induces a voltage in that conductor, causing an electric current to flow. Consequently, when the alternating magnetic field of an eddy current instrument's coil is brought in contact with a conducting test object, a voltage is developed, causing a current to flow in the test object. Thus, electromagnetic induction is considered to be the operating principle of eddy current testing. Joseph Henry also independently discovered electromagnetic induction in the United States at about the same time. In fact, the unit of measure for induction is named after him.

In 1834, Heinrich Lenz stated the principle that defines how the properties of the test object are communicated back to the test system. Lenz's law states that the direction of current flow in the test object will be such that its magnetic field will oppose the magnetic field that caused the current flow in the test object. This means that, in practice, the eddy currents communicate with the test coil by developing a secondary flux that cancels a portion of the coil's flux equivalent to the magnitude and phase of the flux developed by the eddy currents.

The theory describing the chain of events of an eddy current test may thus be fully described by the discoveries of Oersted, Faraday, Henry, and Lenz. The existence of eddy currents themselves, however, was not discovered until 1864. They were discovered by James Maxwell, who is famous for stating the defining equations of electromagnetic theory. The first use of eddy currents for nondestructive testing occurred in 1879 when D. E. Hughes used these principles to conduct metallurgical sorting tests.

8.1.2 Modern Eddy Current Testing

The development of the eddy current method progressed slowly until the late 1940s, when Dr. Friedreich Foerster founded the Institut Dr. Foerster, which made great strides in developing and marketing practical eddy current test instruments. By the late 1960s the Institute had developed a product line covering virtually every application of the eddy current test method and worked with American manufacturers to firmly establish the method in the United States. Two major contributions of Foerster were the development of impedance plane display, which greatly aided in communication of test information to the practitioner, and formulation of the Law of Similarity, which enables the practitioner to duplicate the same eddy current performance under a variety of test situations.

The next major contribution to the advancement of the method, multifrequency testing, was also developed by an equipment manufacturer, Intercontrolle of France, in 1974. Driving the test coil at multiple frequencies helps to overcome what has traditionally been the major limitation of the eddy current method, the fact that the various conditions to which the method is sensitive can vector into a single displayed signal that is difficult to interpret. Originally developed to suppress the display of undesired test variables, multifrequency testing can also optimize an eddy current test for normally conflicting performance variables such as sensitivity and penetration as well as aid in identifying the nature of a particular test response. Multifrequency testing is a very significant innovation that has markedly advanced the state of the art.

The development of microprocessor-based eddy current instruments since the mid-1980s has also enhanced the potential and user-friendliness of the method. It has improved recording capability, provided sophisticated postinspection signal analysis, and has allowed automatic mixing of multifrequency signals. Modern microprocessor-based eddy current instruments offer a breadth of useful features virtually unimaginable in the days of analog equipment. Manufacturers such as Zetek, Hocking, Foerster, Nortec, ETC, and Magnetic Analysis have been important contributors.

In addition to mainstream eddy current testing, more specialized techniques are employed for certain applications. These include flux leakage, remote field eddy current, and modulation analysis inspection. In classifying nondestructive test methods for the purpose of qualifying and certifying test personnel, the American Society for Nondestructive Testing (ASNT) classifies all of these techniques under the umbrella of the Electromagnetic Testing method (ET).

8.1.3 Material Variables Detectable by Eddy Currents

During more than a century of development as a test method, eddy current testing has found application due to its sensitivity to the following variables:

- Conductivity variations
- Detection of discontinuities
- Spacing between test coil and test material (*lift-off* distance)
- Material thickness
- Thickness of plating or cladding on a base metal
- Spacing between conductive layers
- Permeability variations

Eddy current testing is suitable for inspection of the surface and just beneath the surface of conductive materials, volumetric inspection of thin conductive materials, and lift-off measurement to determine thickness of nonconductive materials adhering to or resting on the surface of conductive materials.

8.1.4 Major Application Areas

The versatility of the eddy current method has resulted in broad applications usage. However, the major application areas include the following:

- In-service inspection of tubing at nuclear and fossil fuel power utilities, at chemical and petrochemical plants, on nuclear submarines, and in air conditioning systems
- Inspection of aerospace structures and engines
- Production testing of tubing, pipe, wire, rod, and bar stock

8.2 THEORY AND PRINCIPLES

Eddy current theory is based on the principles of electricity and magnetism, particularly the inductive properties of alternating current. The discussion begins with a review of some basic principles.

8.2.1 Electricity

All matter is made up of atoms, the atom being the smallest unit of any element that retains the properties of that element. The center of an atom, the nucleus, has a positive electrical charge. Orbiting the nucleus and rotating on their own axes are negatively charged particles called electrons. As shown in the illustration of the copper atom (Figure 8-1), orbits of electrons around the nucleus resemble the orbits of planets around the sun in that there can be several orbits, called "shells." However, atomic structure differs from the solar system in that a given shell can contain multiple electrons.

From the perspective of eddy current testing, one is concerned specifically with the outer shell of a material's atoms, because the number of electrons in the outer shell determines whether the material will conduct electricity. The outer shell can contain a maximum of eight electrons, and when the outer shell contains as many as seven or eight elec-

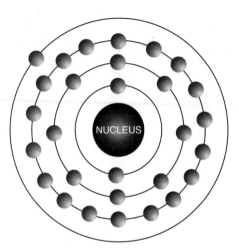

FIGURE 8-1 Copper atom.

trons, the material will not conduct electricity and is called an insulator. However, materials whose atoms have only one, two, or three electrons in the outer shell can conduct electricity and are, in fact, called conductors. Materials whose outer shells contain an intermediate number of electrons are called semiconductors and, although important in the design of computer circuitry, are not significant here.

If undisturbed by outside forces, a conductor's electrons will repeatedly orbit the nucleus. However, when voltage [also called electromotive force (EMF) or potential] is applied to a conductor, its electrons will advance from one atom to the next. That is, there is a flow of electrical charges called current or electricity. Voltage causes electrons to flow because it can attract and repel them; that is, voltage applies polarity to electrons. A battery is an example of a voltage source. Electrons, being negatively charged, will be attracted to a battery's positive terminal and repelled by its negative terminal. As shown in the illustration of a flashlight circuit (Figure 8-2), electrons flow through the bulb's filament from the negative to the positive terminal of the battery.

Although a conductor's atoms will permit current flow when voltage is applied, there is always some opposition to flow, due to the attraction of electrons to their atoms. This opposition varies among the atoms of different materials. Willingness of a test specimen to allow current flow is a key point in eddy current testing, detailed in the following definitions:

- *Conductivity* is the relative ability of a material's atoms to conduct electricity.
- *Resistivity* is the opposition of a material's atoms to the flow of electricity; it is the inverse of conductivity.
- *Conductance* is the ability of a particular component to conduct electricity. Conductance depends on a component's conductivity, length, and cross section.
- *Resistance* is the inverse of conductance. It is the opposition that a particular component offers to the flow of electricity. Like conductance, it depends on a component's conductivity, length, and cross section.

Conductivity is the material property of most interest to us in eddy current testing, whereas resistance is an important element in the display of test information. Material

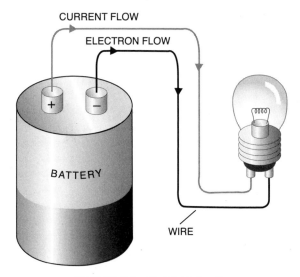

FIGURE 8-2 Flashlight circuit.

conductivities are compared on a scale called the International Annealed Copper Standard (IACS). Pure unalloyed annealed copper at 20°C is the base value on this scale, with a value of 100%. Other materials are assigned a percentage depending on their ability to conduct electricity relative to copper.

Having now identified resistance, as well as voltage and current, these terms can be tied together, showing their units, using the most basic formula of electricity, Ohm's law:

$$I = \frac{V}{R}$$

where
 I = current in amperes
 V = voltage in volts
 R = resistance in ohns

Thus, current flow increases when voltage increases and current flow decreases when resistance increases. There are two types of current: direct current (dc), which flows in only one direction; and alternating current (ac), which continually reverses direction.

8.2.2 Magnetism

Magnetism is a mechanical force of attraction or repulsion that one material can exert upon another. The opposite ends of a magnet exhibit opposing behavior called *polarity*. Thus, the ends of a magnet are called poles—one north and one south.

A magnet has a force field that can be visualized as a number of closed loops that flow through the magnet, travel around the outside of the magnet, and then reenter the magnet at the other end (Figure 8-3). These magnetic loops are called *lines of force* or *flux lines*. The word "flux" literally means "flow" and relates to the fact that the lines of force flow

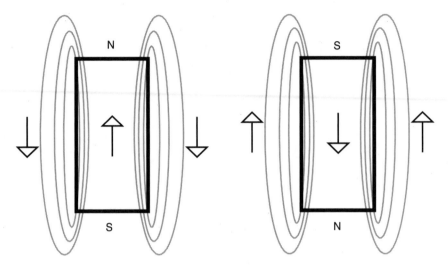

FIGURE 8-3 Magnetic polarity.

from the north to the south pole around the outside of a magnet, and from the south to the north pole within the magnet. Units of measurement and definitions for magnetism include the following:

Magnetic Flux (Φ or phi) is the entire set of a magnet's flowing lines of force.

Flux density (B) is the number of flux lines per unit area, perpendicular to the direction of flow.

The *maxwell* (Mx) is one magnetic field line or line of force.

The *weber* (Wb) is 1×10^8 lines or maxwells.

The *gauss* (G) is one line of force per square centimeter.

The *Tesla* (T) is one weber per square meter.

Field intensity depends on flux density. Flux density is greatest within the core of a magnet and at the poles. Flux density decreases with distance from the magnet according to the inverse square law (Figure 8-4); that is, flux density is inversely proportional to the square of the distance from the poles of the magnet.

When like poles of two magnets are brought together, the magnets push apart as their force fields repel each other. When unlike poles of two magnets are brought together, the magnets attract as the two force fields attempt to combine.

There are two types of magnets: permanent magnets and electromagnets. Permanent magnets are physical materials, having a property called ferromagnetism, which means that they can become magnetized when their *domains* have become aligned (Figure 8-5). Domains are miniature magnets consisting of groups of atoms or molecules present within a material's individual grains.

Permanent magnets were discovered in ancient times and are often produced in bar and horseshoe shapes. The fact that they retain a magnetic field without activation by electrical current is what distinguishes them from electromagnets. *Permeability* is the measure of a material's ability to be magnetized; that is, a material's ability to concentrate

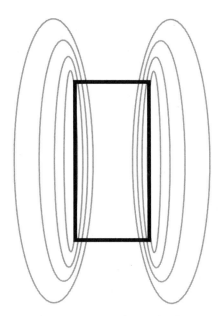

FIGURE 8-4 Flux density distribution.

FIGURE 8-5 Magnetic domains.

magnetic flux. The more flux density obtained from a material by a given quantity of applied magnetizing force, the greater the permeability of that material.

Materials may be classified as *magnetic* or *nonmagnetic*. Although there are three types of magnetic materials—*ferromagnetic, paramagnetic,* and *diamagnetic*—the term "magnetic" usually refers to *ferromagnetic* materials, which have much higher permeability than the other two types. The different magnetic materials may be characterized as follows:

1. Ferromagnetic materials become strongly magnetized in the same direction as the magnetizing field in which they are placed. Their permeability ranges from approximately 50 to more than 100,000. Examples are iron, carbon steel, 400 series stainless steel, and nickel.
2. Paramagnetic materials become slightly magnetized in the same direction as the magnetizing field. Their permeability is slightly more than 1. Examples are aluminum, chromium, platinum, and oxygen gas.
3. Diamagnetic materials become weakly magnetized in the opposite direction from the magnetizing field. Their permeability is slightly less than 1. Examples are copper, gold, silver, and hydrogen.

8.2.3 Electromagnetism

Electromagnetism is the phenomenon whereby the passage of electrons through a conductor causes a magnetic field to develop concentrically around the conductor, perpendicular to its axis. A concentrated magnetic field, similar to that obtained from a bar magnet, can be obtained by winding a conductor into a coil. A coil functioning as an electromagnet is called a solenoid, although an ideal solenoid has a length much greater than its diameter. A solenoid concentrates a magnetic field inside the coil, with opposing poles at each end of the coil, and flux lines completely encircling the loops of the coil. If direct current is applied to the coil, its magnetic field will flow in only one direction and it can perform the same work of attracting ferromagnetic materials as a permanent magnet. In addition, the coil can be wound around a ferromagnetic core for increased field strength.

If the current is alternating, the electromagnetic field will likewise alternate and the coil will exhibit a quality called inductance, L, whose unit is the henry. Inductance is the ability of a conductor to induce voltage in itself or in a neighboring conductor when the current varies.

8.2.4 Permeability

Electromagnets are used to produce permanent magnets. A conducting wire or cable is wound around the ferromagnetic material to be magnetized. Direct current is passed through the conductor, causing it to function as an electromagnet. When the resulting magnetic field enters the material to be magnetized, the material's domains become aligned. The higher the number of turns of wire or cable and the stronger the applied current, the greater the *magnetizing force*.

As stated earlier, permeability defines a material's ability to be magnetized, it's ability to concentrate magnetic flux. Numerical permeability values for different materials, termed relative permeability (μ_r) are stated in comparison to the permeability of air or a vacuum. Permeability can be quantitatively expressed as the ratio of flux density to magnetizing force. Permeability can be a problem in eddy current testing because the relative

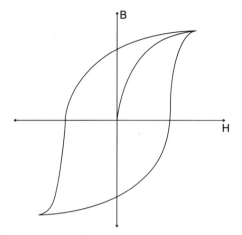

FIGURE 8-6 Hysteresis loop.

permeability of a given ferromagnetic material can vary during testing, causing permeability noise signals that can override the eddy current signals being sought. A *hysteresis loop* (Figure 8-6) is a plot of a material's flux density (B) variations as magnetizing force (H) is varied. By magnetically saturating the test material, permeability becomes constant and eddy current testing can proceed without interference from permeability variations.

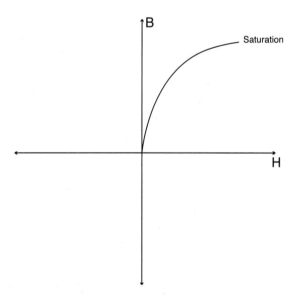

FIGURE 8-7 Magnetic saturation.

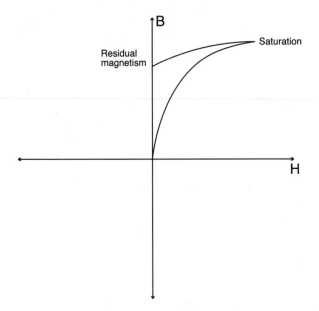

FIGURE 8-8 Residual magnetism.

Saturation (Figure 8-7) occurs at that point on the loop where further increases in magnetizing force do not cause significant increases in flux density.

At the completion of testing, the material will retain a certain amount of *residual magnetism* (Figure 8-8), the amount of flux density remaining in the material after the magnetizing force has been reduced to zero. The residual magnetism must be eliminated by demagnetization, to prevent problems such as the material attracting ferromagnetic debris.

8.3 ALTERNATING CURRENT PRINCIPLES

8.3.1 Sinusoidal Variation

Alternating current flows in a cyclical manner, behavior that is accurately illustrated by a sine curve (Figure 8-9).

The current begins its cycle at zero amplitude and, as time elapses, rises to a peak in one direction, falls back to zero, rises to a peak in the opposite direction, and falls back to zero again to complete the cycle. The end of one cycle is the starting point of the next cycle. One complete 360° cycle is called a *sinusoid.* Activity exhibiting the behavior of a sinusoid is termed sinusoidal.

The 360 degree points of a sinusoid correspond to the 360 degree points of a complete circle. Thus, both a sinusoid and a circle express one complete cycle. However, the sinusoid adds dimensions of amplitude and polarity to the cycle concept. Recall that the distance from the center of a circle to the edge is its radius. The portion of a circle's arc that corresponds to the length of its radius is called a *radian,* which occupies an arc of 57.3°.

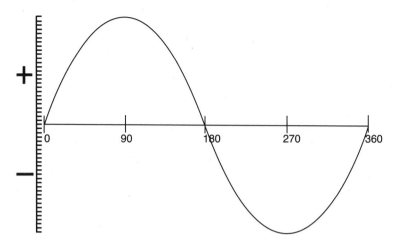

FIGURE 8-9 Sinusoid.

360° divided by 57.3° equals 6.28 or 2π, an important constant in alternating current calculations.

8.3.2 Electromagnetic Induction

Prior to discussing eddy currents, the induction process will first be modeled using a pair of coils as primary and secondary elements of a mutual induction circuit. Sinusoids are included to illustrate time and polarity relationships. The symmetry of two coils aids in explanation of the polarity relationships that occur in mutual induction processes such as eddy current tests. This illustration models an eddy current test in its most basic configuration, where a single coil is the primary circuit and the test material is the secondary. Initially, the primary coil will be examined without the influence of a secondary. The primary coil will be multiturn to represent a typical eddy current test coil. The secondary, when introduced, will be a single-turn coil, which validly represents any eddy current test specimen. It must be remembered that the secondary coil in this discussion represents strictly the test specimen, *not* the second coil of various multicoil configurations that will be described later in this chapter.

The Induction Process

1. An alternating current generator applies alternating voltage to a coil circuit (Figure 8-10a). A portion of this voltage, V_R, is applied across the resistance of the coil wire. The V_R amplitude rises from zero amplitude at zero degrees (Figure 8-10b). The "R" subscript in V_R identifies this voltage, sometimes called "resistance voltage," as a force needed to move current through the resistance of the coil wire. V_R is a component of the circuit's total voltage, V_T, which will be explained later.

2. V_R causes a current, I_P, to flow through the coil (Figure 8-10c), in phase with V_R (Figure 8-10d). The "P" subscript in I_P identifies the coil current as the primary current.

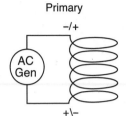
a) AC Generator Applies Voltage

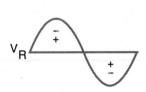
b) "Resistance Voltage" Sinusoid

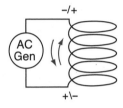
c) Primary Current Flows

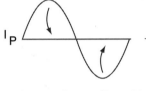
d) Primary Current Sinusoid

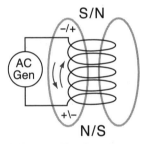
e) Primary Flux Develops

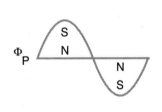
f) Primary Flux Sinusoid

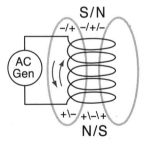
g) Back Voltage Develops

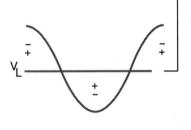
h) Back Voltage Sinusoid

i) Inductive Reactance Occurs

FIGURE 8-10 Self-induction process. (a) AC generator applies voltage. (b) "Resistance voltage" sinusoid. (c) Primary current flows. (d) Primary current sinusoid. (e) Primary flux develops. (f) Primary flux sinusoid. (g) Back voltage develops. (h) Back voltage sinusoid. (i) Inductive reactance occurs.

3. *Electromagnetism* occurs. The alternating current flowing through the coil causes an alternating magnetic field, Φ_P, the primary flux, to develop around the coil (Figure 8-10e), in phase with V_R and I_P (Figure 8-10f).
4. *Self-induction* occurs. Since the coil is standing in the field of its own varying flux, Faraday's law applies and electromagnetic induction is imposed on the coil wire. That is, Φ_P induces an additional voltage, V_L, often called "back voltage," into the coil (Figure 8-10g). The "L" subscript identifies V_L as an "induced voltage." This voltage is separate from the V_R voltage that caused I_P to flow. According to Faraday's law, the quantity of induced voltage is proportional to the rate of flux variation. Since Φ_P is varying the most through the 0°, 180°, and 360° points, and not varying through the 90° and 180° points, the back voltage is induced 90° out of phase (Figure 8-10h) with the coil current and flux.
5. *Inductive reactance* occurs. Since the back voltage is 90° out of phase with the coil current, it will oppose changes in the coil current (Figure 8-10i). In that amplitude change is the very nature of alternating current flow, opposition to change in ac is effectively opposition to flow of ac. This opposition, called *inductive reactance*, X_L, is distinct from resistance, R. Resistance simply opposes flow of current and can occur in either a dc or ac circuit. Inductive reactance, which, strictly speaking, opposes *change* of current flow, can occur only in an ac circuit.

Inductive reactance depends on coil design and test frequency to the extent that, as more flux lines cut across more coil turns per unit time, inductive reactance increases.

The variables influencing inductive reactance are detailed in the following equations:

$$X_L = 2\pi f L$$

$$L = \mu_r \frac{N^2 \times A}{l} \times 1.26 \times 10^{-6}$$

where
X_L = inductive reactance
f = test frequency
L = coil inductance
μ_r = relative permeability of the coil core
N = number of turns
A = cross sectional area
l = coil length

Note. The permeability of air, 1.26×10^{-6}, must be multiplied by an appropriate value for μ_r, which would be 1(one) for an air core coil or some higher value in the case of a coil with a ferromagnetic core.

6. If a secondary circuit is placed in proximity to the primary, mutual induction will occur and Φ_P will induce a voltage into the secondary circuit (Figure 8-11a). This voltage is appropriately called secondary voltage, V_S, and is 180° out of phase (Figure 8-11b) with the inducing primary flux.
7. V_S causes a current, I_S, to flow in the secondary circuit (Figure 8-11c), with the same phase (Figure 8-11d) as V_S. In an actual eddy current test, where the secondary circuit is the test specimen, I_S will be the eddy currents.
8. With current now flowing in the secondary circuit, electromagnetism will again occur

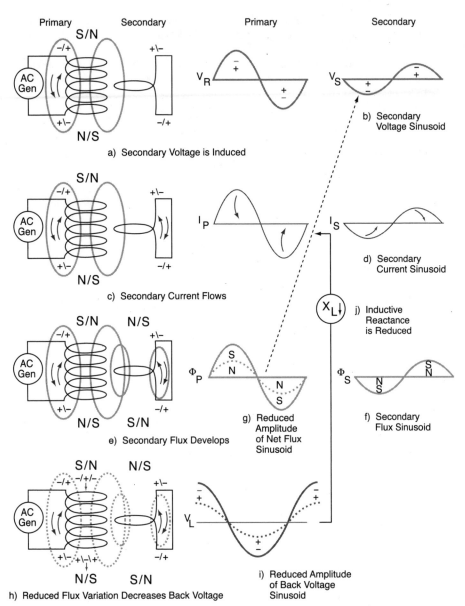

FIGURE 8-11 Mutual induction process. (a) Secondary voltage is induced. (b) Secondary voltage sinusoid. (c) Secondary current flows. (d) Secondary current sinusoid. (e) Secondary flux develops. (f) Secondary flux sinusoid. (g) Reduced amplitude of net flux sinusoid. (h) Reduced flux variation decreases back voltage. (i) Reduced amplitude of back voltage sinusoid. (j) Inductive reactance is reduced.

and a secondary flux, Φ_S, will develop (Figure 8-11e), with its phase (Figure 8-11f) determined by the secondary current.

9. The 180° phase difference between primary and secondary activity indicated in Figures 8-11b, d, and f is the effect of Lenz's law, which states that the induction process in the secondary circuit causes a phase reversal, which results in the secondary flux being opposite in polarity to the primary flux. Due to this state of opposition, the secondary flux will cancel a portion of the primary flux, resulting in an overall decrease in net flux for the two coils. The reduced amplitude of net flux (dashed line in Figure 8-11g) results in a reduced rate of net flux variation. Less flux variation results in reduced back voltage (Figures 8-11h and 8-11i), which results in a reduction of inductive reactance (Figure 8-11j).

Observe that the induction process occurs in a certain order: voltage drives a current, which develops an electromagnetic field, which then induces a voltage to again initiate the process. During an eddy current test, a primary circuit (the test coil) induces eddy currents into a secondary circuit (the test material). Any factors that affect current flow in the secondary circuit, such as primary/secondary coupling or conductance variations, will affect the amplitude of both V_L and of the inductive reactance in the primary circuit.

Variations in the test material change not only the test coil's inductive reactance, but also a quantity called effective resistance. Although the resistance of the coil wire itself does not change, the eddy currents in the test material encounter friction as they circulate, thus dissipating a portion of their energy as heat. That is, the secondary circuit acts as a load on the primary circuit, with electrical energy converting to thermal energy. This energy loss in the circuit is counteracted by an increase in V_R to keep the coil current constant. Thus, both V_R and V_L vary with change of test material properties during an eddy current test.

8.3.3 Signal Output

Voltage Plane
As stated earlier, there is a 90° phase difference between V_R and V_L. These two voltages can be vectorially added to produce a quantity called V_T, which is the total voltage in the primary circuit, the output of the instrument's alternating current generator.

Figure 8-12 illustrates a voltage plane diagram with V_R and V_L values plotted as the base and elevation of a right triangle, and V_T as the hypotenuse. Thus the Pythagorean theorem

$$c^2 = a^2 + b^2$$

may be expressed using voltage values

$$V_T^2 = V_R^2 + V_L^2$$

and restated to solve for V_T by vector addition:

$$V_T = \sqrt{V_R^2 + V_L^2}$$

The basic information available from an eddy current test is the magnitude of V_T and its phase relative to I_P. As shown in the sinusoids of Figures 8-10d and 8-10b, I_P is in phase with V_R. I_P can therefore be placed along with V_L on the horizontal axis of the voltage plane. Although test output could be shown as a sinusoid, indicating variation in magnitude and phase of V_T, a display showing just the tip of the vector arrow as a dot, called "vector point display," provides all necessary information in a simple manner and lends itself especially well to eddy current signal analysis. When the magnitude and phase of V_T

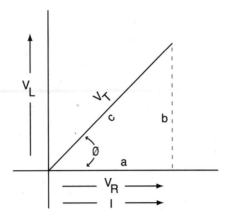

FIGURE 8-12 Voltage plane.

are plotted on the voltage plane, the tip of the V_T vector indicates the magnitudes of V_R and V_L. Figure 8-13 summarizes how the various voltage and impedance components fit into a coil circuit driven by alternating voltage, with the resistive and reactive properties separately identified.

Impedance Plane

Just as V_R and V_L can be combined into V_T, the combined effects of R and X_L on the alternating current in the coil can be expressed as a quantity called impedance. Specifically, *impedance amplitude* (Z) is the magnitude of the vector sum of inductive reactance and resistance, and is the coil's total opposition to current flow. That is:

$$Z = \sqrt{R^2 + X_L^2}$$

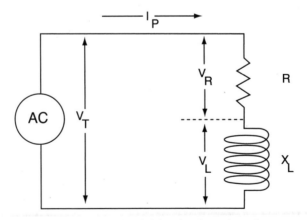

FIGURE 8-13 Coil circuit driven by alternating voltage.

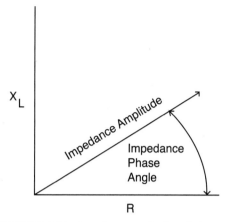

FIGURE 8-14 Impedance amplitude and phase angle.

The *Impedance phase angle* (ϕ), the proportional relationship between inductive reactance and resistance, can be calculated from:

$$\phi = \arctan \frac{X_L}{R}$$

Impedance amplitude and phase angle are illustrated in Figure 8-14.

It is important to understand that there can be only one current in the circuit, flowing through both the resistance and inductive reactance, and influenced by both V_R and V_L. Ohm's law then shows how a voltage plane can be converted into a corresponding impedance plane:

$$R = \frac{V_R}{I}$$

$$X_L = \frac{V_L}{I}$$

$$Z = \frac{V_T}{I}$$

Although eddy current signal variations represent voltage variations as well as impedance variations, the impedance plane is the convention for expressing eddy current signal variations. That is, voltage variations are used to represent impedance variations in the coil. However, before exploring the different patterns of signal variation, it is necessary to examine how eddy currents behave in the test material in order to produce such patterns.

8.4 EDDY CURRENTS

When a test specimen is brought into proximity to the alternating flux field of an eddy current coil, coil flux causes electrons in the specimen to circulate in a swirling eddy-like pattern; hence the term "eddy currents." Eddy current behavior depends on the properties of both the flux and the specimen itself.

8.4.1 Eddy Current Flow Characteristics

Eddy currents have a number of flow characteristics that affect their test performance:

1. They flow only in closed, concentric loops. Their flow paths are circular when unimpeded by the intrusion of material boundaries or discontinuities. Also, the flow paths are parallel to the turns of the bobbin-type coil in shown Figure 8-15 and perpendicular to the axis of the coil's flux field.
2. The orientation of the coil to the test material therefore determines the orientation of the eddy current flow pattern in the test material. Orientation of the coil to the test material can be controlled and varied for optimum results by selection of the proper coil configuration. Several options are shown Section 8.5.2 of this chapter.
3. Discontinuities are detectable by the eddy current method in proportion to the degree to which they disturb the flow pattern. Thus, a discontinuity is least detectable when its longest dimension is parallel to eddy current flow paths (see Figure 8-16a) and most detectable when the longest dimension is perpendicular to the flow paths (see Figure 8-16b). Discontinuities with smaller volumes may not be detectable when oriented parallel to the flow paths. Ensuring that discontinuities of all likely orientations are detectable is an important part of test coil design and selection. Moreover, eddy currents always follow the path of least resistance around nonconducting obstacles, flowing under long, shallow discontinuities and flowing around short, deep discontinuities.

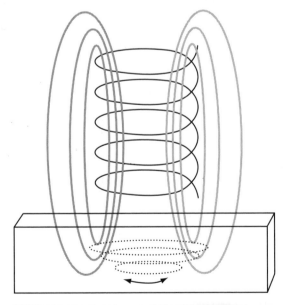

FIGURE 8-15 Bobbin-type coil's flux and eddy currents.

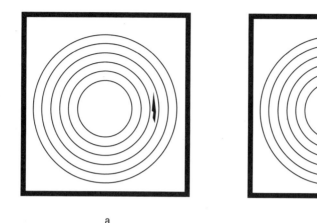

FIGURE 8-16 Discontinuities in eddy current flow patterns. (a) Discontinuity parallel to flow paths. (b) Discontinuity perpendicular to flow paths.

4. Eddy currents behave like compressible fluids. Although the flow paths are circular as long as the eddy currents are undisturbed by nonconducting material boundaries and discontinuities (see Figure 8-17a), the flow paths will distort and compress to accommodate intrusions into their flow (see Figure 8-17b).
5. Since an alternating flux field develops eddy currents, their flow in the test material likewise alternates clockwise and counterclockwise. The frequency of alternation of the eddy currents depends on the frequency of alternation of the flux field.

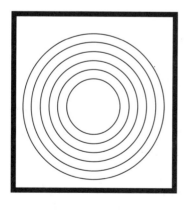 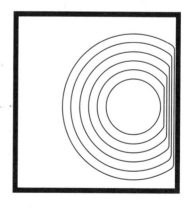

FIGURE 8-17 Effect of material boundaries. (a) Eddy currents undisturbed by material boundaries. (b) Eddy currents compressed by material boundaries .

6. Eddy current density varies in the test material as follows:
    A. Eddy currents exhibit a *skin effect*. That is, current density is maximum at the material surface and decreases exponentially with depth. Thus, in thicker materials, eddy current testing operates only on the outer "skin" of the test material and test sensitivity decreases rapidly with depth. Volumetric tests are possible only in thin specimens.

    Skin depth (δ), also called standard depth of penetration, is defined as the depth at which eddy current density (the portion of electrons active at a particular depth as compared to the material surface) has decreased to $1/e$, where "e" is the so-called natural logarithm, the number 2.71828, a device representing the natural rate of decay for many phenomena. Eddy current density for one, two, and three skin depths calculates as:

    $$1\delta = \frac{1}{e} = \frac{1}{2.71828} = 0.368 = 36.8\%$$

    $$2\delta = \frac{1}{e^2} = \frac{1}{7.38906} = 0.135 = 13.5\%$$

    $$3\delta = \frac{1}{e^3} = \frac{1}{20.08554} = 0.0498 = 5.0\%$$

    Beyond three skin depths eddy current density is too small to provide a displayable signal. (See effective depth of penetration in paragraph 6C below.) Standard depth of penetration in English and metric units for a particular material and test frequency can be calculated as follows:

    $$\delta \text{ (inches)} = 1.98 \sqrt{\frac{\rho}{f \times \mu_r}}$$

    $$\delta \text{ (mm)} = 25 \sqrt{\frac{\rho}{f \times \mu_r}}$$

where:
δ = standard depth of penetration
ρ = resistivity
f = frequency
μ_r = relative permeability

    B. The skin depth formula is truly valid only in the case of infinitely thick test material and large coils. However, at a material thickness of at least five skin depths, where eddy current density is only 0.0067% of surface density, the effect of restriction in thickness is so slight that the material may be considered to be "effectively infinite," rendering the formula accurate enough for practical purposes, providing that coil size is adequate. Conversely, if coil size is adequate but material thickness is restricted, current density at the opposite surface will exceed the calculated values. This is a fortunate circumstance, enhancing the possibility of volumetric inspection on thin-walled specimens.
    C. The extent of a coil's flux field varies with coil diameter such that effective eddy current penetration is approximately limited to the diameter of the coil. Consequently, if the coil is too small, current density at a particular depth will be less than that indicated by the skin depth equation.

    Effective depth of penetration is defined as the depth at which eddy current density decreases to 5% of surface density. This is the minimum eddy current density necessary to develop sufficient secondary flux to change coil impedance by a dis-

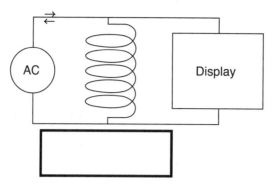

FIGURE 8-18 Voltage causes current to flow.

playable amount. Thus, effective depth will equal three skin depths only where coil diameter is at least as great as three skin depths. Coil diameter, however, is a double-edged sword: as diameter increases, sensitivity to small defects decreases.

7. Eddy currents exhibit a linear phase lag with depth. To visualize the phase lag phenomenon, one may imagine a tall glass of water filled with small ice cubes all the way to the bottom of the glass. If one dips a teaspoon a short distance into the glass and begins to stir, the ice cubes on the surface circulate first and those at greater depths go into motion progressively later as the energy introduced by the spoon proceeds toward the bottom of the glass. In similar fashion, as depth increases, eddy current activity is progressively delayed. Phase lag in the test material proceeds at the rate of one radian (57.3°) per standard depth of penetration. The phase lag signal indicates discontinuity depth and material thickness in eddy current testing.

8.4.2 Eddy Current Test Sequence

The induction process was previously modeled using a pair of coils as primary and secondary circuits. The eddy current test process is now summarized with an actual test specimen as the secondary circuit.

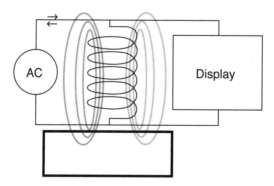

FIGURE 8-19 Electromagnetism.

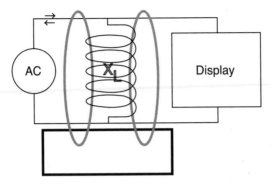

FIGURE 8-20 Varying flux induces back voltage, which causes inductive reactance.

1. The test instrument's AC generator applies an alternating voltage of a certain frequency to the test coil, causing an alternating current to flow through the coil (Figure 8-18).
2. The current in the coil develops a primary magnetic field around the coil (Figure 8-19). The primary magnetic field initiates the following induction processes:
   a. The coil's flux induces a back voltage into the coil, causing inductive reactance (Figure 8-20).
   b. The coil's flux induces a voltage into the test material, causing eddy currents to circulate (Figure 8-21).
3. The eddy currents generate a secondary magnetic field, which reacts with the primary field that the coil is generating (Figure 8-22).

Any changes in the flow of eddy currents will cause changes in the magnetic field that the eddy currents return to the test coil. Any changes in this magnetic field will cause changes in the inductive reactance and effective resistance of the coil, resulting in changes in current flow through the coil.

4. Finally, any changes in current flow through the coil will produce a change in the impedance indication on the instrument's display.

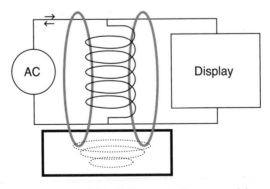

FIGURE 8-21 Eddy currents in test material.

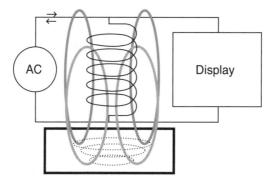

FIGURE 8-22 Secondary flux interacts with primary flux.

8.4.3 Test Performance

Test performance criteria do not seem to be as formally defined for eddy current testing as for ultrasonic testing and radiography. However, the same performance-related terminology can be usefully employed as follows:

- *Sensitivity:* The minimum size of discontinuity that can be displayed from a given material depth. Surface sensitivity is especially important with the eddy current method.
- *Penetration:* The maximum depth from which a useful signal can be displayed for a particular application.
- *Resolution:* The degree to which separation between signals can be displayed.

Test performance depends primarily on material conductivity, permeability, test frequency, coil design, and lift-off. Test frequency and coil design are readily selectable, and are therefore the primary controls over test performance. Guidelines for realizing optimum results for a specific application are given in Section 8.6. The following paragraphs summarize how major test variables affect performance.

1. *Conductivity.* The greater the conductivity of the test material, the greater the sensitivity to surface discontinuities, but the less the penetration of eddy currents into the material. Initially, this reduced penetration may seem contradictory, but is actually quite logical. As the coil's flux field expands, voltage is induced first on the surface and then at increasing depths in the test material. In high-conductivity materials, a considerable eddy current flow and thus a strong secondary flux are developed at the surface. This results in a substantial cancellation of primary flux. Because the primary flux has been greatly weakened, less primary flux is available to develop eddy currents at greater depth.

2. *Permeability.* This variable applies only to ferromagnetic materials. As material permeability increases, noise signals resulting from permeability variations increasingly mask eddy current signal variations. This effect becomes more pronounced with increased depth. Permeability thus limits effective penetration of eddy currents. The problem can be eliminated by magnetically saturating the material. However, the opportunity to saturate is limited by coil/test material geometry.

3. *Frequency.* Eddy current testing is performed within a frequency range of approximately 50 Hz to 10 MHz, although most applications are performed well within the extremes of that range. As test frequency is increased, sensitivity to surface discontinuities increases, permitting increasingly smaller surface discontinuities to be detected. As fre-

TABLE 8-1 Typical Depths of Penetration

Metal	Conductivity % IACS	Resistivity	Permeability	1 KHz	4 KHz	16 KHz	64 KHz	256 KHz	1 MHz
				\multicolumn{6}{c}{36.8% Depth of penetration}
Copper	100	1.7	1	0.082	0.041	0.021	0.010	0.005	0.0026
6061 T-6	42	4.1	1	0.126	0.063	0.032	0.016	0.008	0.004
7075 T-6	32	5.3	1	0.144	0.072	0.036	0.018	0.009	0.0046
Magnesium	37	4.6	1	0.134	0.067	0.034	0.017	0.008	0.0042
Lead	7.8	22	1	0.292	0.146	0.073	0.37	0.018	0.0092
Uranium	6.0	29	1	0.334	0.167	0.084	0.042	0.021	0.0106
Zirconium	3.4	70	1.02	0.516	0.258	0.129	0.065	0.032	0.0164
Steel	2.9	60	750	0.019	0.0095	0.0048	0.0024	0.0012	0.0006
Cast steel	10.7	16	175	0.018	0.0089	0.0044	0.0022	0.0011	0.0006

quency is decreased, eddy current penetration into the material increases. In addition, as frequency is decreased, the speed of coil motion must be decreased in order to obtain full coverage.

The test frequency for obtaining adequate penetration in a given material can be estimated using the skin depth equation or by using a penetration chart plotted from the skin depth equation for various conductive materials. Table 8-1 shows skin depths obtained at various frequencies for a selection of materials. However, because of the number of variables affecting eddy current behavior, this frequency should only be used as a starting point. The optimum frequency is best determined by experimentation.

4. *Coil Design.* Penetration and sensitivity are affected by conflicting requirements for coil geometry. Sensitivity to small surface discontinuities requires that the eddy current field be sufficiently compact so that it will be adequately distorted by the discontinuity. Conversely, penetration requires that the eddy current field extend to the required depth in the test specimen. The rules of thumb are that eddy current penetration is limited to a depth equivalent to coil diameter while sufficient sensitivity requires that coil diameter be limited to the minimum length of discontinuity to be detected.

5. *Lift-off.* Since flux density decreases exponentially with distance from the test coil, the amount of lift-off, or separation between the coil and test specimen, has a significant impact on sensitivity. The closer the coupling between coil and test specimen, the denser the eddy current field that can be developed, and thus the more sensitive the test to any material variable. Conversely, close coupling increases sensitivity to lift-off noise due to causes such as probe wobble.

8.5 TEST EQUIPMENT

Basic eddy current hardware includes instruments, coils and coil fixtures, coil/test specimen transport equipment, recording devices, and reference standards. Test instruments can be either general purpose or designed for a specific application. Coils are usually designed for a particular category of application.

8.5.1 Eddy Current Instruments

A broad variety of eddy current instruments is available for use, from simple to complex. Although these instruments vary greatly in applications flexibility as well as size, most of them operate on similar principles.

In addition to a power supply, all eddy current instruments require at least three circuit elements: AC generator, coil circuit, and processing/display circuitry. The level of flexibility designed into each of these elements generally determines how eddy current instruments differ from each other.

- *AC generators* provide the voltage that drives the coil. They can operate at a single fixed frequency, provide a selection of switchable frequencies, be continuously variable, or even provide multiple frequencies simultaneously. In some instruments, there is adjustment for amplitude of the voltage applied to the coil.
- *Coil circuits* range from designs intended to work with only a single specific coil, a limited range of specified coils, or with virtually any coil configuration available.
- *Displays* can range from single LED and meter readouts to multifrequency presentations on multicolor display screens.

Dedicated Instruments

Dedicated instruments are designed for a specific application and are usually able to perform that application more efficiently than general-purpose instruments. Examples of dedicated instruments are crack detectors, coating thickness gauges, and conductivity meters. Conductivity meters, for example, can give direct readout of conductivity in IACS values. In addition, some crack detection instruments provide lift-off suppression to prevent noise signals caused by variations in coil to test material spacing. When there is sufficient work in a given application to justify investment in a single-purpose instrument, it is likely to be the best choice. However, one must be careful using eddy current meter-type instruments. Because they do not provide the quantity of information available from an impedance plane display, meter-type instruments can mislead less-qualified users. Since meter instruments can display only upscale or downscale deflections, they must be operated so that only one material variable is displayed. However, with impedance plane display instruments, each type of material condition deflects the display dot in a characteristic manner, facilitating separation of variables and interpretation of signals.

Standard Impedance Plane Display Instruments

The AC generator of a standard impedance plane display instrument drives the test coil at only one frequency, which is usually selectable from a wide range of frequencies. These general-purpose instruments can perform an extensive variety of eddy current applications. The ability to view actual impedance plane signals provides the knowledgeable user a great deal of valuable information. Some newer impedance plane instruments have flat displays, offering enhanced portability, such as the unit shown in Figure 8-23. However, test-system-type instruments (Figure 8-24) do not provide portability, but may be expected to operate 24 hours a day to accommodate continuous production at tubing, pipe, rod, or wire mills.

Impedance plane display instruments show variation of both inductive reactance and resistance during testing. Control functions of impedance plane instruments can include, but are not limited to, the following:

- *Frequency:* Adjusts the frequency at which the AC generator drives the test coil
- *Gain (Sensitivity, dB):* Adjusts amplification of the bridge output signal for display (see Mode of Operation)
- *Horizontal/Vertical Dot Position:* Adjusts dot position on the display
- *Phase Rotation:* Rotates the direction of dot deflection
- *Balance (Null, Zero):* Adjusts impedance to be identical on both sides of the bridge

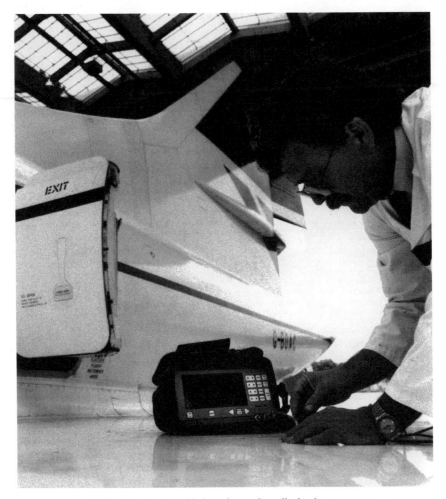

FIGURE 8-23 Portable impedance plane display instrument.

- *Erase (Clear):* Erases the display
- *Gate:* Sensitizes some portion of the display to trigger an alarm
- *Filters:* Prevent display of signal above and/or below a certain frequency range
- *Probe Drive:* Adjusts voltage amplitude applied to the test coil
- *Horizontal and Vertical Display Amplification:* Allows one axis of the display to be expanded relative to the other for signal enhancement

Multifrequency Instruments
Development of multifrequency instruments was one of the most significant advances in the evolution of eddy current testing hardware. These instruments practically eliminate what had been one of the most severe limitations of the method, the fact that signals

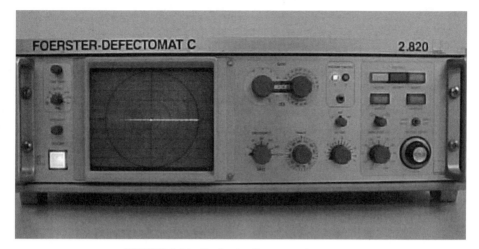

FIGURE 8-24 Production line system instrument.

caused by different material variables can vector into a combined signal that becomes difficult to interpret. In addition, they offer potential for substantial enhancement of performance. Driving the test coil at more than one frequency, multifrequency instruments can not only display the test activity at each frequency separately, but can also show a so-called "mixed output" of different frequency signals subtracted from each other. These capabilities result in the following four advantages:

1. *Suppression of Undesired Variables.* The ability to subtract signals from each other and display the difference as mixed output permits elimination of undesired signals on the display. This feature is the reason why multifrequency instruments were originally developed: to suppress signals from steel supports during inspection of nonferromagnetic tubes, as well as reduce lift-off noise due to probe wobble. A two-frequency instrument can eliminate one source of unwanted signal. Each additional frequency enables the mixing out of an additional type of signal.

2. *Optimization of Normally Contradictory Test Variables.* Use of multiple frequencies allows more than one frequency-dependent performance variable to be optimized simultaneously. For example, during in-service tube inspection using internal coils, a higher frequency provides sensitivity to inner diameter discontinuities, while a lower frequency provides the penetration needed to detect outer diameter discontinuities.

3. *Signal Identification by Pattern Recognition.* A given signal deflection could be caused by a number of detectable conditions. However, each condition exhibits a unique pattern of behavior when viewed over a wide range of frequencies. Multifrequency instruments display this behavior, enhancing the likelihood of identifying the true nature of the signal.

4. *Simultaneous Absolute/Differential Operation.* Some multifrequency instruments have the advantage of allowing a single dual coil assembly to be operated simultaneously in both absolute and differential mode (see Mode of Operation, below), cutting in half the required testing time when the inspection is required to be performed using both of these techniques.

Two types of multifrequency instruments are available: multiplexed and multichannel. Multiplexed equipment operates at only one frequency at a given instant, rapidly switching among the available frequencies. Thus, the test is not being performed simultaneously at all frequencies, although the display gives the illusion that this is the case. Multichannel equipment is the equivalent of having more than one eddy current instrument sharing a single display screen. Early multifrequency instruments required that signal mixing be performed manually by the technician. Recent designs, such as the unit shown in Figure 8-25, perform the mixing automatically.

8.5.2 Test Coils

Eddy current techniques are often classified according to the mode of operation and basic configuration of the test coil assembly. *Mode of operation* determines how the instrument interfaces with the test specimen, such as whether it is comparing coil input from the test specimen to a reference coil (absolute operation) or whether it is comparing coil input from two adjacent portions of the test specimen to each other (differential operation). *Basic configuration* determines how coils are physically packaged to "fit" the test object; that is, whether the coil approaches a portion of the test surface in a probe-like fashion (surface coil), whether it fully encircles the outer circumference of the test object (encircling coil), or whether it passes through the inside of tubular product (internal coil). Coil design, as well as magnitude and frequency of the applied current, all affect the electromagnetic field developed by the coil.

Mode of Operation
With most eddy current instruments, the coil assembly is connected to the instrument via a bridge circuit, as illustrated in Figure 8-26. Bridges are capable of detecting very small

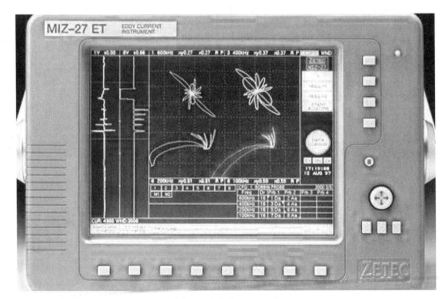

FIGURE 8-25 Multifrequency instrument.

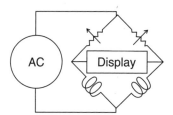

FIGURE 8-26 Bridge circuit.

impedance variations. At the start of the test, the instrument operator balances the bridge to provide a reference signal. During testing, the display provides a readout of bridge imbalance caused by interaction of the coil with the test material.

Absolute coil configurations (Figure 8-27) place a single coil on the test material and employ a second coil, called a *balance load,* remote from the test material to balance the bridge. Absolute coils detect any condition that affects eddy current flow. Although this means that they are capable of detecting any type of condition to which the eddy current method is sensitive, it also means that they are sensitive to potentially unwanted signals such as lift-off and material temperature variations.

Differential coil configurations (Figure 8-28) use a matched pair of coils to perform a comparison. Both coils are coupled to the test material, with one portion of the test material being compared to another. Conditions sensed by both coils are not detected, whereas conditions sensed by only one coil are detected. This has the advantage of suppressing temperature and lift-off variations. Suppression of lift-off helps small discontinuities to be distinguished from lift-off noise. The downside to differential coils is that they provide no signal when a defect condition is simultaneously detected by both coils. Thus, differential coils will only display the ends of long discontinuities; they are not sensitive to gradual discontinuity variations and could ignore a long discontinuity entirely if its ends are very narrow. Differential coil signals are also difficult to interpret: the displayed signal represents the *difference* between two coils' impedances, rather than the impedance of a single coil's interaction with the test material.

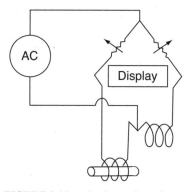

FIGURE 8-27 Absolute coil configuration.

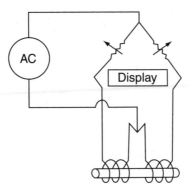

FIGURE 8-28 Differential coil configuration.

External Reference coil configurations (Figure 8-29) combine features of both the absolute and differential modes, placing one coil in contact with the test material and the other coil coupled to a reference standard. This technique provides an indication whenever the test material differs from the standard.

Basic Configurations

Surface Coils. Surface coils are usually designed to be hand-held and are encased in probe-type housings for scanning material surfaces. Surface coils are available in different shapes and sizes to meet different application needs. There is vastly more variety in surface coil design than with encircling and internal coils. Some of them are astonishingly small, wound with wire finer than human hair. Some surface coils can perform a variety of applications, whereas others have been configured to fit a specific size and shape of test specimen. For example, surface probes have been fitted with guides to enable tracing the coil along the edge of turbine blades.

Most surface coils are "bobbin wound" like a spool of thread (Figure 8-30a) and are

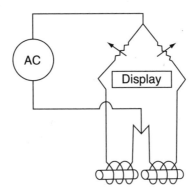

FIGURE 8-29 External reference coil configuration.

designed so the axis of the coil is perpendicular to the surface of the test specimen. Such coils are sensitive to surface cracks and discontinuities that are oriented perpendicular to the test surface; they are generally insensitive to planar subsurface discontinuities. Planar discontinuities can be detected using so-called "horseshoe" or "gap" probes (Figure 8-30b). These probes employ a pair of coils wound on each end of a U-shaped ferrite form so that the flux field flows from one pole of the "horseshoe" to the other and therefore parallel to the test surface. The eddy current field thus flows perpendicular to the test surface, providing sensitivity to planar discontinuities.

Wide surface coils permit rapid scanning and deeper penetration but are less effective at pinpointing the location of small discontinuities. They are often selected for conductivity testing because they tend to average out localized conductivity variations along material surfaces. Conversely, narrow coils are preferred for detecting and pinpointing the location of small surface discontinuities. Because of their smaller diameter electromagnetic fields, narrow coils are less susceptible to edge effect. Surface coils are made in numerous configurations to meet specific application needs. Typical surface coil configurations include the following:

Pencil Probe (Figure 8-31a): Shaped, as its name implies, to be held between the fingers and drawn across the test specimen.

90° Probe (Figure 8-31b): Similar in function to a pencil probe, except that the coil is at a right angle to the probe housing for use where access is limited

Bolt Hole Probe (Figure 8-31c): Designed to fit inside bolt holes with the coil axis perpendicular to the wall of the hole. Manual bolt hole probes are often fitted with retainers so that they can be rotated at a certain depth in the hole and may be fork-shaped to ensure a snug fit. Motorized bolt hole probes are also available. Their output is generally shown on a time base display, allowing the user to determine circumferential position of discontinuities.

Fastener (Doughnut) Probe (Figure 8-31d): A probe designed to fit above the fastener (rivet) holes of an aircraft fuselage. It is used to inspect for cracks around the hole and can be fitted with a clear plastic sight to aid in aligning the probe with the hole.

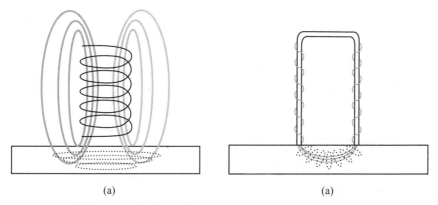

(a)   (a)

FIGURE 8-30 Surface coils: (a) bobbin-wound, (b) horseshoe probe.

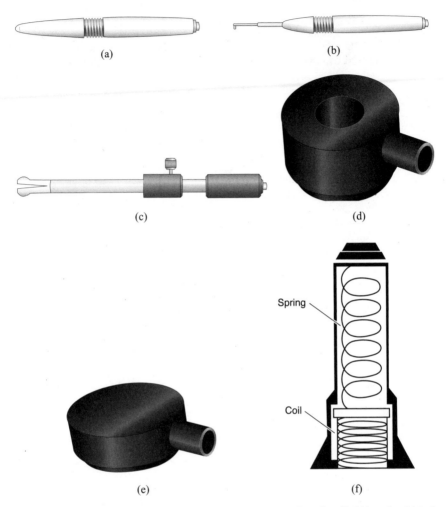

FIGURE 8-31 Typical surface coil configurations: (a) pencil probe, (b) 90° probe, (c) bolt hole probe, (d) fastener (doughnut) probe, (e) pancake probe, (f) spring-loaded surface coil.

Pancake Probe (Figure 8-31e): A low-profile coil generally used for scanning surfaces that have little or no curvature.

Spring Loaded Surface Probe (Figure 8-31f): The coil is mounted like a piston in a cylinder, spring-loaded so that it retracts into an outer housing when pressed against the test surface, thereby minimizing lift-off noise due to probe wobble.

Shielded coils are encased within a cylinder of ferrite, a nonconductive, ferromagnetic material. As shown in Figure 8-32, shielding contains the coil's flux field to prevent interaction with test material boundaries. However, since shielding only operates in the lateral direction, it does not impair penetration.

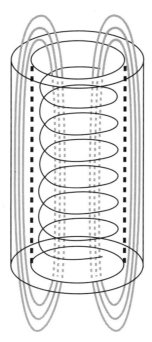

FIGURE 8-32 Shielded coil.

The *cross-axis* coil assembly consists of a pair of adjacent coils interacting with the test material, with the coil axes oriented 90° to each other. Thus, there is sensitivity to defects of all orientations. Cross-axis coils can be placed in a side-by-side configuration as shown in Figure 8-33a, where one coil generates eddy currents parallel to the test surface while the other coil generates eddy currents perpendicular to the test surface. Cross-axis coils can also be wound as a unit with alternate layers wound at 90° angles to each other (x-wound), as shown in Figure 8-33b.

Transmit–receive configurations, such as reflection coils and through-transmission coils, use one coil assembly to induce eddy currents into the test material and a second coil assembly to sense the secondary field.

The *reflection coil* technique, shown schematically in Figure 8-34a, employs two coil assemblies in a single housing, positioned on the same side of the test object. A large single outer coil functions as a transmitter surrounding a pair of smaller, stacked inner coils that form a receiver circuit (Figure 8-34b). The term "reflection" coil results from the fact that the inner coils form a matched pair, wound in opposition. The outer coil induces eddy currents into the test specimen. Secondary flux developed by the eddy currents then induces voltage into the inner coil closest to the test specimen, causing an imbalance in the two-coil receiver circuit, thus providing a signal. Reflection coils have the advantage of being insensitive to temperature drift. They perform well at low frequencies and can function over a broad frequency range.

The *through-transmission* technique (Figure 8-35), positions transmitting and receiving coil assemblies on opposing sides of the test object. It provides a valuable performance advantage in that discontinuities can be detected at greater depths. However, this is

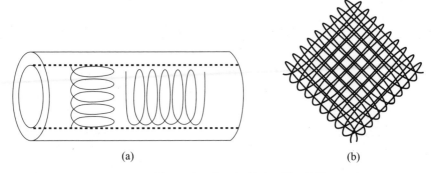

FIGURE 8-33 Cross-axis coils: (a) side by side, (b) X-wound.

offset by the fact that discontinuity depth cannot be displayed because the technique does not provide phase information.

Hall detector. Another transmit–receive technique employs a coil to provide primary flux for generating eddy currents, but uses a solid-state device called a Hall detector as a receiver. In 1879, E. H. Hall discovered that a small voltage is developed across a current-carrying conductor when an inducing magnetic flux is perpendicular to the direction of current flow. Although the voltage was not significant using typical conductors, certain semiconductors develop voltage of magnitude suitable for eddy current test purposes. An ordinary receiver coil operates according to Faraday's law, with induced voltage proportional to the time rate of change of the inducing flux, and therefore depends on the effects of frequency as well as amplitude, with the result that output decreases as frequency decreases. However, a Hall detector responds only to the instantaneous magnitude of the inducing flux, offering good performance at low frequencies. In addition, a Hall detector can be physically quite small.

In addition to the widely known coil types described above, a number of advanced designs have been developed in recent years. A notable contributor in this area is Atomic Energy of Canada, Limited, whose staff has developed a reputation for developing state-of the-art coils for power utility applications.

Encircling coils. Encircling coils (Figure 8-36) completely surround the test material and are normally used for production testing of rod, wire, pipe, tube, and bar stock. Material tested with encircling coils should be centered in the coils by means of guides, so that

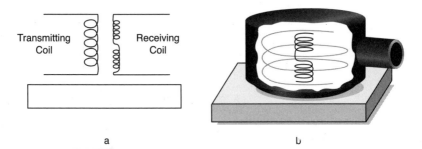

FIGURE 8-34 Reflection coil: (a) schematic diagram, (b) actual configuration.

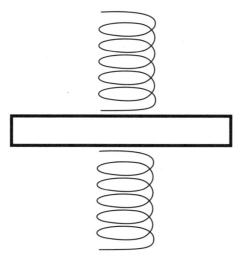

FIGURE 8-35 Through-transmission.

the entire circumference will be tested with equal sensitivity. Production line encircling coils experience more heavy-duty usage than any other type of eddy current coil. Some of them may operate continuously, with product moving through them as fast as 5000 feet per minute. A heavy-duty, production line encircling coil assembly with interchangeable coil size modules is shown in Figure 8-37a. Light-duty, hand-held encircling coil pairs that can be fastened together for differential use or separated for absolute or external reference use are shown in Figure 8-37b.

Because of the "center effect," eddy currents oppose and therefore cancel themselves at the center of solid cylindrical materials tested with encircling coils. Thus, discontinuities located at the center of rods and bar stock cannot be detected with encircling coils. Since encircling coils inspect the entire circumference of the test object, they cannot pinpoint the exact location of a discontinuity along the circumference. So-called *spinning coils* (Figure 8-38), which are actually surface coils that revolve around cylindrical test material, are employed when identification of circumferential location is required in encircling coil applications. Since spinning coils couple to only a limited segment of test material circumference, they are not subject to the center effect. However, spin-

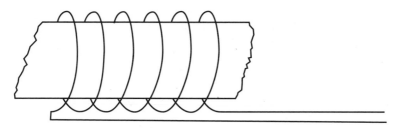

FIGURE 8-36 Encircling coil.

FIGURE 8-37 Typical encircling coil configurations: (a) production line, (b) hand held.

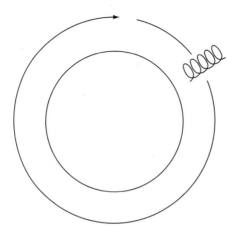

FIGURE 8-38 Spinning coil.

ning coils inspect with a spiral pattern, so their material coverage depends on coil rotation speed versus material transport speed. Care must be taken to ensure adequate coverage.

Internal coils. Internal Coils (Figure 8-39) pass through the core of hollow product and are normally employed for in-service inspection of pipes and tubes. Like encircling coils, standard bobbin-wound internal coils inspect the entire circumference of the test object at one time, but cannot pinpoint the exact location of a discontinuity along the circumference. Again, special designs are available to pinpoint circumferential location of discontinuities. Both manual and automatic means are used to propel internal coils down the length of a long tube. Flexible "u-bend" coil assemblies are available for navigating extreme curvature of tubing. A selection of typical internal coils is shown in Figure 8-40. Inspection of heat exchanger tubing, figure 8-41 is the most common application of internal coils.

8.5.3 Reference Standards

Test calibration or standardization is the process of adjusting the instrument display to represent a known reference standard so that the test can be a comparison between the test

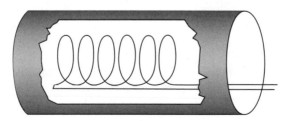

FIGURE 8-39 Internal coil.

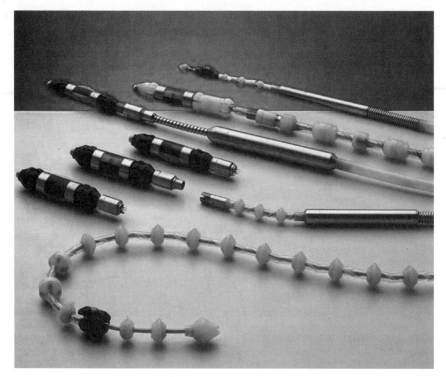

FIGURE 8-40 Typical internal coil configurations.

material and the reference standard. The validity of the test thus depends upon the validity of the reference standard. Moreover, the test system should be checked at regular intervals against the reference standard to ensure that it is operating properly and is still set up correctly for the test being performed. If a variation in instrument performance or setup is discovered, all material tested since the last verification of proper performance and setup should be retested.

Since there is an infinite variety of discontinuity conditions, it is neither possible nor practical to have a set of reference standards so complete as to replicate every possible condition that could be detected during a test. Testing is therefore not a matter of matching each test signal with an identical reference signal. Instead, one obtains practical reference standards that contain a manageable number of representative discontinuity conditions. Signals that vary from these must then be interpreted through techniques such as impedance plane analysis.

The following rules apply to the selection and fabrication of standards:

1. The test standard should be of the same material, with the same wall thickness and configuration, and receive the same processing as the material to be tested.
2. The artificial discontinuities in the standard should model the natural discontinuities expected in the test material. For example:

FIGURE 8-41 Inspection of heat exchanger tubing.

    a. Drilled holes can simulate pits.
    b. EDM notches or saw cuts can simulate cracks.
    c. Thickness reduction in tubing can simulate wear.
    d. Heat Treatment can simulate a conductivity change.
3. Artificial discontinuities in test standards should be sufficiently separated so that their signals will not interfere with each other.
4. Standards should contain no discontinuities other than those intended to produce reference signals.

Figure 8-42 shows tubing standards as well as standards used to calibrate eddy current instruments for detection and evaluation of cracks in material surfaces and inside bolt holes. To augment standards containing artificial discontinuities, it is helpful to build a library of natural discontinuities by obtaining actual specimens removed from

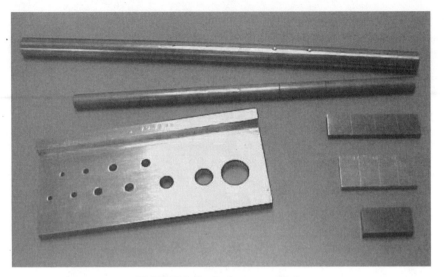

FIGURE 8-42 Reference standards.

service or rejected during manufacture. These are invaluable in sharpening one's interpretive skills.

8.6 EDDY CURRENT APPLICATIONS AND SIGNAL DISPLAY

8.6.1 Eddy Current Display Media

Eddy current testing employs a variety of display media, including:

- Graphic displays, such as cathode ray tubes, liquid crystal screens, and electroluminescent screens
- Analog and digital meter displays
- Simple "go/no-go" displays such as light emitting diode illumination to indicate a deviation of impedance from a reference signal (e.g., detection of a surface crack)
- Recording devices, such as strip charts, magnetic tape, and floppy disks that enable postinspection analysis of test data

Graphic display screens are the prevalent media for obtaining detailed, real-time test information. These displays include:

- A sinusoid showing amplitude and phase variations of total voltage across the coil
- An ellipse that tilts and varies in shape as the test material varies in geometric properties and conductivity
- Deflections of a timebase sweep

- Impedance plane display
- X–Y output of impedance plane display

The sinusoid and ellipse presentations are essentially obsolete, having been replaced by impedance plane displays; but test systems using these outputs may still be in use. The time base is used to show discontinuity location over a defined range of test coil movement, such as the 360° sweep of a coil rotating in a bolt hole; however, time base display is limited in that it indicates signal amplitude but not phase information. Impedance plane display shows complete eddy current signal information and is generally the most useful for signal analysis. The X–Y output of the impedance plane can be sent to various recording and storage media, and is also available on the display screen of some instruments.

8.6.2 Impedance Plane Display

The impedance plane is the key to signal interpretation as well as to obtaining optimum test results. As stated earlier, the impedance plane is a graph of test coil impedance variations with inductive reactance displayed on the vertical axis and resistance on the horizontal. The point on the graph representing the test coil when it is remote from any conductive material is an important display reference position. It is called the "coil in air" or simply "air" point for surface and internal coils and the "empty coil" point for encircling coils (Figure 8-43a). For the remainder of this discussion, the term "air point" will be used, regardless of coil configuration.

As the test coil is brought into proximity with a test specimen, the display dot makes a trajectory from the air point to a point representing the impedance of the test specimen (Figure 8-43b). Each condition that eddy current testing can detect is characterized by a unique position or pattern on the impedance plane, with test variables generally arranged along curves, such as the conductivity curve shown in Figure 8-43c, that are obtained by joining the end-points of the series of lift-off curves. The curves may properly be termed "loci" in that they trace a set of points that represent the range of some variable to which eddy current testing is sensitive.

The impedance plane can be manipulated to advantage by altering parameters such as test frequency and coil diameter. The best presentation is usually a compromise, in that sensitivity, penetration, and resolution are all affected by changes in frequency and coil specifications. By studying impedance graphs and their manipulation, the practitioner can anticipate and optimize test signals for a given application.

Normalization of the Impedance Plane
The sensitivity of the impedance plane to alteration of test parameters does present one problem: changes in these parameters, since they change coil impedance, also change the position and scale of the curves that indicate test material variations. Thus, every change in frequency and/or coil specifications results in a new set of impedance curves. The problem can be illustrated by the effect of frequency variation on inspection of very thin-walled tube with an encircling coil. Figure 8-44a shows that, without the tube inserted, changes in frequency cause changes in inductive reactance values for the coil. That is, five different frequencies result in five different positions for the air point. Figure 8-44b shows curves for variation in conductance from each air point to infinity, with the dashed lines showing the effect of inserting tubes of different conductance into the coil. As frequency is increased, a given tube's impedance takes a more clockwise position on a larger curve. However, the proliferation of curves can be solved with normalization. The impedance plane is normalized by converting the vertical axis from a scale of inductive

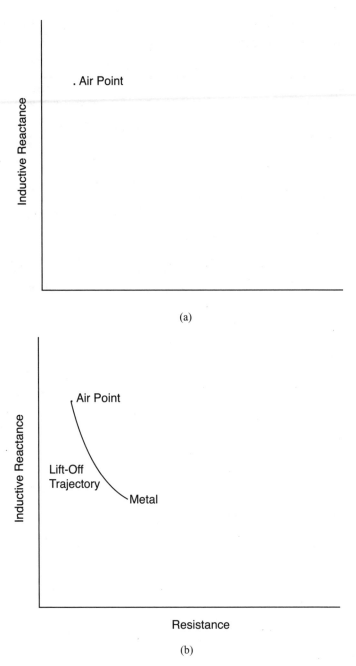

FIGURE 8-43 Impedance plane: (a) air point, (b) lift-off trajectory.

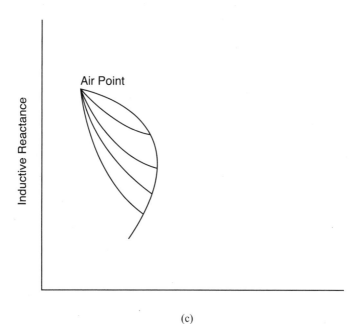

(c)

FIGURE 8-43 (c) Conductivity curve.

reactance of the test coil in ohms to a scale of inductive reactance of the test coil compared to the inductive reactance of the coil's own air point; this is done by dividing actual coil inductive reactance by inductive reactance at the air point. This has the effect of mathematically pulling all of the curves shown in Figure 8-44b into a single curve. Additionally, the horizontal axis needs to be normalized; this is done by subtracting variations in coil wire resistance from total resistance (one is interested in displaying test specimen variations, not coil variations) and dividing the remainder by the inductive reactance at the air point. Thus, the normalized impedance plane would be labeled $\omega L/\omega L_0$ on the vertical axis and $R/\omega L_0$ on the horizontal.

Most of the remaining illustrations will show a normalized impedance plane. However, in some cases it will be more convenient to illustrate the subject matter using a non-normalized impedance plane, with the assumption that frequency and coil properties are fixed.

Effect of Phase Lag on the Conductance Curve

Figure 8-44b showed that impedance traces a semicircular path when conductance in a very thin-walled tube is varied. The semicircular pattern results when the specimen wall is so thin that phase lag is insignificant. Factors that increase current flow result in clockwise movement of the impedance operating point along the semicircle, while those that restrict current flow result in counterclockwise movement of the operating point, as seen in Figure 8-45.

As specimen wall thickness is increased, display of test variables becomes more complicated, with the semicircle evolving into a comma shape and dimensional variations ap-

FIGURE 8-44 Impedance plane before normalization. (a) Movement of air point with change in frequency. (b) Variation of conductance with frequency.

pearing as appendages to the comma curve. The shaded area in Figure 8-46a shows how the semicircular impedance curve changes from a semicircle to a comma shape as a very thin-walled tube inside an encircling coil transforms into a solid bar. The semicircular curve is also valid for very thin-walled tubes inspected by an internal coil. In the case of a solid bar inside a long encircling coil, a "perfect comma" is achieved, with the lower end of the curve intersecting the origin of the graph at a 45° angle. The comma-shaped curve that evolves due to increased wall thickness is normally called a conductivity curve, although variations in frequency and coil diameter also cause movement along this curve on the normalized impedance plane. Internal coils testing thicker tubes, as well as surface coil applications, produce a more flattened impedance curve, similar to that shown in Figure 46b.

Lift-Off Curves
When a test coil is remote from any conductive material, impedance is at a position of high inductive reactance and low resistance. The high value for the air point on the inductive reactance scale occurs because there is no secondary flux available to reduce primary flux; the low value on the resistance scale occurs because the only resistance detected is that of the coil wire.

As the coil approaches a conductive and/or ferromagnetic object, the impedance of

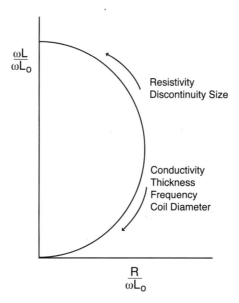

FIGURE 8-45 Effect of test variables on impedance curve operating point.

the coil changes and the display dot moves. If the coil approaches a nonferromagnetic conductive specimen, secondary flux cancels a portion of primary flux, resulting in a decrease in inductive reactance, as shown in Figure 8-47a. The more conductive the test material, the greater the cancellation of primary flux and the further downward the display dot moves. Simultaneously, the specimen acts as a resistive load on the coil and the impedance point advances along the resistance scale. However, if the coil approaches a specimen that is both ferromagnetic and conductive, the specimen's flux adds to the coil's flux and the impedance point moves up both the inductive reactance and resistance scales, as shown in Figure 8-47b. If the material is nonconductive (such as ferrite), the impedance point advances up the inductive reactance axis only, Figure 8-47c, with no movement along the resistance axis. This effect is useful in orienting phase rotation to achieve a vertical deflection for inductive reactance variation. Regardless of direction, the display dot trajectory is the vector sum of movement along both axes and is called a lift-off curve.

Just as the spacing between a surface coil and the test material is called "lift-off"; the spacing between either an internal coil or encircling coil and concentrically positioned test material is called "fill factor". As stated earlier, lift-off is useful for measuring the thickness of nonconductive coatings on metals. Moreover, it can be used to measure the thickness of any nonconductive material that is simply resting on a conductive surface. Fill factor can be used to measure diameter of bars and rods placed inside encircling coils. Sensitivity to lift-off and fill factor depends on flux density and thus decreases as coil to test material distance increases. The decrease in sensitivity is, of course, nonlinear because flux density decreases according to the inverse square law.

The downside of sensitivity to lift-off and fill factor variations is that inadvertent movement of the coil relative to the test material will cause noise signals that can obscure the signals for which the test is being performed. Prevention of noise signals due to lift-

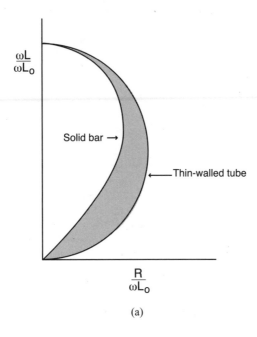

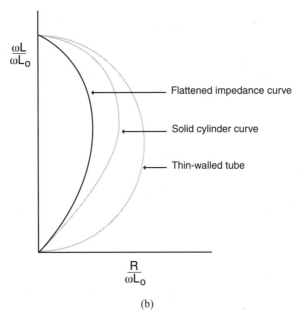

FIGURE 8-46 Variation of impedance curve shape with specimen configuration. (a) Transition of impedance curve from semicircle to comma shape. (b) Impedance curves for internal coil testing thicker tubes and for surface coils.

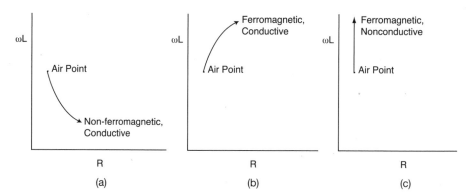

FIGURE 8-47 Lift-off curves. (a) Nonferromagnetic conductive specimen. (b) Ferromagnetic, conductive specimen. (c) Ferromagnetic nonconductive specimen.

off and fill factor variations is an important element in most eddy current inspections. Special techniques, fixtures, and instrument circuit designs have been developed to suppress these signals. When performing lift-off and fill factor applications, it is best to operate at high test frequencies. This reduces penetration into the test material, thus lessening the effects of material variables on test results.

Edge Effect
Just as increasing the spacing between a surface coil and conductive material causes a lift-off curve, moving the coil toward the edge of the material will cause an edge effect trace (Figure 8-48). This results from compression of the field and then reduction of current density as the coil is moved off the edge of the test specimen.

If the eddy current field simultaneously intercepts a discontinuity while approaching the edge, the two conditions will produce a combined response, rather than separate edge and discontinuity indications. Thus, the discontinuity may not be detected. The

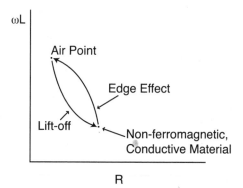

FIGURE 8-48 Edge effect versus lift-off.

problem can be eliminated by scanning the coil parallel to the material edge at a constant distance from the edge to first establish a uniform edge effect indication. Interception of a discontinuity will then cause an additional response. Edge effect is intensified by the wider eddy current fields developed by large diameter coils and lower test frequencies. Use of smaller diameter coils reduces edge effect; use of shielded coils virtually eliminates it.

Surface Coil Impedance Curves for Material and Performance Variables

Although there are some differences among the signals obtained from surface coils, internal coils, and encircling coils, surface coils are ideal for demonstration purposes. This is because they are easily manipulated, are employed for a wide variety of applications, and retain good coupling to the test specimen when their diameter is varied (in contrast to internal and encircling coils, whose diameter must depend on the diameter of the test specimen). The following sections show a variety of material and performance variables using hand-held surface coils.

Conductivity Variation

Figure 8-49a shows the so-called conductivity curve, the locus of the end points of the lift-off curves for all nonferromagnetic, conductive materials. In order for materials to be positioned on this curve, their thickness must be effectively infinite relative to electromagnetic penetration. The counterclockwise extreme of the conductivity curve represents zero conductivity; the clockwise extreme of the curve represents infinite conductivity. Varying test frequency shifts the impedance points for materials of different conductivities along the conductivity curve on the normalized impedance plane. At higher frequency (Figure 8-49b), the impedance points for the various conductivities advance clockwise along the curve, with lower conductivity materials spreading apart while higher conductivity materials compress at the bottom end of the curve. Conversely, at lower frequency (Figure 8-49c), the impedance points move counterclockwise, with higher conductivity materials spreading apart while the lower conductivity materials become compressed at the top end of the curve. Thus, higher frequencies provide greater separation for conductivity tests on lower conductivity materials, while lower frequencies provide greater separation for conductivity tests for high-conductivity materials.

Frequency adjustment also helps separate the lift-off and conductivity variables during conductivity tests, as shown in Figure 8-50. At low frequencies, lift-off curves for low-conductivity materials are almost parallel to the conductivity curve. As frequency is increased, the operating point advances clockwise along the conductivity curve, increasing the angle between the lift-off curve and conductivity curve. Optimum separation is achieved near the so-called "knee" of the conductivity curve. Advancing beyond the knee will lengthen the lift-off curve, increasing sensitivity to lift-off noise caused by probe wobble.

From the standpoint of eddy current testing, conductivity is a material's most important variable. A material must be conductive in order for it to be tested by eddy currents. In addition, conductivity affects test performance. There are two groups of factors that can cause a material's conductivity to vary: those that result in useful applications and those that can sabotage the test process. Factors affecting conductivity variations that can result in test applications include the following:

1. *Variations in chemical composition.* The various metallic elements and alloys can be sorted as long as none of the materials have overlapping conductivities. Certain ranges of copper alloys have overlapping conductivities, for example, and therefore could not be sorted.

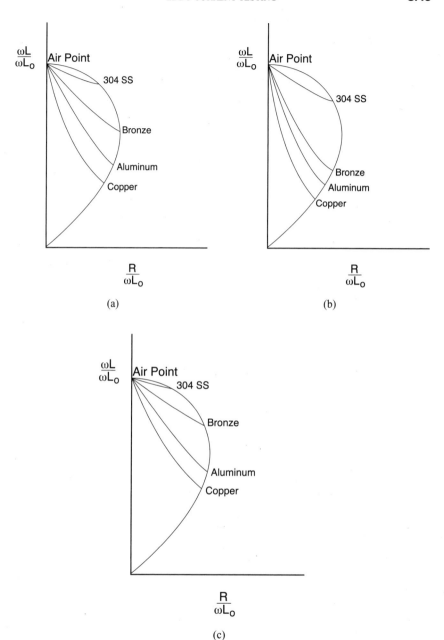

FIGURE 8-49 Conductivity curve: (a) medium frequency, (b) high frequency, (c) low frequency.

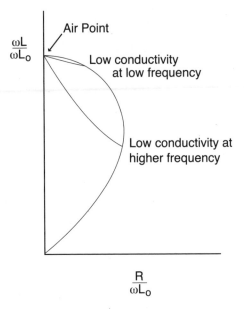

FIGURE 8-50 Effect of frequency on separating signals.

2. *Mechanical processing.* Cold working affects lattice structure, causing minor conductivity changes.
3. *Thermal processing.* Heat treatment causes hardness changes that are detectable as conductivity variations. Because of this, eddy current testing can be used to some extent as a process control for heat treatment.
4. *Variations in thickness of plating or cladding* are a combination of both conductivity and dimensional variables.

Factors affecting conductivity that can negatively impact test performance include:

1. *Material temperature.* As material temperature increases, conductivity decreases. Variations in temperature can be caused by environment, materials processing, and eddy currents themselves. Care must be taken that material temperature does not vary during testing and that reference standards are the same temperature as the test material. Temperature variation during testing can cause the display dot to drift away from the balance point.
2. *Unrelieved residual stress* causes unpredictable conductivity variations. Thus, it is an undesirable variable.

Ferromagnetic Materials

It follows that the effects of ferromagnetic test materials on the impedance of the coil will be similar to those obtained when a ferromagnetic core is used to enhance the flux density of an eddy current surface coil. The coil's alternating electromagnetic field causes the domains of the ferromagnetic material to rotate so that its poles alternate along with the coil, the two fields combining into a field of increased flux density. This increase in primary coil flux density (Φ_P) causes an increase in back voltage, which results in an increase in inductive reactance. The impedance point therefore moves upward on the impedance plane.

The difference between a coil with a ferromagnetic core versus bringing the coil into contact with ferromagnetic test material is that since the core is a permanent part of the coil assembly, it causes the air point to be positioned higher on the inductive reactance axis. Conversely, proximity between the coil and ferromagnetic test material can be varied, with the result that the air point remains lower and inductive reactance increases only as the coil is brought closer to the test material. Moreover, the coil can be brought into proximity to ferromagnetic test materials of various conductivities. Thus, there is theoretically a conductivity curve for every permeability value.

Discontinuity Signal Display

One of the most important applications of eddy current testing is detection and evaluation of cracks and other discontinuities. The impedance plane responds to discontinuities depending on the density and phase lag of interrupted eddy current circulation. Displayed signal amplitude depends on the total quantity of deflected electrons, whose density is decreasing exponentially with increased depth. Displayed phase lag depends on the weighted average of the phase lags of all deflected electrons, whose phase lag is increasing linearly with increased depth. In fact, since a discontinuity of any size will possess some degree of thickness, it will interrupt eddy currents over a *range* of current densities and phase lags, depending on the dimensions, shape, and orientation of the discontinuity.

Table 8.2 shows eddy current density and phase lag at one, two, and three standard depths of penetration. The table indicates that displayed phase lag represents a round trip of lag time from the material surface to the discontinuity and back again to the surface. To obtain meaningful depth information, the skin depth formula must first be employed so that actual depth values can be assigned to 1δ, 2δ, and 3δ.

Subsurface Discontinuities

Assuming that material thickness exceeds five skin depths and that a large-diameter coil is used, theory states that the display of a series of fixed-size voids (see Figure 8-51a) would perform as indicated in Table 8-2 as their depth from the surface varies and produce the display shown in Figure 8-51b.

The lift-off signal represents the test material surface and serves as the 0° reference. Since a so-called "surface void" must necessarily penetrate the surface, it will deflect eddy currents with a phase lag greater than 0°, with it's signal exhibiting some rotation clockwise from the lift-off signal. Similarly, subsurface voids cannot exist only at depths 1δ, 2δ, and 3δ, as indicated in Table 8-2; the voids must have some finite thickness and

TABLE 8-2 Eddy Current Density and Phase Lag for Subsurface Discontinuities

Discontinuity depth	Current density	Phase lag in Material	Phase lag on display
Surface	100.0 %	0.0°	0.0°
1δ	36.8%	57.3°	114.6°
2δ	13.5%	114.6°	229.2°
3δ	5.0%	171.9°	343.8°

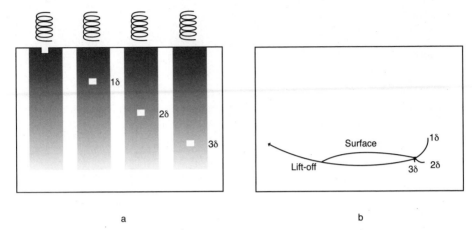

FIGURE 8-51 Subsurface discontinuities: (a) variation in void depth from surface, (b) signal display for voids of different depths.

extend a measurable amount above and/or below the stated depths, with a consequent variation in displayed phase lag.

Changing the frequency affects penetration, sensitivity, and resolution. Figure 8-52 compares display of a series of nine subsurface voids at 300 KHz, 50 KHz, and 1 KHz. At 300 KHz, penetration is just adequate to display the top three subsurface voids well separated at approximately 1δ, 2δ, and 3δ. As frequency is decreased, the signals rotate counterclockwise on the display. Thus, at 50 KHz, deeper voids are displayed as penetration increases. Finally, at 1 KHz, nine voids are now only a small portion of one skin depth, with their signals virtually superimposed and lower amplitude. Penetration has been enhanced at the cost of resolution and sensitivity.

Surface-Breaking Discontinuities

The behavior of surface-breaking discontinuities is more complicated. The key to understanding the difference in the display pattern between a series of subsurface discontinuities of increasing depths and a series of surface-breaking discontinuities of increasing depth is, of course, to consider the effect of each on skin effect and phase lag. The subsurface series represents a void of a *fixed volume* whose position is being shifted down to locations of decreased current density and increased phase lag; the surface series represents a void whose *volume is expanding* into locations of decreased current density and increased phase lag. Thus, in that the surface void disturbs fewer additional electrons each time its depth increases by a fixed amount, its signal amplitude increases by progressively smaller amounts. This likewise causes the weighted average of deflected electrons to similarly increase by smaller amounts, resulting in displayed signal phase correspondingly increasing by progressively smaller amounts.

As with subsurface discontinuities, changes in frequency affect penetration, sensitivity, and resolution. Figure 8-53 compares the behavior of the surface-breaking voids in a 304 stainless steel plate at 100 KHz, 10 KHz, and 1 KHz. At 100 KHz, the eddy currents are concentrated on the surface, giving good resolution of the shallowest void, but there is insufficient penetration to resolve the deeper voids. As frequency decreases, penetration is gained at the expense of surface sensitivity. At 10 KHz, there is sufficient density dis-

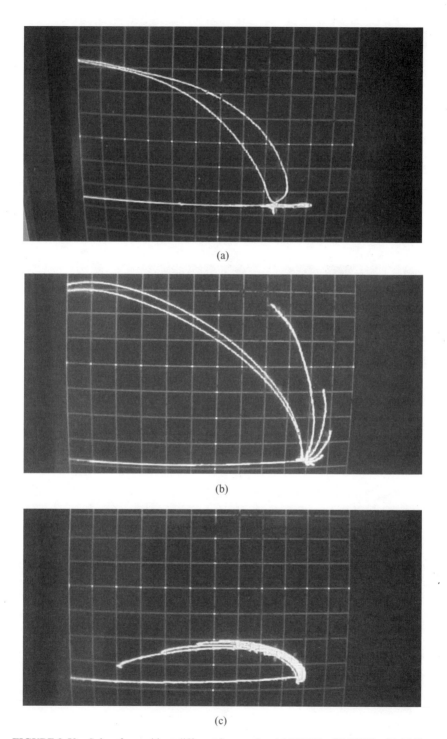

FIGURE 8-52 Subsurface voids at different frequencies: (a) 300 kHz, (b) 50 kHz, (c) 1 kHz.

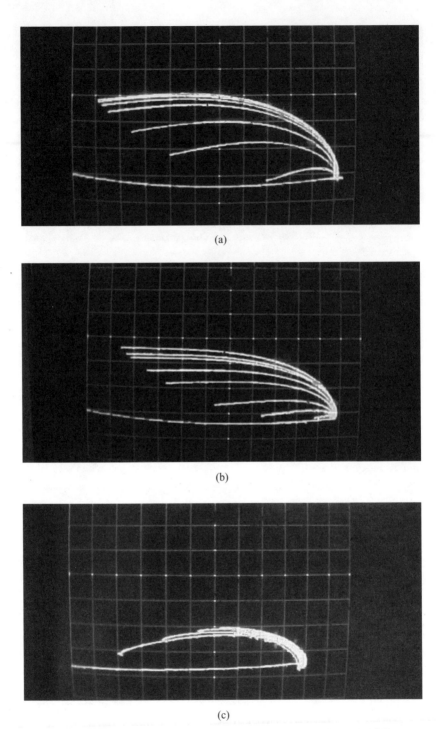

FIGURE 8-53 Surface-breaking discontinuities: (a) 100 kHz, (b) 10 kHz, (c) 1 kHz.

tribution to obtain adequately resolved signals from the shallowest void as well as the deepest. At 1 KHz, decreased surface sensitivity and phase rotation virtually obscure the signal from the shallowest void.

Shape of the Eddy Current Field
Terms such as "skin depth" and "effective depth of penetration" tend to imply that a decrease in current density and phase lag occur only in the downward direction in the test material. However, the coil's flux, which develops the eddy currents, extends in all directions away from the coil. Therefore, the eddy current field is necessarily somewhat ring-shaped, as shown by the field of the surface coil in Figure 8-54a, with current density and phase lag varying laterally as well as downward, indicated by progression from black through increasingly lighter shades of gray in the cross-sectional view of Figure 8-54b. These field properties become obvious when one examines the signal produced by a surface-breaking discontinuity (Figure 8-54c). Close examination shows that the trace initially proceeds vertically, then curves increasingly toward the left, indicating an initially high phase lag that then decreases as the coil approaches the discontinuity.

Display of Discontinuity Orientation
Figure 8-55a shows a portion of an aluminum bar with machined notches at angles decreasing in 10° increments from 90° to 10°. Figure 8-55b shows how phase lag can be used to indicate the varying orientation of these surface-breaking discontinuities. As discontinuity orientation tilts away from the perpendicular, an increasingly open series of loops is traced on the display, indicating a progressively less rapid variation of phase lag. It is also possible to state the direction in which the discontinuity is leaning. If the signal trace advances clockwise as the coil approaches the discontinuity, then the field is being interrupted in an area of lower phase lag, signifying that the discontinuity is being intercepted where it breaks the surface. However, if the trace advances counterclockwise, greater phase lag is initially indicated, the result of the discontinuity being intercepted deeper into the material.

Enhancing Signal Display
In addition to the instrument gain control, which adjusts amplification of bridge output, most impedance plane display instruments provide capability to adjust the ratio of amplification for horizontal versus vertical display of test signals. This feature can be especially useful for improving signal resolution. Figure 8-56a shows a display of surface-breaking notches of increasing depth, with the inevitable reduction of phase separation as notch depth increases. In Figure 8-56b, the vertical amplification has been increased at the expense of the horizontal, stretching apart the signals for improved resolution. The enhancement is accompanied by a corresponding amount of signal distortion, as indicated by the increased curvature of the lift-off trace. The user should be careful to return the horizontal/vertical display amplification to normal settings after completing the application, to preclude undesired distortion during subsequent tests.

Thickness Variation
Thickness variations exhibit the same display behavior as subsurface discontinuities, except that they represent an infinite-size void whose depth is increasing. The phase rotation pattern is the same, but the signal amplitude is greater. Figure 8-57a shows the lift-off curves for a series of aluminum samples of increasing thickness. When the end points of the lift-off curves are joined, a spiral curve is produced. Such a spiral would be traced if the coil were drawn along the length of a wedge covering the same thickness range.

Frequency variation also follows the same pattern as for subsurface discontinuities.

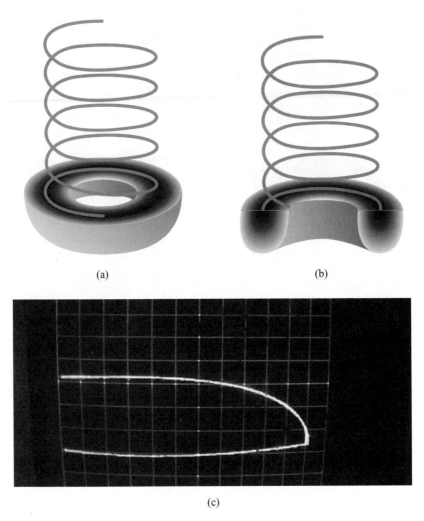

FIGURE 8-54 Effect of field shape on signal display. (a) Eddy current field shape for surface coil. (b) Cross section of eddy current field. (c) Signal for coil approaching surface-breaking discontinuity.

Decreasing frequency improves penetration and rotates signals from greater depth in a counterclockwise direction. Increasing frequency reduces penetration, but improves resolution of signals from thinner materials by rotating the signals clockwise. These effects of frequency variation are illustrated using a series of eleven aluminum steps ranging in thickness from 0.012″ to 0.100″. In Figure 8-57b, at a frequency of 10 KHz, the thinner steps are well separated, but penetration is only sufficient to resolve the first nine steps. When frequency is reduced to 2 KHz (Figure 8-57d), penetration is increased to the point

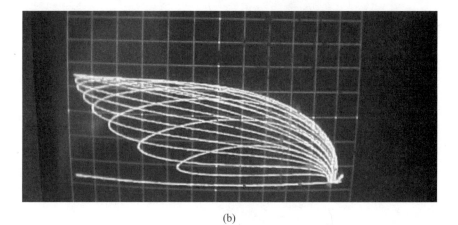

FIGURE 8-55 Display of varying discontinuity orientations. (a) Bar with discontinuities of varying orientations. (b) Signal variation as orientation is varied.

where all eleven steps are resolved, but the reduction in frequency has reduced displayed separation of the thinner steps.

Plating and Cladding
Variation in thickness of one conductive nonferromagnetic material bonded over another is shown in Figure 8-58. Figure 8-58a shows how a wedge of high-conductivity material (copper) bonded over low-conductivity material (304 stainless steel) would be displayed. As thickness increases, a spiral advances from the 304 stainless position on the conductivity curve and reaches the copper position on the conductivity curve when the thickness of the copper becomes effectively infinite and the effects of the 304 stainless can no longer be detected. In effect, a thickness curve for the copper has been grafted onto the conductivity curve, with 1δ, 2δ, and 3δ appearing at their usual positions of phase rotation. Figure 8-58b shows that low-conductivity material bonded over high-conductivity causes the thickness spiral to be inverted.

The display for nonferromagnetic (304 stainless steel) material bonded over ferromagnetic (carbon steel) is shown in Figure 8-59. As thickness increases, a curve proceeds from the impedance plane position for carbon steel to the position for 304 stainless on the conductivity curve for nonferromagnetic materials. The illustration shows the display for

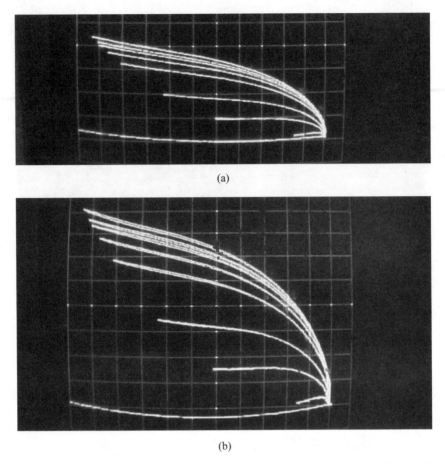

FIGURE 8-56 Signal enhancement using display amplification. (a) Surface-breaking notches without enhancement. (b) Surface-breaking notches with enhancement.

a wedge of 304 stainless positioned on carbon steel, as well as lift-off curves for various thicknesses of 304 stainless bonded over carbon steel.

Spacing between Conducting Materials

The coil's flux can penetrate multiple layers of conducting material. Therefore, it is possible to use eddy currents to measure thickness of adhesives or air gaps between two metal samples, providing that the combination of conductivity, frequency, and coil diameter provides adequate penetration. Figure 8-60 shows the spacing curve traced for a wedge-shaped air space between two layers of aluminum. As the coil is scanned along the upper metal surface, in the direction of increasing gap thickness, a curve is traced from the point of infinite aluminum thickness on the conductivity curve to the point on the aluminum thickness curve that represents the thickness of the upper metal surface. Frequency must be selected so that the position for the thickness of the upper metal layer is far enough up the thickness spiral for a usefully long spacing curve to be traced.

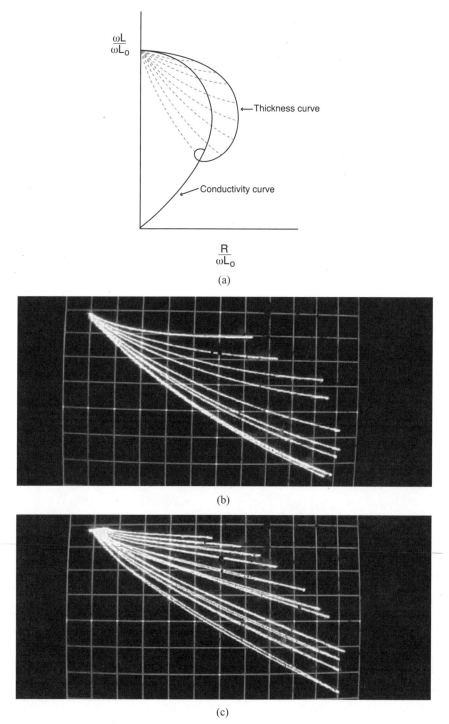

FIGURE 8-57 Thickness testing. (a) Thickness spiral. (b) Thickness signal display at 10 KHz. (c) Thickness signal display at 2 KHz.

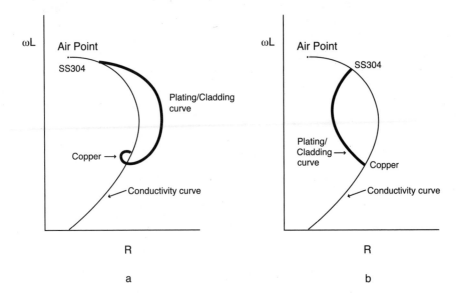

FIGURE 8-58 Thickness of nonferromagnetic plating/cladding over nonferromagnetic base material. (a) High conductivity bonded over low conductivity. (b) Low conductivity bonded over high conductivity.

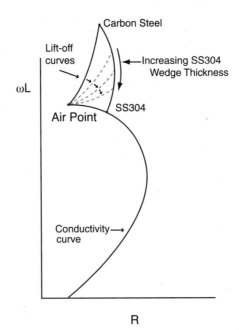

FIGURE 8-59 Thickness of nonferromagnetic plating/cladding over ferromagnetic base material.

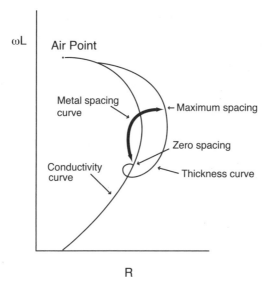

FIGURE 8-60 Metal spacing.

Differential Surface Coil Display
For purposes of convenience and simplicity, all display illustrations in this section have been those of surface coils operating in the absolute mode. Figure 8-61 compares display of absolute (Figure 8-61a) versus differential (Figure 8-61b) surface coil indications for surface-breaking voids of increasing depth. The differential coil pair was oriented so that one coil followed the other as the discontinuities were intercepted. However, if the probe were rotated 90°, so that the coils jointly intercept each discontinuity, the signals would almost completely cancel each other. Thus, while differential coils have the advantage of suppressing unwanted signals such as lift-off noise, the user must be well aware of their characteristics.

Encircling and Internal Coil Display
Figure 8-62 shows signal traces from a series of external circumferential grooves on the same tubing standard intercepted by both encircling (Figure 8-62a) and internal (Figure 8-62b) differential coils. The grooves are detected as surface-breaking by the encircling coil and subsurface by the internal coil.

Law of Similarity
In the discussion of conductivity, it was shown that positioning along the comma-shaped impedance curve advances clockwise as conductivity increases. It was also shown that individual conductivity positions shift along the curve as test frequency is varied. In the case of surface coils, for example, similar shifts also occur with changes in permeability and coil diameter. Thus, a given position along the impedance curve depends on a combination of factors. If positioning along the impedance curve can vary with conductivity, frequency, permeability, and coil diamcter, the question arises as to what a given curve position actually represents. The answer is that a given position along the curve represents

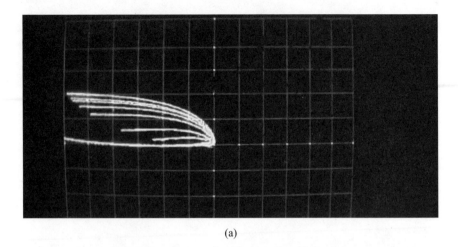

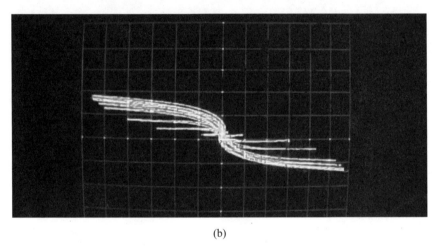

FIGURE 8-61 Absolute versus differential coil systems. (a) Absolute. (b) Differential.

a specific distribution of eddy currents in the test material and that conductivity, frequency, permeability, and coil diameter all affect that distribution. Since distribution of eddy currents affects the major performance criteria—sensitivity, penetration, and resolution—position along the impedance curve largely determines test performance.

Various systems have been devised for assigning reference values for different positions along the impedance curve. The most well-known system is probably the one defined by Foerster for encircling coils enclosing a long bar, the f/f_g ratio. The "f" value represents the frequency at which the test instrument is driving the coil. The f_g value is a mathematically derived reference value known as the *limit frequency,* calculated as follows:

$$f_g = \frac{5.066\,\rho}{\mu_r D^2}$$

EDDY CURRENT TESTING

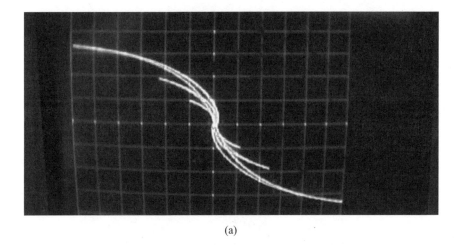

(a)

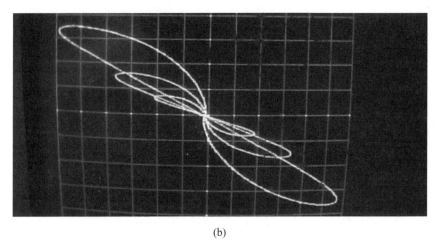

(b)

FIGURE 8-62 Display of signals for encircling and internal differential coils. (a) Encircling coil. (b) Internal coil.

where
f_g = limit frequency (KHz)
ρ = resistivity in microohm-centimeters
μ_r = relative permeability (dimensionless)
D = bar diameter

Ascending f/f_g values are plotted clockwise along the impedance curve, as shown in Figure 8-63. Each position represents a specific eddy current distribution in the test material and, hence, a specific combination of sensitivity, penetration, and resolution. This leads to the concept of the similarity law, which states, in effect, that if test conditions are manipulated to obtain the same f/f_g ratio in two different test objects, performance will be

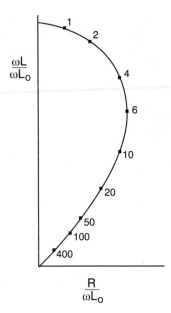

FIGURE 8-63 Impedance curve for f/f_g variations with encircling coil inspecting long bar.

the same for both. For example, a given signal display for an identical discontinuity in both a low-conductivity and a high-conductivity material can be obtained by manipulating test frequency to obtain the same f/f_g ratio for each of them.

8.7 ADVANTAGES AND LIMITATIONS

One of the major advantages of the eddy current method is also one of its most severe limitations. That is, the eddy current method is sensitive to many variables including material conductivity and thickness, size of surface and subsurface discontinuities, thickness of plating or cladding on base metals, spacing between conductive layers, spacing between test coil and test material (lift-off), and permeability variations. However, the major limitation of the eddy current method is that response to these variables is vectorially additive, with the result that when more than one variable is detected by the test coil, all variables combine into a single response that may be difficult to resolve into its separate components. Ability to suppress and render variables separately identifiable is an important element of the user's knowledge. The techniques used to overcome this problem range from relatively simple to complicated and potentially expensive, such as the use of multifrequency instruments.

Additional advantages of the eddy current method include the following:

1. Equipment available for field use has become increasingly portable and lightweight. In addition, many instruments are microprocessor-based, which permits test setups to be saved to memory and test results saved to disk for archiving and analysis.

2. The method is nondestructive. No couplants, powders, or other physical substances

need to be applied to the test material. The only link required between the probe and test material is a magnetic field.

3. Test results are usually instantaneous. As soon as the test coil responds to the test specimen, a qualified user can interpret the results. However, data on a large quantity of test material can be acquired so rapidly, as in the case of in-service tube testing, that it becomes more practical to initially record the data and review it later at a more reasonable pace. Moreover, the use of multifrequency equipment can greatly complicate the amount of data acquired at a given instant and render it overwhelming unless it is recorded and later reviewed as separate elements.

4. Eddy current testing is ideal for "go/no-go" inspections. Audible and visual alarms, triggered by threshold and box gates, as well as gates that can be set to almost any desired shape, are available to automate testing.

5. It is not necessary for the coil assembly to touch the test material. This permits high speed production testing to be done without friction, thus preventing wear of test coils.

6. Eddy current testing is safe; there is no danger from radiation or other such hazards.

7. Material preparation is usually unnecessary and cleanup is not required.

Additional limitations of eddy current testing include the following:

1. The test material must be electrically conductive. However, it is possible to measure the thickness of nonconductive coatings on conducting materials using the lift-off technique.

2. It is difficult to assess subsurface conditions in ferromagnetic materials. Consequently, testing of ferromagnetic materials is limited to detecting surface discontinuities only, unless the material has been magnetically saturated. Magnetic saturation is limited to the testing of geometries that can accommodate saturation coils, primarily encircling coil applications. It is possible to perform inner-diameter inspection of tubing as long as the magnetic field is not so strong as to cause the probe to lock onto the tubing. In addition, magnetically saturated test objects may have to be demagnetized after testing is completed, so that they will not attract ferromagnetic debris.

3. Even on nonferromagnetic materials, the eddy current method has limited penetration, which varies with the material conductivity and test frequency. As a rule, penetration is limited to fractions of an inch in most materials.

4. Inspection speed may have to be limited as a function of test frequency.

5. Much of the eddy current theory is complicated, presenting a challenge to practitioners requiring mastery of the method.

8.8 OTHER ELECTROMAGNETIC TEST TECHNIQUES

In addition to the conventional eddy current testing techniques described above, additional electromagnetic test techniques and equipment have been developed to solve specific applications. Although these techniques cannot be as broadly applied as conventional eddy current testing, they often provide the best solution to a given problem.

8.8.1 Flux Leakage Testing

It was stated earlier that eddy current testing cannot detect subsurface discontinuities in ferromagnetic test material unless the material is magnetically saturated. Saturation, a common technique using encircling coils, is possible but is sometimes inconvenient using internal coils, and is generally impractical using surface coils. Flux leakage testing is used at steel mills and fulfills an important need in the in-service inspection of ferromagnetic tubing.

When a ferromagnetic object that is free of discontinuities is magnetized, an unbroken flux field flows through the object, forming a path between its poles. Flux density decreases toward the outer material surfaces, and a detector of flux leakage would pick up a very weak signal from the surface. However, if the surface is interrupted by a discontinuity, poles are formed at opposite sides of the resulting gap, and flux lines flow externally from the north to south pole. The flux leakage lines are detected by a test probe passing over the discontinuity and a signal is displayed. Disturbance of the material's original flux lines also extends beneath the discontinuity and if the material wall is thin enough, a discontinuity on one wall surface may be detected from the opposite wall. Therefore, it is possible to use internal-type flux leakage probes to detect both inner- and outer-wall discontinuities on tubing.

8.8.2 Remote Field Eddy Current Testing

Remote field eddy current testing is another technique useful for inspection of ferromagnetic tubing. Compared to impedance-type eddy current techniques where detection is limited to surface discontinuities in ferromagnetic materials unless saturation is employed, volumetric inspection of ferromagnetic tubing is possible using remote field testing. This technique responds to material conditions that my not be detectable with flux leakage testing, but this advantage is counterbalanced by possible distortion of the received signal. Although the technique can be used on either ferromagnetic or nonferromagnetic tubing using internal coils, it is not as effective overall as standard eddy current testing, so it is primarily used to obtain penetration in ferromagnetic materials.

Remote field eddy current testing is essentially a geometric variation of familiar send–receive eddy current test techniques, employing a bobbin-wound sending coil that induces eddy currents in the tube wall and a sensor consisting of a single coil or an array. Test frequency is selected such that skin depth is at least as great as tube wall thickness. As usual, eddy current skin effect operates laterally as well as downward from the coil, with eddy currents flowing for some distance in the lateral direction. The eddy currents' secondary flux cancels the coil's primary flux on the inner tube wall near the coil. As distance from the coil increases, this cancellation effect decreases. In the so-called remote field, laterally removed from the sending coil, the sensor can detect phase and amplitude information from the secondary flux developed by the eddy currents. Standard impedance plane display eddy current instruments can be used for remote field inspection. However, the drive voltage for the sending coil must be amplified substantially above the levels normally used for conventional impedance-type eddy current inspection.

8.8.3 Modulation Analysis

Modulation analysis is a production system technique whereby the test material is moved past the coil assembly at a constant high velocity. Impedance amplitude variations are displayed on a strip chart recorder. However, displayed information is selective in that filtration blocks frequencies that contain undesired information from the test material. For example, a high-pass filter would prevent display of the gradual impedance variations characteristic of wall thickness variations, whereas a low-pass filter would block more abrupt impedance variations. Because the amplitude-only display makes it difficult to interpret the type of material variable that causes test signals, material transport speed and filter frequency must be well controlled so that only relevant information is displayed.

8.9 GLOSSARY OF KEY TERMS

Absolute coil technique—An eddy current test in which one arm of the instrument's bridge is connected to a "test coil" that is interfacing with the test material and the opposing arm of the bridge is connected to an electrically similar coil, called a "balance load" or "dummy load," that is remote from the test material.
Alternating current—A current whose direction of flow is periodically reversing.
Atom—The smallest unit of matter; hence, the smallest unit in an electrical conductor. The atom consists of a positively charged nucleus surrounded by one or more "shells" or "orbits" of negatively charged electrons.
Back voltage—Voltage induced into a conductor by means of the varying electromagnetic field of a varying current passing through that conductor.
Bridge circuit—Formally known as a "Wheatstone Bridge," this design is commonly used as the test coil input circuit for eddy current instruments. In this application, it consists of two opposing "arms," each of which contains a coil and is capable of sensing very small impedance differences between the two arms.
Coil in air—A surface coil that is not in proximity to and is therfore unaffected by conducting material.
Conductance—The ability of a particular component to conduct electricity. Conductance depends on a component's conductivity, length, and cross section.
Conductivity—The opposite of resistance, that is, the relative "willingness" of a material to allow the flow of current.
Conductor—A material that contains very few (e.g., one or two) electrons in its outer shell. When there are only a few electrons in the outer orbit, it is easy to cause electrons to move from one atom to the next.
Cross-axis coil—A test coil design in which the coil assembly consists of two coils wound 90° to each other in order to make the coil assembly less directional in sensitivity.
Current—The flow of electrical charges, measured in amperes.
Differential coil technique—An eddy current test in which two opposing arms of the instrument's bridge are each connected to a separate test coil that is interfacing with the test material. Also called the differential self-comparison technique.
Direct current—The flow of electricity in only one direction.
Domains—Miniature magnets consisting of groups of atoms or molecules present within a material's individual grains.
Eddy currents—Circulating electrical currents induced in an isolated conductor by an alternating magnetic field.
Effective depth of penetration—The maximum material depth from which a displayable eddy current signal can be obtained, arbitrarily defined as the depth at which eddy current density has decreased to 5% of the surface eddy current density.
Electricity—The flow of electrons through a conductor from one atom to the next.
Electromagnet—A magnet operating on the principle of electromagnetism, usually consisting of a solenoid coil containing a ferromagnetic core. Electromagnets are distinguished from permanent magnets by the fact that they require electricity in order to operate.

Electromagnetic induction—Relative motion between a magnetic field and a conductor causes a voltage to be induced in that conductor.

Electromagnetism—The phenomenon whereby passage of electrons through a conductor causes a magnetic field to develop concentrically around the conductor, perpendicular to it.

Electron—A negatively charged particle of electrical energy that orbits the atom.

Empty coil—An encircling coil which is not in proximity to and is therfore unaffected by conducting material.

Encircling coil—An eddy current test coil designed so that the test material can pass through its interior wall. It is used primarily for production testing of long, continuous product such as pipe, tube, rod, wire, and bar stock.

External reference coil—A type of absolute coil bridge connection where the bridge's reference coil is sensing a reference standard.

Feed-through coil—See *encircling coil*.

Ferromagnetism—A property whereby materials can become magnetized when their domains have become aligned

Fill factor—The calculated coupling effectiveness between the inner surface of an encircling coil and the outer surface of a specimen enclosed within it or between the outer diameter of an internal coil and the inner surface of a specimen surrounding it.

Flux—See *magnetic flux*.

Flux density—The number of flux lines per unit area perpendicular to the direction of current flow.

Gap probe—See *horseshoe probe*.

Gauss—The unit of measure for flux density. One gauss equals one line of force passing through an area of one square centimeter.

Horseshoe probe—A surface probe configured with the coil wrapped around a horseshoe-shaped permanent magnet, to provide an eddy current field circulating perpendicular to the test surface.

I.D. (inner diameter) coil—See *internal coil*.

Inductance—The property of an electical component, such as an eddy current coil, whereby the component's alternating electromagnetic field induces voltage into the component itself or into a nearby secondary conducting component such as a test specimen.

Inductive reactance—Opposition that induced voltage offers to the alternation of alternating current.

International annealed copper standard (IACS)—A scale used to compare material conductivities with pure, unalloyed, annealed copper at 20°F, as the base value at 100% conductivity. Other materials are assigned percentage values relative to copper.

Impedance—The vector sum of inductive reactance and resistance.

Internal coil—An eddy current test coil designed to pass through the interior walls of tubes and pipes. It is used primarily for in-service inspection of tubing.

Lenz's law—The phenomenon whereby, when a primary circuit induces voltage into a secondary circuit; the direction of current flow in the secondary will be such that the polarity of the secondary flux will be opposite to the primary flux.

Lift-off—The variation in impedance as the distance between a probe coil and a conductor is varied. Lift-off can be a positive effect because it can be used to measure the thickness of nonconductive substances located between the coil and a conductor; but it is often a negative effect because its strong signals can mask the weaker indications of variables that an eddy current examination is intended to measure.

Locus (plural: loci)—A mathematical term indicating a set of points that satisfy a given set of conditions. For eddy current purposes, a set of points that trace a line or curve representing the range of a displayed variable such as conductivity or lift-off.

Magnetic field—A force field radiating from permanent magnets and electromagnets, which decreases in strength inversely as the square of the distance from the poles of the magnet.

Magnetic flux—The lines of force that make up a magnetic field. In the case of an eddy current test coil, which is an electromagnet, the lines of force flow out of one end (pole) of the coil, around the outside of the coil, and back into the other end (pole) of the coil.

Magnetism—The mechanical force of attraction or repulsion that one material can exert upon another.

Normalization—Confirmation to a standard. The impedance plane is normalized to preclude the need to develop separate impedance graphs for each test coil that is used. The vertical axis is normalized by dividing actual inductive reactance of the test coil by inductive reactance of the coil when remote from conductive material. The horizontal axis is normalized by subtracting test coil resistance from total resistance and then dividing the remainder by inductive reactance of the coil when remote from conductive material.

Nucleus—The central portion of the atom, containing protrons, which carry a positive charge and thus balance the negative charge of the orbiting electrons, as well as neutrons, which carry no charge.

Ohm's law—A formula defining the relationship of current, voltage and resistance, as follows:

$$\text{current} = \frac{\text{voltage}}{\text{resistance}}$$

Permanent magnets—Materials that retain their magnetism after they are removed from a magnetic field.

Permeability—The measure of a material's ability to be magnetized; that is, a material's ability to concentrate magnetic flux.

Polarity—The phenomenon whereby opposite ends of a magnet exhibit opposing forces. This is due to the fact that the lines of force flow from the north to the south pole around the outside of a magnet, and from the south to the north pole within the magnet. This causes like poles to repel and unlike poles to attract.

Poles—Nomenclature for the opposite ends of a magnet, one north and one south.

Probe coil—An eddy current test coil, configured in a probe-type housing, designed to be placed in contact with a surface of the test material.

Reflection coil—A coil assembly in which a large outer coil induces primary flux into the test specimen and a pair of stacked inner coils, wound in opposition, sense the secondary flux. The term "reflection" derives from the fact that the inner coils, being wound in opposition, function as electrical mirror images of each other.

Resistance—Opposition to the flow of current. Its unit of measure is the ohm.

Resistivity—The opposition of a material's atoms to the flow of electricity. It is the inverse of conductivity.

Shielded coil—An eddy current surface coil whose turns are surrounded by a cylinder of ferromagnetic material that concentrates the coil's flux field to suppress lateral extension of the field.

Sinusoid—One complete 360° cycle of a sine curve.

Skin depth—See *standard depth of penetration*.

Skin effect—Eddy current density is maximum at the material surface and decreases exponentially with depth. Thus, in thicker materials, eddy current testing operates only on the outer "skin" of the test material. Test sensitivity decreases rapidly with depth, and volumetric tests are possible only on thin specimens.

Standard depth of penetration—The material depth at which eddy current density has decreased to $1/e$ (36.8%) of surface eddy current density.

Self-induction—Induction of voltage into a conductor, such as a test coil, resulting from the coil's own magnetic field expanding and collapsing against the turns of the coil.

Solenoid—A coil of conducting wire whose length substantially exceeds its width.

Surface coil—An eddy current test coil, often configured in a probe-type housing, designed to be placed in contact with a surface of the test material.

Through-transmission—A technique whereby test coils are positioned on opposite sides of the test material, with one coil transmitting and the other coil receiving. This technique increases penetration of eddy currents, but does not provide display of phase information.

Voltage—The polarity applied to electrons, causing current to flow through a conductor. The unit of measure is the volt.

8.10 SUGGESTIONS FOR FURTHER READING

V. S. Cecco, G. Van Drunen, and F. L. Sharp. *Advanced Manual For Eddy Current Method.* Canadian General Standards Board, Ottawa, Canada, 1986.

V. S. Cecco, S. P. Sullivan, J. R. Carter, and L. S. Obrutsky. *Innovations in Eddy Current Testing.* Engineering Technologies Division, Nondestructive Testing Development Branch, Chalk River Laboratories,Chalk River, Ontario, 1995.

R. C. McMaster, P. McIntire, and M. L. Mester. *Nondestructive Testing Handbook,* Second Edition. American Society for Nondestructive Testing, Inc., Columbus, Ohio, 1986.

Robert C. McMaster, *Nondestructive Testing Handbook, First Edition,* American Society for Nondestructive Testing, Inc., Columbus, Ohio, 1959.

D. J. Hagemaier. *Fundamentals of Eddy Current Testing.* American Society for Nondestructive Testing, Inc., Columbus, Ohio, 1990.

ASM International Handbook Committee. Metals Handbook, Ninth Edition, Volume 17, *Nondestructive Evaluation and Quality Control.* ASM International, Metals Park, Ohio, 1989.

Eddy Current Classroom Training Handbook, Second Edition. American Society for Nondestructive Testing, Inc., Columbus, Ohio, 1979.

H. L. Libby. *Introduction to Electromagnetic Nondestructive Test Methods.* Robert E. Krieger Publishing Company, Huntington, New York, 1979.

W. J. McGonnagle. *Nondestructive Testing,* Second Edition. Gordon and Breach, New York, 1961.

B. Hull and V. John. *Nondestructive Testing.* Macmillan Education Ltd., London, England, 1988.

P. E. Mix. Introduction to *Nondestructive Testing.* Wiley, New York, 1987.

D. Lovejoy. *Magnetic Particle Inspection.* Chapman & Hall, London, England, 1993.

B. Grob. *Basic Electronics,* Fourth Edition. Gregg Division, McGraw-Hill, New York, 1977.

CHAPTER 9
THERMAL INFRARED TESTING

1. HISTORY AND DEVELOPMENT

Humans have always been able to detect infrared radiation. The nerve endings in human skin can respond to temperature differences of as little as 0.0162°F (0.009°C). Although extremely sensitive, nerve endings are poorly designed for thermal nondestructive evaluation. Even if humans had the thermal capabilities of a pit viper, which can find its warm-blooded prey in the dark, it is most probable that better heat detection tools would still be needed. Thus, as inventive beings, we have turned to mechanical and electronic devices to allow us to become hypersensitive to heat. These devices, some of which can produce thermal images, have proved invaluable for thermal inspection in countless applications.

Sir William Herschel, who in 1800 was experimenting with light, is generally credited with the beginnings of thermography. Using a prism to break sunlight into its various colors, he measured the temperature of each color using a very sensitive mercury thermometer. Much to his surprise, the temperature increased when he moved out beyond red light into an area he came to term the "dark heat." This is the region of the electromagnetic spectrum referred to as infrared and recognized as the electromagnetic radiation that when absorbed causes a material to increase in temperature.

Twenty years later, Seebeck discovered the thermoelectric effect, which quickly lead to the invention of the thermocouple by Nobili in 1829. This simple contact device is based on the premise that there is an emf (electromotive force) or voltage that occurs when two dissimilar metals come in contact, and that this response changes in a predictable manner with a change in temperature. Melloni soon refined the thermocouple into a thermopile (a series arrangement of thermocouples) and focused thermal radiation on it in such a way that he could detect a person 30 feet away. A similar device, called a bolometer, was invented 40 years later. Rather than measuring a voltage difference, a bolometer measured a change in electrical resistance related to temperature. In 1880 Longley and Abbot used a bolometer to detect a cow over 1,000 feet away!

Herschel's son, Sir John, using a device called an evaporograph, produced the first infrared image, crude as it was, in 1840. The thermal image was caused by the differential evaporation of a thin film of oil. The image was viewed by light reflecting off the oil film. During World War I, Case became the first to experiment with photoconducting detectors (thallium sulfide) that produced signals not as a result of being heated, but by their direct interaction with photons. The result was a faster, more sensitive detector. During World War II, the technology began to expand and resulted in a number of military applications and developments. The discovery by German scientists that cooling the detector increased performance was instrumental in the rapid expansion of the infrared technology.

It was not until the 1960s that infrared thermal imaging began to be used for nonmilitary applications. Although systems were cumbersome, slow to acquire data, and had poor resolution, they proved useful for the inspection of electrical systems. Continued advances in the 1970s, again driven mainly by the military, produced the first portable systems usable for thermal nondestructive testing (TNDT).

These systems, utilizing a cooled scanned detector, proved both rugged and reliable. However, the quality of the image, although often adequate, was poor by today's standards. By the end of the decade, infrared was being widely used in mainstream industry, for building inspections, and for a variety of medical applications. It became practical to calibrate systems and produce fully radiometric images, meaning that radiometric temperatures could be measured throughout the image. Adequate detector cooling, which had previously been accomplished through cryogenics—using either compressed or liquefied gases—was now done much more conveniently using thermoelectric coolers. Less expensive tube-based pyroelectric vidicon (PEV) imaging system were also developed and produced. Although not radiometric, PEVs were lightweight, portable and could operate without cooling.

In the late 1980s, a new technology, the focal plane array (FPA), was released from the military into the commercial marketplace. The FPA employs a large array of thermally sensitive semiconductor detectors, similar to those in charge coupled device (CCD) visual cameras. This was a significant improvement over the single-element, scanned detector and the result was a dramatic increase in image quality and spatial resolution. Arrays of 320 × 240 and 256 × 256 elements are now the norm, and for specialized applications, arrays are available with densities up to 1000 × 1000.

Development of the FPA technology has exploded in the last decade. Both long- and short-wave sensing systems are now available in fully radiometric versions, as are systems with data capture rates as high as 500 frames per second and sensitivities of 0.18°F (0.1°C) or less. While the cost of radiometric systems has not dropped significantly, the quality has increased dramatically. The digital storage of data has allowed 12 and 14-bit digital files; the result is access to far more image and radiometric detail than ever imagined. Figure 9-1, taken with a typical FPA system, shows the clarity of detail available and its appeal for specialized work like this biological study of thermoregulation in ravens. (Because this book is printed with black ink only, much of the information conveyed by the original color images has been lost. Interested readers may obtain the original images from the author's website, www. snellinfrared.com.)

Although not radiometric, pyroelectric FPA systems have also been developed that provide excellent imagery at prices below $15,000. Costs have dropped so much that a fixed-lens pyroelectric system is now being installed in some model year 2001, luxury automobiles to enhance the safety of driving at night.

Concurrently, the use of computers and image processing software has grown tremendously. Nearly all systems commercially available today offer software programs that facilitate analysis and report writing. Reports can be stored digitally and sent electronically over the Internet.

With so many advances being made so rapidly, it is difficult to imagine what is next. Certainly, we will see continued development of new detector systems, most notably the quantum well integrated processing (QWIP) systems, with a promise of higher data capture speeds and greater sensitivities. "Tunable" sensors capable of capturing and processing images at several wavelengths will also become commercially available. Probably the greatest area of growth will be in the development of automated systems in which the infrared instrument, combined with a data processing computer, will capture and analyze data and provide control feedback to a process or system being monitored. Systems are already in use for certain processes in the cement, metals, and chemical industries and have significantly reduced failures and costs, while dramatically increasing quality and profits.

As equipment has improved, so has the need to qualify inspection personnel and stan-

FIGURE 9-1.

dardize inspection methodologies. In the late 1980s, members of the American Society for Nondestructive Testing (ASNT) met to discuss a strategy for having infrared adopted as a standard test method. ASNT, a volunteer professional organization, is responsible for developing personnel qualification programs. Certification, which is conducted by the employer, is based on the practitioner having had the required training and experience, as well as passing written and practical examinations.

In 1992, infrared testing was officially adopted by ASNT and thermographers could for the first time be certified to a widely used, professionally recognized standard. Since that time, a number of large companies in the United States have begun certifying their thermographers in compliance with ASNT recommendations.

The development of inspection standards over the past two decades has also been an interesting process. The American Society of Heating, Refrigeration, and Air Conditioning Engineers (ASHRAE) developed a building inspection standard, which was later modified and adopted by the International Standards Organization (ISO). The American Society for Testing and Materials (ASTM) also developed several standards used for determining the performance of infrared systems, as well as standards for roof moisture and bridge deck inspections. Concurrently, a number of relevant ASTM standards were developed for radiation thermometry, which also has application to infrared thermography. The National Fire Protection Association (NFPA) has developed a standard, NFPA 70-B—the maintenance of electrical systems, which incorporates the use of thermography. Even though inspection standards continue to be developed, there are many applications at this time that lack inspection standards. Without standards, inspection results may lack repeatability, which often leads to misleading, perhaps even dangerous, analysis of data.

Access to the technology has become easy as the value of products has improved. Companies investing in developing solid programs, including inspection procedures and qualified personnel, have a distinct advantage due to access to the remarkable long-term benefits provided by infrared.

2. THEORY AND PRINCIPLES

To use today's infrared equipment, it is essential to understand the basics of both heat transfer and radiation physics. As powerful as modern equipment is, for the most part, it still cannot think for itself. Its value, therefore, depends on the thermographer's ability to interpret the data, which requires a practical understanding of the basics.

Thermal Energy

Energy can be changed from one form to another. For instance, a car engine converts the chemical energy of gasoline to thermal energy. That, in turn, produces mechanical energy, as well as electrical energy for lights or ignition, and heat energy for the defroster or air conditioner. During these conversions, although the energy becomes more difficult to harness, none of it is lost. This is the First Law of Thermodynamics. A byproduct of nearly all energy conversions is heat or thermal energy.

When there is a temperature difference between two objects, or when an object is changing temperature, heat energy is transferred from the warmer areas to the cooler areas until thermal equilibrium is reached. This is the Second Law of Thermodynamics. A transfer of heat energy results either in electron transfer or increased atomic or molecular vibration.

Heat energy can be transferred by any of three modes: conduction, convection, or radiation. Heat transfer by conduction occurs primarily in solids, and to some extent in fluids, as warmer molecules transfer their energy directly to cooler, adjacent ones. Convection takes place in fluids and involves the mass movement of molecules. Radiation is the transfer of energy between objects by electromagnetic radiation. Because it needs no transfer medium, it can take place even in a vacuum.

Transfer of heat energy can be described as either steady-state or transient. In the steady-state condition, heat transfer is constant and in the same direction over time. A fully warmed-up machine under constant load transfers heat at a steady-state rate to its surroundings. In reality, there is no such thing as true steady-state heat flow! Although we often ignore them, there are always small transient fluctuations. A more accurate term is quasi-steady-state heat transfer. When heat transfer and temperatures are constantly and significantly changing with time, heat flow is said to be transient. A machine warming up or cooling down is an example. Because thermographers are often concerned with the movement of heat energy, it is vital to understand what type of heat flow is occurring in a given situation.

Heat energy is typically measured in British thermal units (Btu) or calories (c). A Btu is defined as the amount of energy needed to raise the temperature of one pound of water one degree Fahrenheit. A calorie is the amount of heat energy needed to raise the temperature of one gram of water one degree Celsius. One wooden kitchen match, burned entirely, gives off approximately one Btu or 252 calories of heat energy.

The units to describe the energy content of food are also termed "calories." These calories are actually kilocalories (Kcal or C) and are equal to one thousand calories (c). Various devices, such as a calorimeter or a heat flow meter can measure the flow of heat energy. These devices are not commonly used, except by scientists in research laboratories.

Temperature is a measure of the relative "hotness" of a material compared to some known reference. There are many ways to measure temperature. The most common is to use our sense of touch. We also use comparisons of various material properties, including expansion (liquid and bimetal thermometers), a change in electrical voltage (thermocouple), and a change in electrical resistance (bolometers). Infrared radiometers infer a temperature measurement from detected infrared radiation.

Regardless of how heat energy is transferred, thermographers must understand that materials also change temperatures at different rates due to their thermal capacitance. Some materials, like water, heat up and cool down slowly, while others, like air, change temperature quite rapidly. The thermal capacitance or specific heat of a material describes this rate of change. Without an understanding of these concepts and values, thermographers will not be able to properly interpret their findings, especially with regard to transient heat flow situations. Although potentially confusing, these properties can also be used to our advantage. Finding liquid levels in tanks, for example, is possible because of the differences between the thermal capacitance of the air and the liquid.

Latent Heat

As materials change from one state or phase (solid, liquid or gas) to another, heat energy is released or absorbed. When a solid changes state to a liquid, energy is absorbed in order to break the bonds that hold it as a solid. The same thing is true as a liquid becomes a gas; energy must be added to break the bonds. As gases condense into liquids, and as liquids freeze into solids, the energy used to maintain these high-energy states is no longer needed and is released.

This energy, which can be quite substantial, is called latent energy because it does not result in the material changing temperature. The impact of energy released or absorbed during phase change often affects thermographers. The temperature of a roof surface, for instance, can change very quickly as dew or frost forms, causing problems during a roof moisture survey. A wet surface or a rain-soaked exterior wall will not warm up until it is dry, thus masking any subsurface thermal anomalies. On the positive side, state changes enable thermographers to see thermal phenomena, such as whether or not solvents have been applied evenly to a surface.

Conduction

Conduction is the transfer of thermal energy from one molecule or atom directly to another adjacent molecule or atom with which it is in contact. This contact may be the result of physical bonding, as in solids, or a momentary collision, as in fluids. Fourier's law describes how much heat is transferred by conduction:

$$Q = \frac{k}{L \times A \times \Delta T}$$

where
- Q = heat transferred
- k = thermal conductivity
- L = thickness of materials
- A = area normal to flow
- ΔT = temperature difference

The thermal conductivity (k) is the quantity of heat energy that is transferred through one square foot of a material, which is one inch thick, during one hour when there is a one-degree temperature difference across it. The metric equivalent (in watts) is W/(m × °C) and assumes a thickness of one meter.

Materials with high thermal conductivities, such as metals, are efficient conductors of heat energy. We use this characteristic to our advantage by making such things as cooking pans and heat sinks from metal. Differences in conductivity are the basis for many thermographic applications, especially the evaluation of flaws in composite materials or the location of insulation damage.

Materials with low thermal conductivity values, such as wool, fiberglass batting, and expanded plastic foams, do not conduct heat energy very efficiently and are called insulators. Their insulating value is due primarily to the fact that they trap small pockets of air, a highly inefficient conductor.

The term R-value, or thermal resistance, is a measure of the resistance to conductive heat flow. It is defined, as the inverse of conductivity, or $1/k$. R-value is a term that is generally used when describing insulating materials.

The following are the thermal conductivities and R-values of some common materials:

Material	k (Btu × in/ft² × hr × °F)*	R (ft² × hr × °F/Btu × in)
Extruded polystyrene (1")	0.20	5.0
Fiberglass batts (1")	0.32	3.125
Soft wood (average)	1.0	1.0
Brick (average)	10.0	.1
Concrete (average)	13.0	.07
Steel	314.0	.003
Cast iron	331.0	.0036
Copper	2724.0	.00036

*Metric units for k are W/m × °C.

Another important material property is thermal diffusivity. Thermal diffusivity is the rate at which heat energy moves throughout the volume of a material. Diffusivity is determined by the ratio of the material's thermal conductivity to its thermal capacitance. Differences in diffusivity and consequent heat flow are the basis for many active thermography applications in TNDT.

Convection

Heat energy is transferred in fluids, either gases or liquids, by convection. During this process, heat is transferred by conduction from one molecule to another and by the subsequent mixing of molecules.

In natural convection, this mixing or diffusing of molecules is driven by the warmer (less dense) molecules' tendency to rise and be replaced by more dense, cooler molecules.

Cool cream settling to the bottom of a cup of hot tea is a good example of natural convection. Forced convection is the result of fluid movement caused by external forces such as wind or moving air from a fan. Natural convection is quickly overcome by these forces, which dramatically affect the movement of the fluid. Figure 9-2 shows the typical, yet dramatic, pattern associated, in large part, with the cooling effect of convection on a person's nose.

Newton's Law of Cooling describes the relationship between the various factors that influence convection:

$$Q = h \times A \times \Delta T$$

where
 Q = heat energy
 h = coefficient of convective heat transfer
 A = area
 ΔT = Temperature difference

The coefficient of convective heat transfer is often determined experimentally or by estimation from other test data for the surfaces and fluids involved. The exact value depends on a variety of factors, of which the most important are velocity, orientation, surface condition, geometry, and fluid viscosity.

Changes in h can be significant due merely to a change in orientation. The topside of a horizontal surface can transfer over 50% more heat by natural convection than the underside of the same surface.

In both natural and forced convection, a thin layer of relatively still fluid molecules

FIGURE 9-2.

adheres to the transfer surface. This boundary layer, or film coefficient, varies in thickness depending on several factors, the most important being the velocity of the fluid moving over the surface. The boundary layer has a measurable thermal resistance to conductive heat transfer. The thicker the layer, the greater the resistance. This, in turn, affects the convective transfer as well. At slow velocities, these boundary layers can build up significantly. At higher velocities, the thickness of this layer and its insulating effect are both diminished.

Why should thermographers be concerned with convection? As forced convection, such as the wind, increases, heat transfer increases and can have a significant impact on the temperature of a heated or cooled surface. Regardless of velocity, this moving air has no affect on ambient surfaces. Thermographers inspect a variety of components where an increase in temperature over ambient is an indication of a potential problem. Forced convection is capable of masking these indications.

Radiation

In addition to heat energy being transferred by conduction and convection it can also be transferred by radiation. Thermal infrared radiation is a form of electromagnetic energy similar to light, radio waves, and x-rays. All forms of electromagnetic radiation travel at the speed of light, 186,000 miles/second (3×10^{-8} meters/second). All forms of electromagnetic radiation travel in a straight line as a waveform; they differ only in their wavelength. Infrared radiation that is detected with thermal imaging systems has wavelengths between approximately 2 and 15 microns (μm). Electromagnetic radiation can also travel through a vacuum, as demonstrated by the sun's warming effect from a distance of over 94 million miles of space.

All objects above absolute zero radiate infrared radiation. The amount and the exact wavelengths radiated depend primarily on the temperature of the object. It is this phenomenon that allows us to see radiant surfaces with infrared sensing cameras.

Due to atmospheric absorption, significant transmission through air occurs in only two "windows" or wavebands: the short (2–6 μm) and long (8–15 μm) wavebands. Both can be used for many thermal applications. With some applications, one waveband may offer a distinct advantage or make certain applications feasible. These situations will be addressed in subsequent sections.

The amount of energy emitted by a surface depends on several factors, as shown by the Stefan–Boltzmann formula:

$$Q = \sigma \times \varepsilon \times T^4 \text{ absolute}$$

where
Q = energy transmitted by radiation
σ = the Stefan–Boltzmann constant (0.1714×10^{-8} Btu/hr $\times$ ft^2 $\times$ R^4)
ε = the emissivity value of the surface
T = the absolute temperature of the surface

When electromagnetic radiation interacts with a surface several events may occur. Thermal radiation may be reflected by the surface, just like light on a mirror. It can be absorbed by the surface, in which case it often causes a change in the temperature of the surface. In some cases, the radiation can also be transmitted through the surface; light passing through a window is a good example. The sum of these three components must equal the total amount of energy involved.

This relationship, known as the conservation of energy, is stated as follows:

$$R + A + T = 1$$

where
R = Reflected energy
A = Absorbed energy
T = Transmitted energy

Radiation is never perfectly transmitted, absorbed, or reflected by a material. Two or three phenomena are occurring at once. For example, one can see through a window (transmission) and also see reflections in the window at the same time. It is also known that glass absorbs a small portion of the radiation because the sun can cause it to heat up. For a typical glass window, 92% of the light radiation is transmitted, 6% is reflected, and 2% is absorbed. One hundred percent of the radiation incident on the glass is accounted for.

Infrared radiation, like light and other forms of electromagnetic radiation, also behaves in this way. When a surface is viewed, not only radiation that has been absorbed may be seen, but also radiation that is being transmitted through the target and/or reflected by it. Neither the transmitted nor reflected radiation provides any information about the temperature of the surface.

The combined radiation reflecting from a surface to the infrared system is called its radiosity. The job of the thermographer is to distinguish the emitted component from the others so that more about the target temperature can be understood.

Only a few materials transmit infrared radiation very efficiently. The lens material of the camera is one. Transmissive materials can be used as thermal windows, allowing viewing into enclosures. The atmosphere is also fairly transparent, at least in two wavebands. In the rest of the thermal spectrum, water vapor and carbon dioxide absorb most thermal radiation. As can be seen from Figure 9-3, radiation is transmitted quite readily in both the short (2–6 μm) and long (8–14 μm) wavebands. Infrared systems have been optimized to one of these bands or the other. Broadband systems are also available and have some response in both wavebands.

A transmission curve for glass would show us that glass is somewhat transparent in the short waveband and opaque in the long waveband. It is surprising to try to look thermally through a window and not be able to see much of anything!

Many thin plastic films are transparent in varying degrees to infrared radiation. A thin plastic bag may be useful as a camera cover in wet weather or dirty environments. Be aware, however, that all thin plastic films are not the same! While they may look similar,

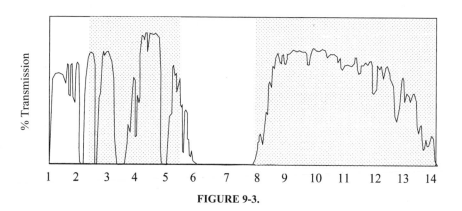

FIGURE 9-3.

it is important to test them for transparency and measure the degree of thermal attenuation. Depending on the exact atomic makeup of the plastic, they may absorb strongly in very narrow, specific wavebands. Therefore, to measure the temperature of a thin plastic film, a filter must be used to limit the radiation to those areas where absorption (and emission) occurs.

The vast majority of materials are *not* transparent. Therefore, they are opaque to infrared radiation. This simplifies the task of looking at them thermally by leaving one less variable to deal with. This means that the only radiation we detect is that which is reflected and absorbed by the surface ($R + A = 1$).

If $R = 1$, the surface would be a perfect reflector. Although there are no such materials, the reflectivity of many polished shiny metals approaches this value. They are like heat mirrors. Kirchhoff's law says that for opaque surfaces the radiant energy that is absorbed must also be reemitted, or $A = E$. By substitution, it is concluded that the energy detected from an opaque surface is either reflected or emitted ($R + E = 1$). Only the emitted energy provides information about the temperature of the surface.

In other words, an efficient reflector is an inefficient emitter, and vice versa. For thermographers, this simple inverse relationship between reflectivity and emissivity forms the basis for interpretation of nearly all of that is seen. Emissive objects reveal a great deal about their temperature. Reflective surfaces do not. In fact, under certain conditions, very reflective surfaces typically hide their true thermal nature by reflecting the background and emitting very little of their own thermal energy.

If $E = 1$, all energy is absorbed and reemitted. Such an object, which exists only in theory, is called a blackbody. Human skin with an emissivity of 0.98, is nearly a perfect blackbody, regardless of skin color.

Emissivity is a characteristic of a material that indicates its relative efficiency in emitting infrared radiation. It is the ratio of thermal energy emitted by a surface to that energy emitted by a blackbody of the same temperature. Emissivity is a value between zero and one. Most nonmetals have emissivities above 0.8. Metals, on the other hand, especially shiny ones, typically have emissivities below 0.2. Materials that are not blackbodies—in other words everything!—are called real bodies. Real bodies always emit less radiation than a blackbody at the same temperature. Exactly how much less depends on their emissivity.

Several factors can affect what the emissivity of a material is. Besides the material type, emissivity can also vary with surface condition, temperature, and wavelength. The emittance of an object can also vary with the angle of view.

It is not difficult to characterize the emissivity of most materials that are not shiny metals. Many of them have already been characterized, and their values can be found in tables such as Table 9-2. These values should be used only as a guide. Because the exact

TABLE 9.2 Emissivity Values*

Human skin	0.98
Black paint (flat)	0.90
White paint (flat)	0.90
Paper	0.90
Lead, oxidized	0.40
Copper, oxidized to black	0.65
Copper, polished	0.15
Aluminum, polished	0.10

*Values will vary with exact surface type and wavelength.

emissivity of a material may vary from these values, skilled thermographers also need to understand how to measure the actual value.

It is interesting to note that cracks, gaps, and holes emit thermal energy at a higher rate than the surfaces around them. The same is true for visible light. The pupil of your eye is black because it is a cavity, and the light that enters it is absorbed by it. When all light is absorbed by a surface, we say it is "black." The emissivity of a cavity will approach 0.98 when it is seven times deeper than it is wide.

From an expanded statement of the Stefan–Boltzmann law, the impact that reflection has on solving the temperature problem for opaque materials can be seen:

$$Q = \sigma \times \varepsilon \times T4 + (\boldsymbol{\sigma \times (1 - \varepsilon) \times T^{4\ \text{background}}})$$

The second part of the equation (in boldface) represents that portion of the radiosity that comes from the reflected energy. When using a radiometric system to make a measurement, it is important to characterize and account for the influence of the reflected background temperature.

Consider these two possible scenarios:

- When the object being viewed is very reflective, the temperature of the reflected background becomes quite significant.
- When the background is at a temperature that is extremely different from the object being viewed, the influence of the background becomes more pronounced.

It becomes clear that repeatable, accurate radiometric measurements can be made only when emissivities are high. This is a fundamental limitation within which all thermographers work. Generally, it is not recommended to make temperature measurements of surfaces with emissivities below approximately 0.50, in other words all shiny metals, except under tightly controlled laboratory conditions. However, with a strong understanding of how heat energy moves in materials and a working knowledge of radiation, the value of infrared thermography as a noncontact temperature measurement tool for nondestructive evaluation is remarkable.

3. EQUIPMENT AND ACCESSORIES

Infrared systems can generally be classified as either thermal imaging systems or point measuring systems. Thermal imaging systems can be further subdivided into quantitative or radiometric (measuring temperatures) and qualitative or nonradiometric systems (thermal images only). Point measuring systems are also often termed "spot" or "point" radiometers and can be either hand-held or fixed mounted.

Qualitative thermal imaging systems are useful when radiometric temperatures are not required. Using these lower-cost instruments, it is possible to make thermal comparisons and thus distinguish thermal anomalies in equipment and materials.

Quantitative thermography involves the use of calibrated instruments. By using correction factors, accurate radiometric temperatures can be determined from the thermal image. Calibration, which is expensive to achieve or maintain, is best made for the specific temperature ranges and accuracy required.

Nonimaging radiometric measuring systems, called point or spot radiometers, are lower-cost instruments, which, although they do not give an image, can be valuable for many types of NDT. For limited areas and pieces of equipment, and when used properly, this equipment can provide accurate, real-time radiometric measurement across a wide range of temperatures. Typically, spot radiometers have emissivity correction capabilities.

Without an image, it can be difficult knowing the exact area the spot radiometer is measuring, possibly resulting in erroneous data. Some models include a laser-aiming device to help aim them. It should be carefully noted, however, that the area being measured is always much larger than the small spot illuminated by the laser. Newer models are available with a projected laser beam that also outlines the exact measurement area, alleviating any misunderstanding.

The waveband response of spot system detectors is typically wide enough to allow them to be filtered to detect narrow wavebands for specialized applications. This is particularly useful for such applications as measuring temperatures of thin plastic films or glass during manufacturing. Spot systems can also be fixed or mounted to constantly monitor process temperatures unattended. The data is typically fed back to a computer or control system. Such thermal data can be correlated not only to temperature, but also thickness, moisture content, material integrity, material type, or parts presence detection. When combined with visual automated systems, a powerful information system is available.

Even though spot radiometers are much less expensive than thermal imaging systems, they should not be thought of as a substitute for an imaging system. Both types of systems have their place in most NDT programs.

A variation of the spot radiometer is the line scanner. It is also very useful for many fixed mounted NDT inspections. This instrument employs a detector or small array of detectors that is then scanned along a line. If the line scanner views a moving object, an image can be created. Line scanners are particularly useful for viewing equipment that is moving at a constant rate because the image is built up one line at a time as the product passes by. Two common applications are viewing the refractory in rotating limekilns or the cross-roll moisture profile on the sheet in a paper machine. Line scanners, which can produce very high-quality images with excellent radiometrics, are less expensive than full imaging systems and lend themselves well to these types of automated inspections.

In the history and development section, it was noted that over the years three main types of infrared imaging systems were developed: scanners or scanning systems, pyroelectric vidicons (PEVs), and focal plane arrays (FPAs). While all three types are still in use today, the overwhelming majority of new equipment sales are the FPA systems.

It is useful to have a general understanding of how imaging systems work. The purpose of the imager is to detect the infrared radiation given off by the object being inspected, commonly called the target. The target's radiosity is focused by the thermally transparent optics onto a detector, which is sensitive to this invisible "heat" radiation. A response from the detector produces a signal, usually a voltage or resistance change, which is read by the electronics of the system and creates a thermal image on a display screen, typically a video viewfinder. In this seemingly simple process, we are able, without contact, to view the thermal image, or thermogram, that corresponds to the thermal energy coming off the surface of the target.

The first infrared imaging systems used a technique called "scanning" to create a full screen image using a single, small detector. A series of rotating or oscillating mirrors or prisms inside the scanner direct the infrared radiation from a small part of the scene being viewed onto the detector. The detector response results in the display of a portion of the field of view. Very quickly, the mirror or prism moves to scan another part of the scene with the resultant display. This process of making a series of single measurements happens thousands of times per second, so that a complete image is built up and viewed in what appears to be "real time," i.e., at 30 frames per second or greater.

For some types of detectors to respond quickly, accurately, and in a repeatable fashion (particularly photon detectors), they must be cooled to extremely low temperatures. Systems with the highest-quality images use liquid nitrogen (LN2) at $-320.8°F$ ($-196°C$). The nitrogen is simply poured into an insulated container, called a "Dewar," that surrounds the detector. Under normal conditions, the detector will stay cold for between 2–4 hours before it must be refilled. Other cooling systems have employed a cylinder of very high pressure (6000 psi) argon gas as a coolant. As it is slowly released, the gas expands and cools the detector to –292°F (–180°C). Obviously, great care must be used in handling cryogenics. Because of these safety problems and the problems of dealing with cryogenic substances in the field, cryogenically cooled systems are seldom used anymore except in laboratory situations. Several types of noncryogenic cooling schemes have replaced them. While noncryogenic devices require a fair amount of power and have a limited life, their benefits make them much more attractive than cryogenic devices.

Historically, pyroelectric vidicon (PEV) designs have offered a lower-cost alternative to scanners, and many are still in use. While PEVs are not radiometric, they can produce a good quality image. In addition, they do not require cooling. The detector in a PEV system is a monolithic crystal that, when heated by the incoming radiation, changes polarity. An electron-scanning gun, similar to those used in older television cameras, reads this change, which then produces an electronic image of the thermal scene. PEVs are slower than scanners in their response by an order of magnitude. PEVs also need to be constantly moved or "panned," or have their signal chopped. At this time, PEV tubes are no longer being manufactured.

Focal plane array (FPA) systems, which use a matrix of detectors to sense the infrared radiation, have a much higher spatial resolution than scanners or PEV systems. Typical detectors have pixel resolutions of 256×256 or 320×240. Resolutions of 512×512 pixels may soon be the norm. Although early FPA systems had to be cooled and were only sensitive in the short waveband, uncooled, long-wave sensing systems are now available.

There are several types of FPA systems. Short-wave instruments typically use photon detectors. These are cooled detectors, which actually count the number of photons received. Photon detecting systems now have very reliable radiometrics and excellent imagery. Long-wave systems use thermal detectors, which do not have to be cooled, although they must be temperature stabilized. While some are classified as "bolometers" and others "pyroelectric" or "ferroelectric," in all cases, thermal detectors are actually heated up by the incoming radiation. In the former this results in a resistance change, whereas in the latter there is a change in polarization. Very reliable radiometrics and excellent imagery is now possible from bolometers, but pyroelectric systems, while producing good quality images, have proven extremely difficult to be used radiometrically.

Infrared systems vary widely in price. At this time, spot radiometer systems start at under $500. Simple pyroelectric FPA instruments are in the range of $10,000 to $20,000. A complete radiometric system with software and normal lens is typically priced between $45,000 and $55,000. Although specifications and configurations vary, all these systems are designed to be used in typical industrial environments.

Infrared imaging systems are generally made up of several common components, including lens, detector, processing electronics, controls, display, data storage, data processing and report generation software, and filter.

Lens

Lenses serve to focus the incoming infrared radiation on the detector. Some infrared lens materials are actually opaque to visible light. Materials commonly used for lenses are germanium (Ge), silicon (Si), and zinc selenide (ZnSe).

Lenses can be of different focal lengths. A normal lens (with a field of view from 16° to 25°, approximately) is useful for most applications. Where inspection space is limited or a wide view required, the use of a wide-angle lens is recommended. Over longer distances, such as when viewing a disconnect switch in a substation, a telephoto lens may be warranted. Lens selection also impacts spatial and measurement resolution.

Coatings used on some lenses may contain small amounts of radioactive material (thorium). When using the system in a nuclear power plant, make sure to test the system for its baseline characteristics prior to entry.

Detector

The radiation is focused on the detector, where it produces a measurable response. Materials commonly used for detectors are platinum silicide (PtSi), mercury cadmium telluride (HgCdTe), and indium antimonide (InSb).

Processing Electronics

The response from the detector is processed electronically to produce a thermal image, a temperature measurement, or both.

Controls

Various controls on the system allow for adjustments to control the input of infrared radiation or the output of data. These typically include adjustments to range and span, thermal level, polarity, emissivity, and other temperature measurement functions.

Display

The processed data is output to a display, either an electronic viewfinder or a liquid crystal display (LCD) screen. Most instruments can display the thermal image in either grayscale or a color scale. The option of displaying the image in color while working in the field is important. Typically, radiometric systems also feature some very powerful analysis functions in the instrument itself, such as spot or area measurement and isotherm display. Again, these features often make field work simpler and more effective and are generally recommended.

Data Storage

Data is typically stored either as a still digital image or as an analog video image. Digital images are stored on floppy discs or PC cards. Digital voice data can also be stored with an image on some systems as well. PC cards, depending on storage capacity, can hold up to 1000 images with accompanying voice data.

It is also possible to output digital data directly to a computer using an RS-232 port, or to control the camera remotely using the same data port. Most systems also include an output for a standard video signal or an "S-video" signal that can be accepted by any compatible videocassette recorder (VCR), either 8 mm or VHS.

Data Processing and Report Generation Software

Most of the software that is available today is both powerful and very easy to use. Digital images are imported into the computer directly from the PC card and may be displayed in grayscale or with a variety of color palettes. Various color palettes can be selected. Adjustments can be made to all radiometric parameters, such as emissivity, background temperature, span, and level. Analysis functions may include spot, area, isotherms, and line thermal measurement, as well as size measurements. Analysis can extend beyond the image by displaying the numerical data in a spreadsheet or in various standard graphical forms such as a histogram.

When an image has been analyzed and processed, notation labels can be added and the image inserted into a report template. From there, the completed report can be sent to a printer, stored electronically or sent via the Internet.

Filters

Many thermographic applications depend upon the use of specialized filters to obtain a useful image or measurement. Before using filters, it is important to know the exact spectral response of the system as determined by the detector and the lens material. Responses within the long or short wavebands can vary from system to system. It is also important to understand how the selected filter interacts with the detector's response.

There are three generic filter designations:

1. High-pass filters, which allow only shorter wavelengths to pass
2. Low-pass filters, which allow only longer wavelengths to pass
3. Band-pass filters, which allow only a designated band of wavelengths to pass

Among the commonly used infrared filters are:

1. Flame filters. These suppress the peak flame radiation by excluding all radiation except that in the 3.8 μm band (for SW systems) or 10.8 μm band (for LW systems). This enables viewing through flames to hot objects beyond.
2. Sun filters. These suppress the undesirable "glint" affects of solar radiation in SW systems by excluding wavelengths below 3 μm.
3. Glass filters:
    A. For short-wave systems one type of filter cuts off wavelengths below 4.8 μm. This allows the system to see the surface of the glass rather than through the glass.
    B. Another glass filter allows a short-wave system to see only through glass by limiting radiation to a 2.35 μm band-pass.
4. Filters for viewing thin plastic films come in various band passes, depending on the exact plastic that is being inspected. Many thin plastic films are very transmissive in both long and short wavelengths. In all cases, the filter limits the radiation detected to that narrow band at which the plastic is opaque (nontransparent). The result is that the surace of the plastic can be seen and measured thermally. For polyethylene, a 3.45 μm band pass filter is used, whereas for polyester a 7.9 μm band-pass filter is used.

To maintain radiometric accuracy, the system must be calibrated for each filter used. Some systems automatically load the correct calibration curves when the filter is installed.

Infrared imaging systems have become much smaller over the past few years. The need to push a cart full of accessory equipment is becoming a thing of the past! With digital data storage densities as high as they are, and with the advent of affordable digital video, data storage is far simpler as well. Several of the instrument manufacturers have developed systems that also allow for bar-code reading. Some systems now allow downloading of images and data from previous inspections onto the imager's PC card, thus allowing for the viewing of archived images in the field.

As with any high-quality measurement tool, it is important to regularly check the calibration of an infrared instrument. It is a good habit to do this before and after each inspection by making a simple check of the temperature of the tear ducts of a person's eyes, which should be between 93–95 °F. For more exacting needs, the use of a calibrated reference blackbody is suggested. This traceable device has a temperature-controllable, high-emissivity target. Set the device for the thermal level at which the work is to be performed, allow it to stabilize, and check the calibration of the infrared instrument. When instruments are out of calibration or in need of a periodic calibration, they should be returned to the manufacturer.

When accurate radiometric measurements are required, additional data may need to be collected at the time the image is captured. Without this data, results may be misinterpreted or incorrect. The following environmental data may be important to collect:

- Ambient air temperature
- Background temperature
- Relative humidity
- Distance to the object being inspected
- Information about wind or other convection
- Impact of the Sun
- Specific operational conditions of the target

In addition, it is necessary to measure or estimate the emissivity of the object. In many instances, especially where safety allows, it may also be important to modify the emissivity of the surface being inspected to make measurements possible.

4. TECHNIQUES

Thermographers have developed a number of very useful techniques for expanding the use of the technology. These techniques can consistently result in higher quality data that is often simpler to gather than in the past. While some development time is often required when applying these techniques, the potential returns usually far outweigh the investment.

Some techniques involve specialized tools or equipment, for instance, lasers used to heat the sample, while others are simply better ways to enhance the flow of heat energy into or out of a sample, such as wetting the surface of a tank to make the level visible.

The basic strategy used in many thermal applications is quite simply termed "comparative thermography." It is no more complex than comparing similar samples or components under similar conditions. When the comparative technique is used appropriately and correctly, the differences between the two (or more) samples will often be indicative of their condition. For this method to be effective, the thermographer must eliminate all but one variable from the comparison. Too often, this simple but essential requirement is not achieved due to the complex circumstances of a test or the poor work habits of the thermographer. As a result, data can be inconclusive or misleading.

As an example, when inspecting three-phase electrical systems, it is very useful to compare the three phases. When the loads are even, which is often the case, all phases will appear thermally similar. However, when the loads are uneven, the phase carrying more load will appear warmer. Unless loads are measured and understood, such a thermogram is without value. Misdiagnosis in such a situation can, at best, result in an embarrassing loss of credibility or, at worst, the loss of valuable equipment.

Prior to inspecting a component or system, it is critical to determine what the normal or desired condition is. What is the "ground truth" or baseline? In some cases it may be easier to determine what the abnormal condition will look like. For flaw detection in aerospace composites, this determination can involve costly theoretical modeling as well as construction and extensive testing of sample test pieces. Understanding a baseline signature may also be simply intuitive. For example, Figure 9-4 shows an outside view of a heated building. The differences in the thermal patterns indicate the location of the framing and cross bracing, the areas that are insulated, and the areas where warm air is exfiltrating along the top of the wall.

In comparative thermography, it is useful to know as much as possible about the sample being tested, such as its construction, basic operations, known failure mechanisms, direction of heat flow, or history. Because this knowledge is often not readily available, the thermographer must become adept at asking clear, simple questions and, even more important, listening carefully to the answers. Many thermographers fail at either one or both of these tasks, and the work quality suffers. Communications skills are as important as technical skills, especially when working with unfamiliar equipment or materials.

A variation of comparative thermography is thermal mapping. Here, temperature distributions over large areas of a sample are compared with other, previously saved images. Baseline thermal mapping is used extensively for inspecting mechanical equipment where the thermal patterns may be complex and the signatures of failure often develop slowly over time. Figure 9-5 shows the degradation over time of a bearing on a motor due

FIGURE 9-4.

FIGURE 9-5.

to misalignment. Again, it is critical to eliminate, or at least understand, all the variables. Because the maintenance of many mechanical systems has a strong time-based component, techniques that trend the changes over time, such as vibration monitoring, are already widely used and accepted. While these thermal mapping techniques can reveal a very accurate and useful temperature trend of past performance, it is important to remember that trending only implies, rather than predicts, the future.

Thermal maps can be made of a material sample and compared to other known references, such as the failure point of a material or the temperature at which contact injury could occur. Thermal mapping has been used with remarkable success in this regard in the automotive industry to validate the temperatures of the sheet metal floor pan. Typically, this temperature data has been gathered via thermocouples, an expensive, time-consuming process that sometimes fails to identify the high-temperature problem areas. Thermal mapping, as can be seen in Figure 9-6, not only verifies that the thermocouple is in the proper location, it also provides a wealth of information about each and every location on the floor pan. These thermal maps can be compared to alarm limits, such as the melting point of the carpet, the ignition temperature of fuel, or the temperature at which a human could be burned.

Thermal mapping has also been used extensively in the electronics industry to look at printed circuit boards (PCBs). Printed circuit boards can be inspected during design, manufacturing, or repair. Typically, the inspection is conducted from the time a cold board is energized until it warms up to steady-state operating temperatures. Image subtraction, another valuable mapping technique, is often used in this application. Specialized image processing software is used to subtract the thermal image of the board being inspected,

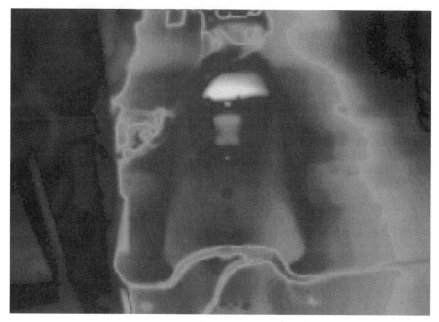

FIGURE 9-6.

pixel by pixel, from a thermal image of a normal or "golden" board. The difference between the two images represents anomalous components or situations.

Manipulation of images using other software routines provides a wide range of techniques for improving thermal sensitivity, spatial resolution, and image contrast. Standard image manipulation routines, such as multiplication, offsetting, and edge enhancement, can be used to beneficially change the mathematical values of an individual or group of pixels. Some experimentation is required to get the best results, but with practice, remarkable detail can often be revealed in otherwise uncooperative images.

These techniques can also be pushed beyond the pictorial qualities of the image. The numerical values of each of the image pixels can be used in a spreadsheet or histogram function for analysis. Because the human eye is limited in what it can distinguish (approximately 256 shades of gray), these numerical analysis routines allow for a much greater depth of analysis. It is possible to distinguish as many as 70,000 discreet thermal levels with the latest infrared systems.

Image averaging, which can be done in the field "on board" many imaging systems, is a particularly powerful routine used to dramatically increase thermal sensitivity. During this routine, a specified number of images over time are averaged together into one image. Although real-time viewing is sacrificed, it is possible to dramatically reduce noise and note very small temperature differences.

One additional software-based comparative thermography technique utilizes either a saturation palette or isotherm function to highlight alarm limits. Data points that are above or below the set point in the image will automatically be portrayed as a different color; for instance, in Figure 9-7 the system has been adjusted so that anything over a pre-

9.20 CHAPTER NINE

(a)

(b)

FIGURE 9-7 (a) Photograph of equipment. (b) Thermograph of (a).

defined temperature will show up as red. As such, it is very easy to note for example, any bearings on a fast moving conveyor that are hotter than normal.

Isotherms can be used in a similar manner to highlight areas either above or below an alarm or indicate areas of similar apparent temperature. Isotherms in thermal images are analogous to elevation contours on a geographical map. They are often used to show an area on the side of a boiler, for instance, that corresponds to degraded refractory insulation. If the boiler operating conditions remain the same, during the next inspection these same isotherm settings will reveal any changes in the damaged area. This information allows for the planning of the material and labor that will be necessary during a repair shutdown.

Another interesting variation on comparative thermography techniques is termed "the null method." The temperature of a material (a fiber leaving a spinneret, for example) is viewed against a blackbody surface of the exact same temperature in the foreground. As it becomes thicker or thinner than specifications allow, the fiber changes temperature and becomes visible against the blackbody. Machine logic can be developed to allow the device to control the fiber-making process.

Many techniques used by thermographers depend upon seeing thermal patterns caused by differences in the rates of conduction, heat capacitance, or thermal diffusivity (a combination of the first two properties). The differences between several materials, or between normal and flawed sections within a given material, become thermally obvious when these techniques are employed. For instance, when inspecting a graphite epoxy honeycomb composite, heat energy is injected into the sample using a high-powered flash of light. The surface absorbs much of the light, converting it to heat, which begins to flow into the material. Where the honeycomb is properly bonded, conductivity is high and the heat continues to flow. Around disbonded areas, the flow of heat energy is stalled by microscopic air gaps. Unable to move forward at the same rate as the bonded area, the heat remains in the vicinity of disbond and causes it to heat up. Using these "active" techniques, a disbond will appear warm when viewed from the heated side.

The use of active thermography has grown exponentially in the past decade, especially in the aerospace industry. High-speed data collection and improved spatial resolution enables us to now view the movement of heat energy in and out of a material. This process is sometimes described as a thermal wave, as the conductive heat movement through the material seems to flow in a wave-like fashion as a sequence of images are viewed rapidly. Thus, it is possible to view a sequence of thermal images—snapshots of heat flow in space and time—that show us not only the location of a flaw but its relative size and depth. Typically, the length of time required until a flaw can be observed is a function of the square of the depth of the flaw. This relationship implies that detectable flaws will be relatively shallow and that contrast will weaken at greater depths. Typically, the radius of the smallest detectable flaw will be at least one to two times larger than its depth. Figure 9-8 shows a graphite epoxy composite test panel that contains a number of flat bottom holes of different sizes drilled to various depths.

Inclusions in a composite material will typically have a different thermal capacitance than that of an unflawed sample. They are thus thermally obvious when the sample is made to change temperature. Active thermography techniques are at this time being refined and promise to become more powerful in the near future.

Flash lamps are the prime heating source for these new thermal wave techniques, but many other heat sources are employed in other applications of active thermography. Heat lamps and hot air guns have been used with great success on materials that are not highly diffusive. Microscopic and hairline cracks can be located by injecting heat into the sample across the face of the sample. As it reaches a crack, which has a higher resistance, the flow of heat energy is reduced. Similar results can be gained by subjecting the sample to

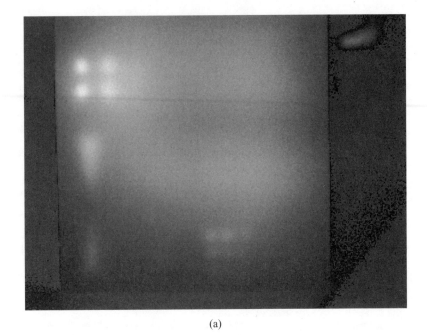

(a)

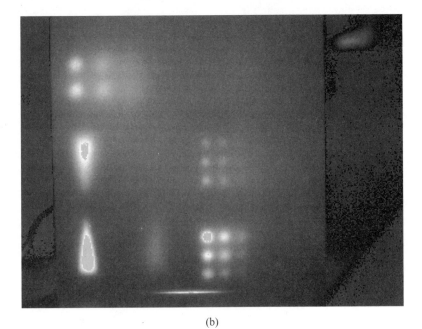

(b)

FIGURE 9-8 (a) Immediately after heat was applied to the surface. (b) 2–3 seconds later, showing heat energy "piling up" over subsurface flaws.

cold; for instance, inserting one end in a cold water bath, which causes heat to flow toward the bath.

On a larger scale, imaging the damage in laminated boat hulls is now common. The hull often consists of inner and outer shells with the space between injected with plastic foam insulation. The insulation bonds to the shells to form a very strong laminate. Unfortunately, impact damage to such structures seriously weakens them but is not readily visible to the eye. If, however, the boat is heated from the inside while in dry dock, the structural anomalies become thermally obvious when viewed from the exterior. Delaminations, which resist the flow of heat energy, will appear cooler, whereas voids, lacking insulation entirely, will usually appear warmer.

The heat of the sun is used as the heat source for many applications including roof moisture inspections, delamination of building facades, fluid levels in outdoor tanks, and the analysis of delaminations in bridge decks and runways. Again, patterns resulting from differential heat flow through the composite allow detection of the defects.

5. VARIABLES

At this time, most applications of infrared thermography require a qualified person to interpret the data. This is largely due to the many variables that are often difficult to understand and sometimes impossible to quantify. For the sake of discussion, in addition to the operator of the instrument, variables can be grouped simply into three categories. These relate to (1) the target, (2) ambient conditions of the system in which the target is operating, and (3) the instrument itself.

Target variables include emissivity, spectral characteristics, temperature, heat transfer relationships in and around the target, thermal capacitance, and diffusivity characteristics. Of these, emissivity is usually the most important. Unless it is very low (below 0.5, approximately), emissivity can be easily characterized and corrected for. As indicated previously, measurements of surfaces with emissivities lower than this are subject to unacceptable error. In an effort to avoid problems regarding absolute measurement of low-emissivity surfaces, many thermographers mistakenly measure the temperature difference between two similar surfaces using an emissivity value of 1.0. The results are nearly always wrong due to the fact that the radiational relationship between compared surfaces is exponential rather than linear.

Some targets, such as glass, thin plastic films, and gases, have transmissivities that vary significantly with wavelength. An understanding of the spectral characteristics of the target, and their relationship to wavelengths detected by the infrared instrument, is thus necessary. The target temperature may also influence emissivity. Polished aluminum, for example, has an emissivity of 0.02 to 0.04 at ambient conditions but at 1000°F (537°C) it is 0.05 to 0.10. The point to remember is to measure or determine emissivity using the conditions which you will encounter at the target.

The target's thermal diffusivity affects how quickly the material changes temperature as well as the shape and intensity of the resulting thermal patterns. Copper has a very high diffusivity, which means that the location of any change in temperature is difficult to pinpoint due to the speed at which heat diffuses through the material.

Among possible variables in ambient conditions are wind speed and direction, solar absorption, radiational cooling, precipitation, surface effects of evaporative cooling, ambient air temperature, background temperature, distance to object, relative humidity, and the presence of emitting/absorbing gases.

The impact of wind-driven convection, in particular, can be profound. Although precise corrections cannot easily be generalized, a commonly used rule of thumb suggests that a 10 MPH wind may reduce the temperature difference (ΔT) from a heated target to an ambient target by approximately one-half; a 15 MPH wind may reduce this ΔT by as much as two-thirds. Be aware that these rules of thumb are based on a very simple analysis; a real life situation will be much more complex and difficult to predict. Temperature data for outdoor electrical inspections are therefore rendered meaningless without local wind speed data.

Precipitation will usually result in evaporative cooling of the surface being inspected, but as freezing occurs—for example, of water absorbed into a masonry surface—latent heat of fusion is released, often with confusing consequences.

Ambient air temperature can add to or subtract from target temperature. An abnormally hot electrical connection, for example, can easily be 100 degrees warmer in the summer than in the winter if no other variables except ambient air temperature change. Especially for low-emissivity targets, a change in the background temperature can be significant. Whenever possible, it is best to have a thermally uniform background free of extremes. If gases with strong spectral emittance/absorptance characteristics, such as CO_2 or water vapor, are present, radiational transfer—and thus radiometric measurements—will be attenuated. Short-wave systems in particular are susceptible to attenuation by the atmosphere and, as a result, correction for relative humidity and distance to object are recommended.

The primary thermal variable relating to the target is whether transient or steady-state heat flow conditions exist. If heat flow is transient, knowing the position in the thermal cycle, as well as the rate of temperature change, is critical to interpreting the thermal image. Regardless of whether the heat flow is steady-state or transient, it is important to know the thermal condition of the target at the time of viewing with respect to the extreme thermal situation that the target may experience. Transfer from a heated building is usually considered steady-state during the late evening and early morning hours, but the effects of the sun can cause heat flow to reverse, even in very cold weather. During these transient situations, thermal patterns can be very confusing. Active thermography—based entirely on variations in transient heat flow—is typically controlled, making its impact much easier to understand.

There are other variables related to the infrared instrument, including the precise waveband detected, thermal sensitivity, detectivity, rate of data acquisition, dynamic range, field of view, spatial and measurement resolution, and system calibration. Short-wave systems are particularly sensitive to problems with solar glint or excessive solar reflection. For extensive outdoor work, long-wave systems are recommended. Long-wave systems, on the other hand, are generally more susceptible to error when used to measure temperatures of very low-emissive surfaces. In general measurements in either waveband produce similar results for temperatures from 14–266°F (–10 to 130°C).

The thermal sensitivity required for some applications is greater than for others. InSb detectors, for example, are favored over less sensitive PtSi detectors for some R&D applications. The detectivity (D^*) of a detector is a measure of its response relative to wavelength. Depending on whether or not this response coincides with the spectral characteristics of the target, one detector may be more or less useful than another. Detectivity also determines in part the rate at which data can be acquired. Most instruments operate at 30 or 60 Hz in order to be compatible with standard video frame rates. Using high-speed computers to store the data, InSb detectors allow for acquisition at rates in excess of 500 Hz over a reduced field of view.

The dynamic range of the acquired data is another important variable. Most of today's focal plane array (FPA) systems can store 12 or 14 bits of data for each image. The result is that a great deal of thermal data is available for analysis beyond the 8 bits that can currently be viewed as an image at any one moment. While it is still possible that data outside these large dynamic ranges may not be acquired, the chances are much less than in the past, when the maximum data acquisition was the 8 bits being viewed.

Using a thermal imager with appropriate resolution is essential. Imagine having to measure a kilometer using a meter stick, or a meter using a car odometer. Each infrared instrument has particular resolution characteristics determined by the detector size and the lens used, and the relationship between the two. On most instruments, the lens can be changed. This results in a change of the field of view (the area viewed by the instrument) as well as a change in spatial and measurement resolution. Spatial resolution—the smallest detail seen—is greater than the measurement resolution, which is the smallest detail that can be measured. It is not uncommon for an inexperienced thermographer to either miss small anomalies or to measure them inaccurately because of a failure to understand resolution. Knowing the specification for the instrument and lens being used, and working within those limits, is critical to accurate work.

Calibration of the instrument is critical to quantitative work. Calibration, usually conducted by the manufacturer, should be for the temperature range in which the instrument will be used. FPA detectors are particularly subject to nonuniform response over the array. Most manufacturers provide for some sort of nonuniformity correction (NUC). The appropriateness of these varies with need.

Where measurement is critical, correction should be at two points in the desired temperature range and should include the lens. Otherwise, a simple one-point correction inside the system is generally adequate. Many instruments make these corrections periodically during normal operations. FPA instruments are also subject to a change in response as the instrument changes temperature. Because they are temperature-stabilized rather than cooled, bolometers, in particular, are susceptible to extreme thermal influences. Where temperature measurement needs are critical, they may need to be allowed to thermally stabilize for up to twenty minutes when ambient conditions change significantly.

Simple field calibration checks are essential for day-to-day accuracy of a thermographer's work. These checks can be as simple as checking the temperature of a human tear duct at instrument startup and shutdown. For more critical work, the inclusion of a thermocoupled surface or a calibrated reference blackbody in the field of view may be warranted. The calibration of an instrument should also be checked against a calibrated reference blackbody on a periodic basis. The exact calibration frequency depends on the requirements of the application. Most manufacturers recommend recalibration on an annual basis. Calibration records should be kept on file.

Of course, the thermographers themselves are probably the greatest variable! Stated simply, the question is "are they qualified?" Qualification is based on appropriate training, experience, and testing. Especially with the industry-wide confusion over the use of the term "certification," variability in personnel qualification is dangerously diverse. Success for many applications also requires the operator to have additional related skills or experience. Regardless of the application, the thermographer must have mastered the basic communications skills of asking questions and listening, without undue "filtering" of the answers.

Today's instruments are so easy to use—literally pressing one button results in a remarkable image—that many operators fail to learn about the many other variables that are not optimized with that single button. Much of thermography is in fact, simply comparative or qualitative work. By comparing the target to another that is similar, it may be possible to minimize or eliminate the effect of one or more of these variables. Clearly, training and experience are fundamental to the process. As an example, it is not uncommon when inspecting a three-phase electrical system to find one phase being warmer than the other two. If the load is balanced, this heating would be considered abnormal, probably associated with an overload condition, an undersized conductor, or possibly heating from a nearby high-resistance connection. But under some conditions, normal loads can also be unbalanced, resulting in an identical thermal pattern. The skill and experience of the thermographer tip the scales in favor of the correct interpretation.

Quantitative thermography requires an even greater understanding of the variables impacting radiometric measurement, as well as a grasp of its limitations. It is vital to determine what margin of error is acceptable before beginning an inspection, and to work carefully to stay within those bounds. Too many thermographers believe that all data output from their instrument is accurate, or that the impact of all variables can be understood and measured. This is simply not true!

Considering the variables that have been discussed, it becomes obvious that there is considerable potential for error in both qualitative and quantitative thermography. Only careful consideration of all variables by a qualified and skilled thermographer can result in good data, and even then it is important to understand—and work within—the margin of error remaining.

6. EVALUATION OF TEST RESULTS AND REPORTING

Much of thermography is still highly dependent on the skill of the thermographer to conduct the inspection correctly, understand the limitations of the test, record all relevant data, and interpret the results properly. As we have seen in the previous section, the number of variables is staggering. Inspection personnel, as a consequence, must be adequately qualified for the task at hand.

It is generally recommended that a Level I qualified thermographer who is working under the supervision of a Level II thermographer collect data. The use of written inspection procedures developed with the support of a Level III thermographer is also critical to success. To write procedures or validate data, it is often necessary to set up other tests. It may be necessary to make emissivity measurements, for instance, of typical components before an inspection methodology can be developed. For NDT of composite materials, it may be necessary to develop costly test pieces to define the limits of detectability.

Procedures should detail the knowledge and skills required to conduct the inspection. As an example, to conduct an electrical inspection, many companies require the thermographer to have a basic working knowledge of electrical systems together with both electrical safety and CPR training. The procedures should also clearly spell out the weather and system conditions necessary for a successful inspection. Any limitations should be clearly spelled out, such as a caution for using short-wave sensing systems outside in the sunshine.

Evaluation of data is also the function of the Level II thermographer. Sometimes additional, supplemental testing using other test methods, as well as specific engineering or Level III support, may be required. On a limited basis, it may be possible for Level I personnel to make "pass/fail" assessments based on clearly written procedures. For instance, when conducting an inspection of boiler refractory, a Level I thermographer can mark refractory as "failed" when the surface temperature exceeds a value set forth in the procedures.

If personnel are qualified and good procedures are followed, evaluation of the data can proceed. A primary challenge is to determine the reliability of measurement, especially for radiometric values. Generally, radiometric measurements are not recommended when emissivity falls below 0.5. Even when qualitatively evaluating low-emissivity materials, one must proceed with caution. System variables, such as load and wind, must also be carefully accounted for, even if it cannot be done precisely.

Even after the data is correctly evaluated, the results must be clearly communicated to others in a report. Part of this process often requires educating the person who is reading the report. If they do not understand the potential problems associated with radiometric temperature measurements or conductive heat flow, for instance, serious misinterpretation of the results is a possibility. Generally, the problem for the thermographer is to have the customer understand the strengths and limitations of the technology or the findings. To what lengths the thermographer should go in the process varies, but is probably directly proportional to his or her investment in having the report make a difference in the inspection process.

Reports can take many forms. As a minimum, the report should include the thermographer's name; the instrument model and serial number(s); and relevant ambient conditions, such as wind speed and direction and ambient temperature. System conditions, such as load, component identification and location information, component emissivity, instrument parameter settings, and, in most instances, a thermal image and matching visual image, should also be included.

These details should be displayed in a way that does not clutter the report, but instead supports the presentation of essential information in an easily understood fashion. The best reports are intuitive in nature with a natural flow of data to support the thermal and visual images.

7. APPLICATIONS

Many of the applications for thermography are not considered mainstream NDE/NDT. Instead, they have come from other areas, such as industrial predictive maintenance and forensic investigation. Regardless of terminology, applications of thermography are nondestructive in nature, allowing the effective evaluation of both materials and systems.

The use of thermography in more traditional NDE/NDT has recently become more widely accepted. Growth has been concentrated in two industries in particular, electronics and aerospace, but new applications are being developed in many other industries as well.

Electronics Industry

In the electronics industry, thermography has become a powerful tool for design and manufacturing of integrated circuit boards (ICBs) as well as, in some cases, their repair. As designers have been pushed to make smaller boards, component density on the boards has increased and so have the problems associated with the heat that they produce. During the design stages, infrared is used to observe the distribution of this heat and its impact on components. The boards are next inspected in situ to evaluate the impact of any further heat flow conditions that might affect their performance.

The emissivities of many of the components on an ICB vary widely, and many are quite low. Two methods have been used successfully to reduce the difficulties of inspection. Boards are first "heat soaked" in an oven. A thermal map of the ICB is created. This map is then compared to the map of the board at operating temperature. The second method involves the application of a relatively high-emissivity, conformal coating to the ICB, thereby allowing direct thermal viewing and analysis.

At the manufacturing stage, infrared can be used to inspect the actual ICB being produced. The board being inspected is energized in a test stand. Its thermal signature is compared to that of a "golden board," meaning an ICB that performs within specifications, both during transient warm-up as well as after it has achieved steady-state operating temperature. Using computerized processing, one image is superimposed on the other, and an image subtraction routine performed. Any anomalous areas are immediately pinpointed. The same technique is used during the repair of more costly boards. The result has been a dramatic reduction in repair costs and an increase in quality assurance.

Aerospace Applications

Nowhere has the use of active thermography grown as it has in the aerospace industry. As more and more composite materials are used, the need to monitor quality of both new and existing stock has also increased. Techniques have evolved that range from simple handheld monitoring to automated inspections using sophisticated computer-based vision systems.

The basic premise of aerospace thermography is that properly laminated or bonded material will have relatively uniform thermal characteristics, including conductivity, capacitance, and diffusivity. Disbonds, delaminations, and inclusions of foreign substances can be located and characterized based on their anomalous thermal signatures. Typically, the component being inspected is heated from one side and viewed from either the same side or the opposite side, depending on the specific requirements. Of paramount importance is that the heat application should be relatively uniform across the area to be inspected. Even heating will promote heat flow through the material, perpendicular to the surface. Active thermography can be accomplished with a variety of heat sources, such as hot air guns, heat lamps, or flash lamps.

Several of today's active systems use a carefully designed system of high-powered xenon flash lamps as a heat source and a computer to capture the thermal data. Flash or thermal wave thermography, as it is often called, is now widely used for production and repair of both commercial and military aerospace components. During flash thermography, the pulse of heat energy is absorbed by the surface of the component. The heat conducts from the surface through the part at a uniform rate until it encounters a discontinuity. Because a delamination or disbond usually has a lower rate of conductivity, the heat over the disbond tends to build up, thus indicating the location, depth, and relative size of the discontinuity.

The wave continues to move into the component, rebounding from discontinuities within the cross section. Data is captured at rates of up to 500 frames per second (60 to 100 frames per second being typical), in order to capture the required detail. Because the entire process takes only a few seconds, the area covered by a single flash pulse—approximately four square feet—can be inspected in a very short time.

While much of this work is still being performed by technicians, new techniques are evolving quickly to utilize robots that can move the flash in a preprogrammed route over the entire aerospace structure. The data is then processed and analyzed by the computer system. Of course, considerable experimentation is required to set up such an operation, but, especially for manufacturing of new components, these automated systems hold great promise.

Also being inspected is moisture intrusion into honeycomb structures. Successful inspections have been conducted right after an aircraft has landed. Because water has a high thermal capacitance, it remains cool (from the high altitude) long after the rest of the plane has reached ambient conditions. Active thermography can also be used to locate moisture. The area of the material with no water intrusion heats evenly, whereas the areas with water appear cooler due to the high thermal capacitance of the water.

Electrical Inspections

Thermography is most widely used for inspecting the integrity of electrical systems. It has the distinct advantages of being fast and noncontact. While heat is not a perfect indicator of all problems in electrical systems, heat produced by abnormally high electrical resistance often precedes electrical failures. Much of electrical thermography work is qualitative, comparing the thermal signature of the similar components. This is quite simple with three-phase electrical systems—the heart of all utility, commercial, and industrial installations—where the phases should almost always appear similar. One or more components at a different temperature (usually warmer) suggest a problem is at hand.

The technique for inspecting electrical systems is quite straightforward. The component is viewed directly while energized. Particular attention is paid to any connection or point of electrical contact. These areas are most susceptible to high resistance and associated heat-related failure. It is possible to inspect a good-sized power substation, containing thousands of points of contact—each a potential problem—in less than an hour. Figure 9-9 shows a classic pattern associated with a loose or corroded connection point. Electrical current imbalances among phases are also readily seen. Often, these are considered normal, such as in a lighting circuit, while in other parts of the electrical system they can result in very costly failures, such as when a phase imbalance is caused by an internal fault in a large motor.

Although the application is straightforward and widely used, it is often used ineffectively. A good electrical systems thermographer must contend with several problems related to the component, the infrared instrument, and the interpretation of the data. Many electrical components have extremely low emissivities, resulting in the inability to accurately measure temperatures. Low emissivity also means that components must be very hot before they radiate enough energy to even be detected. Often, by the time you see a high resistance problem, some damage has already been done. Although it can represent a significant investment, more and more plants are placing high-emissivity targets on these critical components in order to increase the probability of early fault detection and improve the reliability of measurement.

Some thermographers advocate measuring nearby surfaces that have higher emissivities—the electrical insulation, for instance. While this can be useful, care must be taken. Most components have such a high thermal conductivity and diffusivity that a large thermal gradient can exist even over a small distance. This means that the measured temperatures may be much less than those at the site where high-resistance heating is occurring.

Temperature also varies widely as the electrical load on the system changes. The heat output from a high-resistance connection is predictable and can be correlated with changing loads; unfortunately, the temperature change of the connection is much less predictable. All that can be known is that as load increases, the temperature of a connection will increase at a rate that is greater than linear and less than exponential. The National Fire Protection Association (NFPA) standard *70-B, Maintenance of Electrical Systems,* recommends inspections be done with a 40% minimum load and, when possible, at the highest normal load. Even seemingly insignificant problems should be noted when inspecting systems that are temporarily under a light load.

Too many thermographers still try to inspect electrical equipment without opening the enclosures to directly view the components. Except where extenuating circumstances make a direct view impossible or dangerous, components must be viewed directly. Open-

(a)

(b)

FIGURE 9-9 Abnormally high resistance heating in electrical system. (a) Photograph; (b) thermograph.

ing enclosures cannot be done without great care, however, because the thermographer is exposed to the potential for extreme danger, especially from an arc flash explosion. These explosions, which can be triggered simply by opening an enclosure door, can reach temperatures in excess of 15,000°F in less than a half-second.

Every effort should be made to understand and minimize these dangers. NFPA 70-E, which is accepted as law throughout most of the United States, details the steps necessary to protect the thermographer. These include, among others, education and personnel protective equipment (PPE). PPE may include clothing designed to withstand an arc flash, as well as protection for the eyes from the high-energy ultraviolet radiation given off in such an explosion.

Some equipment may be so dangerous to access that other measures are advised. Infrared transparent windows are now being installed in some instances to allow viewing into an enclosure without opening it. Careful placement is required to ensure that all components can be seen. It may also be possible to use equipment that detects airborne ultrasound. Even microarcing in a connection produces a detectable ultrasound signature through a crack or small hole in an enclosure.

Where enclosures cannot be opened easily, such as on an overhead-enclosed bus, the thermal gradient between the problem and the viewed surface is typically very large. A 5°F surface signature on an enclosed bus often indicates that the metal bus bars a few inches away may have already begun to melt! Great care must be used when conducting inspections of unopened enclosures.

For outdoor inspections, wind is a variable that can quickly reduce the value of thermography if it is not used with the greatest of care. Convective cooling by the wind (or inside plants by any forced air currents) will quickly reduce the temperature of a high-resistance connection. What appears to be an insignificant problem during these windy conditions can, in fact, become very significant when the wind speed drops. Many thermographers try to compensate for this by measuring a temperature difference from one phase to another, with the mistaken notion that the effects of the cooling will be neutralized. Such is not the case!

When making delta measurements between like components, one can fail to recognize that the normal phase, operating at only a few degrees above ambient, can never be cooled below ambient temperature. Thus, in the wind, the delta measurement will also seem deceptively small. The results of outdoor inspections when wind speeds are greater than approximately 10 MPH should be viewed only with a great deal of caution. Even "minor" temperature increases may represent extremely serious problems. It is likely that many problems will not even be warm enough to be detectable, much less accurately measured. To be of any value at all, a measurement or estimate of wind speed at the component location must always accompany temperature measurements. The use of Beaufort's Wind Scale, or some other means of estimating local conditions, is recommended.

Two other difficulties crop up during outdoor inspections. The sun can cause heating of any thermally absorbent surface. Dark surfaces in particular can be heated to the point where they are difficult to view with any understanding. Usually, inspecting during the morning hours or on overcast days is sufficient to minimize difficulties. More problematic is using short-wave sensing systems outside in sunshine. Short-wave systems are susceptible to solar "glint," or reflection of the sun's high quantity of short-wave thermal radiation. For extensive outdoor work during sunny periods, long-wave sensing systems will give superior results. Short-wave systems should be used only on a limited basis or, if loads and other conditions allow, on overcast days or at night.

Although many believe otherwise, the job of acquiring reliable thermographic data regarding electrical systems is not easy. Even with good data in hand, too many people use the data incorrectly to prioritize the seriousness of their findings. Temperature is often not a reliable indicator of the severity of a problem. Given the limitations of radiometric measurement on low-emissivity electrical components, the difficulties are further exacerbated.

This reality does not prevent the vast majority of thermographers, or people purchasing thermographic inspections services, from believing that the hotter a finding is, the more serious the problem. Or, more dangerously, they believe that if a finding is not very hot, it is not a problem. Nothing could be less true! The problem shown in Figure 9-10 is an internal fault in a pole-mounted transformer tank; the temperatures seen on the outside of the tank are quite cool compared to the temperature inside at the actual point of high-resistance heating.

The effects of convection on heated surfaces have been previously described. While conducting electrical inspections inside a plant, it is not unusual for warm air to be convected up and over components, causing them to be warmed more than might be expected. High resistance in a loadside (lower) connection of a breaker, for instance, will result in heating, but convection from it may also cause the lineside (upper) connection to appear warm.

When inspecting three-phase electrical systems outside, it is also important to remember that wind (at ambient air temperature) moving over a normal component (also at or near ambient temperature) results in little cooling effect. However, the same wind blowing over a high-resistance, hot connection will dramatically cool it. The result is that when the wind is blowing, the hot problem connection may not be visible or, if it is, it may not appear to be very warm.

FIGURE 9-10.

The seriousness of many findings will not be made obvious by temperature alone. It must be recognized that any components having low emissivity or a high thermal gradient located inside an enclosure subject to convective cooling, or under light load, will probably not appear very warm. In fact, in most cases, they could be extremely unstable. Further, little is understood about the actual failure mechanisms for the diverse collection of electrical components that can be inspected.

Some generalizations are possible, such as the fact that lower melting points will cause aluminum materials to fail before copper alloys. Spring tension, an essential characteristic of many switch and contact components, fails after the material has been at approximately 200°F for a month. The blocked cooling fins shown in Figure 9-11 also indicate a very serious problem. Insulating oils in transformers break down rapidly when subjected to overheating, and this problem will result in overheating when loads are at peak and ambient temperatures are high.

Rather than basing prioritization strictly on temperature, a more useful approach is to look at all the parameters involved and how they interact with each other. This can be done either with a simple approach or a more complex analysis. On the simple end, a weighted prioritization matrix can be established using any parameters deemed significant. These might include, for instance, safety, loading condition, reliability of the radiometric measurement, and a history of similar components. Certain parameters can be given added value or "weight" as a total score is figured. In Table 9-3, the issue of personnel safety is given a weight of twice the other factors. A simple system like this, if used consistently, can become "smarter" with time if a feedback loop is incorporated. Although this approach is particularly useful for electrical components, it can be used for a myriad of mechanical equipment as well. Other factors, values, and weights can be used, depending on the needs of the thermographer.

A more complex analysis may be required in some instances. This might involve full

FIGURE 9-11.

TABLE 9-3 Priority of Parameters

	Greater chance →				
	1	2	3	x	=
Will a failure injure people?			*	2	
Is the component critical?					
Will loads increase?					
Will the wind decrease?					
Is the measurement unreliable?					
If it is reliable, is ΔT great?					
				Total	

*Immediate action required.

engineering studies and incorporate data from other sources or test methods. A volatile gas analysis, for instance, provides excellent indicators of various internal faults in oil-filled equipment such as transformers.

Because the consequences of not doing thermal electrical inspections are so great—production outages, decreased product quality, and the potential for injury or death—care should be taken to structure and implement a program that will achieve optimum results. Companies that do this are able to virtually eliminate unscheduled downtime due to electrical failures.

Mechanical Inspections

Mechanical inspections encompass a diverse variety of equipment. The technology has proven to be an invaluable tool for looking at motors, rotating equipment, steam traps, refractory, and tank product levels, among others. Most of these applications are qualitative in nature, often comparing the current thermal image to a previous one and understanding the cause and extent of any changes. In fact, conducting baseline inspections of new or rebuilt equipment has proven to be of particular value. As with electrical equipment, it is beneficial to establish inspection routes and periodic frequencies based on needs and resources.

Approximately 90% of the electricity used in the United States passes through electric motors. Although there are a number of excellent means to monitor the electrical and mechanical condition of motors, infrared has proven particularly useful as a fast screening device. High resistance, abnormal current flow, or excessive friction produces heat. If problems are indicated by one of these abnormal thermal signatures, other testing is often implemented to determine the exact nature of the problem.

Thus, the abnormal heating of a misaligned motor coupling usually precedes a measurable vibration signature. Left unresolved, the problem will be compounded as it wears on the motor bearing itself. Many problems that are related to the motor's electrical system also have characteristic thermal signatures. An internal short caused by a breakdown of insulation on motor windings will result in an increased temperature internally, as well as on the motor casing.

Heating of bearing surfaces caused by abnormal friction can provide an excellent indicator of the location and nature of many problems for all types of rotating equipment. The pump bearing in Figure 9-12 shows a typical pattern of an overheating bearing, in this

(a)

(b)

FIGURE 9-12 Pump. (a) Photograph; (b) thermograph.

case due to lack of lubrication. As often as not, the problem stems from overlubrication rather than lack of it! Especially for low-speed equipment, such as trolleys on overhead conveyors, where vibration signatures may not be reliable, thermography has proved a boon.

Refractory Inspections

In a variety of applications where high-temperature refractory is used, such as in boilers, rotating kilns, and torpedo cars, infrared can be used to monitor outside skin temperatures. With this information, engineers can validate the condition of the insulation or back-calculate the thickness of remaining refractory. It is also possible to identify any hot spots related to refractory failure. Typically, refractory is inspected regularly with a frequency determined by age and the estimated consequence of failure. Figure 9-13 clearly shows a problem with the refractory in this powerhouse boiler, although the root cause will require additional investigation. Rotary kilns benefit from monthly inspections, whereas torpedo cars may require weekly inspections. Because they are under less stress, boilers are often inspected on an annual basis prior to shutdown. All refractory should also receive a baseline inspection at startup so that trends can be established from consequent inspections.

Steam Traps and Lines

Steam traps are highly susceptible to a "stuck open" failure mode, which can often be detected using infrared testing. Although airborne ultrasound is one of the primary methods of investigation, infrared has the advantage of being fast and able to be used even at a distance. A trap that is functioning properly will typically have a detectable temperature difference between the steam and condensate sides. This difference will temporarily disappear as it cycles. If both sides of the trap are warm, particularly if they are at steam temperatures, the indication is good that there has been a failure.

The inspection of steam lines using thermography is also valuable. Where insulation has failed on aboveground lines, a thermal signature is usually a clear indicator of the location of the problem. This is particularly true outdoors where water has saturated the insulation. Because there are more variables, buried underground lines are more difficult to inspect. Using a map of the system, it is possible to walk a steam line with a good handheld infrared system and locate both steam and condensate leaks. Signatures are generally clearer at night, but variations in ground moisture, bury depth, and soil type must be considered.

Heat tracing (both steam and electric) on process lines can also be inspected, except when insulation levels are very high, but this is rarely the case. More problematic is the use of aluminum or stainless steel sheet metal coverings over the insulation. These make inspections practically impossible due to the low emissivity (and consequent high reflectivity) of the coverings.

Many processes depend on maintaining internal temperature at a precise level. Operating temperatures may fall below these thresholds as a consequence of either the insulation or the tracing failing. In the confectionery and petrochemical industries, in particular, this can lead to blockage of a line as the transported product freezes. When blockages do occur, thermography can be used to help locate them while there is still a thermal difference between liquid and solid.

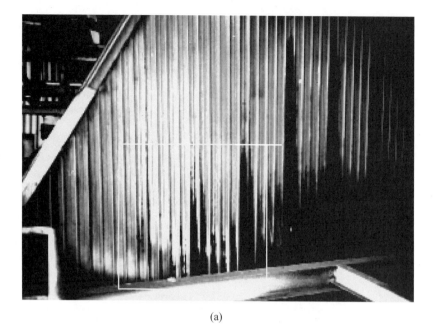

(a)

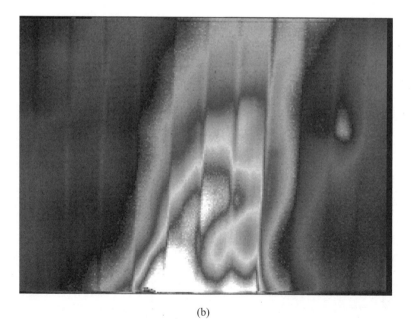

(b)

FIGURE 9-13 Outside wall of boiler. (a) Photograph; (b) thermograph showing problem with refractory.

Tank Level Verification

One of the most surprising, and usually easy, applications for thermography is locating or confirming product levels in tanks and silos. Although most have some sort of instrumentation, it often does not work or readings must be independently confirmed. As with line blockages, the differing thermal capacities of solids, liquids, and gases mean that they will change temperatures at different rates during a transient heat flow cycle. Gases change most quickly. In fact, in large outdoor tanks the sun can cause a detectable thermal change in a matter of minutes. The levels in the tanks in Figure 9-14 are clearly visible. In the late evening or early morning (before sunrise), the pattern would be reversed but the levels would still be visible.

Solid sediments typically change temperature next, as heat is conducted in or out of the material next to the tank wall. The sludge level in a large tank is evident in Figure 9-15. Next, any floating material, such as waxes, can usually be distinguished from the liquid. The liquid, because it may have a higher thermal capacitance and is also subject to convective heat transfer, typically takes the longest to change temperature. For tanks that are indoors, a normal diurnal swing in ambient air temperature is often enough to reveal the levels. Even in air-conditioned spaces, small ambient fluctuations may be enough. The materials inside many tanks are heated or cooled, and thus provide their own thermodynamic, driving force.

Uninsulated tanks and silos are quick to reveal their levels. Where insulation is present, the image may take longer to appear or may require some enhancement. Levels can be enhanced using simple active techniques like applying heat or by wetting the surface to

FIGURE 9-14.

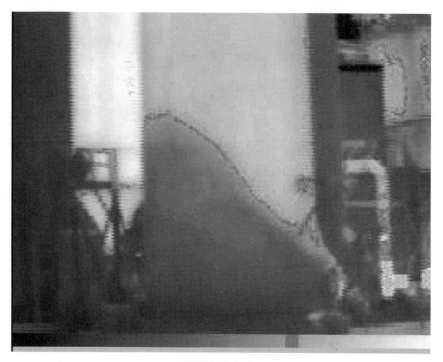

FIGURE 9-15.

induce evaporative cooling. Simply spraying water briefly on a tank is often enough to reveal several of the levels.

Where insulation is covered with an unpainted metal covering, difficulties may arise due to the low emissivity of the surface. Even in this case, however, a high-emissivity vertical stripe, such as paint or a removable tape, can be applied. The levels will be visible on the stripe.

There are numerous other applications for inspecting mechanical equipment thermographically. It is not possible to describe all of them, but it is useful to point out several generalizations. A good baseline inspection is critical. This should be done both at the transition during startup and also during the cool-down period. Be aware that all components of a machine will not go thorough transitional swings at the same rate. Other images should be made when the machine is operating at steady-state temperatures. Again, it is important to understand the impact of variations in conduction and convection and how they affect the surface patterns being seen. Subsequent images taken later on a periodic basis should be compared to this baseline data, noting any changes that may indicate problems. Careful data collection and record keeping is required for these methods to be effective.

Roof Moisture Inspections

The use of infrared roof moisture inspections has grown tremendously over the past decade as a cost-effective maintenance tool. There are literally tens of thousands of acres of flat commercial and industrial roofs throughout the world having a replacement value in the billions of dollars. For a number of reasons related to design, installation, and maintenance, most develop leaks within a year or two. While the damage caused by the actual leak may be substantial, the hidden long-term damage is usually far more costly. From top to bottom, a typical cross section of a flat roof is composed of the membrane, insulation, and a structural deck (see Figure 9-16). Water that enters a roofing system becomes trapped in the roof system, especially in the insulation, resulting in degradation of the system and its premature failure. By locating and replacing the wet roof insulation, subsurface moisture is eliminated, and the life of a roof can be greatly extended beyond the current average. Because water has such a high thermal capacitance, it is slow to change temperature during a transient heat flow cycle. During the daytime, the sun warms a roof. The dry insulation heats faster and to a higher temperature than does the wet insulation. In theory, it would be possible to locate the wet insulation during the daytime, but the sun typically heats the surface of the roof enough to mask these differences. After sunset, however, especially if the night is clear, the roof will begin to cool rapidly. Excessive wind may result in uneven cooling and confusing thermal patterns. At some point in the evening, usually an hour or two after sunset, the dry insulation will become cooler than the wet. It is possible, as this "inspection window" opens, to locate areas of wet insulation by their warmer thermal signature and characteristic patterns. Figure 9-17 shows a typical pattern associated with moisture intrusion into absorbent insulation; water has probably intruded into the system from around the vent pipe seen in the image.

Once patterns are seen, the roof can be inspected quite quickly, marking all wet areas with paint directly on the roof surface. If necessary, the actual presence of moisture in a wet area can be confirmed destructively. It is possible, on a fairly uncluttered roof, to inspect upwards of 500,000 square feet in a night. The inspection window will often remain open long into the night, until it is closed by wind or by the formation of substantial dew.

Typically, radiometric temperatures are not recorded during a roof moisture inspection. A videotape or individual images of all anomalies should be made. Most thermographers work in a black and white or saturation palette with the narrowest of span settings, i.e., high image contrast.

Exactly what pattern will be seen and when depends in large part on conditions and the type of roof insulation. Absorbent insulations, such as fiberglass, wood fiber, and perlite, yield clear signatures. These types of insulation are typically used in built-up roofs. A "board edge pattern," with its characteristic right angles, results because each board of insulation tends to saturate with water before spilling over onto the next one.

On the other hand, nonabsorbent foamboard insulation, which is often used in single-ply roof systems, is extremely difficult to inspect due to the fact that little water is trapped compared to other roof types, although still more than enough to cause degra-

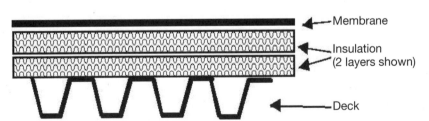

FIGURE 9-16 Cross section of a typical flat commercial roof. Image courtesy Snell Infrared, ©1999.

FIGURE 9-17.

dation. The roofing industry recommends installing an absorbent layer of insulation on top of foamboard, but this is often ignored. The unfortunate consequence is that infrared has limited value as an inspection tool on many foamboard roofs. Many single-ply roofs are also ballasted with pavers or a heavy layer of stone. These, too, can render infrared of little or no value. Clearly, the potential for inspection should be considered at the time of design!

Patterns are also influenced by weather. The roof surface must be dry or else evaporative cooling will prevent it from heating up adequately. If heavy cloud cover forms after sunset, the roof will probably not cool quickly enough for the inspection window to open. Wind in excess of approximately 10 MPH will quickly cool a roof and erase all thermal signatures of value. During cold weather, heat transfer from the building through the roof section can produce detectable thermal patterns in the wet insulation, once again resulting in these sections appearing warm. Very good results can usually be achieved with ambient air temperatures as low as approximately 40°F (4.4°C).

Rooftop conditions themselves also help determine patterns. A roof section that is in the shadow of a higher wall, for instance, will not be heated to the same degree as the unshaded areas of the roof. A west-facing wall, on the other hand, will reradiate its energy to the roof long into the night, preventing it from cooling adequately. Extra gravel or flashing material will stay warmer, sometimes masking over the wet insulation beneath. Where the roof has been previously repaired, differences in the type of insulation may cause variations in the thermal signature. Understanding these influences and the result they have on the thermal patterns is relatively easy for a qualified thermographer.

Roofs should be inspected shortly after installation to establish a baseline signature.

Another inspection would be warranted after any potentially damaging incident, such as a heavy hailstorm or hurricane. When leaks occur, as they are bound to, a quick follow-up infrared inspection will help locate the leak and indicate the extent of the wet insulation needing replacement. It is often possible to inspect a well-designed and maintained roof as infrequently as every 3–5 years.

Great care must be used during roof inspections to prevent accidents and unnecessary injuries. Look for safe access to the roof, and notify all local security officials of the inspection. Never work on the roof alone. Thermographers are most vulnerable because the brightness of their instrument displays prevents their eyes from adjusting to the low light conditions found on most roofs. It is critical to become familiar with, and adhere to, any relevant company or government regulations. Especially when people are within ten feet of the roof edge, it is important to be protected from an accidental fall by an appropriate barrier or fall protection device. Falling over the edge is not the only danger. Even stumbling over a small change in elevation can result in serious injury and a damaged infrared instrument.

A walking rooftop inspection is obviously very labor-intensive, relatively slow, and highly weather-dependent. A great deal of information can be obtained very quickly by performing an aerial infrared roof moisture inspection. Using either a helicopter or fixed-wing aircraft, the instrument can be flown at altitudes between 500 and 1500 feet from the roof, depending on local air traffic requirements and the spatial resolution of the test instrument. Obviously it is important to determine what the size of the smallest detectable area must be. If the conditions are right, data for millions of square feet can be acquired in an evening. Of course, bad weather or roof conditions can result in a costly cancellation of the mission.

Great care must be taken to identify the target roofs. A daytime flyover to conduct a visual preinspection is highly recommended. Most professionals videotape their survey and include some type of on-screen geographical positioning information. Once the data has been collected, analysis and reporting will require additional office time.

Although infrared has some limitations as an inspection tool (mainly that its use is weather-dependent), it is fast, thorough, and reliable. Unlike other methods, it also allows for the inspection of every square foot of roof. On roofs with absorbent insulation, the results of an infrared inspection are extremely accurate, measuring areas of wet insulation within several inches.

The potential savings from infrared roof moisture surveys are astounding. It is not unusual to extend the life of a roof by 50% or more. The life cycle cost drops significantly even when the life is increased by a year or two. As waste disposal costs continue to climb, especially for the hazardous materials that many roofs are constructed of, there is a strong incentive to repair and keep a roof in place. Even when the useful life of the membrane has been reached, it may be possible to install a new membrane over the existing insulation, again at great savings.

Building Energy Surveys

During the "energy crisis" of the late 1970s, thermography was used extensively for inspecting buildings. The technology lends itself to checking for both conduction (insulation) and convection (air leakage) problems as well as moisture intrusion and delamination of facades. When inspecting a building, it is vital to understand exactly how it is constructed. This is often not the same as looking at the drawings for the building. In fact, the thermographer's job is often to find performance flaws in a design or discrepancies between the design and its execution. A simple destructive examination often proves a valuable supplement for the thermographer.

Inspections are also simplest if accomplished during steady-state heat flow situations. Generally, a 20°F (11.1°C) temperature difference from inside to outside is needed for conduction problems to express themselves. Heating of the building's surfaces by the sun, however, can result in a reversal of heat flow and/or confusing patterns. These effects can be noticeable for up to 6 hours after the sun has left a building facade. For this reason, it may be difficult to conduct inspections during sunny days in warm or cold weather.

With optimum conditions, however, missing, damaged, or ineffective insulation, as well as the location of framing, becomes obvious. Patterns vary with the type of insulation and the exact conditions. The conductive inspection is best done from inside the building in order to minimize the effects of wind and sun, and because the inside wall is typically better connected thermally to the outside. If possible, an outside inspection is simpler because views of larger building faces can be achieved.

Any conditioned space is influenced by pressure differences caused by natural or forced convection. These result in air movement, which will often leave a characteristic thermal signature. Excessive air leakage can be identified with thermography, even when temperature differences are only a few degrees. During the heating season, signatures will be seen as cold streaks along the interior building surfaces or warm "blooms" on the outside of the building where heated air is exfiltrating. Some air movement may be evident inside the walls, even interior or insulated exterior walls. By artificially inducing a pressure difference on the building using the HVAC system or a device called a blower door fan, air leakage patterns can be enhanced and to some extent quantified.

Although the inspections of residential-scale buildings are quite straightforward, those of large commercial units can be more complicated, although the returns on the investment, which can be huge, usually warrant a thorough inspection. When possible, such buildings should be inspected during construction as each floor is closed in, insulated, and the relevant finish installed. This allows for design or construction problems to be identified and corrected before the entire building is completed and occupied.

Reducing excessive energy consumption is important, but, generally, a well-planned inspection will also increase occupant comfort, which may also lead to reduced energy use. Several case studies have documented not only an overall reduction in energy use, but more importantly, a reduction in peak use or the size of the system required to condition the space. Other issues may be even more important, such as minimizing unwanted moisture condensation and consequent mold blooms, eliminating the undue buildup of ice on the roof, checking air circulation in conditioned spaces, and preventing the plumbing from freezing.

8. ADVANTAGES AND LIMITATIONS

Previous discussion has revealed many of the advantages and limitations of infrared thermography. In some applications, thermography has proven itself as an inspection technique that is highly effective and very easy to use, whereas other applications may require sophisticated analysis.

Probably the two advantages thermography has over many other inspection techniques is that it is fast and it can create a thermal image. For many applications, an experienced and trained thermographer can make a determination of condition almost immediately upon viewing the thermal image with the right equipment, and good thermographers can create an excellent thermal image in seconds.

Today's equipment is extremely capable of doing amazing things. Radiometers are capable of accurate noncontact temperature measurement. Thermal imaging systems can resolve temperature differences of less than 0.18°F (0.1°C). The spatial resolution of the latest imaging systems provides such detail that many thermographers are forgoing capturing a visual image. In-camera processing allows sophisticated field analysis in a package that weighs less than 5 pounds.

As has been seen, thermography is a very versatile inspection method. The applications where thermography is effective are numerous. The aerospace industry is now augmenting, and in some cases supplanting, ultrasonics and radiography for determining the location of subsurface flaws and inclusions in state-of-the-art composites. Using similar techniques, a roofing contractor can locating areas of wet roof insulation for replacement. From identifying production problems in printed circuit boards, to finding loose connections in electrical systems, to the thermal mapping of complex industrial machines, thermography's list of applications is large and diverse. Where heat is a by-product of a process, or where an object undergoes a thermal cycle, thermography may very well have the ability to provide information about the operation and/or the internal integrity of a component.

The limitations of the technology can be summed up rather quickly. Only the surface of an object can be seen thermally. The thermal pattern is the result of either subsurface differential heat transfer or heat reflecting off the surface. As discussed, some material surfaces are so thermally reflective that they require preparation with a high-emissive coating. If the reflective surface cannot be made more emissive, the subsurface condition may not be resolvable, or the internal temperature will have to be increased until the surface temperature exceeds the minimum detectable temperature difference of the thermal imager. The thermal image must be interpreted. This requires knowledge of the application along with training and experience in thermography.

9. GLOSSARY

Absolute temperature scale—Temperature scales that are measured from absolute zero.
Absolute zero—The point on the Kelvin and Rankin temperature scales that indicates zero. Commonly known as the temperature at which no molecular activity occurs.
Background—The radiating objects that are reflected by a surface to the infrared instrument, usually from "behind" the instrument.
Beaufort's Wind Scale—A simple scale developed by a sea captain named Beaufort that allows one to estimate wind speed based on visual indications, such as the characteristics of the waves or leaves on a tree. It is useful to thermographers who need to estimate wind speed at a specific location.
Blackbody—A surface that absorbs and reradiates all energy incident upon it. Perfect blackbodies do not exist, but surfaces that are close to blackbodies do exist and, if traceable to a standard source, can be used to check the calibration of a system.
Boundary layer—A thin layer of fluid that builds up next to a surface during convection. It reduces heat transfer in proportion to its thickness.
British Thermal Unit (Btu)—A unit of energy defined as the amount of heat required to raise the temperature of a pound of water one degree Fahrenheit at sea level (at standard pressure). A Btu is equal to approximately 1055.06 joules
Calorie—Commonly referred to as the amount of heat needed to raise the temperature of one gram of water one degree Celsius. The modern definition is the amount energy equal to about 4.2 joules. Its symbol is "c" or cal.
Cavity radiator—A hole, crack, scratch, or cavity that will have a higher emissivity that the surrounding surface because reflectivity is reduced. A cavity seven times deeper than wide will have an emissivity approaching 0.98.
CCD—Charge coupled device; one of several types of electronic readout devices used in focal plane array systems.

Coefficient of thermal conductivity—See *Thermal Conductivity*.
Conduction—Heat transfer from warmer (more energetic) to cooler (less energetic) areas in a substance due to the interaction of atoms and molecules. This is the only way heat is transferred in solids. Heat transfer by conduction is also present in fluids (liquids and gasses) when atoms or molecules of different energy levels come in contact with each other.
Conductor—Loosely defined as a material that conducts heat well, usually in comparison with materials that don't conduct well (insulators). Most metals are good heat conductors.
Convection—The movement of fluids in response to a temperature difference.
Density—The mass of a substance per unit volume. Units are pounds per cubic foot.
Electromagnetic radiation—Vibrating electrical and magnetic fields in the form of waves that travel at the speed of light.
Electromagnetic spectrum—Electromagnetic radiation at all possible wavelengths from gamma rays to radio waves.
Emissivity—A property of a material that describes its ability to radiate energy by comparing it to a blackbody (a perfect radiator) at the same temperature. Emissivity values range from zero to one.
Film coefficient—See *Boundary layer*.
First law of thermodynamics—Energy in a closed system is constant; it can't be created or destroyed.
Forced convection—Convection caused by wind, fans, pumps, stirring or some other added force.
Foreground—The radiating objects that are "in front of" the camera and the target.
Fourier's law—The rate equation that describes conductive heat transfer, where energy equals thermal conductivity × area × temperature difference.
FOV—Field of view; a measure of the angular view for a given system and lens combination, usually measured in degrees.
Heat—Also known as thermal energy; energy transferred from regions of higher temperature to areas of lower temperature when a material changes temperature.
Heat of fusion—The latent heat released as a material changes from a liquid state to a solid state or absorbed as it changes from solid to liquid.
Insulator, insulation—Loosely defined as a material that restricts the flow of heat, especially in comparison with materials that conduct heat well (conductors).
IFOV—(Instantaneous field of view.) A measure of the smallest area that can be seen by the system at any one instant, i.e., spatial resolution.
IFOV$_{meas.}$—(Instantaneous field of view—measurement). A measure of the smallest area that can be measured by the system at any one instant. It is a measurement, not a specification of spatial resolution.
InSb—("Ins-bee" or indium antimonide.) A photon detector material with excellent performance in the short-wave band.
Isotherm—A software function that outlines areas of apparent similar temperature or radiosity in the image.
Kilocalories—(One thousand calories.) Commonly used for expressing the energy value of foods. The symbol is Kcal or C.
Kirchhoff's law—For an opaque object, radiant energy absorbed equals radiant energy emitted.
Latent energy—Energy used to make or break the bonds in the phase of a material.
Latent heat of fusion—The energy that is used to create or break the bonds in the solid phase of a material.
Latent heat of vaporization—The energy that is used to create or break the bonds in the gaseous phase of a material.

Long-wave—Radiation with wavelengths between 8–15 μm.
Natural convection—Convection occurring only due to changes in fluid density.
Newton's law of cooling—The rate of heat transfer for a cooling object is proportional to the temperature difference between the object and its surroundings.
Opaque—Nontransparent; $T = 0$.
Phase change—The process that matter goes through when it changes from a solid to a liquid to a gas.
Qualitative thermography—Thermal imaging using nonradiometric equipment or images to compare the radiation coming from various targets without making radiometric measurements of temperature.
Quantitative thermography—Thermal imaging using radiometric equipment (radiometers) or radiometric images to make radiometric measurements of the target temperature.
Quasi-steady-state heat flow—A thermal condition that is assumed to be steady-state for the purpose of analysis.
Radiation—The transfer of heat energy by electromagnetic waves or radiation.
Radiosity—All radiation coming from a surface, including that which is emitted, reflected, or transmitted.
Radiometric—An image or system that is calibrated to infer temperature measurements from the detected infrared radiation.
R-value—The measure of a material's thermal resistance. It is defined as the inverse of thermal conductivity.
Resistance—The measure of a material's ability to resist the flow of energy by conduction. Its value is the reciprocal of its conductivity.
Second law of thermodynamics—Heat cannot flow from a cooler object to a warmer one unless additional work or energy is added. Also stated as "heat cannot be totally changed into mechanical work."
Short-wave—Radiation with wavelengths between 2–6 μm.
Slit response function (SRF)—A test used to determine the smallest object that can be seen or measured by a system.
Solar glint—A phenomenon usually associated with short-wave sensing infrared systems whereby reflections of the sun off any shiny surface are very predominant.
Span—The set of temperature values that can be measured within a preset range. Thermal "contrast."
Spatial resolution—A specification, usually in milliradians (mRad), of the smallest size object that can be seen by the system.
Specific heat—The amount of heat required to raise a unit mass of a given substance by a unit temperature.
Spot radiometer—Nonimaging radiometric device that outputs a temperature or other radiometric measurement; also called infrared thermometer.
Spot size—The size of an area that can be measured at a given distance by a radiometric system.
State change—See *Phase change*.
Steady-state heat flow—A hypothetical thermal condition in which temperature difference across a material or system is unchanging.
Stefan–Boltzmann constant—5.7×10^{-8} W/m²·K⁴
Stefan–Boltzmann law—Total energy radiated by a blackbody surface is proportional to its absolute temperature to the fourth power.
Temperature—The relative measure of hotness or coldness of a material or substance.
Thermal background—The radiation sources "behind" the system that are reflected to the detector.

Thermal capacitance—The ability of a material to store thermal energy. It is defined as the amount to heat energy (in joules) required to raise the temperature of one kilogram of material one degree Kelvin. It is arrived at by multiplying a material's specific heat times its density.

Thermal conductivity——The symbol for thermal conductivity is "k." It is the measure of materials' ability to conduct thermal energy. It is defined as the rate at which heat (Watts) flows through a material of unit area and thickness, with a temperature gradient (Kelvin), over a unit of time. In SI units it is $W/m^2 \cdot K$.

Thermal resistance—The inverse of thermal conductivity. It is the measure of a material's ability to resist the flow of thermal energy.

Thermodynamics—The study of energy, how it changes and how it relates to the states of matter.

Thermography—From the root words for "heat pictures."

Thermograph—A visual picture of thermal data; a thermal image.

Thermographer—A person who is qualified to use thermography equipment.

Transient heat flow—A thermal condition in which the heat flow through a material or system is changing over time.

10. BIBLIOGRAPHY AND REFERENCES

1. *Proceedings of Thermosense.* 1980–1999, SPIE, Bellingham, WA.
2. ASTM Standards (misc.), available from ASTM, 100 Barr Harbor Drive, West Conshohocken, PA 19428-2959; phone 610-832-9500/ fax 610-832-9555.
3. *ISO 6781 Thermal Insulation, Qualitative Detection of Thermal Irregularities in Building Envelopes, Infrared Method.* American National Standards Institute.
4. *Manual for Thermographic Analysis of Building Enclosures (149-GP-2MP).* Canadian General Standards Board, Ottawa, Canada K1A 1G6.
5. *NFPA 70-B, Recommended Practice for Electrical Equipment Maintenance,* and *NFPA 70E, Standard for Electrical Safety Requirements for Employee Workplaces.* National Fire Protection Association, 1995. (NFPA, PO Box 9101, Quincy, MA 02269; 800-344-3555.)
6. *SNT-TC-1A-1996, Guidelines for the Qualification and Certification of Nondestructive Testing Personnel.* American Society for Nondestructive Testing (ASNT), 1711 Arlingate Lane, P.O. Box 28518, Columbus, OH 43228-0518; 800-222-2768.
7. W. L. Wolfe and G. J. Zissis (Eds.). *The Infrared Handbook,* 1993 Edition, ERIM, Ann Arbor, MI ISBN-0-9603590-1-X
8. D. P. DeWitt and G. D. Nutter. *Theory and Practice of Radiation Thermometry.* Wiley, New York, 1988.
9. F. P. Incropera and D. P. DeWitt. *Fundamentals of Heat and Mass Transfer,* 3rd edition. Wiley, New York, 1990.
10. J. M. Lloyd. *Thermal Imaging Systems.* Plenum Press, New York, 1982.
11. E. C. Guyer (Ed.). *Handbook of Applied Thermal Design.* McGraw-Hill, New York, 1989.
12. R. K. Stanley and P. O. Moore. *Nondestructive Testing Handbook, Volume nine: Special Nondestructive Testing Methods.* ASNT, Columbus, 1995.
13. X. P. V. Maldague. *Nondestructive Evaluation of Materials by Infrared Thermography.* Springer-Verlag, London, 1993.

CHAPTER 10
ACOUSTIC EMISSION TESTING

1. HISTORY AND DEVELOPMENT

The word "acoustic" is derived from the Greek word *akoustikos,* which has to do with "hearing." For centuries, the precursor to structural collapse has been sounds that are emitted prior to the failure of a supporting member. A tree branch emits a cracking sound before it actually breaks and stepping onto thin ice produces sounds that warn of impending collapse. Acoustic emission (AE) in this form is to the ears what visual inspection is to the eyes. The analysis of these emissions has become a science in itself.

Acoustic emission testing (AET) has become a recognized nondestructive test (NDT) method commonly used to detect and locate faults in mechanically loaded structures and components. AE can provide comprehensive information on the origination of a discontinuity (flaw) in a stressed component and also provides information pertaining to the development of this flaw as the component is subjected to continuous or repetitive stress.

Discontinuities in components release energy as the component is subjected to mechanical loading or stress. This energy travels in the form of high-frequency stress waves. These waves or oscillations are received with the use of sensors (transducers) that in turn convert the energy into a voltage. This voltage is electronically amplified and with the use of timing circuits is further processed as AE signal data. Analysis of the collected data comprises the characterization of the received voltage (signals) according to their source location, voltage intensity and frequency content.

The major difference between the AE method of NDT and the other NDT methods is that this method is passive, whereas the others, in a sense, are for the most part active. With ultrasonic, radiographic or the other NDT methods, the source of information is derived by creating some effect in or on the material by external application of energy or compounds. AE relies on energy that is initiated within the component or material under test.

The origination of the method is attributed to J. Kaiser in the 1950s. The sounds emitted during crack growth became an issue of scientific investigation during the 1960s. As the technology developed, AE became accepted as a NDT method. Separating the useful information from the background noise was the challenge to the instrument developers. Maturity of the technology led to the ongoing investigation into the micromechanical processes that produce these emissions within various materials.

The technology involves the use of ultrasonic sensors (20 Khz–1 Mhz) that listen for the sounds of material and structural failure. Acoustic emission frequencies are usually in the range of 150–300 kHz, which is above the frequency of audible sound. Crack growth due to hydrogen embrittlement, fatigue, stress corrosion, and creep can be detected and located with the use of this technology. High-pressure leaks can also be detected and isolated. AE technology is also becoming commonly applicable to nondestructive testing for structural integrity of structures made from composite materials.

2. PRINCIPLES OF ACOUSTIC EMISSION TESTING

The AET process is illustrated in Figure 10-1. It begins with forces acting on a body; the resulting stress is the stimulus that causes deformation and with it, acoustic emission. The stress acts on the material and produces local plastic deformation, which is breakdown of the material at specific places. This material breakdown produces acoustic emission: an elastic wave that travels outward from the source, moving through the body until it arrives at a remote sensor. In response, the sensor produces an electrical signal, which is passed to electronic equipment for further processing.

2.1 Acoustic Emission Sources

2.1.2 Stress
As previously mentioned, the AE process begins with stress. Stress is a familiar concept to engineering personnel. It is like an internal force field in a structure that transmits and balances the externally imposed forces (load). Depending on its directional properties, stress may be described as tensile, compressive, bending, shear, or torsion. Stress is measured in pounds per square inch (psi) (kilograms per centimeter^2) To calculate stress, the force (pounds) is divided by the area that carries it (square inches).

Stress can be thought of as a three-dimensional field having different components in different directions at each point in a structure. In response to stress, the structure of the material changes in shape. This change in shape is called "strain." The material deforms elastically and if the stress is high enough, it will deform plastically as well. "Plastic" in this context means "permanent." Plastic deformation involves a permanent change in the relative positions of the atoms in the material structure. On the atomic scale, plastic defor-

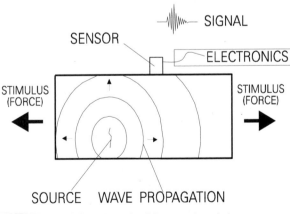

FIGURE 10-1 Schematic of the acoustic emission process.

mation involves the sliding of atomic planes over one another, through the agency of atomic-scale irregularities known as dislocations. The movement of dislocations is the microscopic mechanism that underlies the gross changes in shape that are recognized as yielding, buckling, denting, etc. Acoustic emissions that result from the movement of dislocations, or "strain," have been extensively studied with special laboratory techniques.

Other kinds of permanent deformation take place when materials break and new surfaces are created. On a microscopic scale inside a piece of steel, the materials most likely to break are specks of sulfide, oxide, and carbide, and other nonmetallic materials. The smallest of these items are the carbide "precipitates" scattered within the metal grains, for example, microscopic plates of iron carbide, only a few hundred atoms thick, distributed in "pearlite colonies." These precipitates play a big part in governing the steel's mechanical properties. On a larger scale, there are nonmetallic "inclusions" lying between the metal grains, such as manganese sulfide "stringers" formed during the rolling of the steel plate and slag inclusions introduced during welding. There may also be nonmetallic corrosion products intimately connected to the metal surface. All these nonmetallic components are less ductile than the metallic matrices in which they are embedded. As a result they break more easily when the metal is strained. The breaking of these nonmetallic components is the main source of the acoustic emission observed when crack-free metals are deformed.

When metal is cracked, there is a different kind of acoustic emission source and this one is the most important in nondestructive testing. The occurrence of a crack instantly creates a new surface. This is a major threat to a structure's integrity and is also the best-recognized source of high-amplitude acoustic emissions. Detection of emissions from growing cracks has been the most common goal in the many applications of AE technology. When a surface-breaking crack grows, the structure opens up in response to the applied forces. This is far more serious than the opening of an inclusion, which would tend to have no more than a local effect. Therefore, cracks tend to produce higher amplitude signals that are more readily detectable.

As well as causing large-amplitude AE waves as they progress, cracks produce small-amplitude AE waves from material deformation at the crack tip. Emission can also be produced from the rubbing or "fretting" of crack surfaces as they open and close and grind in response to changing loads. Corrosion products forming on the crack surfaces can enhance this emission, which make the crack even more emissive.

When material in a component deforms in response to any type of loading, the deformation tends to relieve and smooth out the local stresses. This means that after an acoustic emission event has taken place, the elastic energy stored in the stress field will have been reduced; some of it will have been released. The energy released from the stress field is used to create new deformations that will warm the material and produce the acoustic emission. Stated another way, the source of the acoustic emission energy is the energy stored in the elastic stress field produced by the loading of the structure. Acoustic emission is produced at the source, as a short pulse of elastic and kinetic energy that travels through the material as an elastic wave. The theory of frequency spectra shows that, being a short impulse, the wave carries energy at all frequencies from very low to some high upper limit, on the order of 1000 kHz and higher. Experience has shown that high sensitivity to these emissions is most easily achieved by using contact sensors in the upper part of this frequency range, between 100 kHz and 500 kHz.

Some of the lower-frequency emissions (approximately 50 Hz to 15 kHz), if they are loud enough, can he heard. This confirms the idea that the energy of acoustic emissions is spread over a very wide frequency range, and the theory that AE comprises frequencies all the way down to zero is evidenced by the largest acoustic emissions of all, earthquakes, which shake buildings a hundred miles away, at frequencies of a few hertz and less. Finally, the lower-frequency component itself is identical to the permanent change in the stress field, created by the action of the source event.

The amount of acoustic emission energy released, and the amplitude of the resulting wave, depends on the size and the speed of the source event. A strong event produces a greater signal than will a weak event. The theory is that emission amplitude is proportional to the area of the new surface created. A sudden, discrete, crack event will produce a greater signal than will a slow, creeping advance of the crack tip over the same distance. The theory is that emission amplitude is proportional to the crack velocity.

The association between acoustic emission and crack growth has been intensively studied. Processes involving some form of embrittlement, such as hydrogen-induced cracking and stress corrosion cracking, are generally among the better emitters. Ductile processes such as slow fibrous fracture are generally quieter. Weldments are more emissive than parent metal because they are by nature less ductile.

It is useful to distinguish the three different classes of source activity:

1. Primary activity from new, permanent changes in the originally fabricated material. This is typically due to local stresses that are higher than the original stress levels.
2. Secondary activity from materials that were not part of the original fabrication, such as corrosion products.
3. Secondary activities from repetitive processes such as crack surface rubbing (friction) that do not produce new, permanent changes in the material. Secondary activity can be either helpful or a nuisance, depending on the way it is treated. Secondary emission is different from "noise," which is always a nuisance to the AE practitioner.

Noise in AE testing means any unwanted signal and is a major issue in acoustic emission technology. The main types of acoustic noise sources are friction and impact, which can result from many environmental causes. Frictional sources are stimulated by structural loading, which causes movement at movable connectors or loose bolts. Impact sources include rain, wind-driven dust, and flying objects. An intrinsic part of the AE test technique is the ability to eliminate all these noise sources and to focus on what is relevant. Noise is addressed in three ways: (1) by selecting an appropriate test strategy and instrumentation setup; (2) by taking practical precautions on site to prevent noise sources as far as possible; (3) by recognizing and removing noise indications from the recorded data. This last process is the domain of data interpretation.

2.2 Structural Loading and AE Source Activity

Acoustic emissions occur at locations where the local stress is high enough to cause new, permanent deformation. This often happens at stress concentrations, regions where the stress is raised by local geometry. Stress concentrations exist at weld details, changes in section, and structural discontinuities in general; they also exist around cracks and flaws. The stress concentrations at weld details are the reason why fatigue cracks initiate at these locations.

When a material deforms and emits energy, the deformation tends to relieve the high local stresses. Often the load is transferred to some other part of the structure. This has a stabilizing effect. If the structure is unloaded and then reloaded to the same level, the regions that deformed the first time will tend to be stable the second time. Thus, the emission sources will tend not to reemit the second time around, unless the load exceeds the previous maximum.

When a material is loaded (stressed) it changes shape: it stretches, compresses, or shears. The technical term for this change in shape is "strain." The strain has an elastic, reversible component and also (if the load is high enough) a plastic, permanent component.

The elastic component of the strain occurs immediately after the load is applied. The stress/strain field inside the material is quickly redistributed such that all the forces are balanced. Actually, this redistribution takes place at the speed of sound, through the propagation of elastic waves. This is why a body vibrates if a shock force is suddenly applied.

Unlike the elastic component, the plastic component of the strain often takes considerable time to develop. Some of the deformation is immediate but some of it is delayed. Delayed deformation of nonmetallic materials is quite familiar. In time, plastics creep and stretch and wooden beams sag. Steel shows only a trace of this kind of behavior, but acoustic emission is a very sensitive indicator and will often reveal time-dependent behavior that would otherwise go unnoticed.

Figure 10-2 illustrates the characteristic behavior pattern of a newly fabricated component. In this figure, load and AE are both plotted against time. The load is raised and held, then raised and held again. AE is generated during both load rises. During the first load hold, there is no emission. But during the second load hold, the stress is higher. The emission continues for some time into the second hold period, and then the component eventually stabilizes.

Emission that continues during load holds is likely to indicate structurally significant defects. Many test procedures place particular emphasis on emission during load holds. Emission that occurs during rising load on previously unloaded structures is less easy to interpret; it may result from discontinuities. Structurally sound material will also produce emissions while the stress is increased during initial loading. The interpretation of emission during load holds is easier. Another characteristic of structurally significant defects is that they tend to emit on a second loading. If a second loading is carefully monitored, one often sees a little emission before the previous maximum load; not nearly as much as the first time, but not zero either. This emission can be an important indicator of structural instability.

As can be observed in Figure 10-3, emission is plotted directly against load. In this scenario, the load is raised, lowered, raised again to a higher level, lowered, and finally raised to a higher level still. Emission is generated during the first load rise (AB), but as the load is lowered (BC) and raised again (CB) there is no more emission until the pre-

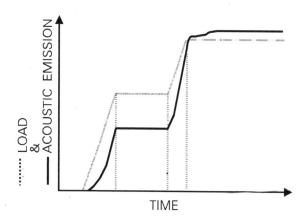

FIGURE 10-2 Emission continuing during load hold indicates instability.

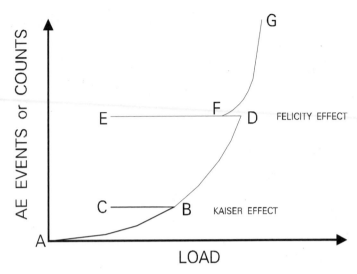

FIGURE 10-3 Emission on repeated loading.

vious load maximum is exceeded. Emission continues as the load is raised further (BD), and stops as the load is lowered for the second time (DE). On raising the load for the last time, a different emission pattern is observed: the emission starts up before the previous maximum load is attained (F). Emission continues as the load is increased (FG). The behavior observed at point B (no emission until previous maximum load is exceeded) is known as the Kaiser effect. The behavior observed at point F (emission at a load below the previous maximum) is known as the Felicity effect. Insignificant flaws tend to show the Kaiser effect while structurally significant flaws tend to show the Felicity effect.

Of special interest is the AE monitoring of structural fatigue. The emission behavior of growing fatigue cracks has been extensively studied. Classic laboratory data is shown in Figure 10-4. This diagram shows both the crack length and the accumulated total of the emission detected. The emission began with crack initiation, and then tracked rather closely with the growth of the crack, increasing as the crack propagated more rapidly toward failure.

The experiment to prove this was carried out with cyclic loading using a fixed load amplitude. It was found that the primary emission from active crack growth occurred only at the peak load levels. In fact, Figure 10-4 shows only the emission that occurred at the peak load levels; secondary emission and noise that occurred at lower load levels were outside the gate threshold level. At first, when the crack was still small, not every cycle produced emissions. However, as the crack approached the critical length for unstable propagation, every cycle produced emissions. This result fits well with the behavior of statically loaded specimens discussed above, in that insignificant flaws tend to show the Kaiser effect, whereas structurally significant flaws tend to show the Felicity effect.

The primary emission from growing fatigue cracks can result from two sources. First, there may be emissive particles, typically nonmetallic inclusions, in the stress-concentrat-

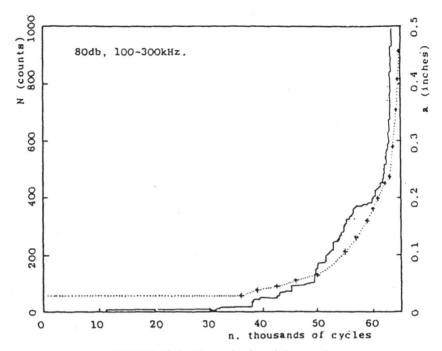

FIGURE 10-4 AE growing from fatigue crack.

ed region near the crack tip. As the crack advances toward these particles, the local stress levels rise, and their breaking will produce a primary emission. The other source is the movement of the crack tip itself. Crack tip movement is typically taking place in a mixed mode: some of the new surface is created by dislocation activity and some of it is created by small-scale cleavage, a sudden separation of the material in a region of local weakness and/or exceptionally high stress. Crack tip movement by dislocation activity is typically not detectable, but cleavage is an abrupt and relatively gross mechanism that produces plenty of AE energy in the normally detectable range.

Secondary activity from crack face friction is also often observed in AE monitoring of fatigue cracks. In constant-cycle fatigue, this activity often produces just the same signal, cycle after cycle, at intermediate load levels. This secondary emission may continue for hundreds or thousands of cycles, then die out only to start again later in the test. The best explanation is that rubbing at rough spots or "asperities" on the crack surface, as indicated in Figure 10-5, produces this. It has also been suggested that the recently created surfaces at the crack tip may stick together then break apart again as the crack tip opens and closes.

Theoretical relationships between AE and crack propagation rates have been developed. Extensive research work has been done on AE from constant-cycle fatigue; however, less work has been done relating to the random-cycle fatigue. Distinguishing between primary and secondary emission is easy in the case of constant-cycle fatigue. In the case of random-cycle fatigue it is not so easy. Crack face movement, either due to friction or to fresh growth, is an undesirable and probably deteriorating condition in the material that should be corrected.

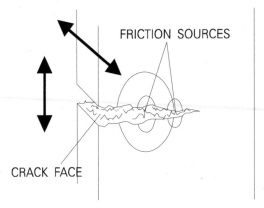

FIGURE 10-5 Crack face rubbing can produce secondary emission.

2.3 The Signal Shaping Chain

The signal shaping chain is shown in Figure 10-6. It has four links: the source, the propagation of the wave, the sensor, and the signal-conditioning electronics. Each link has a controlling influence on the size and shape of the measured signal. The final signal is drastically different in shape from the original motion at the source. An important consideration in discussing the signal shaping chain is the frequency content. All signals can be analyzed as to their "sine wave" frequency components. This is the field of "Fourier" analysis, one of the most powerful tools in the science and technology of signal processing.

The term "frequency" refers to the repetition rate of an oscillation within a given time period, i.e., the number of cycles per second. Anyone who has turned the knob to tune in a car radio has been selecting the frequency that he or she wants to receive. Each radio station broadcasts at a particular frequency. By turning to a station, the receiver becomes sensitive to the desired frequency. An acoustic emission source, however, is not like a radio station radiating just one frequency. It is more like a lightning bolt! When listening to a radio during a thunderstorm, the radio will pick up the lightning discharge anywhere on the AM band. This is known as a "broadband" signal, in contrast with the "narrow band" radio station. Like lightning, the impulsive acoustic emission source radiates energy at many frequencies.

From the fact that the source is broadband and radiating a variety of frequencies, it follows that there is a choice of frequencies to use for sensing the wave and measuring the resulting signal. This is the rationale for selecting the sensor and electronic filtering.

The frequency response of AE sensors can be either broadband or narrow band. Broadband sensors offer higher fidelity, a more faithful rendering of the actual motion of the metal surface at the sensor location; yet, in most practical AE testing, narrow band sensors are preferred. There are several reasons for this. Resonant sensors are generally more sensitive and much less expensive than the broadband types. They have the advantage of operating in a known and well-established frequency band, which can be chosen to optimize system performance when faced with wave attenuation and background noise. Broadband sensors have the potential to deliver extra information, through the use of advanced signal processing techniques; but, unfortunately, the best of this information is at the low end of the spectrum, just where the noise problems are worst. Thus, high-frequency resonant sensors are recommended for most practical structural monitoring.

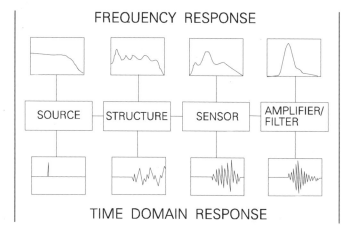

FIGURE 10-6 Signal shaping chain.

Once the resonant frequency sensor has been selected, the frequency bandpass of the amplifier/filter combination in the AE instrument is normally set to match it. Thus, the last two links of the signal shaping chain have been established. Now the signal measurements will be altered only by changes in the first two links. The same kind of source event at different locations can produce different signals; and different kinds of source events at the same location can also result in different signals. To obtain meaningful information from the AE signals, it is necessary to have made a suitable choice of sensor type and frequency filter, as well as to have a working understanding of source behavior and wave propagation effects.

2.4 Propagation of the AE Wave

The AE wave can be visualized by recalling the ripples produced when a stone is dropped into a pond. The ripples spread out from the source and eventually reach the banks, reflecting in a complicated pattern and ultimately dying away. Reacting to these ripples, a floating cork near the bank of the pond will bob up and down for several seconds in a complicated rhythm, even though the impact of the stone was over in a fraction of a second. The same principle applies to the propagation of AE waves. The short pulse from the source is only the beginning of the AE process. The motion at the sensor is quite unlike the original shock at the source. This is the part played by wave propagation in the signal shaping chain. The principal difference between the ripples on a pond and the AE waves in a structure is that the AE process occurs many times faster. The typical AE source motion is finished in a few millionths of a second. The wave takes perhaps a thousandth of a second to reach the sensor and it takes in the order of a hundredth of a second for the motion to die away.

Another difference between the pond and the structure is that wave propagation in a structure is generally more complex. First, the structure has many surfaces that repeatedly reflect the wave. Second, solids (unlike liquids) support shear forces as well as compressional forces. This leads to the existence of several different wave types (modes) that can all be excited simultaneously.

Aspects of the wave propagation process that are particularly important in AE technology areas are:

- Attenuation—the loss of amplitude as the wave travels outward from the source. This is important when detecting waves from distant sources.
- Wave velocity—the speed with which the disturbance travels through the structure is important for some source location techniques. In addition, signal shaping effects have a profound influence on the measured values of the detected AE signal.

2.4.1 Attenuation

As the acoustic wave travels through the structure, its amplitude decreases. This effect is known as attenuation and is illustrated (for a steel plate) in Figure 10-7. In this figure, amplitude is plotted on the Y-axis using the dB_{ae} decibel scale. On this scale, each increment of 20 decibels is a tenfold increase in the signal peak voltage. The graph shows that the peak signal voltage 9 feet from the source is about one-thirtieth of the voltage very close to the source. The dB_{ae} scale is universally used and very convenient because, being logarithmic, it condenses the very wide range of AE signal amplitudes. dB_{ae} is a logarithmic measure of acoustic emission signal amplitude referenced to 1 μV.

Attenuation is due to several factors. In most structures, the important ones are geometric spreading, scattering at structural boundaries, and absorption. When the source is only a few inches from the sensor, as in local monitoring of weld details, the geometric spreading effects are the most important. These effects are therefore especially important for crack detectability. At distances greater than a couple of feet, energy absorption and structural scattering are the most influential properties. The understanding of these effects is also important for identifying extraneous noise.

Geometric spreading effects are fundamental to wave propagation. Basically, the sound wave is "trying" to spread throughout the volume of the structure near to the source; the change in the static stress field and the wave motion are at their maximum.

In theory, for a structure that is large in all dimensions (infinite half-space), a stress

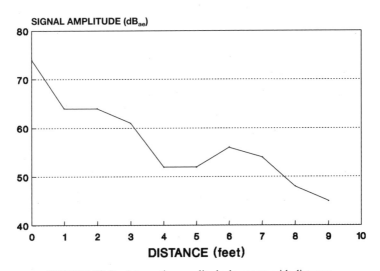

FIGURE 10-7 Attenuation amplitude decreases with distance.

wave would simply spread and continue to attenuate. In realistic structures, the boundaries force the stress wave to remain in a confined space such that attenuation from beam spreading is limited. In a small, well-defined structure such as a rod, the attenuation from geometric beam spreading is minimal and stress waves may travel great distances. The data shown in Figure 10-7 illustrate this relationship (approximately). The curve falls steeply at short distances and more gradually at larger distances. The 30% figure applies to a circular wave front spreading out in a plate-like structure such as a girder web. In a solid medium like concrete, the wave spreads out in all three dimensions and it loses amplitude more rapidly—50% for each doubling of distance. In a rod, the wave is channeled and cannot spread, so attenuation is relatively low.

The second major cause of attenuation is reflection (scatter) at structural boundaries and geometrical discontinuities. Whenever a wave meets a discontinuity, some of the wave energy is reflected. Discontinuities also produce mode conversions. These effects are especially important in the complex geometry of many structures, where there may be changes in direction, connections, stiffeners, and other boundaries along the acoustic path from source to sensor.

The third cause of attenuation is absorption. Here the elastic and kinetic energies in the wave are absorbed and converted into heat by the material through which the wave is passing. Steel absorbs very little at the frequencies used for AE testing. Nonmetallics, in general, and paint, in particular, tend to absorb energy more than steel. Absorption is greater at higher frequencies due to the shorter wavelengths at these frequencies. The distance/amplitude reduction mechanism for absorption is different from the mechanism of beam spread/amplitude reduction. In the case of distance/amplitude, there is a constant number of decibels absorption per foot from the source. This is the same at all distances from the source. For example, if there is a 6 dB reduction after the energy has traveled one foot, there will be an additional 6 dB reduction after the energy has traveled an additional foot, and so on. Thus, whereas energy losses due to beam spreading near the source are high, absorption also accounts for large losses of energy as the wave fronts move further away from the source.

Attenuation measurements are easily made with a simulated AE source. The most widely used simulated AE source is the breaking of a pencil lead pressed against a structural member, as illustrated in Figure 10-8. As the lead is pressed against the structural member, the applied force produces a local deformation that is suddenly relieved when the lead breaks. With good technique, the resulting stress wave is amazingly reproducible.

The Hsu pencil (named after the developer of the technique) and the accessory Nielsen shoe are convenient, inexpensive aids that have been enormously valuable in practical AE testing. The breaking of the lead creates a very short-duration, localized impulse that is quite similar to a natural acoustic emission source such as a crack. Furthermore, the amplitude of the lead break source is well within the range of typical crack sources. The Hsu pencil has become so well accepted as a simulated AE source, that in some procedures for wide-area monitoring, the maximum permissible sensor spacing is based on the ability to detect lead breaking from anywhere in the inspection area.

The usual procedure for developing a graph such as shown in Figure 10-7 is to break lead several times at each of several different distances from a sensor, and then to record the amplitude for each break. The amplitudes for each distance are averaged and, subsequently, the average amplitudes are plotted against distance.

The attenuation curve is an important aid in determining the sensor placements for the specific application. In many acoustic emission applications, the goal of the inspection is to monitor the entire structure. In this case, it is important that all parts of the structure are within detection range of at least one sensor. In global monitoring tests of this kind, the

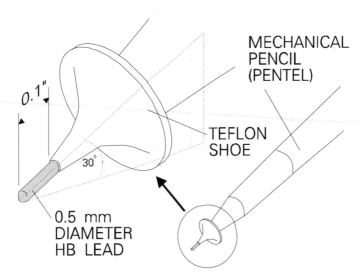

FIGURE 10-8 Hsu–Nielson source.

test procedure typically specifies how the attenuation curve can be used to determine acceptable maximum sensor spacing.

In some tests included in these guidelines, the recommended strategy is limited-area monitoring focusing on a known crack or possibly cracked site. Here the attenuation curve has different applications. It shows how sensitivity to the crack compares with the sensitivity to distant noise sources and whether distant noise sources will reach the sensor or whether they will be attenuated before reaching it. Also, it shows that sensors close to the crack must be placed at similar distances in order to get similar responses.

2.4.2 Wave Velocity

Source location calculations are based on the time of arrival of the wave from the source to the sensors. These arrival times depend on the velocity with which the waves travel from source to sensor. Understanding and predicting wave velocities is an important aspect of the science of physical acoustics.

An important wave mode in AE testing of structures is the Lamb wave. This wave mode is named after Horace Lamb, an English acoustician who in the 1920s developed the mathematical theory of sine waves propagating in finite plates. This kind of theory seeks to describe wave propagation in terms of wave modes—patterns of oscillatory motion that can propagate in a stable way, maintaining their shape as they travel. In plates, Lamb identified two families of wave modes, and he developed equations that described their velocities of propagation. In the first family, the motion is symmetrical about the median plane of the plate, and in the second family it is asymmetrical. The parent members of these families are called the s_0 and a_0 modes respectively. The form of the motion for these modes is shown in Figure 10-9.

The s_0 mode, often called the "extensional mode," is a rippling movement in which the plate is alternately stretching and compressing in the direction of the wave motion. Coupled with this in-plane movement, the sides of the plate are "breathing" in and out symmetrically as the ripples run through the plate. To produce this kind of wave motion, the

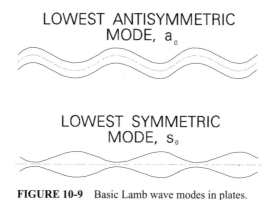

FIGURE 10-9 Basic Lamb wave modes in plates.

exciting force needs to be directed parallel to the plate. A sudden release of in-plane tension will also tend to produce this kind of wave motion.

The a_0 mode, often called the "flexural mode," is a mode in which the body of the plate bends with the two surfaces moving in the same direction. Most of the motion is transverse to the plate; there is relatively little motion in the plane of the plate. Exciting forces perpendicular to the plate produce this kind of wave motion. In addition to this, forces that are parallel to the plate but offset from the center line can produce this wave mode.

The underlying theory of elastic wave modes is the general mathematical theory of "eigenfunctions." According to this theory, wave modes travel independently and do not interfere with one another. Several wave modes can be traveling at the same time in the same material; in this case, the motion at any point in the material is simply the sum of the motions of the various modes. The wave modes are like the raw ingredients of a recipe, which can be combined in various proportions to produce many different flavors. The two Lamb wave modes travel at different velocities. At 300–500 kHz in steel plate thicker than ¼ inch, the a_0 mode travels at 10,000–11,000 ft/sec (3000–3300 m/s). The s_0 mode travels somewhat slower than 10,000 ft/sec. Thus, the a_0 mode will reach the sensor a little earlier. When setting up AE source location systems, the technician usually has to enter into the computer the sensor positions and also the velocity of sound that will be used in the source location calculations. Sensor positions will usually be specified in inches, and in this case the velocity must be entered, correspondingly, in inches per second. A good practical value in AE work on many structures is 120,000 inches per second.

The a_0 Lamb wave mode (flexural mode) is the most important wave mode in acoustic emission testing. It usually produces a higher amplitude wave than the s_0 mode, and in typical structures it also travels faster and therefore arrives at the sensor first. Other members of the two families of Lamb waves (called a_1, s_1, etc.) can travel faster than either the a_0 or the s_0 modes, but their amplitudes tend to be low and they are relatively unimportant.

Lamb waves provide the best wave propagation from the source and at distances many times greater than the plate thickness. Close to the source, i.e., within one or two plate thicknesses, it is better to think in terms of longitudinal and shear waves. These wave types are well known to technicians trained in ultrasonic testing (UT). In summary, a good knowledge of acoustic theory is the key to understanding the variables and evaluating the results of each test or situation.

2.4.3 Signal Shaping Effects

Wave propagation is a tremendously important stage in the signal shaping chain. It is wave propagation that largely determines the signal's size, shape, and shape-dependent signal features such as rise time and duration. Because several successful interpretive techniques are based on signal shape, it is important to have some understanding of these effects.

Figure 10-10 shows a typical record of an AE waveform recorded from a structure. Note the timescale on the X axis, running from –245 microseconds to 1.39 milliseconds (1390 microseconds), a total span of 1635 microseconds. Thus, the wave has occurred and decayed in about a thousandth of a second. In this time, there are several hundred positive and negative oscillations of the signal voltage. Some of the individual oscillations can be recognized as they stand out from the background. The dominant frequency during these oscillations is the resonant frequency of the sensor.

The waveform envelope (overall shape) shows a fast rise and a much slower decay. This is typical of AE signals. The waveform at the start of the time base is from direct waves from the source; the later part of the time base comprises waves that have been reflected back and forth many times before arriving at the sensor. The rising part of the time base is determined by strong reflections from nearby surfaces. The falling part of the timebase is shaped by more remote reflectors and by the acoustic damping in the structure.

2.5 Sensing and Measurement

2.5.1 Sensor Performance and Calibration

Secondary to wave propagation, the next essential stage in the signal shaping chain is the sensor. For sensing AE waves, piezoelectric crystals are used. The Greek word *piezein* means "to squeeze." Piezoelectric materials generate an electrical voltage and a corresponding separation of charge when they are squeezed (deformed). In the AE sensor, the

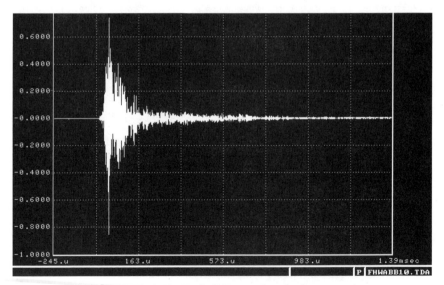

FIGURE 10-10 Typical AE waveform in a structure.

deformation is produced by motion; it is the elastic response of the piezoelectric crystal when it is struck by the incoming stress wave. The piezoelectric element is mounted inside the sensor enclosure, as illustrated in Figure 10-11. The electric voltage is generated by the element material itself. The element does not need an external power supply.

When struck by a sudden impulse, a piezoelectric element will vibrate like a bell, "ringing" at its resonant frequency. In fact, there may be many resonant frequencies all excited together. If shaken by a vibratory motion, a piezoelectric element will produce a corresponding oscillating voltage at the same frequency as the motion. The element has a linear response; if the input motion is doubled, the output voltage will also double.

The ratio (output voltage amplitude)/(input motion amplitude) is a measure of the sensitivity of the sensor. It has been established that this sensitivity depends strongly on the frequency of the motion (i.e., the number of oscillations per second). The sensitivity of the element is greatest at its resonant frequency.

Sensor calibration curves show how sensor sensitivity varies with frequency. An example is shown in Figure 10-12. The shapes of these curves are characteristic of the sensor type and vary widely in their sensitivity and frequency response. For a given sensor type, the calibration curves of individual sensors should be closely matched.

With calibration curves such as that shown in Figure 10-12, the amplitude of the input motion may be specified in units of displacement, velocity (meters per second), or pressure (microbars). In all cases, it refers to the amplitude of a sinusoidal oscillation at the frequency indicated on the X axis.

Calibration curves often have their Y axis (sensitivity) scaled in decibels (dB). Decibels are a relative measure: each decibel corresponds to an increase of 12.2%. Compounding upward, with each decibel being a 12.2% increase, it is easy to show that 6 dB is a doubling and 20 dB is a tenfold increase in sensitivity.

The sensitivity curves on sensor calibration certificates have their Y axis labeled in decibels relative to a stated reference level, such as 1 volt per meter/second. That is, zero dB means a sensitivity of 1 volt per meter/second. As shown on the graph in Figure 10-12, 6dB equals a sensitivity of 2 Volts per meter/second, and so on. Mathematically, the formula is:

$$dB = 20 \log (S/S_{ref})$$

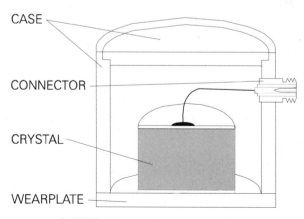

FIGURE 10-11   AE sensor schematic.

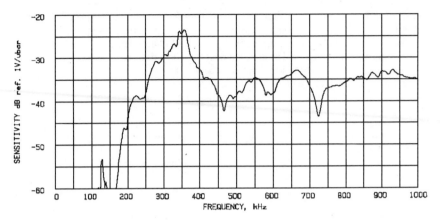

FIGURE 10-12 Typical sensor calibration curve.

where S is the sensitivity and S_{ref} the reference level, both stated in volts per meter/second.

The sensitivity of the sensor depends not only on frequency, but also on the direction of the motion. Unlike accelerometers, which are carefully designed to measure only the component of motion parallel to their axis, AE sensors respond to motion in any direction. Therefore, to understand fully the significance of a sensor calibration curve it is necessary to know about the type and direction of the wave that is being used for the calibration.

In the design of AE sensors for nondestructive testing, the manufacturer's priorities are high sensitivity, well-defined and consistent frequency response, robust performance in the working environment, and immunity to unwanted noise. Unwanted noise can also include electromagnetic interference originating from radio stations, navigation systems, etc. Immunity from this kind of noise was greatly improved with the development of the integral preamp sensor. This type of sensor has a preamplifier built into the housing along with the piezoelectric element. For field testing, it has considerable advantages over earlier sensor types, which required separate preamplifiers to be mounted within a few feet of the sensor.

The sensor must be in good acoustic contact with the structure under test so that it will detect the motion of the AE wave and deliver a strong signal. Coupling and mounting techniques are very important. An acoustic couplant in the form of an adhesive, a viscous liquid, or grease is applied to the sensor face and the sensor is pressed against the structure. The surface on which the sensor is mounted must be smooth and clean; some preparation may be necessary. The sensor must be securely held in place with adhesive, magnetic hold-down, or other means. After mounting, the system performance is verified by simulating an AE signal and checking the system response.

2.5.2 *Signal Conditioning, Detection, and the Hit Process*

The signal produced by the sensor is amplified and filtered, detected and measured. Amplifiers boost the signal voltage to bring it to a level that is optimum for the measurement circuitry. Along with several stages of amplification, frequency filters are incorporated into the AE instrument. These filters define the frequency range to be used and attenuate low-frequency background noise. These processes of amplification and filtering are called "signal conditioning." They "clean" the signal and prepare it for the detection and measurement process (evaluation).

After conditioning, the signal is sent to the detection circuit. This is an electronic comparator, a simple circuit that compares the amplified signal with an operator-defined threshold voltage. The principle is illustrated in Figure 10-13.

Whenever the signal voltage rises above the threshold voltage, the comparator generates a digital pulse. The first pulse produced by the comparator marks the start of the "hit." This pulse is used to trigger the signal measurement process. As the signal continues to oscillate above and below the threshold level, the comparator generates additional pulses. While this is happening, electronic circuits are actively measuring several key features of the signal. In time, the amplitude of the signals reduce to a point where there are no more threshold crossings. After the predetermined time (the "hit definition time") has passed without any further pulses from the comparator, the system determines that the event has ended. The hit process is then brought to a close. Control circuitry terminates the measurement process and passes the results to a microprocessor. Finally, the measurement circuitry is reset and rearmed ready for the next event.

2.5.3 Signal Features and Their Measurement

Measurements are made of the relevant features of the signal such as the amplitude, duration, signal energy, and counts. These features and their relationship to the detection threshold are shown in Figure 10-14. This diagram shows a typical AE signal as a voltage–time curve, just as it would appear on an oscilloscope screen. A typical signal lasts less than a thousandth of a second.

"Amplitude", in AE testing, refers to the largest voltage present in the signal waveform. It is one of the most important measures of signal height. It is fundamental because for a signal to be detected, its amplitude must exceed a predetermined threshold. Amplitude is usually measured in dB_{ae}, a decibel scale running from 0 to 100. 0 dB_{ae} is defined as an amplitude of one microvolt at the preamplifier input.

"Duration" is the length of time from the first threshold crossing to the last, measured in microseconds (millionths of a second). The relationship between duration and amplitude tells the user about the signal's shape: whether it is a short sharp "click," or a long drawn-out "scrape."

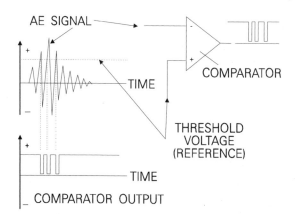

FIGURE 10-13 Signal detection by comparison with threshold.

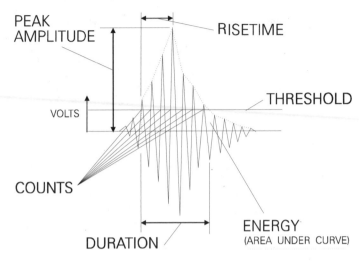

FIGURE 10-14 Key signal features.

"Signal energy" is the area under the voltage–time envelope, i.e., the area under the dotted line shown in Figure 10-14. This is another important measure of signal size and is the most widely used measure of AE activity. When a structure produces many emissions in response to loading, the energies of the individual signals can be added to produce a total amplitude. Of all the techniques that have been used to describe emission quantity in a single number, this has been the most successful.

"Counts" are the comparator output pulses corresponding to the threshold crossings. A single hit may provide only a few counts or it might furnish hundreds of counts, depending on the size and shape of the signal. For the electronics designer, this is the easiest measurement to make, and in the early years of AE, "counts" were the most common way to describe and report AE quantities. During the 1980s, energy replaced counts as the preferred measure of AE activity. However, counts are still useful for data interpretation; used in conjunction with amplitude or duration, they can give valuable information on signal shape.

Several other signal features may be measurable, depending on the equipment available, but those noted above are the most widely used. A block diagram for typical instrumentation used for making these measurements is shown in Figure 10-15.

As well as measuring the features of the individual signals, the AE instrument usually measures also the times at which they are detected and the environmental variables that may be causing the activity. The broader aspect of instrument architecture is discussed in the next section.

2.6 Instrument Architecture

The most common overall design for AE instruments is the so-called "hit-based" architecture. Typical AE activity consists of a series of distinct signals. These signals occur at irregular time intervals and have widely varying shapes and sizes. Hit-based architecture is designed for the efficient measurement and recording of this kind of activity.

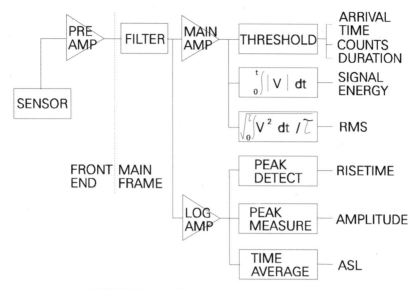

FIGURE 10-15 Signal measurement block diagram.

In hit-based architecture, the system remains dormant until a signal is detected by the threshold circuitry, as discussed in Section 2.5.2. The signal is measured and a microprocessor stores a data record containing the time of detection and the results of the measurements made. After this, the measurement circuitry is reset, ready for the next hit. Thus, the information passed to the computer consists of a series of "hit data sets." Each hit data set takes up typically 30 bytes of storage space. Ideally, each hit data set will correspond to the detection of one AE event (e.g., microscopic crack jump) in the stressed material of the structure under test. The test data builds up in computer memory, or is converted into displays or written to disk or some other medium according to the system design. Environmental data such as pressure, force, displacement, etc., may also be measured at the same time, and the results may be written along with the signal data.

From these data received, the computer can generate many different kinds of displays, both during data acquisition and also at any time afterward by replay of data files written to a disk or other storage medium. Representative architecture for a hit-based system is shown in Figure 10-16. This diagram shows a channel with amplification, filtering, detection, measurement of signal features, and the transfer of data and timing information to data storage and processing circuitry.

A major advantage of hit-based architecture is that it delivers a very detailed description of the emission while making very economical use of data storage space. The storing of hit data sets optimizes the use of storage media and gives the greatest scope for data interpretation and evaluation. The data may be retrieved weeks or even years later for evaluation. The test record is available in the same form as the day it was stored. The permanence of the data record and its ready availability for reanalysis is one of the true advantages, not only of this particular architecture but also of AE testing as a whole.

Hit-based architecture is aimed at comprehensive data acquisition and versatile analysis capability. There are other kinds of architecture with different goals, as well as variations within the broad class of hit-based architecture. Some instruments sacrifice perform-

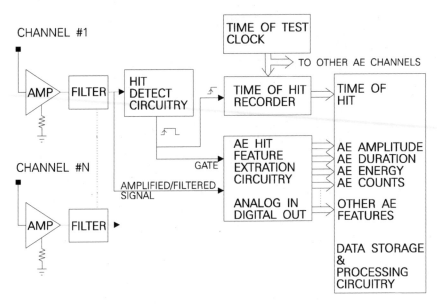

FIGURE 10-16 Instrument functional diagram.

ance in favor of portability. Some systems sacrifice versatility in favor of automatic operation optimized for specific applications. There are simple systems that accumulate AE energy or counts in hardware counters and output a corresponding voltage for plotting on a chart-type recorder. There are dedicated industrial instruments in which the computer prints a predefined set of graphs showing various aspects of the AE data but does not produce a permanent record of every hit. Instruments exist with special alarms or interpretive software. Both research laboratories and commercial vendors have implemented many different AE instrumentation ideas through the years. Out of several hundred diverse AE instrument designs that have been constructed, only a few have been successful enough to warrant production in substantial quantities. A state-of-the-art AE instrument is shown in Figure 10-17. This is a complete digital 52 channel AE system.

2.7 Multichannel Source Location Techniques

2.7.1 Introduction

AE instrumentation can include many channels. A single AE event can be detected on several channels, producing a hit on each one. These hits will occur in very quick succession, typically all falling within a few hundred microseconds as the wavefront ripples through the sensor array. By comparing the arrival times at different sensors, we can find out about the location of the source. This capability is one of the most useful features of AE testing. When AE is used to inspect a large structure with just a few widely spaced sensors, source location techniques can be used to pinpoint the exact regions that should be inspected later, using other NDT methods.

When AE is used to determine if a known flaw is active, source location techniques can be critical in enabling the user to discriminate successfully between the known flaw

FIGURE 10-17 State-of-the-art acoustic emission system.

and extraneous noise sources. Figure 10-18 shows more details of the basic concept. As the wave spreads out from the source, it reaches first the nearest sensor, S2, then the farther sensors, S3 and S1. As the resulting signals cross the detection thresholds on each channel, "hits" are produced on the corresponding channels in the measurement instrument. The first hit, on sensor S2, occurs at time r_2/c after the source event, where r is the distance from source to sensor and c is the wave velocity discussed in Section 2.4.2. The second and third hits occur at times r_3/c and r_1/c after the source event, respectively. Note that the instrument has no direct awareness of the time of the source event. It can measure only the times of the resulting hits; but from the differences between those hit times, it can calculate both the location and the time of the source event.

There are a variety of practical techniques derived from this general concept. The intention might be to focus on one or more predefined areas of interest, to discriminate against one or more predefined noise sources, or to set about the test without preconceptions and take an approach using a location-type overview of all sources on the entire structure. There are techniques based on computation from measured time differences and there are also simplified techniques that only consider the sequence of the hits, without even measuring the actual time differences. There are also techniques based on two, three, four, or more sensors with computations oriented to linear, planar, solid, and complex structural geometries.

With AE instruments, techniques are executed sometimes in hardware, sometimes in acquisition-time software or posttest software, and sometimes according to the user's choice. Three major approaches—computed source location, zone location, and guard techniques—are explained in the sections below.

2.7.2 Computed Source Location

Two of the most widely used location techniques are known as linear location and planar location. The principle of linear location is illustrated in Figure 10-19, which shows the time of arrival difference as a function of the source position. When the source is at the midpoint, the wave arrives at sensors S1 and S2 simultaneously and the time of arrival difference is zero:

$$\Delta T = 0$$

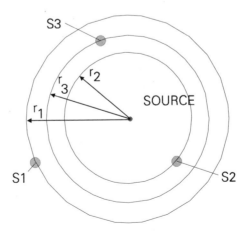

FIGURE 10-18 Basic principle of AE source location.

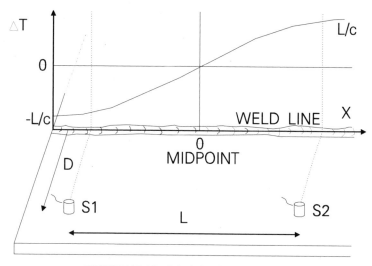

FIGURE 10-19 Principle of linear location.

ΔT ("delta T") is the algebraic symbol that stands for the time difference between hits. The greater the distance between the source and the midpoint, the greater the delta T. The relationship is linear:

$$\Delta T = \frac{2X}{c}$$

Where X is the distance of the source from the midpoint, and c is the velocity of sound discussed in Section 2.4.2. This relationship holds so long as the source is between the sensors. However, if the source is beyond one of the sensors, the delta T has a constant value of L/c, where L is the distance between sensors. Thus, a source between the sensors can be pinpointed, but not a source beyond a sensor. Note also in the above equation that an accurate calculation of the source location requires both an accurate measurement of the delta T, and an accurate knowledge of the wave velocity c. If the delta T differs from true because of some accident of wave propagation, the calculated location will be in error.

If it is desired to use two sensors to locate sources along a weld in a plate but the sensors are offset a distance D from the weld line, a situation results as is depicted in Figure 10-20. The relationship between source position and delta T is quite similar to the one shown in Figure 10-19 but the straight lines have become curves because the sensors are offset from the weld line.

In this case, it is possible, in principle, to pinpoint a location on the weld line beyond the sensors as well as between them. The greater slope of the curve in the region between the sensors indicates that the technique will be most accurate and reliable in the central region. Beyond the sensors, a small change (or error) in the measured delta T produces a large change (or error) in the calculated source position. In other words, the technique is not as reliable.

The examples in Figures 10-19 and 10-20 illustrate how, with two sensors (one delta T), it is possible to locate the X position of a source along a line. With three sensors, (two

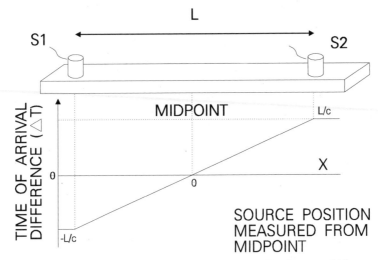

FIGURE 10-20 Source location with two sensors offset from a weld line.

delta T's), it is possible to locate the X and Y positions of a source on a plane. This is the situation illustrated in Figure 10-18. There are many mathematical approaches to source location on a plane, some using three sensors and some using four sensors.

Just as with linear location, the best accuracy for planar location is achieved when the source is well within the sensor array, rather than outside of it. In fact, with only three sensors on a plane, the source location mathematics is ambiguous over most of the plane. For many points outside the triangle, it is possible to find matching points inside the triangle that will produce the same delta T's. The resulting regions of ambiguous location are shown in Figure 10-21 for an array of three sensors in an equilateral triangle.

Location tends to be ambiguous when the first arrival occurs long before either of the other arrivals. When a three-sensor array is used for monitoring a specific area of interest, it is clearly important to place the area of interest in the area of unique location, not in the area of ambiguous location. Alternatively, the problem of ambiguous location can be resolved by the use of an extra delta T from a fourth sensor.

There are many mathematical techniques for accomplishing source location on a plane. Some use three sensors and some use four. Several exact analytical solutions have been described, some slow and clumsy, some fast and sophisticated. Approximation solutions have also been widely used. These can offer a good combination of speed, accuracy, and robustness when faced with the option of inferior data. These include iterative approaches, least-squares solutions, and linear and quadratic approximations.

The accuracy of the computed source location is limited more by wave propagation factors than by the mathematical approaches. The conventional approaches all assume that the wave travels directly from source to sensor with a well-defined velocity. In practice, reflections, multiple wave modes, and other propagation effects that produce uncertainties in the effective velocity, as well as scatter in the experimental data, can upset this assumption. The result is that computed source location using conventional approaches is very accurate in some situations but prone to serious error in others. Structural geometry and operating frequency govern whether conventional source location techniques will work well or not.

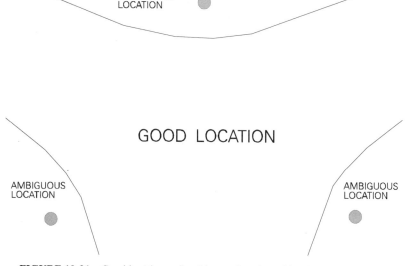

FIGURE 10-21   Good location and ambiguous location with a three-sensor array.

With computed source location, several hits are needed or there can be no computation. Smaller events that only hit one sensor are liable to be ignored. This can be dangerous, especially in wide-area monitoring. Overreliance on computed source location has sometimes led users to ignore important data and to draw the wrong conclusions about structural integrity. Therefore, computed source location must be used wisely. The sensor spacing must be properly adapted to the wave attenuation on the particular structure being monitored, and to the desired sensitivity.

2.7.3   Zone Location and Guard Techniques

A simpler type of source location is known as zone location. Here, conclusions are drawn from the hit sequence alone, without actually measuring the delta T's. This development began with the idea of guard sensors, which is illustrated in Figure 10-22. The technique was conceived in the early 1960s as a way to record data from a limited area of interest, while rejecting noise from outside.

A "data" sensor D is placed on the area of interest surrounded by several guard sensors G. AE waves from the area of interest will hit the data sensor before hitting any of the guards. Waves from outside will hit at least one of the guards before hitting the data sensor. Based on this, it is easy to reject the outside noise. The concept can be implemented in several different ways. Sometimes this process is conditioned by hardware circuitry in the instrument or by software between the detecting and the recording of the hits. In these cases, the noise signals are not recorded. Sometimes this process is performed by posttest analysis on systems that record all hits, including hits on guards. Posttest processing is safer, but the files are longer and more work is necessary during the analysis of the data. As a variant on the guard sensor technique, guard sensors can be placed on known noise sources and data sensors can be placed around the rest of the inspection area.

The idea of first-hit zone location was a further development of the guard concept,

10.26 CHAPTER TEN

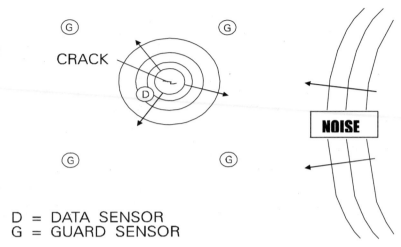

FIGURE 10-22 Guard sensor concept.

conceived in 1972. This is a technique for wide-area monitoring. Many sensors monitor the inspection area, and a determination is made as to which channel detects the wave first. If the wave is traveling uniformly, the first-hit sensor must be closer to the source than any other sensor. By paying attention to the first hit sensor, good information is gained about the location of the source.

With first-hit zone location, the inspection area can be divided into zones, each zone being centered on one sensor. If a particular sensor is the first one to be hit, it can be concluded that the source lies somewhere in that sensor's zone.

Figure 10-23 shows the pattern of zones for a simple layout of sensors on a plane. The zone boundaries are straight lines that evenly divide the spaces between pairs of sensors. In geometric terms, they are the perpendicular bisectors of the lines joining neighboring sensors. Figure 10-24 shows the expansion of this concept into the three-dimensional structure.

Data from structural tests is often displayed by channel, e.g., a graph of hits versus channel or a graph of a particular channel's activity. The practical value of first-hit zone location is that it allows the user to remove second-hit and third-hit information from the data, so that each channel shows only the emission that actually originated within its zone. This substantially sharpens and cleans up the data.

As with the guard technique, the first-hit zone location technique can be implemented by hardware circuitry, by software at the time of data acquisition, or by software during posttest analysis. Details of the implementation differ depending on the specific instrument design.

Zone location can be extended to take into account the second and third hits, if there are any. With the knowledge of the channel that received the second hit, the source location can be narrowed down to a small segment of the primary zone. With the knowledge of the channel that received the third hit, the source location can be narrowed down still further. However, these techniques are seldom implemented.

A benefit of the zone location approach is that only a single hit is needed for the event to be admitted into the analysis process. In the preceding discussion of computed source

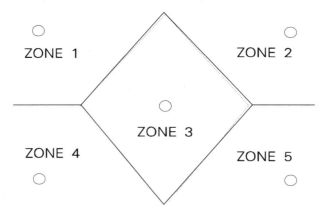

FIGURE 10-23 First-hit zone location.

location, it was pointed out that several hits are needed to give the delta T's required for computation. Compared with this, zone location has, in effect, a higher sensitivity and none of the detected events are ignored.

2.8 Data Displays

Becoming familiar with the AE data displays and learning how to read them is a very important part of the AE user's training. Even those who are familiar with other NDT meth-

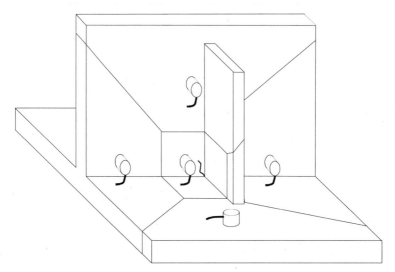

FIGURE 10-24 Zone location on a structure.

10.28 CHAPTER TEN

ods may find AE displays strange at first because they are basically graphs of numerical data, rather than the visual images and displays that are the working tools of some other NDT methods.

AE data displays can be grouped into four categories:

1. Location displays
2. Activity displays
3. Intensity displays
4. Data quality (crossplot) displays

Location displays are widely used. However, other kinds of displays are very valuable for background information on-site, for investigative work, and for data analysis, especially in the more difficult situations. Therefore, with training and practice, the AE practitioner becomes familiar with all the display types. Location displays show where the emission is coming from. When computed location techniques are being used, the typical location display is a map of the inspection area. An example is shown in Figure 10-25. The user defines the X–Y coordinate frame for the map as the software is set up. The sensor numbers are shown in the appropriate places on the map. When planar source location is being used, the location of each event is plotted as a dot on the map. If a source emits repeatedly, the dots form a cluster, as shown in Figure 10-25. The user's eye is drawn naturally to the clusters on the screen that correspond to the most emissive sources.

When linear location is being used, the display is a histogram with the span between the sensors laid out along the X axis. This is illustrated in Figure 10-26. The X axis is divided into a number of segments or "bins," typically 100. Each located event is assigned to the bin that corresponds to its location. The number of events in each bin is indicated by the height of the histogram bar. When viewing this display, the user's eye is drawn to the highest peak or the highest concentration of indications. This is where the majority of the emission activity originates.

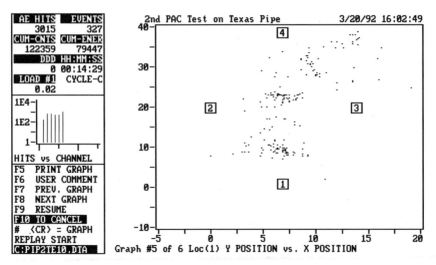

FIGURE 10-25   Planar source location display.

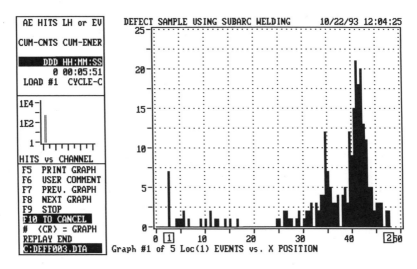

FIGURE 10-26 Linear location display.

A third kind of location display is a histogram of events as related to channels. This is used for first hit zone location, where each event is assigned to the channel that detected it first, and later hits on other channels are disregarded. This type of display is illustrated in Figure 10-27. The height of the bar for each of the five channels shows the number of events detected by that channel first. Channel 4 shows the most activity, with 1700 events hitting it first in one hour of monitoring. In this case, the crack was between channels 1 and 2; channels 3, 4, and 5 were guard sensors used to screen out extraneous noise.

There are variations on the display shown in Figure 10-27. For example, the total AE signal energy from each zone could be plotted on the Y axis instead of the total number of events. This would provide greater prominence to the zones that produced high-energy

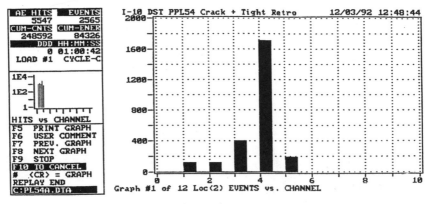

FIGURE 10-27 Zone location display.

hits. The second major class of displays is activity displays. These include displays that have time on the X axis (see Figure 10-28). The displays show when the AE activity occurred and how much activity there was. As in Figure 10-26, the X axis is divided into bins. In this example, there are 100 bins from left to right, so each bin represents a time slice of 36 seconds. On the Y axis, the height of each histogram bar shows how much emission occurred during that time slice. As shown, there typically are detected hits, located events, or total energy shown. In older reports the Y axis often showed AE counts, i.e., threshold crossing counts, as shown in Figure 10-14.

The Y axis scaling can be linear, as in Figure 10-28, or logarithmic, as in Figure 10-29. Looking at the linear plot, a few time slices with very high activity stand out prominently from a low-level background.

In the logarithmic plot, the scaling reduces the visual impact of these very active time slices, but provides more information about the fluctuating lower-level activity in between. In the logarithmic Y axis scaling of Figure 10-29, 1E1 stands for 10^1 which is 10, 1E2 stands for 10^2, which is 100, 1E3 stands for 10^3, which is 1000, and so on. $1E^0$ stands for 1. Logarithmic axes can cover a large span. A 24-hour logarithmic plot would allow comparison in the same glance of large amounts of AE produced in "rush-hour traffic" and a very small amount produced in the "middle of the night." With linear axes, very small amounts may not even show above the baseline.

In becoming familiar with logarithmic plots, it is important to learn how to read the scale and how to interpret the display. In the lower part of the display, the patterns of rising and falling activity may be informative, but it must always be remembered that the actual level is low. In the upper part of the display, the level of activity is drastically greater and even a small increase in the height of the histogram bar means thousands of additional emissions detected.

Another form of activity display is shown in Figure 10-30. Instead of showing the amount of activity in each time slice, this display shows a running total of all the activity detected since the start of the test. This is the best display for measuring the average emission rate as well as for observing the total emission quantity.

The third major class of displays is "intensity displays." These provide statistical information about the size of the detected signals. In many cases, large signals are likely to

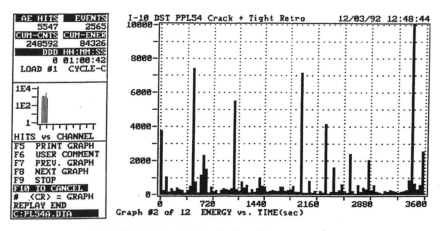

FIGURE 10-28 Activity display: AE rate versus time.

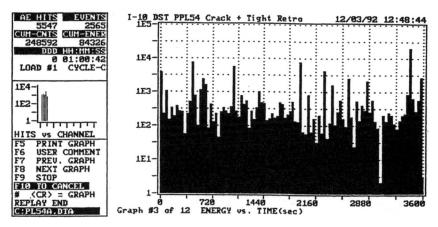

FIGURE 10-29 Activity versus time, logarithmic scale.

be more important than small signals. Furthermore, a given total amount of AE signal energy might be originating from a few large signals or from many small signals; it could be important to know about these origins. These are the issues that are addressed by intensity displays.

The best-known intensity display is the "amplitude distribution" plot, which shows how many of the hits were large and how many were small. This display exists in two forms. The "differential" form, illustrated in Figure 10-31, is a histogram with amplitude on the X axis. The X axis is divided into a hundred 1 dB slices. The heights of the bars

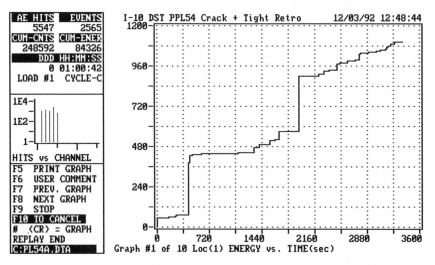

FIGURE 10-30 Cumulative activity display.

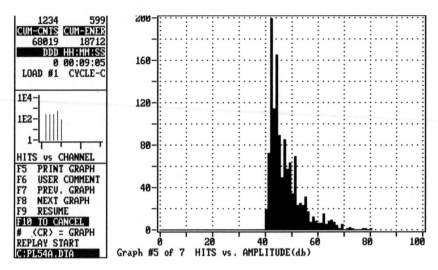

FIGURE 10-31 Differential amplitude distribution.

show how many hits were recorded at each 1dB level. Typically, there are many low-amplitude signals and fewer high-amplitude signals. Naturally, there are no measured signals with amplitudes less than the detection threshold, so the display is blank to the left of the threshold dB value.

The cumulative form of the amplitude distribution display is a line graph, presenting the number of hits that were greater than the X axis amplitude. This form is derived from the differential display by scanning it from right to left and accumulating the contents of the bins. The cumulative display derived from Figure 10-31 is shown in Figure 10-32. Both differential and cumulative forms are often displayed with a logarithmic scale on the Y axis. This makes it easier to assess the high-amplitude activity even when the low-amplitude hits are much more numerous.

The shapes of the amplitude distributions often give valuable clues about the source mechanisms that are producing them. For example, growing cracks often give straight lines when a logarithmic scale is used on the Y axis. If the plot is far from linear, it is probably not a growing crack. Figure 10-32 is a typical example of a straight-line amplitude distribution. Another type of intensity display is the so-called "energy account," illustrated in Figure 10-33. This shows whether the total signal energy is coming mostly from low-energy events or mostly from high-energy events. The X axis shows the energy of the individual hits, on a logarithmic scale from 1 to 100,000, divided into 50 bins. The height of each bar shows the total energy from all the hits that fall into that bin.

Different emission source mechanisms cause amplitude distributions and energy accounts to show as different shapes. These displays can help to identify the mechanisms that are working, and can help to tell the difference between flaw-related AE and extraneous noise. The fourth and last major class of AE displays is crossplot displays. These are used during interpretation to assess data quality. The term "crossplot" implies a plot in which each hit gives one point on the display, showing the cross-relationship between two measured signal features. The best-established crossplots are "duration versus amplitude," "counts versus amplitude," and "counts versus duration." A duration/amplitude crossplot is illustrated in Figure 10-34.

ACOUSTIC EMISSION TESTING   10.33

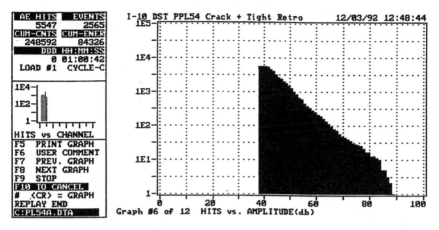

FIGURE 10-32   Cumulative amplitude distribution.

Larger signals typically have higher amplitudes and also more counts (see Figure 10-14). Therefore, the data in the counts/amplitude crossplot tend to fall into a diagonal band running from lower left to upper right. Along with this basic tendency there are subtle variations that can provide additional information about the shape of the signal. Shape, in turn, depends on the source mechanism.

Short, sharp sources, such as crack growth, cause short-rise, quick-decay signals, whereas long, drawn-out source processes, such as frictional sliding, show as slow-rise, slow-decay signals. This is indicated in Figure 10-35.

A crack-growth signal and a friction signal might have the same amplitude, but the

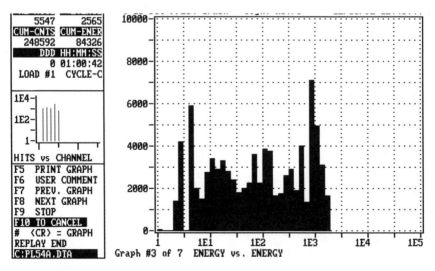

FIGURE 10-33   Energy account.

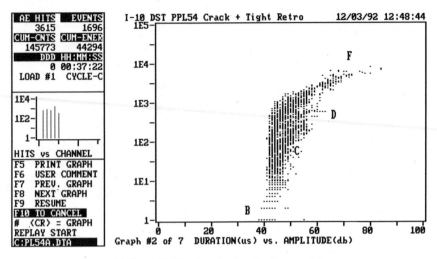

FIGURE 10-34 Duration/amplitude crossplot.

friction signal would have a much longer duration. This provides a technique for recognizing signals from friction and distinguishing them from signals originating from crack growth. Figure 10-34 shows three distinct bands of data. There is a large, broad band at the top, a short, tight band at the bottom, and a third lying in between. These bands correspond to different signal shapes. Signal shape is influenced not only by source mechanism but also by wave propagation, so the analysis of this kind of data is quite demanding. It is not immediately obvious whether the different bands correspond to different channels, to different source locations, or to different source mechanisms. Various techniques of advanced data analysis and signal processing can be used to address these issues.

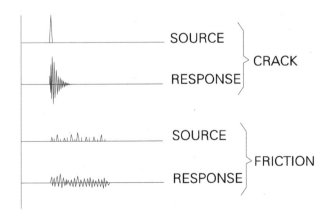

FIGURE 10-35 Signal shapes depend on source mechanism.

2.9 Interpretation and Evaluation

In nondestructive testing, data interpretation and evaluation proceed as shown in Figure 10-36. ASTM has developed the following definitions:

Indication. Response or evidence of a response in a nondestructive test.

Interpretation. The determination of whether indications are relevant, nonrelevant, or false.

Evaluation. The determination of the significance of relevant indications.

False (referring to an indication). Obtained through improper technique or processing.

In AE testing, the most basic kind of indication is simply a hit (the data record produced after the signal has crossed the threshold). Relevant indications are hits produced by the crack. Nonrelevant indications include hits produced by sources outside the inspection area. False indications include hits due to "echoes," which can occur if the instrument is badly set up.

In AE as in all other NDT methods, interpretation occurs before evaluation. Sometimes, interpretation is performed explicitly and deliberately, as a well-defined step in a documented data analysis process. In other instances, interpretation is made more implicitly and seems little more than a common-sense effort to extract the most out of the available data.

When conducted explicitly, the first task in data analysis is to identify any nonrelevant and false indications in the data file, and then to filter them out. A new data file is written that contains only the relevant data, and running this filtered file through an evaluation program performs the evaluation. This process is used in several AE applications where the techniques have been standardized and a clear-cut procedure has been written. A good example is the AE testing of railroad tank cars. In this and other applications, the duration–amplitude crossplot shown in Figure 10-34 is a key tool in the interpretation process. The use of this crossplot can be included as part of a well-defined procedure, and technicians using it have a common understanding and basis for discussion, refinement, and ongoing development.

Evaluation procedures have been codified for several major AE applications. The best way to arrive at documentable evaluation procedures is to obtain a broad and representa-

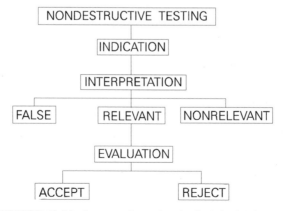

FIGURE 10-36 Interpretation and evaluation of NDT data.

tive range of field experience, following up the AE tests with other NDT inspections to characterize the flaw conditions as thoroughly as possible. By documenting conditions and results with the use of a database, data can be accumulated for use in future analyses. This has been done for several leading AE applications, such as pressure vessels and tanks, railroad tank cars, jumbo tube trailers, and aircraft fuselages.

Evaluation sometimes hinges on located clusters of events, sometimes on total emission quantities, and sometimes on emission in specific parts of the test, such as high-stress load holds. Often, the data is evaluated in accordance with several applicable acceptance criteria. Comparison of AE data with specific acceptance criteria serves as a screening process. If the structure meets the acceptance criteria, it is deemed acceptable. If not, the procedure specifies either another stage of AE data analysis or verification of the AE data with other nondestructive testing methods.

In less standardized applications, interpretation and evaluation are often conducted more implicitly. In interpretation, the technician uses commonsense observations to remove false indications (noise), or to focus on the parts of the test that are determined to contain the most meaningful information. The final evaluation tends to emerge from a personal blending of data analysis with the experience, commonsense observation and technical intuition. These can be subjective processes and the degree of success depends very much on the skills of the individual technician. These skills are difficult to teach and communicate, and the processes actually used by an individual will change over the years as interests and preferences shift and as new ideas are considered. This style of operation is best seen as a necessary prelude to the development of standardized procedures that can be used widely and reliably.

Evaluation criteria are only meaningful when they are related to a specific instrumentation setup and a specific load stimulus. In testing pressure vessels, tanks, etc., the load stimulus can be numerically specified, controlled, and measured by the technician.

3. ADVANTAGES AND LIMITATIONS OF ACOUSTIC EMISSION TESTING

In contrast with most other NDT methods, in AE testing the discontinuity itself is the releaser of energy, making its own signal (in response to *stress*). AE testing detects *movement* (other methods detect *geometric discontinuities*).

Advantages of AE testing:

- AE can be used in all stages of testing including:
    - Preservice (proof) testing
    - Inservice (requalification) testing
    - On-line monitoring of components and systems
    - Leak detection and location
    - In-process weld monitoring
    - Mechanical property testing and characterization
- Material anisotropy is good
- (Less) geometry sensitive
- Less intrusive
- Global monitoring

- Real-time evaluation
- Remote scanning
- Performance/price ratio

Limitations of AE testing:

- Repeatability: Acoustic emission is stress unique and each loading is different.
- Attenuation: The structure under test will attenuate the acoustic stress wave.
- History: Tests are best performed if the loading history of a structure is known.
- Noise: Acoustic Emission can be subject to extraneous noise.

The advantages and disadvantages of AE testing can be summarized as in the following table:

Acoustic emission	Most other NDT methods
Discontinuity growth/movement	Discontinuity presence
Stress, damage-related	Shape-related
Material anisotropy is good	Material anisotropy is bad
(Less) geometry sensitive	(More) geometry sensitive
Each loading is unique	Inspections are readily repeated
Less intrusive	More intrusive
Global monitoring	Local scanning
Principal limitations: attenuation, history dependence, and noise	Principal limitations: access, geometry, and dependence on discontinuity orientation and proximity to surface

4. GLOSSARY OF ACOUSTIC EMISSION TERMS

This glossary is presented in two forms. First, selected key terms are listed in logical order to show how standard terminology is used to describe the AE process from source to data storage. Second, a more complete glossary is presented in alphabetical order. Definitions for items marked with an asterisk are drawn from ASTM E 1316, and in some cases, reflect minor changes.

Key Terms in Logical Order

Acoustic emission (AE)—elastic waves generated by the rapid release of energy from sources within a material.
Event (AE)—a local material change giving rise to acoustic emission.*
Source—the physical origin of one or more AE events.
Sensor—a device containing a transducing element that turns AE wave motion into an electrical voltage.
Signal—the electrical signal coming from the transducing element and passing through the subsequent signal conditioning equipment (amplifiers, frequency filters).
Channel—a single AE sensor and the related equipment components for transmitting, conditioning, detecting, and measuring the signals that come from it.

Detection—recognition of the presence of a signal (typically accomplished by the signal crossing a detection threshold).
Hit—the process of detecting and measuring an AE signal on a channel.
Signal features—Measurable characteristics of the AE signal, such as amplitude, AE signal energy, duration, counts, and risetime.
Hit Data Set—the set of numbers representing signal features and other information, stored as a result of a hit.
Event data set—the set of numbers used to describe an event, pursuant to data processing that recognizes that a single event can produce more than one hit.

Terms in Alphabetical Order

Acoustic emission (AE)—elastic waves generated by the rapid release of energy from sources within a material.
Activation (AE)—the onset of AE due to the application of a stimulus such as force, pressure, heat, etc.
Activity* (AE)—a measure of emission quantity, usually the cumulative energy count, event count, ringdown count, or the rates of change of these quantities.
Amplitude (AE)—the largest voltage peak in the AE signal waveform; customarily expressed in decibels relative to 1 microvolt at the preamplifier input (dB_{ae}).
Amplitude distribution*—a display of the number of AE signals at (or greater than) a particular amplitude, plotted as a function of amplitude.
Attenuation—loss of amplitude with distance as the wave travels through the test structure.
Burst emission*—a qualitative description of the discrete signal related to an individual emission event occurring within the material.
Channel (AE)—a single AE sensor and the related equipment components for transmitting, conditioning, detecting, and measuring the signals that come from it.
Continuous emission*—a qualitative description of the sustained signal level produced by rapidly occurring acoustic emission events.
Counts—the number of times the AE signal crosses the detection threshold. Also known as "ringdown counts" and "threshold crossing counts."
dB_{ae}—a unit of measurement for AE signal amplitude A, defined by A (dB_{ae}) = 20 log V_p where V_p is the peak signal voltage in microvolts referred to the preamplifier input.
Detection (AE)—recognition of the presence of a signal (typically accomplished by the signal crossing a detection threshold).
Event* (AE)—a local material change giving rise to acoustic emission.
Event data set (AE)—the set of numbers used to describe an event, pursuant to data processing, that recognizes that a single event can produce more than one hit.
Event description—a digital (numerical) description of an event, comprising one or more signal descriptions and/or information extracted from them or calculated from them.
Event energy* (AE)—the total elastic energy (in the wave) released by an acoustic emission event.
Felicity effect*—the presence of AE at stress levels below the maximum previously experienced. (The reverse of the Kaiser effect.)
Frequency—for an oscillating signal or process, the number of cycles occurring in unit time.
Guard sensors—sensors whose primary function is the elimination of extraneous noise based on arrival time differences.

Hit (AE)—the detection and measurement of an AE signal on a channel.
Hit data set—The set of numbers representing signal features and other information, stored as a result of a hit.
Intensity* (AE)—A measure of the size of the emission signals detected, such as the average amplitude, average AE energy, or average counts.
Kaiser effect*—The absence of detectable acoustic emission at a fixed sensitivity level, until previously applied stress levels are exceeded.
kHz—Kilohertz, an SI unit of frequency, 1000 cycles per second.
Location*—relating to the use of multiple AE Sensors for determining the relative positions of acoustic emission sources.
Noise—nonrelevant indications; signals produced by causes other than AE, or by AE sources that are not relevant to the purpose of the test.
Parametric inputs—environmental variables (e.g., load, pressure, temperature) that can be measured and stored as part of the AE signal description.
Risetime—the time from an AE signal first threshold crossing to its peak.
Sensor (AE)—a device containing a transducing element that turns AE wave motion into an electrical voltage.
Signal (AE)—the electrical signal coming from the transducing element and passing through the subsequent signal conditioning equipment (amplifiers, frequency filters).
Signal description—the result of the hit process: a digital (numerical) description of an AE signal and/or its environmental context.
Signal features—measurable characteristics of the AE signal, such as amplitude, AE energy, duration, counts, and risetime, that can be stored as part of the AE signal description.
Signal strength (AE)—the strength of the absolute value of a detected AE signal. Also known as "relative energy," "MARSE," and "signal strength."
Source (AE)—the physical origin of one or more AE events.
Source energy (AE)—the total energy (of all forms) dissipated by the source process.
(Primary) zone—the area surrounding a sensor from which AE can be detected and from which AE will strike the sensor before striking any other sensors.

5. REFERENCES

1. A. A. Pollock, "Inspecting Bridges with Acoustic Emission," Guidelines prepared for the U.S. Department of Transportation and Federal Highway Administration (FHWA), June 1995. Technical Report No. TR-103-12 6/95. Physical Acoustics Corp., Princeton Junction, NJ 08550.
2. R. K. Miller (Ed.), *Nondestructive Testing Handbook,* 2nd edition, Volume 5 "Acoustic Emission Testing." 1987. American Society for Nondestructive Testing, Columbus, OH.

BASIC METRIC–ENGLISH CONVERSIONS

METRIC TO ENGLISH

To convert	Into	Multiply by
Celsius	Fahrenheit	9/5 and add 32
Centimeters	Inches	0.3937
Cubic centimeters	Cubic inches	0.06102
Cubic meters	Cubic feet	35.31
Grams	Ounces	0.03527
Kilograms	Pounds	2.205
Kilometers	Miles	0.6214
Liters	Gallons (U.S.)	0.2642
Meters	Feet	3.281
Meters	Yards	1.094
Millimeters	Inches	0.0393
Square centimeters	Square inches	0.1550
Square kilometers	Square miles	0.3861
Square meters	Square feet	10.76
Square meters	Square yards	1.196

ENGLISH TO METRIC

To convert	Into	Multiply by
Cubic feet	Cubic meters	0.02832
Cubic inches	Cubic centimeters	16.39
Fahrenheit	Celsius	5/9 after subtracting 32
Feet	Meters	0.3048
Gallons (U.S.)	Liters	3.785
Inches	Millimeters	25.40
Inches	Centimeters	2.540
Miles	Kilometers	1.609
Ounces	Grams	28.35
Pounds	Kilograms	0.4536
Square feet	Square meters	0.0929
Square inches	Square centimeters	6.452
Square miles	Square kilometers	2.590
Square yards	Square meters	0.8361
Yards	Meters	0.9144

INDEX

A

Absolute coil technique, 8.67
Absolute temperature scale, 9.44
Absolute zero, 9.44
Acoustic emission, 1.6
Acoustic emission testing, 1.15, 10.1–10.39. *See also* AE
  acoustic emission sources, 10.2
    stress, 10.2
    structural loading, 10.4
  advantages, 10.36
  attenuation, 10.10
  computed source location, 10.22
  data displays, 10.27
    activity displays, 10.28
    crossplot displays, 10.28
    data quality displays, 10.28
    intensity displays, 10.28
    location displays, 10.28
  disadvantages, 10.37
  evaluation, 10.35
  guard techniques, 10.25
  history and development, 10.1
  indication, 10.35
  instrument architecture, 10.18
    hit-based, 10.18
  interpretation, 10.35
  measurements, 10.17
    amplitude, 10.17
    counts, 10.18
    duration, 10.17
    signal energy, 10.18
  multichannel source location techniques, 10.20
  principles, 10.2
  propagation of the AE waves, 10.9
  sensing and measurement, 10.14
  sensor performance and calibration, 10.14
  signal conditioning, 10.16
  signal detection, 10.16
  signal features, 10.17
  signal shaping chain, 10.8
  signal shaping effects, 10.14
  the hit process, 10.16
  wave velocity, 10.12
  zone location, 10.25
Adhesion, 4.32
AE, 1.7, 1.15
Allotropic transformation, 2.12
Alloys, 2.2
  ingots, 2.2
Alternating current (AC), 5.51
American Industrial Radium and X-ray Society, 1.16, 1.23, 6.7. *See also* American Society for Nondestructive Testing
American Petroleum Institute, 3.23
American Society for Nondestructive Testing, 1.16, 1.23 3.1, 6.7. *See also* ASNT
  Recommended Practice No. SNT-TC-1A, 3.1
American Society of Mechanical Engineers, 3.1

American Society of Mechanical
    Engineers *(continued)*
  Boiler and Pressure Vessel Code,
    Section V—Nondestructive
    Examination, 3.1
  Section XI—Rules for Inservice
    Inspection of Nuclear Power Plant
    Components, 3.1
Amplitude, 10.38
Amplitude distribution, 10.38
Angle beam, 7.111
Angle of field, 3.54
Angstrom, 3.54, 4.32
Annealing, 2.23
Anode, 6.68
Anodizing Process for Inspection of
    Aluminum Alloys and Parts, 4.2
ANSI N45.2.6, 3.1
Area amplitude response curve, 7.112
Artifact, 2.23, 6.68
A-scan, 7.111
A-Scan presentation, 7.111
ASNT, 1.23
ASNT Level III Certificate Program,
    1.25
Aspect ratio, 3.54
Attenuation, 7.112
Audigage, 1.6

B

Back echo, 7.112
Back reflection, 7.112
Back surface, 7.112
Back voltage, 8.67
Background, 4.32, 5.51, 9.44
Beam spread, 7.112
Beaufort's wind scale, 9.44
Bell-makers, 1.3
Betz, C., 4.2
Billet, 2.23
Binary eutectic, 2.9
Black light, 4.7, 4.32, 5.51
    filter, 4.32
Blackbody, 9.44
Blacksmiths, 1.3, 1.5

Bleedout, 4.32
Blister, 2.23
Blotting, 4.32
Blowhole, 2.23
Boilers, 1.5, 1.6, 1.8, 1.9, 1.10
Borescopes, 3.3, 3.13, 3.54
    fiberoptic, 3.13, 3.15
    lens-optic, 3.15
    miniborescope, 3.15
Boundary layer, 9.44
Brady, E., 4.4
Brazing, 2.14, 2.24
Bridge circuit, 8.67
Brightness, 3.54
British thermal unit (Btu), 9.44
Brittle cracking, 2.24
Brittle fracture, 2.20
Brittleness, 2.24
B-scan presentation, 7.112
Bubbler, 7.112
Burned-in image, 3.54
Burning, 5.51
Burr, 2.24
Burst, 2.24
Burst emission, 10.38

C

Calorie, 9.44
Capillary action, 2.24
Carrier, 4.32
Carrier fluid or liquid, 5.51
Cast structure, 2.24
Casting, 2.4
    cold shut, 2.7
    cracking, 2.4
        hot cracking, 2.6
        solidification shrinkage cracking, 2.6
    discontinuities, 2.4
        cracking, 2.4
        microporosity, 2.4
        nterdendritic porosity, 2.4
        pipe, 2.4
        porosity, 2.4
        slag, 2.4
        voids, 2.4

gas porosity, 2.6
grains in, 2.4
inclusions, 2.7
microporosity, 2.6
pipe, 2.4
porosity, 2.6
scabs, 2.7
seams, 2.7
shrinkage porosity, 2.4, 2.6
shrinkage voids, 2.4, 2.6
solidification shrinkage, 2.4
Casting shrinkage, 2.24
Cavity radiator, 9.44
CCD, 3.3, 3.54
Charged-coupled devices, 3.3, 3.15, 3.21, 3.54. *See also* CCD
Charpy test, 1.18, 2.24
Chatter, 2.24
Checks, 2.24
Circumferential magnetization (circular magnetization), 5.51
Cleavage, 2.24
Code, 3.54
Coefficient of thermal conductivity, 9.45
Cohesion, 4.32
Coil in air, 8.67
Coil method, 5.52
Cold shut, 2.7, 2.24
Cold Working, 2.24
Collimator, 6.68, 7.112
Color, 3.5, 3.54
  brightness, 3.5
  hue, 3.5
  saturation, 3.5
Columnar crystal structure, 2.24
Composite viewing, 6.68
Conduction, 9.45
Conductor, 9.45
Cone, 3.54
Constant potential x-ray unit, 6.68
Contact testing, 7.112
Continuity, 2.1
Continuous emission, 10.38
Contrast, 3.4, 3.54, 4.33
Contrast ratio, 4.33
Contrast sensitivity, 6.68

Convection, 9.45
Corrosion, 2.24, 3.54
Corrosion fatigue, 2.24
Corrosion resistance, 1.18
Corrosion-induced discontinuities, 2.15
  hydrogen embrittlement, 2.17
  intergranular attack, 2.17
  intergranular stress corrosion cracking, 2.17
  pitting, 2.15
  stress corrosion cracking, 2.17
Couplant, 7.112
Cracking, 2.10
  fatigue cracking, 2.11
  solidification cracking, 2.10
Cracks, 2.10, 2.24
Crater, 2.24
Crazing, 2.19
Creep, 2.19
Creep cracks, 2.24
Creep voids, 2.24
Crevice corrosion, 2.24, 3.54
Critical angle, 7.112
Cross-axis coil, 8.67
C-scan, 7.112
Curie, M., 1.6, 7.1
Curie, P., 1.6, 7.1
Curie, 6.1
Current induction method, 5.52

D

DAC curve, 7.112
Dead zone, 7.112
Decarburization, 2.24
Defect, 2.24
deForest, A. V., 1.6, 1.11, 4.2–4.3
Delay line, 7.112
Delay sweep, 7.112
Demagnetization, 5.52
Demagnetizing coil, 5.52
Dendrite, 2.24
Densitometer, 6.69
Density, 9.45
Depth of field, 3.54
Depth of focus, 3.54

Destructive testing, 1.17
　benefits, 1.19
　characteristics measured, 1.18
　instruments, 1.18
　limitations, 1.20
Developer, 4.33
　aqueous, 4.33
　dry powder, 4.33
　liquid film, 4.33
　nonaqueous, 4.33
　soluble, 4.33
Differential coil technique, 8.67
Diffuse indications, 5.52
Digital cameras, 3.21
Direct current (DC), 5.52
Direct viewing, 3.54
Discontinuities, 2.1–2.2, 2.11
　categorization, 2.1
　changes in microstructure, 2.1
　chemical segregation, 2.1
　cracks, 2.1
　definition, 2.1
　evaluation, 2.1
　folds, 2.1
　fretting, 2.1
　galling, 2.1
　gouges, 2.1
　inclusions, 2.1
　　slag, 2.2
　lamellar tearing, 2.3
　laps, 2.1
　notches, 2.1
　origin, 2.1
　plastic deformation, 2.15
　primary processing, 2.2
　scratches, 2.1
　secondary processing, 2.2
　sharp angles, 2.1
　slag, 2.12
　voids, 2.1
　welding, 2.11
　welding undercut, 2.1
Distal end, 3.54
Distance amplitude, 7.112
Distance linearity range, 7.113
Doane, F. B., 1.6, 1.11, 4.2
Drain time, 4.33

Drake, H. C., 1.6, 1.11
Dry particle technique, 5.52
Drying oven, 4.33
Drying time, 4.33
Dual search unit, 7.113
Ductility, 1.18, 2.24
Dustproof construction, 3.54
Dust-tight construction, 3.54
Duty cycle, 6.69
Dwell time, 4.33
Dynamic range, 7.113

E

Eddy current testing, 1.15, 8.1–8.70. *See also* ET
　advantages, 8.64
　alternating current principles, 8.10
　　sinusoidal variation, 8.10
　　electromagnetic induction, 8.11
　　induction process, 8.11
　coil design, 8.24
　　cross-axis, 8.33
　　encircling, 8.34
　　Hall detector, 8.34
　　internal, 8.37
　　reflection, 8.33
　　shielded, 8.32
　　through-transmission, 8.33
　　transmit–receive, 8.33
　conductance, 8.4
　conductivity, 8.4
　detectable material variables, 8.3
　　conductivity variations, 8.3
　　discontinuities, 8.3
　　lift-off distance, 8.3
　　material thickness, 8.3
　　permeability variations, 8.3
　　spacing between conductive layers, 8.3
　　spacing between test coil and test material, 8.3
　　thickness of plating or cladding on a base metal, 8.3
　diamagnetic materials, 8.8
　differential surface coil display, 8.61

INDEX

display media, 8.40
  analog and digital meters, 8.40
  cathode ray tubes, 8.40
  graphic display screens, 8.40
  liquid crystal screens, 8.40
eddy currents, 8.17
  characteristics, 8.18
electricity, 8.3
electromagnetism, 8.8
encircling and internal coil display, 8.61
ferromagnetic materials, 8.8
field intensity, 8.6
flux density, 8.6
history and development, 8.1
impedance plane, 8.16
impedance plane display, 8.41
  conductivity variation, 8.48
  discontinuity signal display, 8.51
  display of discontinuity orientation, 8.55
  edge effect, 8.47
  effect of phase lag on the conductance curve, 8.43
  effect of plating and cladding, 8.57
  effects of ferromagnetic test materials, 8.50
  enhancing signal display, 8.55
  lift-off curves, 8.44
  normalization, 8.41
  shape of the eddy current field, 8.55
  spacing between conducting materials, 8.58
  subsurface discontinuities, 8.51
  surface coil impedance curves, 8.48
  surface-breaking discontinuities, 8.52
  thickness variations, 8.55
instruments, 1.6, 1.15, 8.24
  control functions, 8.25
  dedicated, 8.25
  multifrequency, 8.26
  standard impedance plane display, 8.25
law of similarity, 8.61
lift-off, 8.24
limitations, 8.64
magnetic flux, 8.6

magnetism, 8.5
major application areas, 8.3
  aerospace, 8.3
  bar stock, 8.3
  in-service inspection, 8.3
  pipe, 8.3
  rod, 8.3
  tubing, 8.3
  wire, 8.3
paramagnetic materials, 8.8
permeability, 8.8
probes, 8.31
  90°, 8.31
  bolt hole, 8.31
  fastener (doughnut), 8.31
  pancake, 8.32
  pencil, 8.31
  spring loaded surface, 8.32
recording devices, 8.40
reference standards, 8.37
remote field eddy current testing, 8.66
resistance, 8.4
resistivity, 8.4
signal output, 8.15
test coils, 8.28
  basic configurations, 8.30
  surface coils, 8.30
test equipment, 8.24
  AC generators, 8.25
  coil circuits, 8.25
  displays, 8.25
test performance criteria, 8.23
  penetration, 8.23
  resolution, 8.23
  sensitivity, 8.23
test sequence, 8.21
theory and principles, 8.3
voltage plane, 8.15
Eddy currents, 1.6
Effective depth of penetration, 8.67
Electric current innduction, 1.6
Electric Power Research Institute, 3.1
Electricity, 8.3
Electromagnetic induction, 1.6, 8.68
Electromagnetic radiation, 9.45
Electromagnetic spectrum, 9.45
Electromagnetism, 8.1, 8.68

Elongation characteristics, 1.18
Emissivity, 9.45
Empty coil, 8.68
Emulsification time, 4.33
Emulsifier, 4.33
   hydrophilic, 4.33
   lipophilic, 4.33
Encircling coil, 8.68
Endoscope, 3.3, 3.54
Entrant surface, 7.113
Equilibrium phase diagram, 2.9
Equivalent penetrameter sensitivity, 6.69
Erdman, D. C., 1.6
Erosion, 3.54
ET, 1.5, 1.7, 1.15, 1.23, 8.2
Etching, 4.33
Event data set, 10.38
Event description, 10.38
Event energy, 10.38
Exfoliation, 2.25
Explosion-proof construction, 3.54
External reference coil, 8.68

F

False indication, 4.33
Far field, 7.113
Faraday, 1.15, 8.1
Farraday, M., 1.6
Farrow, C., 1.6
Fatigue, 2.25
Fatigue cracking, 2.18
   thermally induced, 2.18
Fatigue life, 1.18
Feed-through coil, 8.68
Felicity effect, 10.38
Ferromagnetism, 5.52, 8.68
Fiber camera, 3.15
Fiber optics, 3.54
Field of view, 9.45
Fill factor, 8.68
Film coefficient, 9.45
Film contrast, 6.69
Film speed, 6.69
filters, 9.15
Firestone, F., 1.6, 1.15

First law of thermodynamics, 9.45
Flakes, 2.25
Flaw characterization, 7.113
Fluorescence, 5.52
Flux, 5.52, 8.68
Flux (luminous), 3.55
Flux density, 5.52, 8.68
Flux field penetration, 5.52
Flux leakage testing, 8.65
Focal spot, 6.69
Foerster, F., 1.6, 8.2
Fog, 6.69
Folds, 2.25
Foot-candle, 3.55, 4.33
Forced convection, 9.45
Foreground, 9.45
Forging, 2.15
Forging discontinuties, 2.25
Fourier's law, 9.45
Fovea centralis, 3.55
Fracture, 2.25
Frame, 3.55
Fretting, 2.25
Full-wave rectified current (FWDC), 5.52

G

Gamma radiography, 1.6, 6.69
Gamma rays, 6.8
Gap probe, 8.68
Gas holes, 2.25
Gas porosity, 2.25
Gate, 7.113
Geometric anisotropy, 2.3
Geometric discontinuities, 2.21, 2.25
   excessive undercut, 2.23
   fretting, 2.23
   galling, 2.23
   gouges, 2.22
   notches, 2.22
   scoring, 2.23
   scratching, 2.23
   sharp radius, 2.23
Geometric distortion, 3.55
Geometric unsharpness, 6.69
Gouge, 2.25

Grain, 2.25
Grain boundary, 2.25
Graininess, 6.69
Gross porosity, 2.25
Guard sensors, 10.38

H

Half-life, 6.69
Half-value layer (HVL), 6.69
Half-wave rectified current (HWDC), 5.52
Hardness, 1.18
Hardness test, 1.18
Heat of fusion, 9.45
Heat-affected zone, 2.25
Herschel, J., 1.6
Herschel, Sir William, 1.6, 9.1
Hit, 10.39
Hit data set, 10.39
Horizontal (hum) bars, 3.55
Horseshoe probe, 8.68
Hot cracks, 2.25
Hot tear, 2.25
Hughes, E. E., 1.6, 1.15
Hydrogen embrittlement, 2.18, 2.25
Hysteresis loop, 5.52

I

I.D. (inner diameter) coil, 8.68
Image, 3.55
Image burn, 3.56
Image quality indicator (IQI), 6.69
Immersion rinse, 4.33
Immersion testing, 7.113
Impact resistance, 1.18
Impedance, 7.113, 8.68
Inclusions, 2.3, 2.25
    elongated, 2.3
    stringers, 2.3
Incomplete fusion, 2.25
Incomplete penetration, 2.25
Indication, 7.113
Indium antimonide, 9.45
Inductive reactance, 8.68

Industrial radiography, 1.6
Infrared image, 1.6
Inherent discontinuity, 2.25
Inhomogeneities, 2.2
Initial pulse, 7.113
Instantaneous field of view, 9.45
Intensifying screen, 6.69
Intensity, 10.39
Intensity threshold, 3.4
Interlaced scanning, 3.55
Internal coil, 8.68
International Annealed Copper Standard
    (IACS), 8.68
Intrinsically safe construction, 3.54
IQI sensitivity, 6.69
Isotherm, 9.45
Isotope, 6.69

J

Jaeger (J) test, 3.5

K

Kaiser effect, 10.39
Kaiser, 1.6, 1.15, 10.1
Kinetic energy, 6.69
Kirchhoff's law, 9.45
Knerr, H. C., 1.6

L

Lack of fusion, 2.11, 2.25
Lamb wave, 7.113
Lamellar tearing, 2.3, 2.25
Lamination, 2.25
Lap, 2.26
Latent energy, 9.45
Latent heat of fusion, 9.45
Latent heat of vaporization, 9.45
Latent image, 6.69
Leak Testing (LT), 1.25
Lenz, H., 8.2
Lenz's law, 8.68

INDEX

Lester, H. H., 1.6, 1.10
Lift-off, 8.68
Linear indication, 5.52
Location marker, 6.70
Longitudinal magnetization, 5.52
Longitudinal wave, 7.113
Lumen, 3.55

M

Macroshrinkage, 2.26
Magnetic domains, 5.52
Magnetic field, 8.69
Magnetic field detection, 1.6
Magnetic flux, 8.69
Magnetic particle inspection, 5.52
Magnetic particle test, 1.6
Magnetic particle testing, 1.11, 5.1–5.53.
    See also MT
  advantages, 5.50
  amperage selection, 5.40
  applications, 5.48
    aircraft and aerospace, 5.50
    automotive, 5.50
    bar stock, 5.49
    bearings and bearing races, 5.49
    billets, 5.49
    blooms, 5.49
    castings, 5.49
    construction, 5.50
    cylinders, 5.49
    defense, 5.50
    discs, 5.49
    forgings, 5.49
    gears, 5.49
    ingots, 5.49
    inherent discontinuities, 5.48
    nuclear, 5.50
    nuts and bolts, 5.49
    petrochemical, 5.50
    plate, 5.50
    primary processing, 5.48
    rod, 5.49
    secondary processing, 5.48
    service, 5.49
    shafts, 5.49
    sheet, 5.49
    shipping, 5.50
    slabs, 5.49
    transportation, 5.50
    tubular products, 5.50
    welding, 5.49
    welds, 5.49
    wire, 5.49
  classification of indications, 5.44
  control of magnetization, 5.16
  demagnetization, 5.22
  depth limitations, 5.45
  detection media, 5.21
  detection of discontinuities, 5.13
  development, 5.2
  distorted fields, 5.13
  electromagnetic yoke, 5.26
  equipment, 5.24
    AC demagnetizing coils, 5.30
    black lights, 5.28
    Burmah castrol strips, 5.29
    flux direction indicators, 5.28
    mobile units, 5.26
    pie gauge, 5.28
    portable units, 5.26
    quantitative quality indicator (QQI),
      5.30
    stationary units, 5.24
  evaluation of test results, 5.44
  false indications, 5.45
  history, 5.1
  leakage fields, 5.13
  limitations, 5.51
  nonrelevant indications, 5.45
  options, 5.40
  part geometry, 5.16
  permanent magnets, 5.27
  quality control, 5.37
  relevant indications, 5.45
  reporting, 5.44, 5.46
  techniques, 5.30
    central conductor technique, 5.33
    coil technique, 5.33
    color contrast, 5.36
    current flow techniques (direct), 5.32
    dry continuous, 5.33, 5.34
    dry residual, 5.33, 5.35

fluorescent, 5.36
magnetic flow techniques (indirect), 5.31
magnetization, 5.31
wet continuous, 5.33, 5.36
wet residual, 5.33, 5.36
theory and principles, 5.2
electromagnetic field direction, 5.10
flux density, 5.3
magnetic field, 5.3
magnetic forces, 5.3
magnetic permeabilit, 5.4
magnetic reluctance, 5.5
magnetic saturation, 5.6
magnetization, 5.10
magnetizing force, 5.3
polarity, 5.3
systeresis, 5.6
types of electrical current, 5.17
altenating current (AC), 5.17
direct current (DC), 5.17
electrical power, 5.18
full-wave rectification, 5.18
half-wave rectification, 5.17
variables, 5.39
visual appearance, 5.45
Magnetic poles, 5.52
Magnetic writing, 5.52
Magnetism, 5.1, 5.52, 8.69
principles, 5.3
Magnetizing force, 5.53
Material density, 6.70
Maxwell, 8.2
Mehl, R. F., 1.6, 1.13
Mercury vapor lamp, 5.53
Metal path, 7.113
Metallic glass, 2.4
Metallurgical notch, 2.26
Metals, 2.2
alloys, 2.2
primary production, 2.2
slags, 2.2
Microfissure, 2.26
Microsegregates, 2.26
Microshrinkage cracking, 2.26
Microshrinkage porosity, 2.26
Miniature cameras, 3.16, 3.18

Mode, 7.114
Mode conversion, 7.114
Modulation analysis, 8.66
MT, 1.4, 1.7, 1.11, 1.23
Multiple back reflections, 7.114

N

Natural convection, 9.46
NDE, 3.1
NDT, 1.1, 1.3, 1.4, 1.5, 1.6, 1.7, 1.9, 1.11, 1.15, 1.16, 1.22, 3.1
Near field, 7.114
Neutron radiography, 1.25
Newton's law of cooling, 9.46
Noise, 7.114
Nondestructive evaluation, 1.1–1.26. *See also* NDE
concerns regarding, 1.2
Nondestructive examination , 1.1
Nondestructive inspection (NDI), 1.1
Nondestructive testing, 1.2–1.26. *See also* NDT
benefits, 1.20
conditions for effective testing, 1.21
approved procedures, 1.21
documentation, 1.21
equipment, 1.21
personnel, 1.22
testability, 1.21
history, 1.3
chronology, 1.6
Genesis, 1.3
Second World War, 1.4
limitations, 1.21
Nonmetallic inclusion, 2.26
Nonrelevant indication, 4.33
Normal incidence, 7.114
Normalization, 8.69
Notch, 2.26

O

Object to film distance, 6.70
Oersted, H. C., 8.1

INDEX

Ohm's law, 8.69
Oil and whiting technique, 1.5–1.6, 4.1
Operationally induced discontinuities, 2.18
  brittle fracture, 2.20
  crazing, 2.18
  creep cracking, 2.19
  fatigue cracks, 2.18
  macroscopic plastic deformation, 2.18
Ore, 2.2
Overemulsification, 4.33
Overremoval, 4.33

P

Parametric inputs, 10.39
Particle mobility, 5.53
Penetrant comparator block, 4.34
Penetrant technique, 1.6
Penetrant test, 1.6
Penetrant testing, 1.13, 4.1–4.34. *See also* PT
  advantages, 4.32
  application of developer, 4.17
  applications, 4.29
  development, 4.1
  development time, 4.17
  early techniques, 4.3
  environmental considerations, 4.14
  equipment, 4.6
    black light, 4.7
    in-line test systems, 4.6
    light intensity meters, 4.8
    portable kits, 4.6
    test panels, 4.8
  equipment checks, 4.30
  evaluation, 4.27
  fluorescent, 4.4
  history, 4.1
  interpretation, 4.18
  lighting, 4.14
  limitations, 4.32
  material checks, 4.29
  oil and whiting method, 4.1
  overwashing, 4.3
  penetrant application, 4.15

penetrant materials, 4.9. *See also* Penetrants
  cleaners, 4.9
  developers, 4.13
  emulsifiers, 4.12
  precleaners, 4.9
  solvent removers, 4.13
  solvents, 4.9
penetrant removal, 4.17
postcleaning, 4.18
precleaning, 4.15
principles, 4.4
procedures, 4.14
quality control, 4.29
surface condition considerations, 4.14
system checks, 4.30
techniques, 4.18
  Technique I Process A, 4.19
  Technique I Process B, 4.20
  Technique I, Process C, 4.23
  Technique I, Processes D, 4.20
  Technique II Process A, 4.25
  Technique II Process B, 4.25
  Technique II, Process C, 4.26
temperature, 4.14
theory, 4.4
variables, 4.18
water-removable technique, 4.3
water-washable technique, 4.3
Penetrants, 4.3
  fluorescent, 4.4
  postemulsifiable, 4.34
  solvent removable, 4.34
  visible, 4.4, 4.34
  water-removable, 4.3, 4.34
  water-washable, 4.3
Penetration depth, 7.114
Peripheral vision, 3.55
Permanent magnets, 8.69
Permeability, 5.53, 8.69
Personnel, 1.22
  certification, 1.22
  qualifications, 1.22
Phase change, 9.46
Phase diagram, 2.26
  binary, 2.26
  ternary, 2.26

Pitting, 2.26
Pixel, 3.55
Plastic deformation discontinuities, 2.15
  bursts, 2.15
  folds, 2.15
  laps, 2.15
  scale pits, 2.15
  seams, 2.15
  splits, 2.15
Polarity, 8.69
Postcleaning, 4.34
Postemulsification, 4.34
Precleaning, 4.34
Pressure vessels, 1.5
Pressurized construction, 3.54
Primary processing discontinuities, 2.26
Primary radiation, 6.70
Primary zone, 10.39
Probe, 7.114
Probe coil, 8.69
Prods, 5.53
PT, 1.7, 1.13, 1.23, 4.1
Pulse length, 7.114
Pulse-echo, 1.6
Pulse-echo technique, 7.114
Purged construction, 3.54

Q

Qualification programs for NDT personnel, 1.22
  ASNT Level III Certificate Program, 1.25
  AET, 1.25
  ASNT/ANSI-CP189, 1.25
  Central Certification Program (ACCP), 1.26
  ET, 1.25
  ISO 9712, 1.26
  Level III, 1.25
  LT, 1.25
  MILSTD-410, 1.26
  MT, 1.25
  NAVSEA 250-1500, 1.26
  NRT, 1.25
  overview, 1.25
  PT, 1.25
  RT, 1.25
  TIR, 1.25
  UT, 1.25
  VA, 1.25
  VT, 1.25
  Canadian Certification System (CSGB), 1.27
  Certification Scheme for Welding and Inspection Personnel (CSWIP), 1.26
  Personnel Certification in Nondestructive Testing (PCN), 1.26
  SNT-TC-1A, 1.23
    benefits, 1.24
    Level I, 1.23
    Level II, 1.23
    Level III, 1.23
    limitations, 1.24
    overview, 1.23
Qualitative thermography, 9.46
Quantitative thermography, 9.46
Quasi-steady-state heat flow, 9.46

R

Radiation, 9.46
Radiographic contrast, 6.70
Radiographic quality, 6.70
Radiographic sensitivity, 6.70
Radiographic testing, 6.1–6.70. *See also* RT
  accessories, 6.22
    collimators, 6.24
    densitometers, 6.24
    film and cassettes, 6.22
    film badges, 6.59
    film hangers, 6.24
    high-intensity film illuminators, 6.24
    lead numbers and letters, 6.22
    lead screens, 6.22
    penetrameters, 6.22
  advantages, 6.58

Radiographic testing *(continued)*
  applications, 6.54
    aerospace, 6.57
    food products, 6.58
    law enforcement and security, 6.57
    medicine, 6.57
    nonmetals, 6.57
    objects of art or historic value, 6.58
    petrochemical, 6.57
    power generation, 6.57
  characteristics of radiation, 6.11
  density, 6.47
  equipment, 6.17
    betatrons, 6.17, 6.19
    gamma ray equipment, 6.21
    ionization chambers, 6.60
    linear accelerators, 6.18, 6.19
    photoluminescent glasses, 6.60
    processing equipment, 6.24
    thermoluminescent dosimeters, 6.60
    Van de Graaff generators, 6.18, 6.19
    x-ray equipment, 6.21
    x-ray tubes, 6.17
  evaluation, 6.50, 6.52
    casting discontinuities, 6.53
    cold shuts, 6.54
    concavity, 6.53
    convexity, 6.53
    cracks, 6.52
    fas voids and porosity, 6.54
    geometric conditions, 6.53, 6.54
    hot tears and cracks, 6.53
    inclusions, 6.52
    incomplete penetration, 6.52
    lack of fusion, 6.52
    misrun, 6.54
    overreinforcement, 6.53
    porosity, 6.53
    shrinkage, 6.53
    slag and sand inclusions, 6.53
    undercut, 6.53
    underfill, 6.53
    unfused chaplets, 6.54
  false indications, 6.51
  film artifacts, 6.51
  film processing, 6.44
  gamma radiography, 6.19
  gamma rays, 6.8
  history and development, 6.1
  image quality, 6.33
  limitations, 6.59
  postprocessing artifacts, 6.52
  principles of, 6.17
  processing artifacts, 6.52
  radiographic film, 6.44
  responsibility of the interpreter, 6.54
  safety considerations, 6.59
  shielding, 6.59
  techniques and procedures, 6.39
  theory and principles, 6.10
  variables, 6.25
  x-rays, 6.1
Radium, 1.6
Range, 7.114
Raster, 3.55
Rayleigh wave, 7.114
Reference block, 7.114
Reference standard, 3.55
Reflection, 3.55
Reflection coil, 8.69
Reflector, 7.114
Relevant indication, 7.114
Reluctance, 5.53
Remote viewing, 3.55
Residual magnetism, 5.53
Resistance, 8.69, 9.46
Resistivity, 8.69
Resolution, 3.55
  horizontal, 3.56
  vertical, 3.56
Resolution threshold, 3.56
Resolving power, 3.56
Retained image, 3.56
Retentivity, 5.53
Retina, 3.56
Risetime, 10.39
Rod, 3.56
Roentgen, W. C., 1.6
Roentgen, 1.10, 6.1
RT, 1.4, 1.7, 1.23, 6.1
R-value, 9.46

S

Saxby, S. H., 1.6
Scab, 2.7
Scale pits, 2.26
Scratches, 2.26
Search unit, 7.114
Second law of thermodynamics, 9.46
Secondary processing discontinuities, 2.26
secondary radiation, 6.70
Segregation, 2.26
Self-induction, 8.70
Sensitivity, 7.114
Service discontinuities, 2.26
Shear wave, 7.114
Shielded coil, 8.69
Signal description, 10.39
Signal features, 10.39
Signal strength, 10.39
Simple magnifier, 3.56
Sinusoid, 8.69
Skin depth, 8.69
Skin effect, 8.69
Skip distance, 7.114
Slag, 2.10, 2.26
Slag inclusions, 2.2
　concentrations, 2.2
　localization, 2.2
Slit response function (SRF), 9.46
SNT, 1.23
SNT-TC-1A, 1.24, 1.25
Society for Nondestructive Testing, 1.23. *See also* SNT
Sokolov, S. Y., 1.6, 7.2
Solar glint, 9.46
Solenoid, 8.70
Solvent remover, 4.34
Sonic technique, 1.3
Sound path, 7.114
Source energy, 10.39
Source to film distance, 6.70
Spatial resolution, 9.46
Specific heat, 9.46
Specification, 3.56
Sperry Rail Service, 1.11
Sperry, E., 1.6, 1.11

Split, 2.26
Spot radiometer, 9.46
Spot size, 9.46
Sproule, D. O., 1.6, 7.2
Standards, 3.56
Standard depth of penetration, 8.69
Steady-state heat flow, 9.46
Stefan–Boltzmann law, 9.46
Step wedge comparison film, 6.70
Stockman, L., 4.4
Stringers, 2.26
Subject contrast, 6.70
Supersonic reflectoscope, 1.16
Surface coil, 8.70
Switzer, J., 1.13, 4.2
Switzer, R., 1.13, 4.2

T

TAM panel, 4.34
Target, 6.70
Tensile strength, 1.18
Tensile testing machine, 1.18
Tesla, 5.53
Thermal background, 9.46
Thermal capacitance, 9.47
Thermal conductivity, 9.47
Thermal infrared testing, 9.1–9.47
　advantages, 9.43
　applications, 9.27
　　aerospace, 9.28
　　building energy surveys, 9.42
　　electrical inspections, 9.29
　　electronics industry, 9.27
　　mechanical inspections, 9.34
　　refractory inspections, 9.36
　　roof moisture inspections, 9.39
　　steam traps and lines, 9.36
　　tank level verification, 9.38
　boat hulls, 9.23
　controls, 9.14
　cracks, 9.21
　data processing and report generation software, 9.14
　data storage, 9.14

Thermal infrared testing *(continued)*
  detectors, 9.14
  displays, 9.14
  equipment and accessories, 9.11
    bolometers, 9.13
    cooling systems, 9.13
    cryogenically cooled systems, 9.13
    Dewar, 9.12
    focal plane arrays, 9.12
    focal plane arrays, 9.13
    line scanner, 9.12
    noncryogenic devices, 9.13
    photon detectors, 9.12
    point radiometers, 9.11
    pyroelectric systems, 9.13
    pyroelectric vidicons, 9.12
    spot radiometers, 9.11
    thermal detectors, 9.13
  evaluation of test results and reporting, 9.26
  history and development, 9.1
  inclusions, 9.21
  instruments, 9.24
    calibration, 9.25
    detectivity, 9.24
    dynamic range, 9.24
    field of view, 9.24
    rate of data acquisition, 9.24
    spatial and measurement resolution, 9.24
    system calibration, 9.24
    thermal sensitivity, 9.24
  lenses, 9.13
  limitations, 9.43
  processing electronics, 9.14
  steady-state heat flow, 9.24
  techniques, 9.16
    comparative thermography, 9.17
    image averaging, 9.19
    isotherms, 9.21
    null method, 9.21
    thermal mapping, 9.17
  theory and principles, 9.4
    conduction, 9.5
    convection, 9.6
    emissivity, 9.10
    latent heat, 9.5
    radiation, 9.8
    thermal energy, 9.4
  transient heat flow, 9.24
  variables, 9.23
    ambient air temperature, 9.23
    background temperature, 9.23
    diffusivity, 9.23
    distance to object, 9.23
    emissivity, 9.23
    evaporative cooling, 9.23
    heat transfer relationships, 9.23
    instrument-related, 9.24
    precipitation, 9.23
    presence of emitting/absorbing gases, 9.23
    radiational cooling, 9.23
    relative humidity, 9.23
    solar absorption, 9.23
    spectral characteristics, 9.23
    temperature, 9.23
    thermal capacitance, 9.23
    wind speed and direction, 9.23
    wind-driven convection, 9.23
Thermal resistance, 9.47
Thermography, 1.6, 1.16
Through-transmission, 8.70
Through-transmission technique, 7.114
Tip, 3.56
TIR, 1.7
Title 10, Code of Federal Regulations, Part 50, 3.1
Toughness, 1.18
Trace, 3.56
Transducer, 7.114
Transient heat flow, 9.47
Transmitted film density, 6.70
Tube current, 6.70

U

Ultrasonic flaw detectors, 1.6
Ultrasonic test method, 1.6
Ultrasonic testing, 1.6, 7.1–7.115. *See also* UT
  advantages, 7.110
  amplitude control linearity, 7.67

applications, 7.107
  bond testing, 7.108
  elevated temperature, 7.108
  fluid level measurement, 7.109
  hydrogen embrittlement, 7.108
  inaccessible components, 7.109
  liquid flow rate, 7.109
  stress analyses, 7.109
bandwidth, 7.45
bubblers, 7.48
calibration, 7.59
  amplifier, 7.62
  time base, 7.59
decibel system, 7.28
digital instrumentation, 7.31
  attenuator, 7.33
  controls, 7.31
  delay, 7.31, 7.33
  gain, 7.31, 7.33
  range, 7.31
  sweep, 7.31, 7.33
discontinuity detection, 7.108
discontinuity length, 7.94
displays, 7.35
distance amplitude correction (DAC), 7.63
distance and area amplitude blocks, 7.63
dual transducers, 7.46
economic factors, 7.107
electronic gates, 7.34
equipment, 7.29
  clock, 7.30
  display, 7.30
  pulser, 7.30
  receiver/amplifier, 7.30
  selection, 7.33
  transducer, 7.30
evaluation of test results, 7.105
factors affecting sound waves, 7.9
  absorption, 7.25
  beam characteristics, 7.18
  beam spread, 7.26
  couplant, 7.13
  factors affecting sound waves, 7.19
  interference, 7.19
  reflection, 7.11

reflective mode conversion, 7.16
refraction and mode conversion, 7.13
scatter, 7.26
velocity, 7.9
wavelength, 7.10
focused transducers, 7.47
frequency selection, 7.34
gain, 7.63
gating, 7.88
history, 7.1
horizontal linearity, 7.67
inspection techniques, 7.68
  beam path distance, 7.75
  contact scanning using angle beam shear waves, 7.73
  contact scanning using compression waves, 7.68
  full skip, 7.75
  half skip, 7.75
  immersion, 7.86
  Lamb wave, 7.84
  multiple transducer, 7.81
  pulse-echo, 7.68
  rod and pipe, 7.80
  tandem, 7.84
  through-transmission, 7.81
interface triggering, 7.88
laser-generated ultrasound, 7.45
limitations, 7.110
maximum amplitude system, 7.95
nature of sound waves, 7.4
noncontact methods, 7.43
  EMAT, 7.43
normalization, 7.87
offset method for generating shear waves, 7.88
periodic motion, 7.5
phased array transducers, 7.48
reject, 7.35
reporting, 7.106
sizing procedure, 7.92
sound vibrations, 7.7
sound wave propagation, 7.7
  compression wave, 7.8
  Lamb wave, 7.8
  longitudinal wave, 7.8
  plate wave, 7.8

Ultrasonic testing *(continued)*
   Rayleigh wave, 7.8
     shear wave, 7.8
     surface wave, 7.8
   squirters, 7.48
   storage memory, 7.35
   system checks, 7.59, 7.66
   techniques, 7.52
     calibration, 7.52
     discontinuity sizing, 7.91
     distance, gain, size (DGS) technique, 7.65
     intensity drop, 7.92
     standard angle beam transducer, 7.98
     time of flight diffraction (TOFD), 7.99
     tip diffraction, 7.97
   theory and principles, 7.4
   thickness measurement, 7.107
   through-thickness dimension, 7.92
   transducers, 7.37
     angle beam shear wave, 7.56
     characteristics, 7.53
     compression wave, 7.53
     piezocomposite, 7.41
     problems, 7.51
     PVDF, 7.43
   transfer corrections, 7.65
   ultrasonic extensiometers, 7.109
   variables, 7.103
     attenuation, 7.103
     contact pressure, 7.104
     couplant, 7.104
     crequency, 7.104
     dendritic structures, 7.104
     diameter changes, 7.104
     gain, 7.105
     grain size, 7.104
     resolution, 7.104
     surface condition, 7.104
     temperature, 7.103
     transducer frequency, 7.105
     velocity, 7.103
   vibration, 7.5
   wheel transducers, 7.50
Ultraviolet light, 5.53
UT, 1.7, 1.23

V

Vee path, 7.114
Vibration analysis (VA), 1.25
Video, 3.56
Video cassette recorders, 3.19
Video tape recorders, 3.19
Videoscope, 3.56
Viscosity, 4.34
Visibility, 3.56
Vision, 3.56
Visual acquity, 3.5
Visual acuity, 3.5
Visual acuity, 3.56
Visual angle, 3.57
Visual detection of discontinuities, 3.23
   arc strikes, 3.35
   arc welding, 3.27
   ball and plug valves, 3.40
   brazing, 3.27
   burn-through, 3.31
   butterfly valves, 3.41
   castings, 3.26
   chaplets, 3.26
   chills, 3.26
   cold shuts, 3.26
   concavity, 3.31
   cracking, 3.30
   cracks, 3.25
   crater cracks, 3.33
   diaphragm valves, 3.41
   drawing, 3.26
   excessive reinforcement, 3.31
   extruding, 3.26
   forging, 3.25
   forging laps, 3.25
   gas, 3.26
   gate valves, 3.39
   globe valves, 3.40
   hot tears, 3.26
   inclusions, 3.26
   incomplete penetration, 3.31
   Ingots, 3.24
   inherent discontinuities, 3.23
   insufficient leg, 3.31
   insufficient throat, 3.31
   leakage, 3.41

metal joining processes, 3.27
misalignment, 3.31
misrun, 3.26
nonmetallic inclusions, 3.30
overlap, 3.31
piercing, 3.26
porosity, 3.28
rolling, 3.25
scabs, 3.26
scoring, 3.26
seams, 3.25
service-induced discontinuities, 3.36
    abrasive wear, 3.36
    adhesive wear, 3.37
    corrosion, 3.37
    crevice corrosion, 3.38
    erosion corrosion, 3.38
    erosive wear, 3.36
    fatigue cracking, 3.38
    fretting wear, 3.37
    galvanic corrosion, 3.37
    gouging wear, 3.37
    grinding wear, 3.37
    uniform corrosion, 3.37
    wear, 3.36
slugs, 3.26
soldering, 3.27
spatter, 3.35
stringers, 3.25
swing check valves, 3.40
undercut, 3.31
underfill, 3.31
weld craters, 3.31
Visual examination, 3.1
Visual field, 3.57
Visual perception, 3.57
Visual testing, 1.2–1.3, 3.1–3.57. *See also*
    Visual testing of discontinuities,
    VT
    acceptance criteria, 3.48
    accessories, 3.9
    advantages, 3.49
        photography, 3.49
    applications, 3.22
    ASME categories, 3.1
        VT-1, 3.1
        VT-2, 3.1
        VT-3, 3.1
        VT-4, 3.1
    brightness, 3.6
    boilers, 3.44
    CCD based records, 3.50
    cleanliness, 3.6
    codes, 3.23
        ANSI B31.1—Power Piping and
            B31.7—Nuclear Power Piping,
            3.23
        API 1104, 3.23
        ASME Boiler and Pressure Vessel
            Code, 3.23
    color, 3.5
    direct, 3.1
    distance, 3.8
    environmental factors, 3.7
    equipment, 3.9
        borescopes, 3.13
        charged coupled device, 3.15
        delivery systems, 3.52
        depth indicators, 3.12
        digital camera, 3.21
        fiber camera, 3.15
        gauges, 3.13
        light sources, 3.10
        magnifier, 3.10
        measuring devices, 3.11
        micrometers, 3.12
        miniature camera, 3.16
        monitors, 3.20, 3.53
        optical comparator, 3.12
        straightedge scale, 3.12
        vernier caliper, 3.12
        video tape recorder, 3.19
        videocassette recorder, 3.19
    evaluation, 3.45
    fiberoptic cameras, 3.51
    film-based records, 3.49
    heat exchangers, 3.44
    human eye, 3.4, 3.9
    light, 3.5
    lighting requirements, 3.5
    limitations, 3.49
    miniature cameras, 3.51
    nuclear reactor vessels, 3.43
    object factors, 3.3

Visual testing *(continued)*
  of component supports, 3.42
    corrosion, 3.43
    fatigue, 3.43
    hangers, 3.42
    inadequate construction practices, 3.42
    overload, 3.42
    physical damage, 3.42
    pipe, 3.42
    restraints, 3.42
    snubbers, 3.42
    spring can, 3.42
  of pressure vessels, 3.43
  perception, 3.8
  photographic cameras, 3.49
  photographic techniques, 3.50
  physiological factors, 3.7
  psychological factors, 3.7
  records, 3.49
    CCD-based, 3.50
    film-based, 3.49
  reflectance, 3.7
  remote, 3.1
    borescopes, 3.1
    endoscopes, 3.1
    fiberscopes, 3.1
    video technology, 3.1
  reporting of test results, 3.45
  reports, 3.46
  service-induced discontinuities, 3.36
    corrosion, 3.37
    erosion corrosion, 3.38
    fatigue cracking, 3.38
    leakage, 3.41
    wear, 3.36, 3.37
  shape, 3.7
  size, 3.7
  specifications, 3.23
  standards, 3.23
    ASME, 3.23
  steam generators, 3.44
  storage tanks, 3.43
  surface condition, 3.6
  techniques, 3.22
  temperature, 3.7
  texture, 3.7
  theory and principles, 3.3
  video borescopes, 3.51
  visual angle, 3.8
Voids, 2.26
VT, 1.4, 1.7, 3.1

W

Water path, 7.115
Waterproof construction, 3.57
Watertight construction, 3.57
Weber, 5.53
Wedge, 7.115
Weld undercut, 2.26
Welding discontinuities, 2.11, 2.26
  cracking, 2.12
  deformation, 2.12
  delayed cracking, 2.13
  excess concavity, 2.14
  fracture, 2.12
  hydrogen cracking, 2.13
  lack of penetration, 2.14
  slag, 2.12
  solidification cracks, 2.12
  underbead cracking, 2.12
  undercut, 2.13
Wetability, 4.34
Wheel search unit, 7.115

X

X-radiography, 1.4
X-ray techniques, 1.10
X-rays, 1.6, 1.10, 6.1

Y

Yield point, 1.18

Z

Zuschlag, T., 1.6